Smart Innovation, Systems and Technologies

Volume 22

For further volumes:
http://www.springer.com/series/8767

Anne Håkansson · Mattias Höjer
Robert J. Howlett · Lakhmi C. Jain
Editors

Sustainability in Energy and Buildings

Proceedings of the 4th International Conference on Sustainability in Energy and Buildings (SEB'12)

Volume 1

Springer

Editors
Professor Anne Håkansson
The Royal Institute of Technology - KTH
Kista
Sweden

Professor Mattias Höjer
Centre for Sustainable Communications
KTH Royal Institute of Technology
Stockholm
Sweden

Professor Robert J. Howlett
KES International & Bournemouth University
United Kingdom

Professor Lakhmi C. Jain
School of Electrical and Information Engineering
University of South Australia
South Australia
Australia

ISSN 2190-3018 ISSN 2190-3026 (electronic)
ISBN 978-3-642-36644-4 ISBN 978-3-642-36645-1 (eBook)
Printed in 2 Volumes
DOI 10.1007/978-3-642-36645-1
Springer Heidelberg New York Dordrecht London

Library of Congress Control Number: 2013932298

Printed on acid-free paper

Springer is part of Springer Science+Business Media (www.springer.com)

Preface

The volume of Smart Innovations, Systems and Technologies book series contains the proceedings of the Fourth International Conference on Sustainability in Energy and Buildings, SEB12, held in Stockholm, Sweden, and is organised by KTH Royal Institute of Technology, Stockholm, Sweden in partnership with KES International.

The International Conference on Sustainability in Energy and Buildings is a respected conference focusing on a broad range of topics relating to sustainability in buildings but also encompassing energy sustainability more widely. Following the success of earlier events in the series, the 2012 conference includes the themes Sustainability, Energy, and Buildings and Information and Communication Technology, ICT.

SEB'12 has invited participation and paper submissions across a broad range of renewable energy and sustainability-related topics relevant to the main theme of Sustainability in Energy and Buildings. Applicable areas include technology for renewable energy and sustainability in the built environment, optimisation and modeling techniques, information and communication technology usage, behaviour and practice, including applications.

This, the fourth conference in the SEB series, attracted a large number of submissions all around the world, which were subjected to a two-stage review process. With the objective of producing a high quality conference, papers have been selected for presentation at the conference and publication in the proceedings. The papers for presentation are grouped into themes. The papers are included in this proceedings.

Four prominent research professors gave interesting and informative keynote talks. Professor Göran Finnveden, Professor in Environmental Strategic Analysis and Vice-President for sustainable development at KTH Royal Institute of Technology, Stockholm, Sweden gave a talk entitled "Sustainability Challenges for the Building Sector". Professor Per Heiselberg Professor at the Department of Civil Engineering at Aalborg University, Denmark, gave a talk about "Buildings – both part of the problem and the solution!". Professor Guðni A. Jóhannesson, Professor in Building Technology, Director General of the Icelandic National Energy Authority, Iceland, and chair of IPGT the International Partnership for Geothermal Technologies spoke on the topic of "Meeting the challenges of climatic change - the hard way or the clever way". Professor Lynne A. Slivovsky, Associate Professor of Electrical and Computer Engineering at California

Polytechnic State University, San Luis Obispo, California, USA spoke on the topic of "The Questions That Keep Me Up At Night".

Thanks are due to the very many people who have given their time and goodwill freely to make the SEB'12 a success. We would like to thank KTH Royal Institute of Technology for their valued support for the conference. We would also like to thank the members of the International Programme Committee who were essential in providing their reviews of the conference papers, ensuring appropriate quality. Moreover, we would like to thank invited session chairs for their hard work and providing reviewers of the conference papers and upholding appropriate quality. We thank the high profile keynote speakers and panellists for agreeing to come and provide very interesting theme and talks, as well as, inform delegates and provoke discussions. Important contributors to SEB'12 are the authors, presenters and delegates without whom the conference could not have taken place, so we offer them our thanks for choosing SEB'12 conference. We also like to thank the panellists for coming and discussing " Panel: Sustainability: Current and Future with focus on Energy, Buildings and ICT"

The KES International, KES Secretariat staff and the local organising committee worked very hard to bring the conference to a high level of organisation, and we appreciate their tremendous help and we are thankful to them. Finally, we thank City Hall, City of Stockholm Sweden for hosting the reception for the SEB'12 conference and Quality Hotel Nacka for accommodating the conference 3 – 5 of September 2012 International Conference on Sustainability in Energy and Buildings.

We hope that readers will find SEB'12 proceedings interesting, informative and useful rescource for your research.

Assoc Professor Anne Håkansson
SEB'12 General Chair
KTH Royal Institute of Technology,
Sweden

Professor Mattias Höjer
SEB'12 General Co-chair
KTH Royal Institute of Technology,
Sweden

Professor Robert J. Howlett
Executive chair
KES International & Bournemouth University,
United Kingdom

Organization

Honorary Chairs

Göran Finnveden	Strategic Environmental Analysis KTH Royal Institute of Technology, Sweden
Björn Birgisson	Division of Highway Engineering KTH Royal Institute of Technology, Sweden
and	
Professor Lakhmi C. Jain	University of South Australia, Australia

General Chairs

Assoc Prof. Anne Håkansson	Communication Systems KTH Royal Institute of Technology, Sweden
and	
Professor Mattias Höjer	Centre for Sustainable Communications KTH Royal Institute of Technology, Sweden

Executive Chair

Professor Robert J. Howlett	Executive Chair, KES International Bournemouth University, United Kingdom

Program Chairs

Ronald Hartung	Computer Sciences and Mathematics Franklin University, Columbus, Ohio, USA

and

Mark Smith
Communication Systems
KTH Royal Institute of Technology, Stockholm, Sweden

Publicity and Support Chairs

Bernhard Huber
Centre for Sustainable Communications (CESC)
KTH Royal Institute of Technology, Stockholm, and Sweden

Dan Wu
Communication Systems
KTH Royal Institute of Technology, Stockholm, Sweden

Nils Brown
Centre for Sustainable Communications (CESC)
KTH Royal Institute of Technology, Stockholm, Sweden

and

Esmiralda Moradian
Communication Systems
KTH Royal Institute of Technology, Stockholm, Sweden

Faye Alexander
Conference Operations, KES International, UK

Shaun Lee
Information Systems Support, KES International, UK

and

Russ Hepworth
Business Development Manager, KES International, UK

Invited Sessions Chairs and Workshop

Smart Buildings, Smart Grids
Chair: Dr Catalina Spataru

Assessment and Monitoring The Environmental Performance of Buildings
Chair:Dr John Littlewood

Methodology for Renewable Energy Assessment
Chair: Dr Rainer Zah

Improving Office Building Energy Performance
Chair:Dr Emeka Osaji

Multi-Energy Sources
Chair: Prof Aziz Naamane

Technologies and Applications of Solar Energy
Chair: Prof Mahieddine Emziane

Energy Planning in Buildings and Policy Implications
Chair: Dr Eva Maleviti

Sustainable Energy Systems and Building Services
Chair: Prof Ivo Martinac

Sustainable and healthy buildings
Chair: Prof Jeong Tai Kim

Co-chair: Prof. Geun Young Yun

Special selection
Chair: Prof Robert Howlett

REQUEST Workshop

Invited Keynote Speakers

Professor Göran Finnveden	KTH Royal Institute of Technology, Stockholm, Sweden
Professor Per Heiselberg	Aalborg University, Denmark,
Professor Guðni A. Jóhannesson	Icelandic National Energy Authority, Iceland,
Professor Lynne A. Slivovsky	California Polytechnic State University, San Luis Obispo, California, USA

International Programme Committee

Meniai Abdeslam-Hasse	Université Mentouri de Constantine, Algeria
Nora Cherifa Abid	Aix-Marseille University, France
Vivek Agarwal	Indian Institute of Technology Bombay, India
Abdel Aitouche	Lagis Hei, France
Nader Anani	Manchester Metropolitan University, UK
Naamane Aziz	Marseilles Aix Marseille Universite (AMU), France
Messaouda Azzouzi	Ziane Achour University of Djelfa, Algeria
Brahim Benhamou	Cadi Ayyad University of Marrakech, Morroco
Frede Blaabjerg	Aalborg University Inst. Of Energy Technology, Denmark
Saadi Bougoul	Université de Batna, Algerie
Mohamed Chadli	University of Picardie Jules Verne, France
Christopher Chao	Hong Kong University of Science and Technology, China
Zhen Chen	Heriot-Watt University, Scotland
Derek Clements-Croome	Reading University, UK
Gouri Datta	Deshbandhu College, Kalkaji, University of Delhi, India and Stromstad Academy, Stromstad.
Mohamed Djemai	Université de Valenciennes et du Hainaut Cambrésis, France
Philip Eames	Loughborough University, UK
Mahieddine Emziane	Masdar Institute of Science and Technology, Abu Dhabi
Luis Fajardo-Ruano	Uumsnh, Morelia, Mexico
Antonio Gagliano	University of Catania, Italy
Oleg Golubchikov	University of Birmingham, UK
Ahmed Hajjaji	University of Picardie Jules Verne, France
Abdelaziz Hamzaoui	University of Reims Champagne Ardenne, France
Sture Holmberg	Royal Institute of Technology (KTH) Stockholm, Sweden
Robert J. Howlett	Bournemouth University, UK
Bin-Juine Huang	National Taiwan University, Taipei, Taiwan
Kenneth Ip	University of Brighton, UK
Hong Jin	Harbin Institute of Technology, China
Roger Kemp	Lancaster University, UK
Sumathy Krishnan	North Dakota State University, USA
Angui Li	Xi'an University of Architecture & Technology, China
Soren Linderoth	Technical University of Denmark
John Littlewood	Cardiff Metropolitan University, UK
Nacer Kouider M'Sirdi	Laboratoire des Sciences del'Information et des Systèmes, France
Noureddine Manamani	University of Reims, France

Ahmed Mezrhab	University Mohammed 1, Oujda Morocco
Behdad Moghtaderi	University of Newcastle, Australia
Roger Morgan	Liverpool John Moores University, UK
Mostafa Mrabti	Universite Sidi Mohamed Ben Abdellah, Fes, Morocco
Rui Neves-Silva	Universidade Nova de Lisboa FCT/UNL, Portugal
Emeka Efe Osaji	University of Wolverhampton, UK
Frederici Pittaluga	University of Genova, Italy
Giuliano C Premier	University of Glamorgan, UK
Abdelhamid Rabhi	MIS Amiens, France
Ahmed Rachid	University of Picardie Jules Verne, France
Enzo Siviero	University IUAV of Venice, Italy
Shyam Lal Soni	Malaviya National Institute of Technology, Jaipur, India
Catalina Spataru	UCL Energy Institute, Uk
Alessandro Stocco	University of Nova Gorica and partner Progeest S.r.l. of Padua, Italy
Lounes Tadrist	Polytech.univ-mrs, France
Dario Trabucco	IUAV University of Venice, Italy
Mummadi Veerachary	Indian Institute of Technology, Delhi, India
Wim Zeiler	TU Eindhoven, Faculty of the Built Environment, Netherlands
Mohcine Zouak	USMBA FST, Morocco
Rainer Zah	Life Cycle Assessment & Modelling Group, Empa, Switzerland

Keynote Speakers

We are very pleased to have acquired the services of an excellent selection of keynote speakers for SEB'12. These speakers gave a view about technological and scientific activities, relating to sustainability in energy and buildings, taking place in various areas of the world.

Professor Göran Finnveden

KTH Royal Institute of Technology, Sweden

Sustainability Challenges for the Building Sector

Abstract:

The building and real estate management sector is responsible for a significant part of the environmental impacts of our society. The sector's contribution to the threat of climate change for production of heat and electricity for the buildings are of special importance. It is important to consider the full life-cycle of buildings and also consider

production and transportation of building materials, construction and waste management. In Sweden, emissions of gases contributing to climate change from heating of buildings have decreased during the last decades as results of strong policy instruments. One the other hand emissions from other parts of the life-cycle of buildings have increased, illustrating the need to have a wide systems perspective in order to avoid sub-optimizations. It is also important to consider other environmental threats such as the use of hazardous chemicals, air quality, generation of waste and impacts on ecosystems from production of building materials as well as on building sites.
The building sector has a large potential to reduce its environmental footprint. Many of the most cost-efficient possibilities for mitigation of climate change are related to the building sector. Governmental policies are important for changes to be made. Voluntary instruments such as building rating tools may have an additional role. The ICT-sector may have one of its largest potentials in contributing to a more sustainable society in the building sector. Because of the long life-time of buildings, we are now constructing the future environmental impacts. When looking for cost-efficient solutions, we must therefore also consider the future cost-efficiency. In the presentation also social aspects of sustainability will be discussed including possibilities for the building sector to contribute to a better health and reduced health inequalities.

Biography:

Göran Finnveden is Professor in Environmental Strategic Analysis and Vice-President for sustainable development at KTH Royal Institute of Technology, Stockholm, Sweden. He is a M.Sc. in Chemical Engineering 1989, PhD in Natural Resources Management, Associate Professor in Industrial Ecology 2003 and full Professor 2007. His research has focused on environmental systems analysis tools such as Life Cycle Assessment, Strategic Environmental Assessment and Input-Output Analysis. It has included both methodological development and case studies. Application areas include buildings, energy systems, information and communication technologies, infrastructure and waste management. He has also worked with environmental policy in areas such as environmental policy integration, integrated product policy and waste policy. He is a currently a member of the Scientific Advisory Council to the Swedish Minister of the Environment, an expert in the governmental commission on waste management and a member of the board of directors of the Swedish Waste Nuclear Fund. According to Scopus he has published more than 60 scientific papers and is cited nearly 2000 times.

Professor Per Heiselberg

Aalborg University, Denmark

Buildings – both part of the problem and the solution!

Abstract:

Energy use for room heating, cooling and ventilation accounts for more than one-third of the total, primary energy demand in the industrialized countries, and is in this way a major polluter of the environment. At the same time the building sector is identified as providing the largest potential for CO2 reduction in the future and many countries across the world have set very ambitious targets for energy efficiency improvements in new and existing buildings. For example at European level the short term goal has recently been expressed in the recast of the EU Building Performance Directive as "near zero energy buildings" by 2020.
To successfully achieve such a target it is necessary to identify and develop innovative integrated building and energy technologies, which facilitates considerable energy savings and the implementation and integration of renewable energy devices within the built environment. The rapid development in materials science, information and sensor technology offers at the same time considerable opportunities for development of new intelligent building components and systems with multiple functions.
Such a development will impose major challenges on the building industry as building design will completely change from design of individual components and systems to integrated design of systems and concepts involving design teams of both architects, engineers and other experts. Future system and concepts solutions will require that building components must be able fulfill multiple performance criteria and often contradictory requirements from aesthetics, durability, energy use, health and comfort. A key example of this is building facades that instead of the existing static performance characteristics must develop into dynamic solutions with the ability to dynamically adjust physical properties and energetic performance in response to fluctuations in the outdoor environment and changing needs of the occupants in order to fulfill the future targets for energy use and comfort. Buildings will also be both consumers and producers of energy, which creates a number of new challenges for building design like identification of the optimum balance between energy savings and renewable energy production. The interaction between the energy "prosuming" building and the energy supply grid will also be an important issue to solve.
The lecture will address and illustrate these future challenges for the building sector and give directions for solutions.

Biography:

Per Heiselberg is Professor at the Department of Civil Engineering at Aalborg University, Denmark. He holds a M.Sc. and a Ph.D. in Indoor Environmental Engineering. His research and teaching subjects are within architectural engineering and are focused on the following topics:

- Energy-efficient building design (Net zero energy buildings, design of low energy buildings - integration of architectural and technical issues, modelling of double skin facades, night cooling of buildings and utilization of thermal mass, multifunctional facades, daylight in buildings, passive energy technologies for buildings, modeling of building energy use and indoor environment)

- Ventilation and air flow in buildings (Modelling and measurements of air and contaminants flows (both gas- and particles) in buildings, ventilation effectiveness, efficient ventilation of large enclosures, numerical simulation (computational fluid dynamics) of air and contaminant flows as well as modeling of natural and hybrid ventilation)

Per Heiselberg has published about 300 articles and papers on these subjects.
Currently, Per Heiselberg is leading the national strategic research centre on Net Zero Energy Buildings in Denmark (www.zeb.aau.dk). The centre has a multidisciplinary research approach and a close cooperation with leading Danish companies. He has been involved in many EU and IEA research projects in the past 20 years. He was the operating agent of IEA-ECBCS Annex 35 (1997-2002) and IEA-ECBCS Annex 44 (2005-2009), (www.ecbcs.org). Presently he is involved in ECBCS Annexes 52, 53 and 59.

Professor Guðni A. Jóhannesson

Icelandic National Energy Authority, Iceland

Meeting the challenges of climatic change - the hard way or the clever way

Abstract:

We may not agree on how the possible CO2 driven scenarios of climate change in the future may look like but we all can agree that the anthropogenic increase in CO2 levels in the world atmosphere exposes humanity to higher risks of changes in the environment than we want to face in our, our children's or their children's lifetime.
It is evident that we are now facing a global challenge that we are more often dealing with by local solutions. Our guiding rule is that by saving energy we are also mitigating greenhouse gas emission. Also if we are using renewable energy and substituting fossil fuels we are also moving in the right direction. There are however important system aspects that we should be considering.
The first one if we are using the right quality of energy for the right purpose. A common example is when high quality energy such as electricity or gas is used directly to provide domestic hot water or heat houses instead of using heat pumps or cogeneration processes to get the highest possible ratio between the used energy and the primary energy input.
The second one is if we are obstructing necessary structural changes that could lead to a more effective energy system globally. We have big reserves of cost effective renewable energy sources, hydropower and geothermal energy around the world that are far from the markets and would therefore need relocation structural changes in our industrial production system to be utilized.
The third aspect is if we are using our investments in energy conversion and energy savings in the best way to meet our climatic goals or if we are directed by other hidden agendas to such a degree that a large part of our economical input is wasted.
It is evident that the national and local strategies for energy savings are closely linked to other strategic areas such as industrial development, household economy, mobility.

Also a necessary precondition for investment is that the nations maintain their economic strength and their ability to develop their renewable resources and to invest in new more efficient processes.
The key to success in mitigating the climatic change is therefor to create a holistic strategy that beside the development of technical solutions for energy efficiency and utilization of renewable energy also considers the local and global system aspects. With present technologies for energy efficient solutions, proper energy quality management r and with utilization of cost effective renewable energy sources we have all possibilities to reduce energy related the global CO2 emissions to acceptable levels.

Biography:

Professor Guðni A. Jóhannesson is born in Reykjavik 1951. He finished his MSc in Engineering physics in 1976, his PhD thesis on thermal models for buildings in 1981 and was appointed as an associate professor at Lund University in 1982. He was awarded the title of doctor honoris causae from the University of Debrecen in 2008 and the Swedish Concrete Award in 2011. From 1975 he worked as a research assistant at Lund University, from 1982 as a consultant in research and building physics in Reykjavik and from 1990 as a professor in Building Technology at KTH in Stockholm and from 2008 an affiliated professor at KTH. His research has mainly concerned the thermodynamical studies of buildings, innovative building systems and energy conservation in the built environment. Since the beginning of 2008 he is the Director General of the Icelandic National Energy Authority which is responsible for public administration of energy research, energy utilization and regulation. He was a member of the The Hydropower Sustainability Assessment Forum processing the Hydropower Sustainability Assessment Protocol adopted by IHA in November 2010 and presently the chair of IPGT the International Partnership for Geothermal Technologies.

Professor Lynne A. Slivovsky

California Polytechnic State University, USA

The Questions That Keep Me Up At Night

Abstract:

This keynote will provide an opportunity for reflection on the work we do. We're here talking about energy and sustainability but we're also talking about a different way of living. We, as a technical field, a society, a world, are on a path of profound technological development. What does it mean to educate someone to contribute to this world? To have a technical education? What does it mean to live in this world? And is it possible that we as designers, innovators, engineers, and scientists can consider these questions in our day-to-day work?

Biography:

Lynne A. Slivovsky (Ph.D., Purdue University, 2001) is Associate Professor of Electrical and Computer Engineering at California Polytechnic State University, San Luis Obispo, California, USA. In 2003 she received the Frontiers In Education New Faculty Fellow Award. Her work in service-learning led to her selection in 2007 as a California Campus Compact-Carnegie Foundation for the Advancement of Teaching Faculty Fellow for Service-Learning for Political Engagement. In 2010 she received the Cal Poly President's Community Service Award for Significant Faculty Contribution. She currently oversees two multidisciplinary service-learning programs: the Access by Design Project that has capstone students designing recreational devices for people with disabilities and the Organic Twittering Project that merges social media with sustainability. Her work examines design learning in the context of engagement and the interdependence between technology and society.
Panel: Sustainability: Current and Future with focus on Energy, Buildings and ICT

Panel:

Sofia Ahlroth, Working Party on Integrating Environmental and Economic Policies (WPIEEP), Swedish EPA
Magnus Enell, Senior Advisor Sustainability at Vattenfall AB, Sweden
Göran Finnveden, Professor, KTH Royal Institute of Technology, Sweden
Danielle Freilich, Environmental expert at The Swedish Construction Federation (BI), Sweden
Catherine.Karagianni, Manager for Environmental and Sustainable Development at Teliasonera, Sweden
Örjan Lönngren, Climate and energy expert, Environment and Health Department, City of Stockholm, Sweden
Per Sahlin, Simulation entrepreneur, Owner of EQUA Simulation AB, Sweden
Mark Smith, Professor, KTH Royal Institute of Technology, Sweden
Örjan Svane, Professor, KTH Royal Institute of Technology, Sweden
Olle Zetterberg, CEO, Stockholm Business Region, Stockholm, Sweden

Contents

Volume 1

Session: Sustainability in Energy and Buildings

Short Papers

Session: Smart Buildings, Smart Grids

Invited Sessions

Volume 2

Session: Methodology for Renewable Energy Assessment

Session: Improving Office Building Energy Performance

Session: Multi-Energy Sources

Session: Technologies and Applications of Solar Energy

Session: Energy Planning in Buildings and Policy Implications

Session: Sustainable Energy Systems and Building Services

Session: Sustainable and Healthy Buildings

Chapter 1
Transformational Role of Lochiel Park Green Village

Stephen Robert Berry

Barbara Hardy Institute, University of South Australia
Mawson Lakes, SA 5095, Australia

Abstract. Energy use in housing has a significant negative impact on the environment. The South Australian Government responded to concern for anthropogenic greenhouse gas emissions by creating a model green village of near zero carbon homes in a near zero carbon impact estate. The creation of the Lochiel Park Green Village challenged actors from industry and government to set objectives, performance targets and regulatory guidelines outside existing institutional and professional norms. Evidence collected through a series of interviews has found that industry has responded to their involvement in the development by shifting away from some dominant technologies, practices and beliefs, and embracing new tools, construction practices and technologies, and policy makers have used the experience to consider new standards of building performance. Using a multi-level socio-technical framework this paper demonstrates how structural change at the regime level has come from the experience of actors at the niche level. The creation of the Lochiel Park Green Village has allowed many organisations to gain a more detailed and practical understanding of sustainable housing, and has given organisations the confidence to change industry practices, government policies, and regulatory standards.

1 Introduction

New homes, like all products, embody the complex interaction of economic, political, technical and professional influences (Bijker & Law 1992). Our buildings are the result of many social, economic and technical influences, such as the cultural institutions that shape communities, the technology norms applied by industry, the education and training of actors that meet market demand for new housing, and the experience of actors that demand and use new residential products (Guy 2006). The interaction of many actors, institutions, rules and technologies over many years has brought us to the point where developed nations create high energy use housing that has a significant negative impact on the local environment and the global climate (Organisation for Economic Co-operation and Development 2003).

Housing in developed nations such as Australia has evolved to the extent that households expect high levels of thermal comfort control, expect electricity to be readily available at the flick of a switch, and expect hot water on demand, irrespective of the environmental impact of making that energy available. Housing expectations and cultural norms, embedded institutional construction practices, regulatory

A. Håkansson et al. (Eds.): *Sustainability in Energy and Buildings*, SIST 22, pp. 1–13.
DOI: 10.1007/978-3-642-36645-1_1

standards and dominant technologies have evolved in parallel without due consideration of the resultant ecological footprint (Crabtree 2005).

The Australian building sector constructs around 140,000 new dwellings per year according to the prevailing economic conditions (Australian Bureau of Statistics 2010). Each new dwelling that uses more energy than it produces increases the need for additional electricity generation capacity and associated energy supply infrastructure, and adds to global greenhouse gas emissions.

Since the 1980s consecutive Australian governments have recognised the need to address anthropogenic climate change. This policy evolution has been shaped by the dominant liberalism political economy theory advocating competitive and free markets, where the role of government is to steer the economy by the development of a legal and institutional framework within which the market operates and address market imperfections with market-based responses (Davidson 2011). As such, the Australian Government does not control national emissions directly, but instead sets targets, encourages action by other actors, supports the creation of institutions, enacts laws, and provides market-based incentives that facilitate change.

Throughout the development of Australian Government climate change policy, the residential sector has been highlighted as an important and integral part of the strategy to reduce Australia's greenhouse gas emissions (Australian Minerals and Energy Council 1990; Council of Australian Governments 1992; Australian Greenhouse Office 1998; Council of Australian Governments 2009). This policy imperative has manifested as minimum energy efficiency standards introduced into the Building Code of Australia approximating 4 NatHERS stars in 2003, 5 stars in 2006 and 6 stars in 2010, where 0 NatHERS Stars represents a home frequently thermally uncomfortable and 10 Stars represents a home that is thermally comfortable in that local climate without the need for additional heating or cooling. At 6 Stars, this level of thermal comfort related building energy performance remains far below the regulatory standards set for equivalent climates in the United Kingdom and the United States of America (Horne et al. 2005).

Concern for anthropogenic greenhouse gas emissions has lead governments in Europe, North America, Australia and Asia to consider regulating building energy and carbon performance approximating net zero energy or carbon (Kapsalaki & Leal 2011). In the United Kingdom, the national government has set the path to net zero carbon for new residential buildings by 2016 (Department of Communities and Local Government 2006). Similarly in Australia, policy makers have suggested developing a pathway to net zero carbon buildings by 2020 (Department of Climate Change and Energy Efficiency 2010).

During this policy debate on the best approach to reduce the carbon impact of homes, the South Australian Government chose to showcase the cutting edge of sustainable urban development by creating a model green village of national and international significance (Bishop 2008). This case study of the Lochiel Park Green Village demonstrates the role niche applications can play in transforming the residential market.

2 Socio-Technical Transitions

To move to a new housing paradigm of significantly lower environmental impact will require a substantial shift away from dominant technologies, practices and beliefs (Moloney, Horne & Fien 2010). Change will be necessary within existing institutions and professional norms, technologies that are currently considered innovative will need to become standard systems, and lifestyles are likely to be altered (Brown & Vergragt 2008).

Any change towards sustainability, especially one as large as a move to net zero carbon homes, will face many barriers (Lowe & Oreszczyn 2008; Crabtree & Hes 2009; Geels 2010; Moloney, Horne & Fien 2010). To understand the change process, the role of the various actors and institutions, and the barriers faced, it is valuable to apply a suitable analytical framework. In this paper a multi-level socio-technical framework is used to describe how structural change comes from the interplay between diversity creation at the niche level, the selection environment at the regime level, and changes on the landscape level (Smith 2003; Geels 2004, 2010).

Any transition from the status-quo, represented by the interconnection between technologies, institutions and society which has reached a level of stability, will encounter resistance due to the dominant regime of existing technological production practices, organisational structures and cultural values, and the beliefs of various actors (Hoogma, Weber & Elzen 2005). Sunk investments, not just in built infrastructure but in behavioural patterns, regulatory structures, organisational systems, policy frameworks and the like, mean that transitions challenge technological, institutional and social vested interests (Unruh 2000).

Studies covering various industries have found that niche situations, created by pioneering organisations, can lead to radical change (Kemp, Schot & Hoogma 1998; Unruh 2002; Smith 2007; Verbong & Geels 2007). Smith (2007) argues that the creation of niche situations provides the opportunity for new ideas, artefacts and practices to evolve without the pressure and overwhelming influence of the current regime. These case studies have illustrated that some changes proven at niche situations have been adopted by organisations, changed standard behaviour patterns, altered policy frameworks, infiltrated regulatory structures, and delivered transitions which have become the new status-quo.

In this paper we will see that the Lochiel Park case study demonstrates how, as a response to the landscape level impact of climate change, the creation of a niche development, removed to a large extent from the constraints of the current norms, has started to transform the local building market towards higher levels of environmental sustainability.

3 Creation of Lochiel Park Green Village

Many processes, policies, and interactions between actors, have led to the concept of the Lochiel Park Green Village (Bishop 2008), and many other interactions, processes and tools have resulted in the design and construction of housing which is significantly more environmentally sustainable than the typical product previously offered (Saman et al. 2011b).

In 2001, the Land Management Corporation (LMC), an agency of the South Australian Government responsible for the creation and implementation of residential, commercial and industrial development projects, purchased 15 hectares of surplus government land at Lochiel Park (Land Management Corporation 2005), approximately eight kilometres North-East of the Adelaide central business district. Originally intended to be disposed to the commercial property market as a standard broad-acre land sale of 150 residential allotments, the change of state government in 2002 introduced new policy objectives including an increased interest in delivering environmental outcomes (Donaldson, Bishop & Wilson 2008).

By 2003, policy agreement had been reached within the South Australian Government to limit development to only 4.25 of 15 hectares, retaining the rest as open space by incorporating an urban forest and substantial wetlands (Donaldson, Bishop & Wilson 2008).

In 2004, the Premier of South Australia, Mike Rann declared the project's intent:

> "I want South Australia to become a world leader in a new green approach to the way we all live. The Lochiel Park Development will become the nation's model 'Green Village' incorporating Ecologically Sustainable Development (ESD) technologies."

Given explicit direction from the State Premier to develop a niche urban development of world standing, an advisory panel of state and local government officials and community representatives was established to define the project objectives.

The project's objectives were established as (Land Management Corporation 2005):

Green Village

Develop the land as a model 'Green Village' of national significance incorporating a range of best practice sustainable technologies, which will serve as a model for other urban developments

- Create a 'showcase' for ESD
- Raise environmental awareness
- Foster a culture of sustainability

Urban Consolidation

To achieve Government urban consolidation objectives, develop a high quality, medium density, master planned residential project, incorporating a diversity of housing product

- Capitalise on a surplus Government land asset
- Develop innovative, acceptable and desirable design solutions for urban consolidation projects

Expand Housing Choice

Provide a mix of housing types and meet a range of housing opportunities with products that have broad market acceptance in regard to both design and cost

- Provide housing choice
- Seek affordable housing solutions

Excellence in Urban Design

Achieve excellence in urban design and innovative built form outcomes through an integrated approach to development providing a high level of residential amenity

- Create a model for future urban infill projects
- Push the boundaries of urban design and built form

Enhance Biodiversity

Complement and enhance the biodiversity of the adjoining open space areas, through the minimisation of impacts from the residential development on the surrounding environment

- Create a sustainable urban area that minimises environmental impacts though the management of water, energy and waste
- Use appropriate landscaping to complement the adjoining natural open space areas

Open Space Planning

Facilitate the planning and development of the open space areas, incorporating an urban forest and other active and passive recreation

- Provide a well-planned open space area that contributes to the increased biodiversity of the area
- Consult with key stakeholders and the community regarding open space outcomes

Integrate with Surroundings

Ensure appropriate linkages and integration with surrounding land uses to promote community interaction and a sense of place and belonging

- Complement and integrate with adjoining pedestrian/cycle trails
- Use urban design techniques to reduce the potential for the creation of an insular development

Financial Return

- Achieve an acceptable return on investment for Government
- Ensure the project is economically sustainable and provides an appropriate return on investment to Government

These objectives demonstrated government policy interest across a relatively broad range of areas including environmental sustainability, social sustainability, urban form, transport, industry development and economics. The development of Lochiel Park was seen by the South Australian Government as more than the building of another standard housing estate, but rather, it was seen as an opportunity to foster a culture of environmental sustainability within the house building industry and the wider community, establishing new standards in urban form and house design, while at the same time delivering an economic return for the government.

Leadership from the highest level of State Government allowed many organisations, from government, academia and industry, the freedom to consider innovations outside of dominant technologies, practices and beliefs, and to establish a new set of rules, tools and practices that would shape the development.

4 Setting New Standards

The development of high level objectives provided guidance but little numerical detail of appropriate performance targets. To enable the development of specific numerical targets the government commissioned a benchmarking exercise which examined approaches taken in other major international and national niche green residential developments (Bishop 2008; Donaldson, Bishop & Wilson 2008). Areas of interest included water, energy, waste, biodiversity, transport, sustainable design and built form, landscaping, community and information technology.

The initial detailed performance targets developed for the sustainability framework were set against a baseline developed from Australian Bureau of Statistics data for South Australian households in 2004, stating:

LMC's Green Village will connect environmental, social and economic principles and create a showcase sustainable development that:

- Reduces potable water consumption by 50%;
- Requires plumbed rainwater tanks for every home;
- Reduces open space irrigation demand by 100%;
- Cleans urban stormwater and treats for re-use;
- Reduces energy consumption by 50% to 60%;
- Requires solar water heaters and photovoltaic cells in every home;
- Uses solar lighting for public open space;
- Offsets community CO_2 emissions via urban forest & wetland plantings;
- Provides access to smart metering to monitor resource usage;
- Applies sustainable building design for energy efficiency;
- Requires every dwelling meets eight star energy ratings;
- Recycles 100% of all building waste;
- Encourages reduction of 30% in solid waste production;
- Encourages composting of 100% of organic waste;
- Investigates opportunities for community gardens;
- Ensures all dwellings have optimal solar access;
- Enables a 200 year building life span;
- Establishes the priority of pedestrians in shared used zones;
- Selects building materials based on whole of life cycle costs;
- Investigates shared transport systems (eg community vehicles);
- Provides extensive open space relative to residential footprint;

- Provide educational material on local heritage (including natural, European and Aboriginal) and sustainability principles;
- Ensures all households are within 200 metres of public open space;
- Uses endemic and native plants in reserves & open space; and
- Enables a connected community through provision of broadband Internet facilities and community web portal.

These detailed targets were refined in consultation with the various expert groups (Donaldson, Bishop & Wilson 2008), with final design overarching targets set at a reduction of:

- 66% energy used
- 74% greenhouse gas emissions
- 78% potable water use

The various elements of the detailed and overarching targets would be delivered through two main paths: (a) the creation of community level infrastructure; and (b) the creation of individual houses. This meant that meeting the targets would require the adoption of sustainability principles and practices by many different contractors and organisations, cooperating to deliver a single coherent sustainable development.

Many of the community infrastructure related targets were translated by consultant experts into physical plans such as the Development Master Plan and a Community Development Plan. The Development Master Plan set out the physical layout of the development, establishing links to the river and walking/cycling paths, local transport connections, plus landscaping and stormwater harvesting opportunities. The Community Development Plan focused on creating a physical and social environment that should lead to environmentally beneficial behavioural change, reduced crime and an enhanced sense of community. Functional plans were also generated for the associated stormwater treatment wetlands.

Expert consultative groups for energy and water were used to translate the original targets into rules and guidance materials that could communicate the desired sustainability principles and practices to be applied to each house. These were published in the Urban Design Guidelines (Land Management Corporation 2009), which spell out those actions that are mandatory and a further set of 'advisory' actions that are encouraged to facilitate improvements in lifestyle, amenity and sustainability. The engagement of consultative groups allowed a 'reality check' interaction between the target setters and the local industry and academic experts, whereby specific technical issues could be addressed through revision or re-interpretation of targets.

The final Urban Design Guidelines established a set of performance requirements designed to create near zero carbon homes in a near zero carbon impact development, and were published in a form that communicated the intent equally to the building industry and prospective households. The minimum requirements included:

- 7.5 NatHERS Stars thermal comfort
- Solar water heating, gas boosted
- 1.0kWp photovoltaic system for each 100m^2 of habitable floor area
- High star rated appliances

- Low energy lighting (CFLs & LEDs)
- Ceiling fans in all bedrooms and living spaces
- Rainwater harvesting feeding the hot water system
- Greywater harvesting feeding toilet flushing

The Guidelines established a new set of rules, calling for practices sometimes outside existing institutions and professional norms, requiring the application of technologies and systems uncommon to the mainstream building industry, and the consideration of new performance indicators bringing new concepts to building design and construction practices.

Upon the release of the Urban Design Guidelines, housing product was offered to the open market in 2007 with construction of the first homes beginning in 2008. The majority of the 106 housing allotments were sold by late 2011, and around half of the homes had been completed and occupied by mid-2012.

5 Major Barriers

A number of industry experts, policy makers and members of the Lochiel Park community were interviewed to ascertain the types of impacts made from the creation of the green village and the barriers to its development. These semi-structured 'face-to-face' interviews used a common set of open-ended questions to draw on the experiences and perceptions of those playing a key role in the Lochiel park development.

Earlier it was discussed that any process of socio-technical transition will be subject to barriers associated with the lock-in of dominant technologies, practices and beliefs. The following paragraphs note some of the barriers found within policy organisations, regulatory systems, utilities, the building industry, the real estate industry, and the households as users of the end building product, that were identified from the interviews.

The interpretation of development objectives into quantified performance targets has challenged the embedded policies and processes of policy organisations, particularly because agencies were relatively inexperienced at delivering sustainable housing. Policy makers interviewed noted that institutionalised conservatism and concern for delivering affordable housing, allowed industry to push back on specific desired environmental actions, albeit without a loss of overall environmental outcome. For example: the original 8 Star NatHERS target for thermal comfort was revised in the Urban Design Guidelines to a 7.5 Star target; and the photovoltaic requirement was re-interpreted from a mandatory 1.5 kWp requirement for each dwelling, to a minimum 1.0 kWp capacity for each $100m^2$ of habitable floor area. The reduction in the NatHERS requirement reflected the perceived limit of cost effective thermal comfort using standard construction techniques. The change in the photovoltaic requirement reflected a more economically equitable target which increased system costs in line with expected increased energy demand for larger homes.

Both building industry professionals and policy makers indicated there were perceived technical barriers associated with going beyond the minimum requirements of the Building Code of Australia, as the mainstream building industry was not experienced with designing and building sustainable homes with high levels of thermal comfort and the incorporation of solar thermal and solar photovoltaic technologies. The relatively high sustainability requirements are thought to have frightened some builder companies away from direct involvement, and in the end having only two volume builders providing housing product probably meant a loss in product diversity and consumer choice.

The building industry was reluctant to change some construction practices, and raised issues of risk premiums impacting housing affordability and market acceptance. For example: reverse brick veneer construction was encouraged but resisted by the industry, which successfully negotiated performance standards sufficiently low to continue using standard construction techniques.

Similarly, interviewees noted the real estate industry was reluctant to change their practices, and maintained a mindset that consumers were not interested in environmentally sustainable features. For example: the real estate industry was comfortable communicating properties using traditional house valuation metrics (no. of bedrooms, no. of bathrooms, floor area, car spaces, air-conditioning etc), rather than communicate unfamiliar features of sustainable homes (size of photovoltaic array, higher thermal comfort, lower energy costs, etc).

The local electricity infrastructure provider demonstrated an institutional barrier to recognising the benefit of an expected lower average and peak load for the green village. The privately owned utility installed standard infrastructure, even though the lower load and self-sustaining nature of the homes should have required a less costly level of infrastructure.

Some interviewees suggested that government policy organisations were slow to react and develop appropriate policy settings to support the development. For example: the delay in communicating the solar rebate policy caused some uncertainty in the economics of photovoltaic systems, and discouraged some households from increasing their systems above the minimum requirement.

Institutional barriers were found across many government agencies because 'nation leading' often meant doing things differently. For example: the transport department allocated the standard 50km/h speed limit to the development when an active walking/cycling community would have been better supported with a much lower speed limit.

An economic barrier to the delivery of the development was the process of allocating economic costs. There are many aspects of the Lochiel Park development that meet wider government policy objectives (water, waste, biodiversity, energy), yet the full costs of delivery was allocated to the project and not to the agencies responsible for determining the policy. For example: the district stormwater recycling scheme provides a wider community benefit to the City of Campbelltown and the users of the River Torrens, yet all costs are allocated to Lochiel Park and are passed to the home buyers.

An interesting barrier to the consistent achievement of energy and water savings has been the interaction between households and technologies. Many of the households moving into the development were not highly technologically literate and have been unable to operate and maintain the solar technologies to their full potential (Saman et al. 2011a). And although an interactive energy and water use feedback monitor was installed in each house to allow households to track and modify their behaviour, the inability of households to operate the feedback and solar systems has limited the monitor's effect.

6 Lochiel Park's Impacts

The impact of niche events can be considerable and wide ranging. Earlier it was noted that new ideas, technologies, artefacts and practices that evolve at niche events may be adopted at regime level leading to a more environmentally sustainable outcome (Kemp, Schot & Hoogma 1998; Unruh 2002; Verbong & Geels 2007). The Lochiel Park case study points to a number of impacts perceived by policy makers and industry professions which have led to changes in processes, skill sets, practices, knowledge, and policies, suggesting some change in the incumbent socio-technical regime.

The sustainability transitions literature describes how niche developments can provide vital education and training opportunities that facilitate changes to existing institutional and professional norms (Hoogma, Weber & Elzen 2005; Geels 2010; Moloney, Horne & Fien 2010). The evidence provided by interviewees attests to a similar experience for the Lochiel Park case study.

Building industry professionals suggested that several large building companies and their associated tradespersons have gained substantial experience designing and building sustainable homes at Lochiel Park, and that the experience has led to the adoption of new practices and new understanding in the installation of technologies such as solar water heaters, photovoltaics and LED lighting. The result is a section of the local building industry now confident they can design and construct product attractive to the new home market at 7.5 Stars, with lighting densities less than $3.5 w/m^2$, and with local electricity generation. Involvement in Lochiel Park has also given building designers experience in using sustainability performance assessment tools.

Some of these experiences have been shared with the broader building industry through education and training sessions held at Lochiel Park or presented by those involved with the development. Training sessions for the Housing Industry Association's GreenSmart program have been held on site to allow participants to gain practical knowledge of sustainable housing. The Green Plumbers and Green Painters programs are amongst the many other groups that have held training courses at Lochiel Park.

Policy makers such as planners and local government officials have participated in guided tours of Lochiel Park, and university training programs have incorporated site visits for architecture, renewable engineering and property courses. Students at primary and high school have also been exposed to new concepts in urban sustainability during informative site visits.

Lochiel Park has demonstrated to policy makers, the building industry and the real estate industry that narrow, small lot development, which increases urban density whilst lowering the development's ecological footprint, is technically feasible and attractive to the local housing market. The development density of 25 dwellings per hectare is nearly double that traditionally applied locally to new estates (Bishop 2008).

Policy interviewees have suggested that the success of delivering Lochiel Park has given government the confidence to move policy goalposts, with some agencies establishing sustainability requirements above industry norms or regulatory standards. For example: all 60 display homes at the 2300 dwelling Light's View development in Adelaide (Land Management Corporation 2010) have a minimum requirement of 7 NatHERS Stars, well above the Building Code of Australia requirement of 6 Stars, and all dwellings on the estate are required to have either solar thermal water heating or solar photovoltaic electricity generation.

The Lochiel Park Sustainability Centre, a public information centre, has allowed the government to communicate concepts of sustainability for estate design, house design and lifestyle to the broader community. Thousands of people have visited the Sustainability Centre and learnt about many different aspects of sustainable living.

Lochiel Park, particularly through the development of the community garden and residents association, has demonstrated that community spirit can be achieved through empowering people to engage with their local community, even on new estates.

Greater understanding by house purchasers of sustainability opportunities has been achieved through the practical demonstration of actions and by information dissemination. Policy makers have suggested that the creation of Lochiel Park has encouraged aspirational thinking by other policy makers and the greater community on the potential for environmentally sustainable communities.

An unusual aspect of the development is a commitment to monitor and analyse the energy and water use of all Lochiel Park homes for at least five years. This means that the Lochiel Park Green Village has the potential to provide a legacy of data to inform future policy targets.

The Zero Carbon Challenge (Land Management Corporation 2011), a house design competition for the final residential allotment released at Lochiel Park, has enabled further interaction between architects, builders, engineers and student researchers on creating a net zero carbon home. The competition winner is contractually obligated to build the house and make it available as a display home for a set period.

A number of interviewees noted that by demonstrating industry can build high performing houses, the creation of the Lochiel Park Green Village has influenced policy makers to consider net zero carbon as a future regulatory standard.

The practical experience of creating the niche sustainable housing development has allowed many organisations to interact with new performance standards, construction techniques, and technologies. These interactions have facilitated changes to industry practices, government policies, regulatory standards, and the way the community understands housing.

7 Conclusion

Concern for anthropogenic greenhouse gas emissions has lead governments to consider regulating building energy and carbon performance. Responding to this exogenous environmental impact the South Australian Government chose to showcase the cutting edge of sustainable urban development by creating a model green village of near zero carbon homes in a near zero carbon impact estate.

The creation of the Lochiel Park Green Village challenged many organisations to set objectives, performance targets and regulatory guidelines outside existing institutional and professional norms, thus requiring a shift away from dominant technologies, practices and beliefs. At this early stage in the Lochiel Park niche some evidence is emerging of impacts that if sustained and absorbed into the mainstream may lead to wider environmental outcomes.

By applying a multi-level socio-technical framework we can see how structural change at the regime level has come from new tensions at the landscape level and the interplay of actors at the niche level. The creation of the Lochiel Park Green Village has allowed many organisations to gain a more detailed and practical understanding of sustainable housing. This experience has given organisations the confidence to change industry practices, government policies, and regulatory standards.

References

Australian Bureau of Statistics, Building Approvals 8731.0, Commonwealth of Australia, Canberra (2010)

Australian Greenhouse Office, The National Greenhouse Strategy, Australian Greenhouse Office, Canberra (1998)

Australian Minerals and Energy Council, Energy and the Greenhouse Effect, Commonwealth of Australia, Canberra (1990)

Bijker, W.E., Law, J.: Shaping technology/building society: studies in sociotechnical change. MIT Press, Cambridge (1992)

Bishop, A.: Lochiel Park Case Study, CSIRO (2008), http://yourdevelopment.org/casestudy/view/id/7 (viewed February 7, 2012)

Brown, H.S., Vergragt, P.J.: Bounded socio-technical experiments as agents of systemic change: The case of a zero-energy residential building. Technological Forecasting and Social Change 75(1), 107–130 (2008)

Council of Australian Governments, National Greenhouse Response Strategy, Commonwealth of Australia, Canberra (1992)

Council of Australian Governments, National Strategy on Energy Efficiency, Commonwealth of Australia, Canberra (2009)

Crabtree, L.: Sustainable housing development in urban Australia: Exploring obstacles to and opportunities for ecocity efforts. Australian Geographer 36(3), 333–350 (2005)

Crabtree, L., Hes, D.: Sustainability uptake on housing in metropolitan Australia: an institutional problem, not a technological one. Housing Studies 24(2), 203 (2009)

Davidson, K.: A Typology to Categorize the Ideologies of Actors in the Sustainable Development Debate. Sustainable Development (2011)

Department of Climate Change and Energy Efficiency, Report of the Prime Minister's Task Group on Energy Efficiency, Commonwealth of Australia, Canberra (2010)

Department of Communities and Local Government, Building a Greener Future: Towards Zero Carbon Development, Department of Communities and Local Government, London (2006)

Donaldson, P., Bishop, A., Wilson, M.: Lochiel Park - A nation leading green village. Eco City World Summit, San Francisco (April 2008)

Geels, F.W.: From sectoral systems of innovation to socio-technical systems: Insights about dynamics and change from sociology and institutional theory. Research Policy 33(6-7), 897–920 (2004)

Geels, F.W.: Ontologies, socio-technical transitions (to sustainability), and the multi-level perspective. Research Policy 39(4), 495–510 (2010)

Guy, S.: Designing urban knowledge: Competing perspectives on energy and buildings. Environment and Planning C: Government and Policy 24(5), 645–659 (2006)

Hoogma, R., Weber, M., Elzen, B.: Integrated Long-Term Strategies to Induce Regime Shifts towards Sustainability: The Approach of Strategic Niche Management, In: Weber, M., Hemmelskamp, J. (eds.) Towards Environmental Innovation Systems, pp. 209–236. Springer, Heidelberg (2005)

Horne, R., Hayles, C., Hes, D., Jensen, C., Opray, L., Wakefield, R., Wasiluk, K.: International comparison of building energy performance standards. RMIT Centre for Design, Melbourne (2005)

Kapsalaki, M., Leal, V.: Recent progress on net zero energy buildings. Advances in Building Energy Research 5(1), 129 (2011)

Kemp, R., Schot, J., Hoogma, R.: Regime shifts to sustainability through processes of niche formation: The approach of strategic niche management. Technology Analysis and Strategic Management 10(2), 175–195 (1998)

Land Management Corporation, Lochiel Park Green Village Development Project: Submission to Public Works Committee, Land Management Corporation, Adelaide (2005)

Land Management Corporation, Lochiel Park Urban Design Guidelines, Land Management Corporation, Adelaide (2009)

Land Management Corporation, Light's View, Land Management Corporation, Adelaide (2010), http://www.lightsview.com.au (viewed June 13, 2012)

Land Management Corporation, Zero Carbon Challenge, Adelaide, (2011), http://www.lmc.sa.gov.au/zerocarbonchallenge (viewed June 8, 2012)

Lowe, R., Oreszczyn, T.: Regulatory standards and barriers to improved performance for housing. Energy Policy 36(12), 4475–4481 (2008)

Moloney, S., Horne, R.E., Fien, J.: Transitioning to low carbon communities-from behaviour change to systemic change: Lessons from Australia. Energy Policy 38(12), 7614–7623 (2010)

Organisation for Economic Co-operation and Development, Environmentally sustainable buildings: challenges and policies, Organisation for Economic Co-operation and Development, Paris (2003)

Saman, W., Babovic, V., Whaley, D., Liu, M., Mudge, L.: Assessment of Electricity Displacement due to Installation Parameters of Solar Water Heaters, University of South Australia, Adelaide (2011a)

Saman, W., Mudge, L., Whaley, D., Halawa, E.: Sustainable Housing in Australia: Monitored Trends in Energy Consumption. Sustainability in Energy and Buildings, Marseilles, France, p. 247 (2011b)

Smith, A.: Transforming technological regimes for sustainable development: A role for alternative technology niches? Science and Public Policy 30(2), 127–135 (2003)

Smith, A.: Translating sustainabilities between green niches and socio-technical regimes. Technology Analysis and Strategic Management 19(4), 427–450 (2007)

Unruh, G.C.: Understanding carbon lock-in. Energy Policy 28(12), 817–830 (2000)

Unruh, G.C.: Escaping carbon lock-in. Energy Policy 30(4), 317–325 (2002)

Verbong, G., Geels, F.: The ongoing energy transition: Lessons from a socio-technical, multi-level analysis of the Dutch electricity system (1960-2004). Energy Policy 35(2), 1025–1037 (2007)

Chapter 2
Evaluation and Validation of an Electrical Model of Photovoltaic Module Based on Manufacturer Measurement

Giuseppe M. Tina and Cristina Ventura

Abstract. The analysis of the performance of a photovoltaic (PV) array needs basically the reporting the real working conditions to a reference condition of irradiance and temperature. Normally it is used the Standard Test Conditions (STC). Then the corrected I-V curves can be compared and an analysis of the performances can be carried out. In this context this paper proposes an analytical model to evaluate the energy performance of a PV module. The proposed model is based on some data provided by the manufacturer of the module in STC conditions. The photovoltaic module used as test-bed in the experiments gives the possibility to have the six terminals of the three strings forming the module, that normally are connected in series. This is very useful in the case of shading or disuniform radiation. The model is validated with numerical examples, and tested using both measured and estimated data relative to each single string and their connection in series and parallel. Results show how the parameters extraction depends on the measured value of the maximum power points, if measures are not accurate the analytic model here implemented can not converge to a feasible solution.

1 Introduction

Nowadays there are many efforts to increase the yearly energy production of a PV plant. Mainly it depends on design choices and construction solutions, but on the other hand the initial efficiency must be kept as high as the initial value of the PV plants in order to insure the goodness of the investment. In this context many efforts are focused on developing monitoring tools that, starting from a suitable model of the PV plant, can detect not only a specific difference between the expected efficiency and the measured one, but also a trend over the time that could indicate aging phenomena. These tools are mainly based on proper models of both PV module and PV strings. Actually in literature there are many models but very often they have been developed for PV cells mainly for testing aims. Then these models have been extended to PV modules and a few to PV arrays. The passage of the model from cell to module is critical as the presence of the layers of materials (e.g. Glass, EVA, Tedlar), that forms a PV module, causes some uncertainties, such as the real irradiance that strikes the cells inside a PV module and also the value and distribution of the temperature on the cells. The general approach to assess the electrical performance of a PV system is

A. Håkansson et al. (Eds.): *Sustainability in Energy and Buildings*, SIST 22, pp. 15–24.
DOI: 10.1007/978-3-642-36645-1_2

based upon the capability of analytically describing the I–V characteristic of the photovoltaic component for each operative temperature and solar radiation. Traditionally, the analytical models used in the study of these phenomena evaluate the behaviour of the PV cell by assimilating it into an equivalent electrical circuit that includes some non-linear components [13]. These electrical equivalent circuits are based on some unknown parameters, from three to seven depending on the complexity of the model; the most common electric circuit is known in literature as RP-model, which consists of a current generator, a diode, a series resistant and a shunt resistant. The characteristic equation of this equivalent circuit contains five independent parameters, for this reason it is also called "*five parameters model*". This model offers a good compromise between simplicity and accuracy and has been applied widely [3]. The five parameters can be evaluated by means of either numerical methods, that minimize the difference between a measured I-V curve and the one calculated by the model, or just using the technical data provided by that manufacturer of the PV module. However, due to the transcendental nature of the current equation for PV module, significant computation effort is required to obtain all the five model parameters [10]. For example, in [9] they have analyzed the development of a method for the mathematical modeling of PV arrays. In order to improve the accuracy of the method, analytic solutions [2, 6, 7, 8] and intelligence algorithms [4, 11, 12] are applied to deduce the all parameters. Moreover, recently various high accuracy algorithms techniques have been reported, such as particle swarm optimization (PSO) [15], differential evolution (DE) [10], genetic algorithm (GA) [13] and pattern search (PS) [1].

The aim of this paper is to develop a five parameter model that allows to evaluate the I-V curve of a photovoltaic model starting from data provided by the manufacturer relative to STC conditions. This model has been used since it has been successfully applied and it seems to give good approximations of the I-V curve [5, 14]. In this case, a module that allows to have as output the six terminals of the three strings forming the module, that normally are connected in series, is used to test and verify the adopted solution. The manufacturer provides data relative to each single string and their connection in series and parallel. These data are compared with data evaluated using the model here proposed.

2 Five Parameter Model

The five parameter model is relative to the equivalent circuit representative either a PV cell or a PV module, shown in Fig. 1. It is a complete circuit where both the sources of power losses are used, R_S and R_{SH}.

Application of Kirchoff's Current Law on the equivalent circuit results in the current flowing to the load:

$$I = I_{PH} - I_D - I_{SH} \tag{1}$$

If the diode current and the current through the shunt resistance (I_D and I_{SH}, respectively) are expanded, Eq. 2 is obtained.

$$I = I_{PH} - I_o \cdot \left(e^{\frac{V + I \cdot R_S}{\eta \cdot V_t}} - 1 \right) - \frac{V + I \cdot R_S}{R_{SH}} \tag{2}$$

where $V_t = m \cdot k \cdot T/q$, m is the number of cells connected in series, *k* is the Boltzmann's constant, *T* the absolute temperature and *q* the electronic charge.

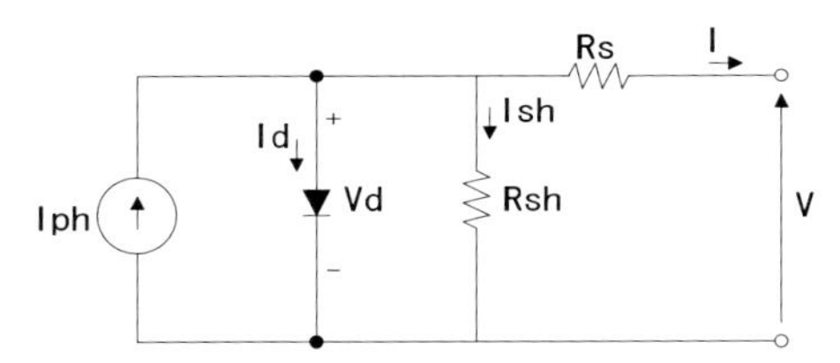

Fig. 1. Equivalent circuit representing the five-parameter model

The characteristic equation of the equivalent circuit contains five parameters: I_{PH} that represents the light current (A), I_0 that represents the diode reverse saturation current, η that represents the ideality factor, R_S that represents the series resistance and finally R_{SH} that represents the shunt resistance. In general, these five parameters are functions of the solar radiation incident on the cell and cell temperature. Reference values of these parameters are determined for a specified operating condition such as STC. Three current–voltage pairs are normally available from the manufacturer at STC: the short circuit current, the open circuit voltage and the current and voltage at the maximum power point. A fourth piece of information results from recognizing that the derivative of the power at the maximum power point is zero. Therefore, to calculate the five parameters, Eq. 1 has to be calculated in the following three points: open circuit "OC" (Eq. 3), short circuit "SC" (Eq. 4) and maximum power point "MP" (Eq. 5).

$$0 = I_{PH} - I_o \cdot \left(e^{\frac{V_{OC}}{\eta \cdot V_t}} - 1 \right) - \frac{V_{OC}}{R_{SH}} \tag{3}$$

$$I_{SC} = I_{PH} - I_o \cdot \left(e^{\frac{I_{SC} \cdot R_S}{\eta \cdot V_t}} - 1 \right) - \frac{I_{SC} \cdot R_S}{R_{SH}} \tag{4}$$

$$I_{MP} = I_{PH} - I_o \cdot \left(e^{\frac{V_{MP} + I_{MP} \cdot R_S}{\eta \cdot V_t}} - 1 \right) - \frac{V_{MP} + I_{MP} \cdot R_S}{R_{SH}} \tag{5}$$

Differentiating Eq. 2 with respect to V gives:

$$\frac{dI}{dV} = - \frac{\frac{I_o}{\eta \cdot V_t} \cdot e^{\frac{V + I \cdot R_S}{\eta \cdot V_t}} + \frac{1}{R_{SH}}}{\frac{I_o \cdot R_S}{\eta \cdot V_t} \cdot e^{\frac{V + I \cdot R_S}{\eta \cdot V_t}} + \frac{R_S}{R_{SH}} + 1} \tag{6}$$

Calculating Eq. 2 at the maximum power point, we have:

$$\frac{dP}{dV} = I + V \cdot \frac{dI}{dV} \tag{7}$$

$$I_{MP} - V_{MP} \cdot \left(\frac{\frac{I_o}{\eta \cdot V_t} \cdot e^{\frac{V_{MP}+I_{MP} \cdot R_S}{\eta \cdot V_t}} + \frac{1}{R_{SH}}}{\frac{I_o \cdot R_S}{\eta \cdot V_t} \cdot e^{\frac{V_{MP}+I_{MP} \cdot R_S}{\eta \cdot V_t}} + \frac{R_S}{R_{SH}} + 1} \right) = 0 \tag{8}$$

Considering and manipulating Eq. (3), (4) and (5) it is possible to express the parameters I_{PH}, I_0 and R_{SH} in function of R_S and η:

$$R_{SH} = -R_S + \frac{V_{MP} \cdot \left(e^{\frac{I_{SC} \cdot R_S}{\eta \cdot V_t}} - 1 \right) \cdot \left(e^{\frac{I_{SC} \cdot R_S}{\eta \cdot V_t}} - e^{\frac{V_{OC}}{\eta \cdot V_t}} \right) - V_{OC} \cdot \left(e^{\frac{I_{SC} \cdot R_S}{\eta \cdot V_t}} - e^{\frac{V_{MP}+I_{MP} \cdot R_S}{\eta \cdot V_t}} \right) \cdot \left(e^{\frac{I_{SC} \cdot R_S}{\eta \cdot V_t}} - 1 \right)}{I_{SC} \cdot \left[\left(e^{\frac{V_{MP}+I_{MP} \cdot R_S}{\eta \cdot V_t}} - 1 \right) \cdot \left(e^{\frac{I_{SC} \cdot R_S}{\eta \cdot V_t}} - e^{\frac{V_{OC}}{\eta \cdot V_t}} \right) - \left(e^{\frac{V_{OC}}{\eta \cdot V_t}} - 1 \right) \cdot \left(e^{\frac{I_{SC} \cdot R_S}{\eta \cdot V_t}} - e^{\frac{V_{MP}+I_{MP} \cdot R_S}{\eta \cdot V_t}} \right) \right] - I_{MP} \cdot \left(e^{\frac{I_{SC} \cdot R_S}{\eta \cdot V_t}} - 1 \right) \cdot \left(e^{\frac{I_{SC} \cdot R_S}{\eta \cdot V_t}} - e^{\frac{V_{OC}}{\eta \cdot V_t}} \right)} \tag{9}$$

$$I_{PH} = \frac{V_{OC} \cdot \left(e^{\frac{I_{SC} \cdot R_S}{\eta \cdot V_t}} - 1 \right) - I_{SC} \cdot (R_S + R_{SH}) \cdot \left(e^{\frac{V_{OC}}{\eta \cdot V_t}} - 1 \right)}{R_{SH} \cdot \left(e^{\frac{I_{SC} \cdot R_S}{\eta \cdot V_t}} - e^{\frac{V_{OC}}{\eta \cdot V_t}} \right)} \tag{10}$$

$$I_O = \frac{R_{SH} \cdot (I_{PH} - I_{SC}) - R_S \cdot I_{SC}}{R_{SH} \cdot \left(e^{\frac{I_{SC} \cdot R_S}{\eta \cdot V_t}} - 1 \right)} \tag{11}$$

Substituting these equations in (8) we have an equation in two variables R_S and η. Calculating the absolute minimum of the obtained function, it is possible to find the five parameters. The main point is that this approach has the advantage to use input data that are always provided by the manufacturers such as: V_{OC}, I_{SC}, V_{MP}, I_{MP}, irradiance and the PV cell temperature. In particular these data refer to STC conditions. In this case these parameters are specified with a lowercase "ref ", as follows: I_{PH_ref}, I_{0_ref}, $\eta_{\_ref}$, R_{S_ref} and R_{SH_ref}.

3 Experimental Setup

For the experiments, data relative to the basic PV module SG Mono GF245F, manufactured by SUNEL, has been used. This module is composed by 60 monocrystalline silicium cells. The advantage of using this module is that there is the possibility to have the six terminals of the three strings forming the module, that normally are connected in series. This is very useful in the case of shading or disuniform radiation. Having the outputs of the three strings, in fact, it is possible to obtain three maximum power points, each one relative to each sub-string. Moreover, it is possible to have different configurations connecting the strings in series or in parallel. Moreover, two solutions have been considered: mono junction-box and multi junction-box, shown in Fig. 1 and Fig. 2, respectively. While

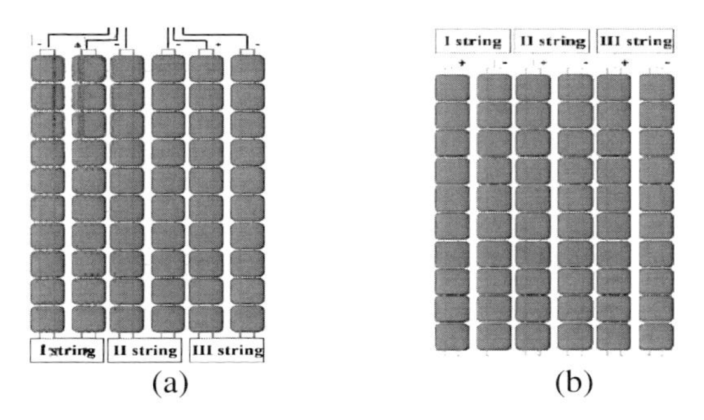

Fig. 2. Connection of PV module "SG Mono GF245F" manufactured by SUNEL; (a) mono junction-box solution; (b) multi junction-box solution

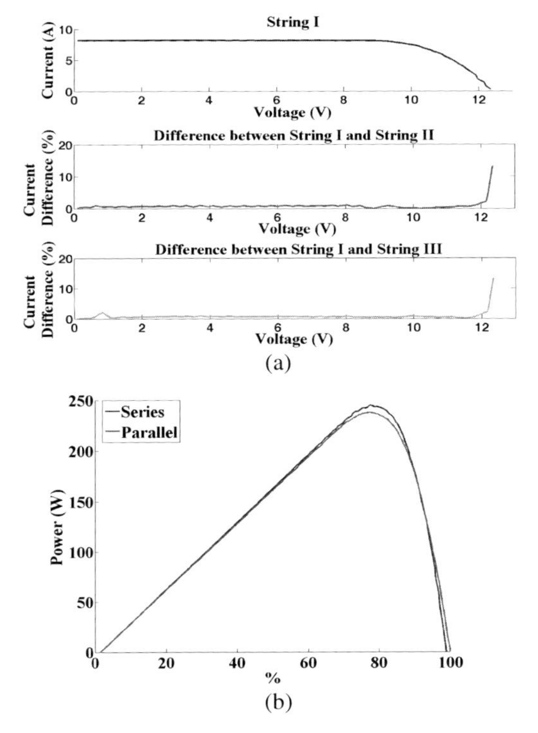

Fig. 3. Characteristics of the module SG Mono GF245F using a mono junction-box; (a) I-V curve of the string I and the difference between the current of string II and string III compared to the current of string I, on equal voltage; (b) Power curve considering series and parallel connections, in the x axis there is the percentage of the voltage relative to V_{OC}

Fig. 3 shows the I-V curve calculated in STC of each of the three strings and their connection in series and in parallel considering the mono junction-box solution, while Fig. 4 shows results obtained using a multi j-box solution. The I-V curves of the three sub-strings are compared, in particular the current of the string II and string III are compared with the current of the string I considering on equal voltage. The power of the series and parallel connections are also compared considering the percentage of the voltage respect to V_{OC}.

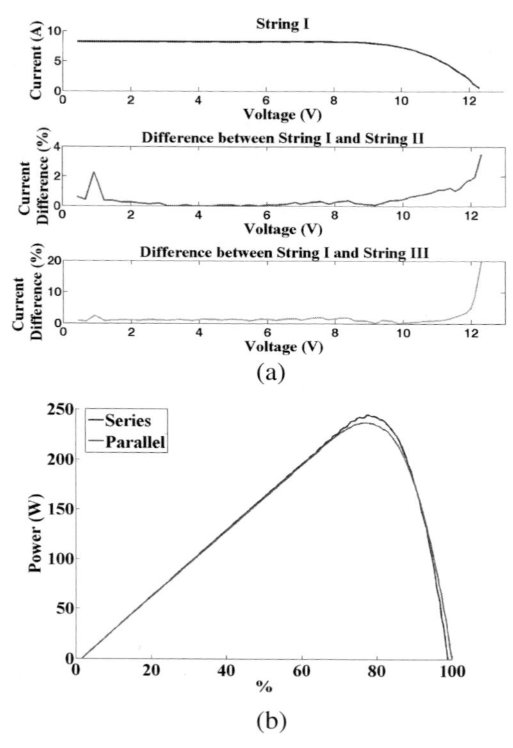

Fig. 4. Characteristics of the module SG Mono GF245F using a multi junction-box; (a) I-V curve of the string I and the difference between the current of string II and string III compared to the current of string I, on equal voltage; (b) Power curve considering series and parallel connections, in the x axis there is the percentage of the voltage relative to V_{OC}

4 Experimental Results

The five parameter model has been used starting from some data provided by the manufacturer of the module. In particular, the manufacturer provides the values of: V_{OC}, I_{SC}, V_{MP} and I_{MP} in STC relative to each string and the connection in series and parallel of the three strings, for both mono and multi junction-box solutions. These data have been obtained using a Solar Array Simulator PASAN class A. Data used in the model are reported in Table 1.

Using the five parameter model proposed in this paper, the parameter R_{S_ref} and $\eta_{\_ref}$ can be calculated and then the values of I_{PH_ref}, I_{0_ref} and R_{SH_ref} can calculated using Eq. 9, Eq. 10 and Eq. 11, starting from the values of V_{OC}, I_{SC}, V_{MP} and I_{MP}.

The model has been validated using numerical example. Fixing the values of the five parameters and starting from Eq. 3, Eq. 4 and Eq. 5, the values of V_{OC}, I_{SC}, V_{MP} and I_{MP} have been calculated and then the obtained values have been used to calculate I_{PH_ref}, I_{0_ref}, $\eta_{\_ref}$, R_{S_ref} and R_{SH_ref} using the method here implemented. Table 2 shows the obtained results, that demonstrate that the method here implemented can calculate the five parameter with a good approximation.

Applying the method here presented to the data obtained using the Solar Array Simulator relative to the "SG Mono GF245F" manufactured by SUNEL, we have noted that calculating the five parameters, their values depends on Maximum Power Point, and therefore the algorithm cannot converge if the values of V_{MP} and I_{MP} are not accurate. Therefore, an analysis of the I-V curves obtained starting from different values of V_{MP} and I_{MP} has been done. These values have been calculated varying the measured values of V_{MP} and I_{MP} (Table 3) of ±0.05 with a step of 0.0025, obtaining 1681 combinations of V_{MP} and I_{MP} values. Using these values the five parameters have been evaluated and the I-V curve has been estimated.

Table 1. Parameters of the PV module used in the numerical simulations, in STC (irradiation = 1kW/m2, temperature = 25°C)

		$V_{OC,ref}$ (V)	$I_{SC,ref}$ (A)	$V_{MP,ref}$ (V)	$I_{MP,ref}$ (A)
Mono Junction-box	I String	12.3047	8.1934	9.6191	7.6953
	II String	12.3535	8.1494	9.6191	7.7246
	III String	12.3047	8.1152	9.6191	7.7637
	Series	37.2070	8.1445	29.1504	7.8857
	Paral.	12.4512	24.2725	9.6680	22.9980
Multi Junction-box	I String	12.3535	8.1982	9.4238	7.9395
	II String	12.3535	8.2129	9.7656	7.6563
	III String	12.3535	8.1982	10.0586	7.4707
	Series	37.1582	8.1933	29.1015	7.9003
	Paral.	12.5098	24.3164	9.6826	23.0420

Table 2. Numerical example of the implemented method: T = 48.3°C, V_{OC} (V) = 18.7, I_{SC} (A) = 3.22, V_{MP}(V) = 14.6 and I_{MP}(A) = 2.91

	I_{PH_ref}	I_{0_ref}	$m{\cdot}\eta_{\_ref}$	R_{S_ref}	R_{SH_ref}
Starting value	3.2205	5.7782·10-7	46	0.3597	972.6
Estimated value	3.2208	3.9426e-007	42.4677	0.4201	905.5186

Then the error between the measured I-V curve and the estimated one has been evaluated using the normalized Root Mean Square Error (nRMSE) as main measure. The normalized Root Mean Square Error (nRMSE) is a non-dimensional form of the error:

$$nRMSE = \frac{\sqrt{\frac{1}{N} \cdot \sum_{i=1}^{N} (y_i - x_i)^2}}{\sqrt{\frac{1}{N} \cdot \sum_{i=1}^{N} (x_i)^2}} \tag{12}$$

where N is the number of data, the variable y_i represents the estimated value, while x_i represents the measured one.

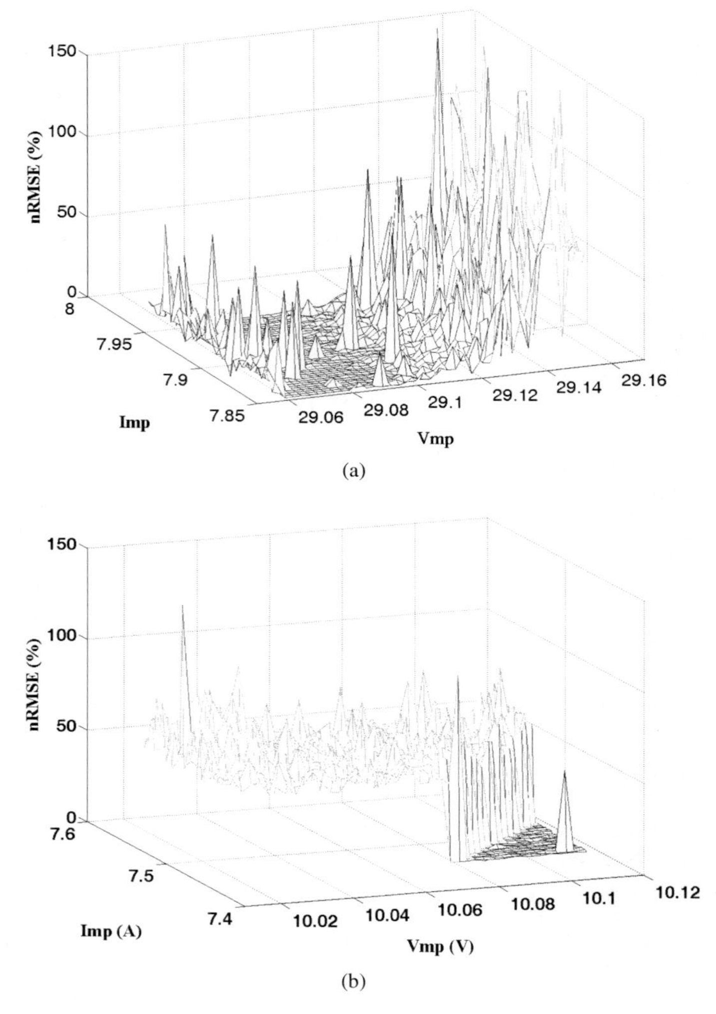

Fig. 5. Error between the measured I-V curve and the estimated one considering different combinations of V_{MP} and I_{MP} values. Results are referred to (a) the series connection of the multi-junction solution; (b) the string III of the multi-junction solution.

Fig. 5 shows two example of the error between the measured I-V curve and the estimated one, considering different combinations of V_{MP} and I_{MP} values. Results are referred to the series connection of the multi-junction solution (Fig. 5.a) and the string III of the multi-junction solution (Fig. 5.b). As it possible to note, the values of nRMSE obtained using the five parameters calculated for each combination of V_{MP} and I_{MP} decrease in some areas of the graphics. That means that the correct Maximum Power Point is situated in that area.

Therefore, it is possible to assert that the measured I-V curve is coherent: if the values of voltage and current in the case of open circuit voltage, short circuit current and maximum power point satisfy the implemented model: if I_{PH_ref}, I_{0_ref}, and R_{SH_ref}. satisfy the analytical equations; and if exists a numerical solution of $\eta_{\_ref}$ and R_{S_ref} that is coherent with the physic significance of these parameters.

5 Conclusion and Future Works

In this context, an analytical model to evaluate the energy performance of an I-V characteristic of a photovoltaic module is proposed. The model is based on some data provided by the manufacturer of the module. It has been validated using numerical examples, and results show how it is used to estimate the I-V curve of a photovoltaic module with a good precision. The photovoltaic module used during the experiments gives the possibility to have six output terminals, relative to each of the three substrings forming the module. Therefore, the manufacturer provided us data relative to the three remarkable points of the I–V curve of the practical array: open circuit, maximum power, and short circuit in STC relative to each string and the connection in series and parallel of the three sub-strings, measured using a Solar Array Simulator. These data have been used to evaluate the I-V curves relative to each case and then these results have been compared with the measured ones and the errors have been calculated using the Root Mean Square Error (nRMSE) as main measure, and tested using both measured and estimated data relative to each single string and their connection in series and parallel. Results show how the parameters extraction depends on the measured value of the maximum power points, if measures are not accurate the analytic model here implemented cannot converge to a feasible solution. Therefore, it is possible to assume that the values of voltage and current in the case of open circuit voltage, short circuit current and maximum power point satisfy the implemented model: if I_{PH_ref}, I_{0_ref} and R_{SH_ref}. satisfy the analytical equations and if exists a numerical solution of $\eta_{\_ref}$ and R_{S_ref} that is coherent with the physic significance of these parameters.

The I-V curve varies with irradiance and cell temperature so, as matter of fact, an analysis of dependence of the parameters on operating conditions is needed.

References

[1] AlHajri, M.F., El-Naggar, K.M., AlRashidi, M.R., Al-Othman, A.K.: Optimal extraction of solar cell parameters using pattern search. Renewable Energy 44, 238–245 (2012)

[2] Bouzidi, K., Chegaar, M., Bouhemadou, A.: Solar cells parameters evaluation considering the series and shunt resistance. Solar Energy Materials and Solar Cells 91, 1647–1651 (2007)

[3] Carrero, C., Amador, J., et al.: A single procedure for helping PV designers to select silicon PV modules and evaluate the loss resistances. Renewable Energy 32(15), 2579–2589 (2007)

[4] Celik, A.N.: Artificial neural network modelling and experimental verification of the operating current of mono-crystalline photovoltaic modules. Solar Energy 85, 2507–2517 (2011)

[5] Chan, D.S.H., Phang, J.C.H.: Analytical methods for the extraction of solar-cell single- and double-diode model parameters from I-V characteristics. Electron Devices 34, 286–293 (1987)

[6] Chegaar, M., Azzouzi, G., Mialhe, P.: Simple parameter extraction method for illuminated solar cells. Solid-State Electronics 50, 1234–1237 (2006)

[7] Chen, Y., Wang, X., Li, D.: Parameters extraction from commercial solar cells I–V characteristics and shunt analysis. Applied Energy 88, 2239–2244 (2011)

[8] Garrido-Alzar, C.L.: Algorithm for extraction of solar cell parameters from I–V curve using double exponential model. Energy Efficiency and the Enviroment 10, 125–128 (1997)

[9] Gradella, V.M., Gazoli, J.R., Filho, E.R.: Comprehensive approach to modeling and simulation of photovoltaic arrays. IEEE Transactions on Power Electronics 20(24), 1198–1208 (2009)

[10] Ishaque, K., Salam, Z.: An improved modeling method to determine the model parameters of photovoltaic (PV) modules using differential evolution (DE). Solar Energy 85, 2349–2359 (2011)

[11] Ishaque, K., Salam, Z., Taheri, H., Shamsudin, A.: A critical evaluation of EAcomputational methods for Photovoltaic cell parameter extraction based on two 1 diode model. Solar Energy 85, 1768–1779 (2011)

[12] Karatepe, E., Boztepe, M., Colak, M.: Neural network based solar cell model. Energy Conversion and Management 47, 1159–1178 (2006)

[13] Moldovan, N., Picos, R., et al.: Parameter extraction of a solar cell compact model usign genetic algorithms. In: Spanish Conference on Electron Devices, CDE 2009 (2009)

[14] Phang, J.C.H., Chan, D.S.H., Phillips, J.R.: Accurate analytical method for the extraction of solar cell model parameters. Electronics Letters 20, 406–408 (2010)

[15] Wang, K., Ye, M.: Parameter determination of Schottky-barrier diode model using differential evolution. Solid-State Electronics 53, 234–240 (2009)

Chapter 3
Evolution of Environmental Sustainability for Timber and Steel Construction

Dimitrios N. Kaziolas[1], Iordanis Zygomalas[2], Georgios E. Stavroulakis[3], Dimitrios Emmanouloudis[1], and Charalambos C. Baniotopoulos[2,4]

[1] Technological Educational Institute of Kavala, Greece
{dnkazio,demmano}@teikav.edu.gr
[2] Institute of Metal Structures, Department of Civil Engineering, Aristotle University of Thessaloniki, Greece,
izygomal@civil.auth.gr
[3] Institute of Computational Mechanics and Optimization, Department of Production Engineering and Management, Technical University of Crete, Chania, Greece
gestavr@dpem.tuc.gr
[4] Chair of Sustainable Energy Systems, School of Civil Engineering, The University of Birmingham, Edgbaston B15 2TT, Birmingham, United Kingdom
ccb@civil.auth.gr

Abstract. The movement for sustainable development aims at the optimization of the whole of human activity in terms of environmental, economic and social impact. The aim of the present paper is the examination of the content and evolution of environmental sustainability in order to identify the key implications and requirements regarding timber and steel structures, two fields with significant potential in terms of sustainability. The conclusions drawn include the identification of issues such as raw materials, the construction stage of a project and waste management and their potential influence on the environmental sustainability of timber and steel construction.

1 Introduction

The term 'sustainable development' is nowadays used at various instances, sometimes with a varying capacity and context. There are cases in which it is used as means of attracting attention or promoting a proposal or agenda without actual reference to its principles or application. Construction is one of sectors where construction product manufacturers often present products that are 'sustainable' without always taking into account any specific measures or processes. Although these cases should be regarded as the exception to the rule, it is nevertheless necessary to examine the specific content of environmental sustainability and its requirements and implications for the construction sector. In such a way, the quality and sustainability of construction products and services will be higher, thus rendering construction a more efficient and sustainable sector.

A. Håkansson et al. (Eds.): *Sustainability in Energy and Buildings*, SIST 22, pp. 25–33.
DOI: 10.1007/978-3-642-36645-1_3

1.1 Purpose of Sustainable Development

In order to identify the requirements regarding sustainability that are related to timber and steel construction, it is necessary to examine the purpose for which the movement for sustainable development was created. One of the first documented descriptions of the conditions surrounding the origin and introduction of the concept of sustainable development can be found in the book 'The Limits to Growth' published in 1972 (Meadows et al. 1972). In it, the issue of the consequences of human activity to the environment, in light of finite resources available on the planet, was approached. A number of international organizations such as the United Nations (1972) also highlighted the importance of environmental issues thus laying the groundwork upon which the concept of sustainable development was formed and a number of environmental protection agencies were established.

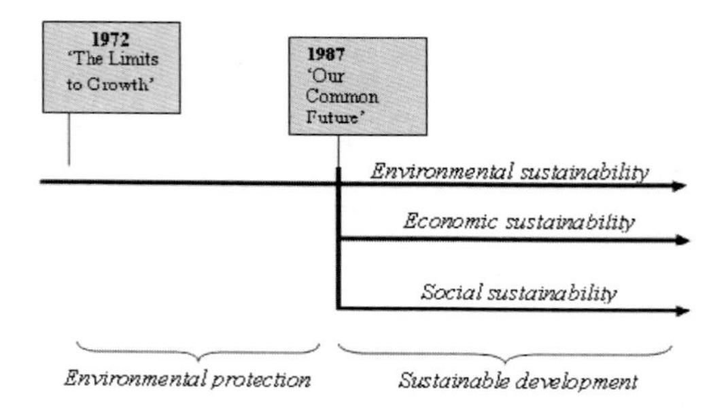

Fig. 1. Evolution of the concept of sustainable development

However, the content of sustainable development today has come to include an even broader range of factors which are not all related to the environment (Figure1). The most quoted definition of sustainable development being about "meeting the needs of the present without compromising the ability of future generations to meet their own needs" (United Nations 1987) confirms that its purpose is a much more general one, to ensure the future of next generations. This general goal includes both environmental issues and also social and economic ones that together provide a more complete description of the aforementioned 'needs' of future generations.

2 Environmental Sustainability

The identification of the three dimensions of sustainability led to the observation that environmental issues should be considered as a part of sustainable development and not as its whole content. The term 'environmental sustainability' is therefore since used to refer to the issues regarding the environment within the broader movement for sustainable development. Several events have occurred and key documents have been published regarding its context and characteristics. In order to identify its implications and requirements regarding timber and steel structures, it is necessary to examine these events and document their meaning and influence on these construction sectors.

2.1 Key Events and Documents

The first time environmental issues were discussed on a global level was at the United Nations Conference on the Human Environment which was held in Stockholm in 1972. The outcome of this event included the adoption of a significant amount of documents, including the 'Declaration on Human Environment', an action plan consisting of 109 recommendations and a resolution on institutional and financial arrangements (Momtaz 1996). The Declaration on the Human Environment is aimed at the preservation and enhancement of human environment and among the issues it refers to are the consideration of environmental consequences, the earth's capacity to produce renewable resources, pollution and the importance of research (Vasseur 1973).

A document that recreates much of the momentum of the United Nations Stockholm conference was the report entitled 'Our Common Future' which was published in 1987 by the United Nations World Commission on Environment and Development. It highlighted environmental issues and dealt with environment and development not as two separate aspects but as a single issue. Instead on focusing on the continuous environmental decay and pollution, the report put forth the possibility of economic growth based on sustainable policies that 'expand the environmental resource base' (United Nations 1987). It also highlighted the urgent need for decisive political action and promoted cooperation between people to both sustainable human progress.

The next significant event regarding the evolution of environmental sustainability is the United Nations Conference on Environment and Development which was held in Rio de Janeiro in 1992. The conference led to the formation of a number of influential documents and actions, among which were the Rio Declaration on Environment and Development and Agenda 21. The Rio Declaration consists of 27 principles intended to guide future sustainable development around the world. The very first principle states that 'human beings are at the centre of concerns for sustainable development' (United Nations 1992), while other principles refer to the need for the integration of environmental protection into development and the use of environmental impact assessment for decision-making. Agenda 21 is a specific action plan which refers to actions that have to be implemented to address current issues and future challenges. It includes the participation of a variety of organizations, from United Nations to national governments.

Another relative document of importance is the Kyoto Protocol adopted on 11 December 1997 and entered into force on 16 February 2005. It is an international agreement linked to the United Nations Framework Convention on Climate Change that set binding targets for 37 industrialized countries and the European community for reducing greenhouse gas (GHG) emissions.

All these events and publications, and many others, had a very influential role in establishing environmental issues as top priorities among agendas referring to future policies and actions. Despite the fact that some of the goals set within the scope of these events -or of other similar ones of acknowledged importance- still require strong efforts, their effect on human activity cannot be questioned. It can therefore be concluded that the current time period should be regarded as remarkably different from previous ones, in that the requirements associated with sustainable development have become the focal point in all business sectors. For an organization to remain competitive and secure a place in tomorrow's market, it will have to incorporate sustainability into the supply chain. This is emphasized in sectors that have already been identified as in need of radical transformation, such as construction.

2.2 Environmental Sustainability in Construction

It has been documented a number of times that the construction sector is responsible for the consumption of vast amounts of raw materials and energy, while it also produces significant quantities of waste. In a report published by a European Commission research project (2003) it is stated that construction activities are responsible for the consumption of more raw materials by weight (about 50%) than any other sector, while the waste streams generated from construction were also found to be the largest. Another EU report states that more than 50% of all materials extracted from earth are transformed into construction materials and products (European Commission 2007).

European statistics (Eurostat 2011) show that in 2008, construction was the economic activity that was responsible for 32.9% of the total waste generation in the EU-27. Statistical data for waste in the EU also show that the construction activity associated with the largest production of non-recyclable waste is onsite construction. The materials documented were mainly concrete, bricks and other materials found in traditional building techniques (European Commission 2010). These materials are frequently used in construction projects and although characterized by a number of advantages regarding the traditional requirements –such as stability and mechanical resistance- they do not perform adequately in terms of environmental sustainability. Concrete is characterized by a very low capacity for recycling which leads to the vast amounts of construction waste documented after the decision for demolition of a construction project has been made.

The same applies to waste generated by the manufacturing of construction materials. Again, the increased use of concrete has proportionally increased the quantities of the relevant construction materials which often cannot be recycled or reused at the end of the project's life cycle.

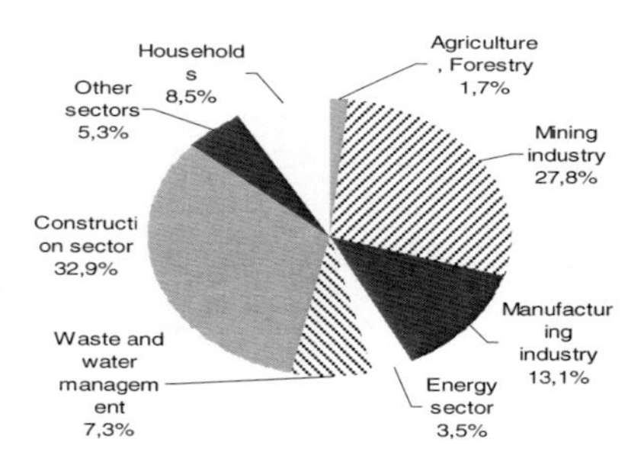

Fig. 2. Total waste generation in the EU-27 by economic activity in 2008 (Eurostat 2011)

It is therefore observed that traditional construction materials and techniques such as concrete do not hold adequate potential in terms of minimization of environmental impact. On the contrary, steel and timber construction present increased opportunities in regard to environmental sustainability, mainly due to the properties of the materials and their potential for recycling and reuse. In regard to the waste generated within the EU-27 by the manufacturing of wood and wood products, an extremely low amount of non-recyclable waste was documented for year 2008 (European Commission 2010), which clearly highlights the potential of the sector in terms of environmental sustainability.

3 Timber and Steel Construction

Timber and steel construction constitute two sectors of construction that are characterized by increased potential in terms of environmental sustainability. This potential can be attributed to the differences that separate the two techniques from traditional ones which are primarily based on concrete. Since concrete is a material that cannot be efficiently recycled or reused, its use leads to the generation of large amounts of waste. On the contrary, wood and steel can be retrieved and recycled or reused after the decision for demolition has been made.

3.1 Raw Materials

Having examined the content and evolution of environmental sustainability it is possible to identify the implications and requirements that are related to timber and steel construction. One of the major issues is raw materials. Although the protection of the environment was introduced on a very general level, it has nowadays become closely associated with the urgency to minimize the use of natural resources and thus decrease the amounts of extracted raw materials. Since both timber and steel construction projects utilise significant large amounts of materials, it is necessary to find ways to optimize their use, along with the use of energy or other related environmental inputs.

The capacity of both wood and steel for recycling and reuse enables the minimization of newly extracted raw materials required for a construction project. To fully exploit the recycling and reuse options, establishing a well-organized and efficient network of collection and sorting points would prove most beneficial. Wood, steel and iron that can be used for the production of new construction materials is a key ingredient of the sustainable nature of the two sectors as it eventually leads to the minimization of newly extracted quantities of raw materials.

The influence of the appropriate raw materials in terms of environmental impact was highlighted by Zygomalas and Baniotopoulos (2011) who calculated the air emissions to the environment that are caused for the acquisition of 1 kg of hot-rolled structural steel members intended for use in steel construction projects in Greece (Table1). Three acquisition routes were examined which cover the local manufacturing or the steel members, their import as ready-to-use products from foreign suppliers and the inland processing of imported semi-finished steel products such as steel billets.

The results show that the imported products that also include the transport to Greece are responsible for significantly higher air emissions, while the steel members manufactured locally cause the lowest emissions across all three categories.

Table 1. Air emissions caused by the acquisition of 1 kg of hot-rolled structural steel members for use in steel construction in Greece (Zygomalas and Baniotopoulos 2011)

Substance outputs (Emission to air)	Unit	Locally manufactured	Imported products	Inland processing of imported semi-finished products
Carbon dioxide (CO_2)	kg	0,267208	0,419745	0,298384
Carbon dioxide, fossil (CO_2)	kg	1,090667	1,307253	1,648536
Carbon monoxide (CO)	kg	0,003653	0,001258	0,000706
Dinitrogen monoxide (N_2O)	kg	2,2E-05	2,13E-05	2,29E-05
Hydrogen Chloride (HCl)	kg	0,000224	8,25E-05	0,000121
Hydrogen Sulphide (H_2S)	kg	5,56E-06	1,93E-05	2,49E-05
Lead (Pb)	kg	4,64E-07	3,86E-06	4,69E-06
Mercury (Hg)	kg	5,15E-08	1,69E-06	1,72E-06
Methane (CH_4, fossil)	kg	0,001244	0,003783	0,004824
Nitrogen oxides (NO_x)	kg	0,00172	0,005041	0,004763
Non-methane volatile organic compounds (NMVOC)	kg	0,000618	0,0018	0,001374
Particulates, < 2.5 um ($PM_{2,5}$)	kg	0,000544	0,000787	0,001032
Particulates, < 10 um, mobile & stationery (PM_{10})	kg	4,17E-06	0,000237	0,000129
Sulfur dioxide (SO_2)	kg	0,004007	0,003095	0,004032
Sulfur oxides (SO_x)	kg	9,81E-05	0,000802	0,000476
Zinc (Zn)	kg	5,84E-07	1,88E-05	1,93E-05

3.2 Construction Time

Another basic requirement concerning environmental sustainability and timber and steel structures is the minimization of the use of resources during the construction stage. The more time and resources required for the completion of a project, the more increased the environmental impact of the whole project. It is therefore necessary to shorten project delivery times. This can be achieved and further improved for timber and steel construction, since both techniques allow for the formation of off-site elements that are only assembled onsite. This advantage has to be exploited to the highest degree since it directly decreases the total amount of time required for the completion of a timber or steel structure.

3.3 Waste Management

Environmental sustainability has shown that project designers should not concentrate only on the initial phases of a construction project but also the last. In the age of rapid economic transformations it is very common for the service life of a construction project to be finished well before its scheduled ending. New owners, new functions and new uses are all reasons for interrupting the service life of a structure and proceeding with interventions or even reconstruction activities. In view of environmental sustainability such options increase the environment impact caused since large amounts of waste are generated and new materials are required. It is therefore crucial to be able to recycle or reuse the retrieved construction materials from the existing building during its demolition. Wood and steel are indeed characterized by this potential and can subsequently minimize the amount of waste which cannot be utilized and has to be sent to landfills. This can be observed in a study conducted by Zygomalas et al. (2009) in which the environmental impacts caused during the life cycle of a simple steel shed were calculated. According to their findings, one third of the environmental burden caused by the construction of the steel shed was deducted from the total life cycle impact due to the environmental benefit from the recycling of retrieved materials (Figure3).

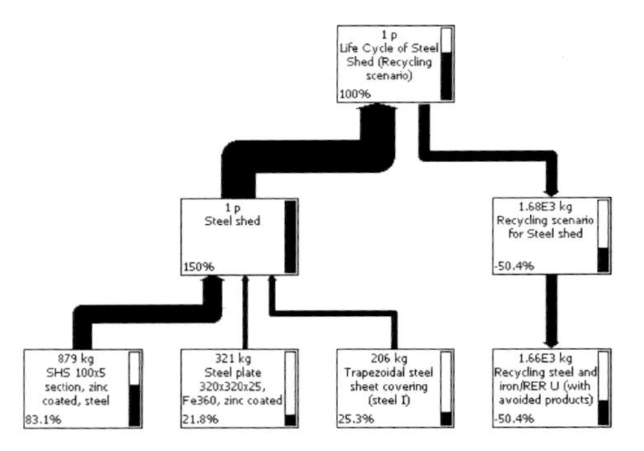

Fig. 3. Illustration of environmental impact caused during the life cycle of a simple steel shed construction (Zygomalas et al. 2009)

3.4 Quantitative Evaluation of Sustainability

Sustainability involves a lot of uncertain information and vague declarations. Therefore suitable modeling tools must be used. Fuzzy inference seems a suitable environment for the evaluation and ranking of various alternatives and finally for the choice of the best ones. Fuzzy sets and their calculus, fuzzy logic and fuzzy inference have been based on the seminal works of L. Zadeh. These techniques allow for the usage of imprecise information and linguistic reasoning, both including vagueness and uncertainty, in classification, optimization and control. Applications of fuzzy evaluation on the environmental sustainability assessment of various activities have been published recently (Phillis et al. 2001, Amindoust et al. 2007, Pislaru et al. 2011). The evaluation of environmental sustainability of structures taking into account their life cycle has not been considered till now, to the best of our knowledge. Research work in this direction is planned for the near future from our group.

4 Conclusions

The concept of environmental sustainability was first introduced as an urgent requirement for the protection of the environment against the consequences of human activity in view of finite resources. In the coming years, environmental issues gained increasing attention and were being prioritized in international agendas for future policies. Ultimately, the overarching goal of sustainable development was set to ensure that the future of the next generations would not be compromised. Environmental sustainability was defined as one of the three dimensions of sustainable development, the remaining two being economic and social sustainability. A number of key events and publications contributed to the development of the concept and led to its current establishment.

The construction sector can play a critical role in ensuring sustainable development, since it is responsible for the consumption of about 50% of raw materials within the EU and generates the largest waste streams as well. It was found that most of the materials in these waste streams are coming from onsite construction based on traditional materials such as concrete. In this view, steel and timber construction were found to present increased environmental sustainability potential, due to the properties of the materials and their capacity for recycling and reuse.

The main implications and requirements for timber and steel construction that were identified concern raw materials, the construction stage of a project and the end of its life cycle. The urgency to minimize the use of natural resources and thus decrease the amounts of extracted raw materials can be balanced with a well-organized and efficient network of collection and sorting points which ultimately enables the minimization of newly extracted raw materials required for a construction project. Environmental impact during the construction stage can also be minimized through the shortening of project delivery times, as timber and steel construction both allow for the formation of off-site elements that are only assembled onsite. At the end of a construction project life cycle and after the decision for its demolition has been made, timber and steel also provide increased potential in terms of environmental sustainability. The recycling and reuse of materials retrieved can minimize the amount of waste which cannot be utilized and has to be sent to landfills.

In this respect, it can be argued that timber and steel construction do not actually produce ‘waste’ with the traditional meaning of the term but rather collected quantities of raw materials which can be used for the manufacturing of new wood and steel construction products.

Acknowledgments. This research work is co-funded by the European Union (European Social Fund) and National (Greek) Resources – **ARCHIMEDES III.**

References

Amindoust, A., Ahmed, S., Saghafinia, A., Bahreininejad, A.: Sustainable supplied selection: A ranking model based on fuzzy inference system. Applied Soft Computing 12, 1668–1677 (2012)

European Commission, Purchasing Guidelines for Green Buildings - Background Document, Work package 13. EU-Research Project RELIEF funded under the 5th Framework Programme - Key action "City of tomorrow and Cultural Heritage", European Commission, Directorate General for Research (2003)

European Commission, Accelerating the Development of the Sustainable Construction Market in Europe. Report of the taskforce on sustainable construction. Composed in preparation of the Communication "A Lead Market Initiative for Europe" COM 860 final (2007), http://ec.europa.eu/enterprise/policies/innovation/policy/lead-market-initiative/files/construction_taskforce_report_en.pdf (accessed on April 5, 2012)

European Commission, FWC Sector Competitiveness Studies N° B1/ENTR/06/054 – Sustainable Competitiveness of the Construction Sector, Final report. Directorate-General Enterprise & Industry (2010), http://ec.europa.eu/enterprise/sectors/construction/files/compet/sustainable_competitiveness/ecorys-final-report_en.pdf (accessed on April 5, 2012)

Eurostat, Waste statistics (2011), http://epp.eurostat.ec.europa.eu/statistics_explained/index.php/Waste_statistics (accessed on April 5, 2012)

Meadows, D.H., Meadows, D.L., Randers, J., Behrens III, W.W.: The Limits to Growth, Commissioned by the Club of Rome. Universe Books (1972)

Momtaz, D.: The United Nations and the protection of the environment: from Stockholm to Rio de Janeiro. Political Geography 15(3-4), 261–271 (1996), ISSN 0962-6298, doi:10.1016/0962-6298(95)00109-3
http://www.sciencedirect.com/science/article/pii/0962629895001093

Phillis, Y.A., Andriantiatsaholiniaina, L.A.: Sustainability: an ill-defined concept and its assessment using fuzzy logic. Ecological Economics 37, 435–456 (2001)

Pislaru, M., Trandabat, A., Avasilcai, S.: Environmental assessment for sustainability determination based on fuzzy logic model. In: 2nd International Conference on Environmental Science and Technology IPCBEE, vol. 6. IACSIT Press, Singarpore (2011)

SAFE Sustainability Assessment by Fuzzy Evaluation, http://www.sustainability.tuc.gr/

United Nations, Report of the United Nations Conference on the Human Environment, Stockholm, Sweden, June 5-16 (1972), http://www.unep.org/Documents.Multilingual/Default.asp?DocumentID=97 (accessed on March 28, 2012)

United Nations, Our Common Future, Report of the World Commission on Environment and Development (1987), http://www.un-documents.net/wced-ocf.htm (accessed on April 02, 2012)

United Nation, Rio Declaration on Environment and Development (1992), http://www.unep.org/Documents.Multilingual/Default.asp?documentid=78&articleid=1163 (accessed on April 02, 2012)

Vasseur, E.: United Nations Conference on the Human Environment: Stockholm. Water Research 7(8), 1227–1233 (1973), ISSN 0043-1354, doi:10.1016/0043-1354(73)90077-8
http://www.sciencedirect.com/science/article/pii/0043135473900778

Zygomalas, I., Efthymiou, E., Baniotopoulos, C.C.: On the Development of a Sustainable Design Framework for Steel Structures. Transactions of FAMENA 33(2), 23–34 (2009)

Zygomalas, I., Baniotopoulos, C.C.: Extending an existing LCI database to include the complete range of steel members acquisition. In: Proceedings of the 2nd International Exergy, Life Cycle Assessment and Sustainability Symposium, Nisyros island, Greece, June 19-21, pp. 641–648 (2011)

Chapter 4
Using the Energy Signature Method to Estimate the Effective U-Value of Buildings

Gustav Nordström, Helena Johnsson, and Sofia Lidelöw

Division of Structural and construction engineering, Luleå University of Technology, Sweden SE-971 87 LULEÅ Sweden

Abstract. The oil crisis of the 1970s and the growing concern about global warming have created an urge to increase the energy efficiency of residential buildings. Space heating and domestic hot water production account for approximately 20% of Sweden's total energy use. This study examines the energy performance of existing building stock by estimating effective U-values for six single-family houses built between 1962 and 2006. A static energy signature model for estimating effective U-values was tested, in which the energy signature was based on measurements of the total power used for heating and the indoor and outdoor temperatures for each studied house during three winter months in northern Sweden. Theoretical U-values for hypothetical houses built to the specifications of the Swedish building codes in force between 1960 and 2011 were calculated and compared to the U-values calculated for the studied real-world houses. The results show that the increasingly strict U-value requirements of more recent building codes have resulted in lower U-values for newer buildings, and that static energy signature models can be used to estimate the effective U-value of buildings provided that the differences between the indoor and outdoor temperatures are sufficiently large.

Keywords: Energy signature, building energy use, houses in cold climate, average U-values.

1 Introduction

Residential buildings account for about 40% of Sweden's total energy use, with space heating and hot water alone being responsible for around 20% of the total (Energimyndigheten 2012). There is therefore great interest in improving building performance. A building's performance is dependent on its design and the way in which it was built, and so building codes can be seen as tools that can be used to increase systematic efficiency and to mandate improvements in the thermal properties of new buildings. Building performance is also sensitive to users' behavior and consumption of building services (Guerra-Santin and Itard 2010; Lundstrom 1986). Since the establishment of the first Swedish national building code in the 1950s, factors such as the oil crisis of the 1970s and the growing awareness of the problems posed by global warming have increased the demand for buildings with high energy performance. This has, in turn, led to changes in the way we build our houses. Since 1920, the

A. Håkansson et al. (Eds.): *Sustainability in Energy and Buildings*, SIST 22, pp. 35–44.
DOI: 10.1007/978-3-642-36645-1_4

dominant housebuilding technique used in Sweden has been to construct a light timber-framed structure with evenly spaced studs. In more recent years, it has become common to use thicker walls with one or two horizontal layers of studs (Nordling and Reppen 2009). Figure 1.1 shows the maximum average U-values [W/m²°C] permitted under Swedish building codes from the 1960s onwards. To facilitate comparisons between codes from different years, all U-values were calculated for a hypothetical house with a floor area of 16x10 m, an inside height of 2.5 m, a window area of 19.5 m² (15%), and an envelope area of 450 m². The building codes of 1960 and 1980 specified maximum U-values for each building part such as the walls, roof and floor, while the building code of 1988 imposed a cap on the average U-value of the building envelope. Conversely, both the current code and that issued in 2006 stipulate an upper limit on the energy usage per square meter of heated floor area [kWh/m²], with the limit values for houses with electrical heating (excluding household electricity) being different to those for houses with non-electrical heating. Thus the U-values for 2006 and 2011 in Figure 1.1 also include ventilation losses in the house whereas the limits stipulated between 1960 and 1988 relate to transmission losses alone.

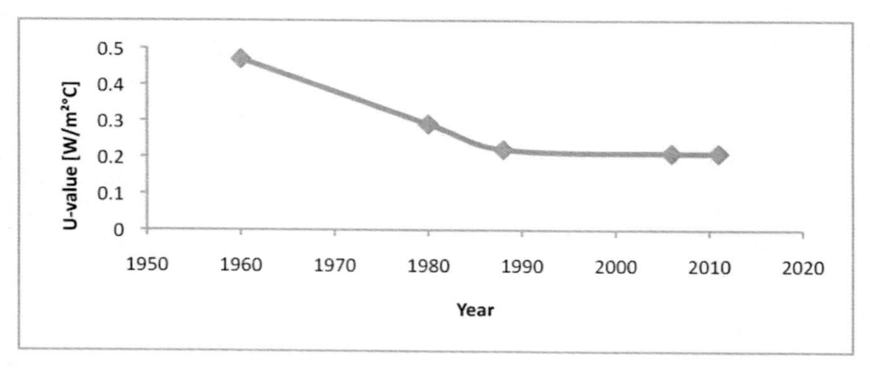

Fig. 1. Caps on average U-values mandated by Swedish building codes between 1960 and the present day

The case study discussed herein was conducted to develop a method based on a static energy signature model for estimating effective U-values for wooden single-family homes in a cold climate. A number of houses constructed between 1962 and 2006 were investigated; the nominal U-values that would be expected for such buildings are shown in Figure 1. The developed method was used to determine whether the changes in the national building code over the years have yielded practical increases in building performance.

2 Theory

2.1 Energy Signature

Energy signature models estimate the overall power loss (K_{TOT}) as a function of the difference between the indoor (T_i) and outdoor (T_e) temperatures, as described by Sjögren (2007) and Hammarsten (1987):

$$K_{TOT}(T_i - T_e) = P_H + P_G - P_{DYN} \quad (1)$$

Here, K_{TOT} is the sum the of heat losses from the building due to transmission and ventilation, P_H is the power supplied for heating, P_G is power gained at no cost ("free power"), and P_{DYN} is power that is dynamically stored and released. If heat is supplied via a district heating system, P_H is given by the total amount of power drawn from the district heating system (P_{DH}) minus that used for heating water (P_{DHW}) and that lost from the system (P_L):

$$P_H = P_{DH} - P_{DHW} - P_L \quad (2)$$

The free power (P_G) consists of power gained from insolation (P_{SUN}), household electricity (P_{HE}), household operating electricity (P_{BE}) and heat generated by the actions of the building's occupants (P_P):

$$P_G = P_{SUN} + P_{HE} + P_{BE} + P_P \quad (3)$$

K_{TOT} can then be described as:

$$K_{TOT}(T_i - T_e) = P_{DH} - P_{DHW} - P_L + P_{SUN} + P_{HE} + P_{BE} + P_P - P_{DYN} \quad (4)$$

Where K_{TOT} equals $\mathbf{P_{TOT}}$:

$$K_{TOT}(T_i - T_e) = P_{TOT} \quad (5)$$

3 Method

3.1 Monitored Buildings

The case study was based on measurements conducted in six inhabited detached single family houses in the city of Luleå, Sweden. The city has a subarctic climate with a yearly average temperature of 1°C (SMHI 2011). All of the houses have wooden structures, wooden or brick façades and are connected to the district heating system of the city. The houses were constructed in 1962, 1967, 1983, and 1987, with the last two having been built in 2006. As such, they were constructed under a range of different building codes, using various techniques. Some of the more important properties of the studied buildings are presented in Table 1. The study focused on measurements acquired during the winter of 2011-2012 between November 2011 and February 2012. Large variations in outside temperature are common during this period, which creates differences in temperature between indoor and outdoor air (ΔT) of 15°C to 50°C; large ΔT values make it easier to accurately estimate K_{TOT}.

Table 1. Key properties of the six single-family houses examined in this work

House	1	2	3	4	5	6
Stories*	1.5**	1	1.5	1	1	1.5
Year of construction	1962	1967	1983	1987	2006	2006
Insulation thickness [mm]						
-Wall	70	130	180	245	215	215
-Roof	200	145	270	435/290	450	315
-Floor	0	50	200	240	190	200
Envelope area [m²]	382	298	454	459	523	596
Heated floor area [m²] house/garage	210	100	168/29	174/32	142/38	197/45
Inhabitants adults/children	3 (2/1)	3 (2/1)	2 (2/0)	4 (2/2)	3 (2/1)	4 (2/2)

*1.5 denotes one and half storey houses that have a finished attic.
**This house also has a basement that extends beneath the entire building.

3.2 Input Data

A Saber measurement system (KYAB, Sweden) was installed in each monitored house. The system consists of a measurement unit connected to the Internet that collects all data sampled by the sensors in each house. Sampling was conducted once per minute and the measurements were later converted to daily averages.

3.2.1 Temperature

T_i and T_e were measured using temperature sensors (one indoors and one outdoors) that were connected to the Saber unit via a cable. The sensors were factory calibrated to read temperatures of -40°C to +80°C with an accuracy of ±0.1°C. Indoor sensors were placed in a bedroom, hallway or living room away from heat sources and not in direct sunlight. Outside sensors were placed on the building façades in locations that would minimize the level of incident sunlight.

3.2.2 Power Supplied for Heating

P_{DH} and P_{DHW} were measured and separated by the Saber unit using a previously established (Yliniemi 2007) and experimentally verified (Yliniemi, et al. 2009) method of estimation. The Saber unit was also used to collect data on the amount of power drawn from the district heating system, which was gathered via the infrared (IR) port on the existing meter in each house. The Saber unit recorded the total amount of energy drawn from the district heating system for space and water heating. Individual residential houses connected to district heating systems have no ability to store heat from the system, and so any power required for space heating or hot water is supplied on demand. Space heating generates a steady baseline use of power from the district heating system, with hot water usage generating additional spikes in usage on top of this (Yliniemi et al. 2009). It was assumed that there were no losses from the system (i.e. the value of P_L was zero) because all measurements of P_{DH} and P_{DHW} were conducted indoors, and all losses that occur inside the building envelope are assumed to contribute to the heating of the house. This means that all production and transportation losses occur outside of the measurement zone.

3.2.3 Gained Free Power

P_{SUN} was assumed to be zero because the measurements were conducted during the months of November-February, during which the level of insolation in northern Sweden is very low (Sjögren et al. 2007). Luleå has about 80 hours of sunshine in October, about 10 in December, and about 60 in February (SMHI 2012). P_P was estimated to 71W at low activity and 119W at higher activity per person (Sauer et al. 2001). Each person was assumed to spend sixteen hours per day in their house, during which they would be highly active for 8 hours and less active for the remaining eight. According to Petersson (2010) 20% of the power used to produce hot water and 75% of the electrical power usage can be considered as heat gains. Because the Saber unit recorded P_{DH} and P_{DHW} separately, a new term (P_{HW}) was introduced to denote the heat gains from water heating. Electricity use was measured using existing electricity meters in each house, which were connected to the Saber unit via a pulse detector. Since P_{HE} and P_{BE} could not be separated, a single term denoting heat gains from electricity was introduced (P_E) and calculated as 75% of the total electrical energy used.

3.2.4 Dynamically Stored/Released Power

P_{DYN} was assumed to have negligible effect on the results. According to Hammarsten (1987), the dynamics of the building can be neglected if twenty-four hour averages are used as input data for the energy signature model as was the case in this work. The assumption that dynamics can be ignored is strengthened by the fact that the case study houses are made of wood, which is a light construction material that stores relatively little energy.

3.2.5 Determination of P_{TOT}

Using the above assumptions, P_{TOT} was calculated for each house.

$$P_{TOT}=P_H+P_{HW}+P_E+P_P \tag{6}$$

3.3 Data Management and Evaluation

All collected data were downloaded from the Saber units' web server and imported to Excel. The measured outdoor temperatures at the studied houses were compared to detect any sudden peaks or drops that might indicate malfunctions or the introduction of external heat sources. The indoor temperatures were also checked to ensure that they remained relatively steady and contained no peaks due to the introduction of new heat sources. P_{TOT} was calculated using equation 6 and was plotted as a function of the difference between the outdoor and indoor temperatures (ΔT) for each house. Linear regression was then used to calculate K_{TOT} from the slopes of the plots. The K_{TOT} values obtained in this way were divided by the area (A) of the envelope for the corresponding building from Table 1 to give a set of estimated average U-values.

4 Results and Analysis

Figure 4 shows a plot of P_{TOT} against ΔT for houses 1-6 over the period between November and February. House 1 changed owners in the beginning of December, which meant that the measuring systems had to be removed. There is thus no data from house 1 for the period between December and February, which were the coldest months of the measurement period and thus had the highest ΔT values. Figure 2 shows the calculated U-values for the period between November and February as a function of the year in which the houses were constructed. No usable temperature data was obtained for house 4 during January. To compensate for this, the indoor temperature for January was assumed to be equal to the average for the preceding months and the outdoor temperature was assumed to be equal to the outdoor temperature measured at a house located approximately 1km away from house 4 on the same river. The average difference between the measured temperatures for houses 3 and 4 was 0.44°C, with a peak difference of 5.2°C during one day. There were some problems with the temperature measurements in house 5 that made some of the expected indoor and outdoor temperature measurements unavailable. The missing indoor values were replaced with the average measured indoor temperature over the entire experimental period. In addition, the Saber unit for this house suffered from a driver malfunction that caused it to stop recording the outdoor temperature below 0°C. The outdoor temperature for house 5 was therefore assumed to be identical to that for house 6, which is located about 100m away from house 5. The average difference between the measured outdoor temperatures for houses 5 and 6 was 0.84°C, with a peak difference of 2.83°C. Due to the missing data, house 1 was also investigated for the time period August to November and house 2 was used as reference house for this period (Figure 2). Based on this time period house 1 has a calculated U-value of 0.58 $W/m^2°C$ and house 2 a calculated U-value of 0.50 $W/m^2°C$. Due to a lack of data from December and January, the ΔT span for house 1 is rather narrow, which makes the accuracy of the regression and the derived U-value uncertain ($R^2 = 0.55$). The calculated U-value for house 1 differs from that for the other house constructed in the 1960s by an unreasonable amount, Figure 3.

Table 1 shows that house 1 is less well-insulated than house 2, suggesting that its U-value should be higher. A more realistic U-value (0.58 $W/m^2°C$) for house 1 with a better model fit ($R^2 = 0.83$) was achieved when considering data from the period between August and November (Figure 2). It should however, be noted that the U-value for house 2 calculated based on data gathered between August and November was lower than that calculated for the period between November and February. This implies that the U-value of house 1 for the August – November period would also be lower than that which would have been measured at higher ΔT values. The goodness of fit of the regression for house 2 was high ($R^2 = 0.97$), which implies that the assumption of a linear relationship between the power expended on heating and the temperature difference is valid. House 3 has the lowest measured U-value of all those measured in this work (0.26 W/m² °C) and a good model fit ($R^2 = 0.96$).

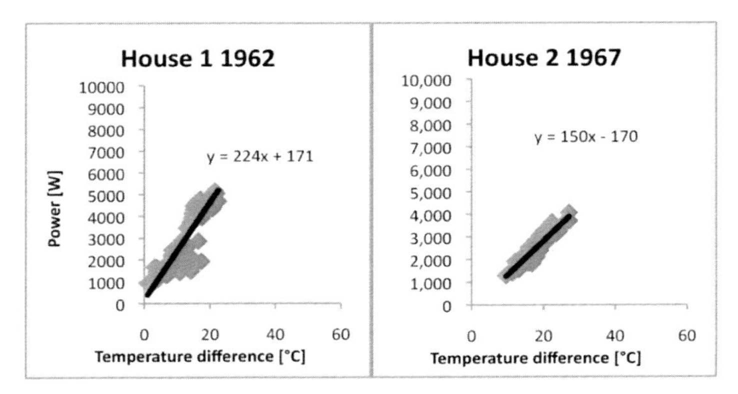

Fig. 2. The total power (P_{TOT}) plotted against the temperature difference (ΔT) for houses 1 and 2 between August and November

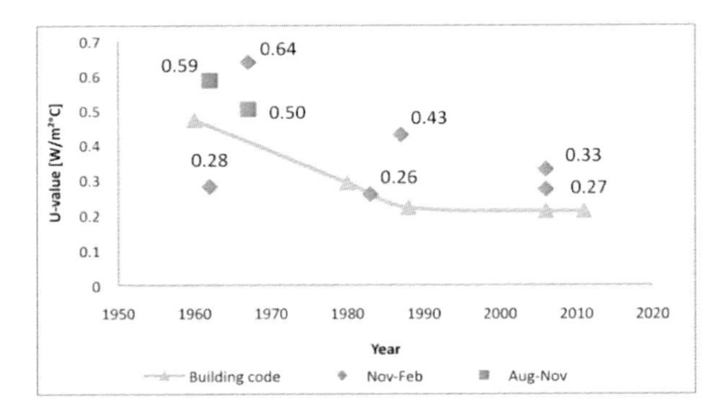

Fig. 3. U-values plotted against the year of construction of the studied houses superimposed on the maximum average U-values permitted by the corresponding building codes as shown in Figure 1

The U-value calculated for house 4 is rather high and there is considerable uncertainty in the corresponding dataset due to the problems encountered in recording its indoor and outdoor temperatures. Because of this the curve in Figure 4 has more variation than the others and the regression analysis yielded an R^2 value of only 0.86. The calculated U-value for house 5 (built in 2006) was 0.27 W/m²°C which is close to that required by the 2006 building code. The model fit for this house was also very good ($R^2 = 0.95$). The calculated U-value for house 6 was 0.34 W/m²°C, which is above the limit specified in the building code. The relatively poor model fit of the regression ($R^2 = 0.91$) for this house, together with the wide spread of measured temperatures is probably due to the residents' use of a stove.

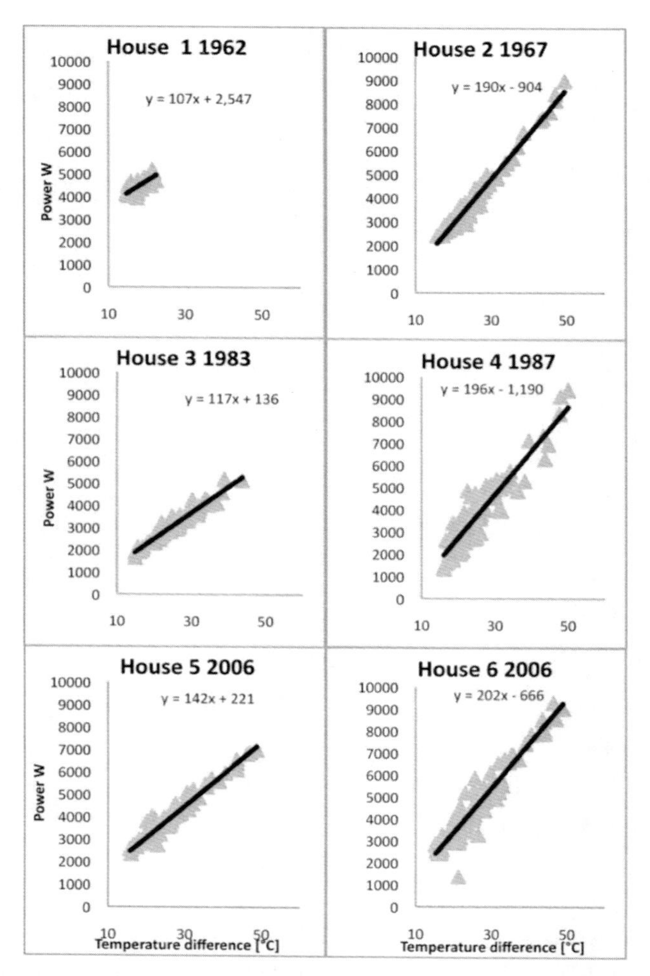

Fig. 4. The total power usage (P_{TOT}) as a function of the inside-outside temperature difference (ΔT) for houses 1-6 between November and February

5 Discussion

5.1 Effect of Building Regulations on Energy Performance

The increasingly rigorous U-value requirements specified by the Swedish building codes over time are reflected in lower U-values for newer houses. However, the results presented in Figure 3 show that the U-values of all houses considered in this work were above those required by the building codes in force when they were constructed; in this context it should be noted that only transmission losses were considered in the requirements of building codes prior to 2006. The gap between the values specified in building codes and those achieved in practice is illustrated by house 2 (built in 1967), which has an effective U-value of 0.64 W/m²°C compared to the upper limit of 0.47 W/m²°C specified in the building code of the time for transition losses alone. Another example is House 4, which was advertised as a "low energy house" at the time of construction and was expected to meet or only modestly exceed the requirements set out in the building code in force when it was constructed. Table 1.1 shows that it has a level of insulation comparable to or better than that found in newer houses. However, its effective U-value was actually higher than that for house 3 (Figure 3), which implies that the measured P_{TOT} was affected by some factor that the energy signature model did not account for.

5.2 Applicability of Static Energy Signature Models in a Cold Climate

The energy signature method provides a straightforward way of estimating a house's effective U-value. Because the method is most reliable when the difference between indoor and outdoor temperatures is large, it is advantageous to estimate the U-value under cold conditions. The method is sensitive to interference from external heat sources such as fireplaces. In addition, the reliability of the energy signature method for houses with basements has yet to be explored. Both Hammarsten (1984) and Westergren et al. (1999) have compared the performance of static and dynamic models for the energy signature method and have discussed the problems associated with the dynamics of heat exchange in buildings. Hammarsten recommends using daily or weekly averages for static models, while Westergren et al. recommended only weekly averages. Daily averages were used in this work, which produced realistic estimated U-values. As such, our results support the recommendation of Hammarsten (1984). User behavior was not considered in this work, but its effects can be appreciated by comparing houses 3 and 4. House 3 is occupied by a childless middle aged couple whereas house 4 belongs to a middle aged couple with two sport-playing teenagers. This probably explains why the electricity and hot water use of house 4 were three and fifteen times greater, respectively, than those in house 3 between November and February. To complement this case study, a theoretical calculation of the average U-values for the studied properties is planned, together with a survey of the user behavior of the inhabitants of the case study houses.

Acknowledgment. Financial support from the European regional development fund via the Interreg IVA North program is gratefully acknowledged.

References

Sauer, H.J.J., Howell, R., Coad, W.J.: ASHRAE handbook, fundamentals 2001. American Society of Heating, Atlanta (2001)

Energimyndigheten 2012 Energiläget 2011. Eskilstuna

Guerra-Santin, O., Itard, L.: Occupants' behaviour: Determinants and effects on residential heating consumption. Build Res. Inf. 38, 318–338 (2010), doi:10.1080/09613211003661074

Hammarsten, S.: Estimation of energy balances for houses. Dissertation, Royal Institute of Technology (1984)

Hammarsten, S.: A critical appraisal of energy-signature models. Appl. Energy 26, 97–110 (1987), doi:10.1016/0306-2619(87)90012-2

Lundstrom, E.: Occupant influence on energy consumption in single-family dwellings. D5: Swedish Council for Building Research, Stockholm (1986)

Nordling, L., Reppen, L.: Så byggdes villan: Svensk villaarkitektur från, till, Forskningsrådet Formas, Stockholm (2009)

Petersson, B.: Tillämpad byggnadsfysik. Studentliteratur, Lund (2010)

Sjögren, J.U., Andersson, S., Olofsson, T.: An approach to evaluate the energy performance of buildings based on incomplete monthly data. Energy Build 39(8), 945–953 (2007), doi:10.1016/j.enbuild.2006.10.010

Sjögren, J.U.: Energy performance of multifamily buildings: Building characteristics and user influence. Licentiate thesis, Umeå University (2007)

SMHI, Sun statistics (2012), http://www.smhi.se/klimatdata/meteorologi/stralning/1.3054 (accessed February 13, 2012)

SMHI, Temperature statistics (2012), http://www.smhi.se/klimatdata/meteorologi/temperatur (accessed September 15, 2011)

Westergren, K., Högberg, H., Norlén, U.: Monitoring energy consumption in single-family houses. Energy Build 29, 247–257 (1999), doi:10.1016/S0378-7788(98)00065-6

Yliniemi, K.: A device and a method for measurement of energy for heating tap water separated from the building's heating energy-usage. Patent no. PCT/SE2006/001250 World Intellectual Property Organization (2007)

Yliniemi, K., Delsing, J., van Deventer, J.: Experimental verification of a method for estimating energy for domestic hot water production in a 2-stage district heating substation. Energy Build 41, 169–174 (2009), doi:10.1016/j.enbuild.2008.08.008

Chapter 5
Two Case Studies in Energy Efficient Renovation of Multi-family Housing; Explaining Robustness as a Characteristic to Assess Long-Term Sustainability

Vahid Sabouri and Paula Femenías

Department of Architecture, Chalmers University of Technology, Gothenburg, Sweden
vsabouri@gmail.com, Paula.femenias@chalmers.se

Abstract. This study addresses two energy efficiency (EE) approaches to renovation of multi-family housing in Sweden aiming at a better understanding of robustness as a building characteristic especially in terms of energy performance of buildings and indoor air quality (IAQ). Gårdsten (Solar houses) and Brogården (passive houses) have been analyzed using an analytical framework. Adaptability, Redundancy, preference for passive techniques, users control over IAQ, transparency of systems to users and maintenance facility have been considered as the main criteria for robustness analysis and the performance of cases has been studied in relation to major factors likely to face uncertainties such as household appliances, occupant behavior, maintenance support, energy sources, technical systems, envelope quality and climatic conditions.

Keywords: Robustness, Energy efficiency, Sustainable buildings, multi-family housing, Renovation.

1 Introduction

1.1 Sustainable Development and Sustainable Building

According to the most often-quoted definition of sustainable development in The Brundtland Report, sustainable development is development that meets the needs of the present without compromising the ability of future generations to meet their own needs... The concept by its definition encompasses a wide range of domains and thus many various fields of science. As it has been stressed by Roggema (2009), global warming and change of climate is likely to be the most critical problem of the 21st century. Tin (2008) explains that changes in the climate are happening faster and stronger than expected. This means that it is not possible to predict the future and we might face significant uncertainties. Thus, adaptation to changes seems inevitable. According to Roggema (2009), the best strategy is to get ready for facing the worst case scenario to be able to deal with probable serious changes in close future.

Sustainable building is a term which is usually used to stress the objectives of sustainable development in relation to building activities and the built environment

A. Håkansson et al. (Eds.): *Sustainability in Energy and Buildings*, SIST 22, pp. 45–57.
DOI: 10.1007/978-3-642-36645-1_5

(Femenías 2004). Since buildings are responsible for environmental issues such as CO2 emissions and consequently climate change, sustainable buildings are characterized partly by having less impact on the environment. However, the other side could be how well these buildings withstand the environmental conditions and adapt to future situations. Although there is no globally accepted definition for sustainable building as Femenías (2004) explains, for implementation of sustainable building it is suggested to consider several factors including a life-cycle systemic approach for different stages from planning to maintenance and even demolition of buildings in order to prolong the life span of the design and make it more flexible and adaptable. Thus, regarding sustainability in the field of architecture, buildings should be designed in a way that they are capable of dealing with unforeseen situations.

1.2 Problem Statement

Nowadays, there is a growing tendency towards strict energy consumption targets in building codes and energy efficiency standards (Simm et al. 2011). Thus, during the past recent years there have been efforts to design and build energy efficient buildings or renovate the existing housing stock into energy efficient housing in order to optimize buildings' energy consumption. Though in some cases the results have been reported to be satisfactory, in some other cases energy efficiency measures have been vulnerable to factors such as aging, maintenance requirements or user behavior etc.

The performance of some systems in a real life situation is not the same as their expected performance on the drawing board or in the test chamber (Leyten et al. 2005). This discrepancy between the predicted design performance and what will happen during the real life operation of a building can considerably influence the energy efficiency (EE) objectives of the building (Simm et al. 2011). Among other reasons, poor assumptions regarding the performance of the building and installations during modeling which can mislead the designers in their approach and occupant behavior could be mentioned as two commonly cited causes for such a performance gap (Simm et al. 2011 with refer to Raslan et al. 2009, Masoso 2009 and Torcellini et al. 2004). Gonzalez (2011) states that measures improving energy efficiency do not always result in the anticipated energy savings since part of the savings might be offset through other mechanisms. This phenomenon is referred to as the *rebound effect* in literature and energy efficiency debates and is partly caused by overestimating energy-saving potentials and underestimating saving costs. Such misestimation is mostly due to disregarding the impact of user behavior (Haas and Biermayr 2000). Especially if the use of any type of energy and natural resource or other inputs such as labor is considered, the system sometimes deviates from its efficient use of energy or economic objectives to a large extent. As for HVAC systems which are in a close relation to energy savings in a building, some factors including sensitivity to aberration from design assumptions, unfeasible maintenance requirements and lack of transparency to occupants and building management account for such vulnerability of measures and goals (Leyten and Kurvers 2005). Furthermore, technically sophisticated systems are more likely to be fragile due to their dependency on technology (Leyten and Kurvers 2005) and could easily affect energy efficiency of buildings. Consequently, in achieving sustainable architecture, energy efficient buildings which are dependent on sensitive measures are not desirable results, especially in a long-term perspective, and designers should plan for buildings with reduced vulnerability.

1.3 Methods

The approach undertaken in this study is based on qualitative methods. According to the explanation of Femenías (Femenías 2004), the process can be seen as abduction which is a kind of approach in between deduction and induction. According to the classification of Groat and Wang (Groat and Wang 2002), the methods used for this research are mainly literature review and case studies. However, since to collect data for case studies, different articles and brochures have been studied and methods such as interviews and study visits have also been conducted, it could be considered as combined strategies.

2 Robust Design

2.1 Main Concept

In a scientific approach, a correct understanding of a concept entails high perception of that concept which is not attained unless one is perfectly acquainted with its definition in that field of science, since considering just the lexical meaning, a word can be variously interpreted. Although there have been attempts to define concepts such as *Robustness* and *Robust design* in industrial science and the fields related to technical products or socio-technical systems, it seems there is still no comprehensive agreed definition for these terms in the scientific terminology. On the other hand, in some cases the term *Robustness* might be compared with conceptions such as *reliability*, *Durability* and *Dependability* on one side and *Stability*, *Resilience* and *adaptability* on the other side. Here the notion is investigated in two different but at the same time similar areas which would be related to the field of architectural design.

Robustness in Technical Systems (Andersson 1997). Andersson (1997) tries to clarify the difference between reliability engineering and robust design by providing the definitions of reliability, availability, durability and dependability and how they are all associated with the larger image of Quality. Eventually, Andersson formulates his definition of robustness in technical systems based on the model of technical process already presented by Hubka and Eder. Since the technical system is considered as the main operator of the technical process, the aim of robust engineering is to design systems in which unexpected secondary inputs cannot excessively affect the performance of the system and the result of the process. According to Andersson *If a technical system maintains a stated performance level of its properties in spite of fluctuations in primary and secondary inputs, the active environment, the operands and in human operation, then the system is robust* (Andersson 1997, P 282).

Robustness Engineering vs. Reliability Engineering. Andersson believes that the main difference between reliability engineering and robust engineering lies in the assumed conditions for the performance of a system. As the definition of reliability implies, reliability engineering deals with some anticipated conditions in a known environment including a set of usual expected variations, while in robustness engineering the system should be able to handle unusual unexpected situations and rare events.

Robustness in Socio-technical Systems (Pavard et al. 2006). In the article by Pavard et al. robustness of socio-technical systems is mainly studied by means of comparing the differences between *regulation*, *resilience* and *robustness* within the theoretical framework of complex systems. From this point of view three types of engineering for complex systems has also been presented; Classical engineering, resilience engineering and robustness engineering. *Intuitively, a robust system is one which must be able to adapt its behavior to unforeseen situations, such as a perturbation in the environment, or to internal dysfunctions in the organization of the system, etc* (Pavard et al. 2006, P 2). In order to better clarify the differences between these notions, three types of regulations are presented:

a) Classic regulations which aim to maintain a constant control over the behavioral variables of the system to guarantee the stability of the system's behavior.
b) Structural regulations which are able to adjust the structure of the system to the new situation by self-adaptation in order to preserve the function of the system.
c) Emergent and self-organized regulations that let the system to govern itself in an emergency situation by self-organization and in association with its environment.

This point of view for managing complex socio-technical systems is followed by introducing three required types of engineering:

1) Classical engineering which is characterized by functional stability and anticipating probable situations. This approach aims for *stable organization.*
2) Resilience engineering which is characterized by uncertain situations and reduced anticipation of the system's behavior. This approach aims for dynamic *reorganization.*
3) Robustness engineering which is characterized by emergent functionalities and no anticipation of further situations. This approach aims for *self-organization.*

Robustness Engineering vs. Resilience Engineering. As implicitly explained, the major difference between these two notions is that resilience engineering deals with undesired situations which are still possible to be anticipated and although changes might happen in the organization of the system, the aim is to preserve a certain result and keep the function of the system alive. This approach by its nature considers the system clearly separated from its environment. However, in robustness engineering, which deals with non-deterministic emergent situations in complex systems, firstly it is not possible to ensure that the function of the system or its subsets will be maintained and secondly the system is not assumed as a distinct entity since there might be a close interaction between the system and its environment and they could be tightly associated.

2.2 Robustness and Architectural Design

A building is a complex system, comprising several technical and socio-technical subsystems each of which might be subjected to the concept of robustness and could effectively influence the robustness of the whole system. Nevertheless to study robustness of a building as a set of interconnected systems there are major factors which could be generally considered in the design approach. These factors could be categorized as follows. However, since the focus of this study is on the energy efficiency measures, part 3 which is more relevant to this discussion will be further investigated.

1. Robustness of building's physical structure

-Main structure -Materials and installations

2. Robustness of user's comfort and satisfaction

-Indoor environmental quality -Functionality of spaces

-Aesthetic features

3. Robustness of feasible operation and maintenance

-Maintenance facility -Energy efficiency

Robustness of Maintenance Facility. Ease of care and maintenance of buildings is practically and economically influenced by design of details and implementation of different methods (Bokalders and Block 2010). IEQ of a building encompasses IAQ as well as thermal comfort, health, safety, quality of potable water and other issues such as lighting, acoustics, ergonomics and electromagnetic frequency levels. It would not be an exaggeration if one says that the success or failure of a building lies on its IEQ. To quote Chris Alexander (Brand 1994) more money should be spent on the basic structure and ceaseless adjustment and maintenance than on finishing. In order for robustness of facile maintenance and care in ecologically constructed buildings, they should be considered in the early planning phase. An important issue regarding maintenance of building services is that they should be easily accessible and adjustable as well as adaptable to new technologies and energy suppliers. According to Brand (Brand 1994) if the systems are too deeply embedded in the construction they cannot be easily replaced and this has caused many buildings to be demolished earlier than their efficient lifetime.

Robustness of Building's Energy Efficiency. Although the cost of energy in the future is still questionable, buildings with high energy consumption are very likely to be unacceptable in a few decades at least due to their ecological issues. Thus, it is of utmost importance that buildings are constructed with robust EE measures, since this will not only make their maintenance economically affordable in the whole building's lifetime but also prevent extra costs in the future to improve their efficiency by applying alternative solutions. According to researches done in Delft University of Technology (Linden 2007), regarding IAQ and EE, robustness of buildings can generally be improved through *user-oriented* and *climate-oriented* design approaches. Considering the definition of robustness, EE measures in a building should be insensitive to changes in the situation or the active environment. In their article, Leyten and Kurvers (2005) refer to the definition of robustness of a technique in statistics which is *the ability of a certain technique to deliver accurate results, although its assumptions are violated* and analogously formulate a definition for robustness of a building and an HVAC system as *the measure by which the building or the system lives up to its design purpose in a real life situation.* Furthermore, in a comparison between low-tech and high-tech solutions, less technically complicated buildings are often more robust (Leyten and Kurvers 2005). This means that the measures should not be dependent on sophisticated technical solutions. Juricic (2011) explains that complex building systems such as mechanical ventilation or active cooling systems are very likely to cause high energy consumption or lack of thermal comfort to users due to lack of transparency which leads to misuse. Moreover, experiences indicate that people would prefer buildings without cooling but with operable windows to those with fixed windows and cooling systems and they even accept temperatures higher than the comfort range in the former case (de Dear et al. 1997). Leyten et al. (2009) stress that user control over IEQ such as control over natural ventilation, temperature, sun shading and artificial lighting increases robustness of buildings. This is because users get the opportunity to adapt IEQ to their specific personal preferences and probable malfunctioning of the building will be compensated. Juricic (2011) points to *redundancy of systems and multiplicity of functions* as a building characteristic which helps its robustness. Sussman (2007) has developed a metaphor explaining such a concept in natural systems such as in human body where several functions are fulfilled by different organs or some other organs might be adapted to achieve the goal in case of failure in the main organs. According to Juricic (2011) *one system, several functions* would be the worst case while *several systems, one function* could be considered as the best case.

3 Case Studies

For this study two cases have been selected which are both well-known demonstration projects of multi-family housing renovation in Sweden. Gårdsten in Gothenburg and Brogården in Alingsås have been retrofitted both with the main focus on the energy performance of the building but with two different approaches to energy efficiency.

3.1 Solar Houses, Gårdsten, Gothenburg

Solar houses1 (Solhus1) is the renovation of 255 apartments comprising 10 buildings (3 high-rises and 7 low-rises) in West Gårdsten which was initiated at the time for the call for targeted projects for the THERMIE program in 1996 (Dalenbäck 2007). The renovation project started in early 1998 and was finalized in 2001.

Energy Efficiency Measures. In the high-rises, the original flat roofs have been covered by extra insulation on top and a shed roof facing south with integrated solar collectors for preheating domestic hot water was added on top of the building. In each block, storage tanks have been placed in the basement of 6-story building and preheated water (heated almost 35% by solar energy) is stored there to be distributed to all the apartments and the common laundry in the block. The supplementary heat is provided through district heating system. Furthermore, the previous open balconies to the south have been repaired and enclosed with glazed panels. The apartments are supplied with fresh air preheated by sunlight through these glazed balconies. Fresh air enters the living rooms and bed rooms adjacent to these balconies through air inlets designed in the windows and balcony doors. The exhaust air is directed out from the existing exhaust system in kitchens and bathrooms on the northern part of the flats (Dalenbäck 2007). Despite of low investment incentives for insulation of all external walls due to low energy costs, the gables in the high-rise buildings were insulated. Existing laundry rooms, located in the basement of the high-rises were replaced with new laundry rooms, designed in the ground floor of these buildings. The new laundries were equipped with energy efficient washing and drying machines, connected to the domestic hot water system in the basements to save electricity for water temperatures below 50°C. Moreover, communal greenhouses have been built on the ground level of these buildings, adjacent to the new laundry rooms along more than half the length of the building to the south. In all buildings the inner window panes of the existing double glazed windows have been replaced with new low-emission panes (Dalenbäck 2007). All apartments have been equipped with energy efficient household appliances and all households have been provided with individual metering systems for water, electricity and space heating in their flats (Nordström 2005).

However, in the low-rise buildings the existing ventilation systems were equipped with a heat recovery installation and the flat roofs were covered by external thermal insulation. One of the low-rises has a unique design in this project. The external walls to the east, north and west of this building have been covered with an extra layer of thermal insulation and a cavity has been created between the original walls and this new layer. These walls are not only protected from outdoor cold weather but also warmed up by circulation of heated air in this gap. The air is heated through solar collectors vertically installed and integrated to the southern façade of this building (Gårdstensbostäder 2010).

3.2 Passive Houses, Brogården, Alingsås

Brogården consisted of 299 apartments in sixteen 3-story buildings originally constructed in the early 1970s as part of *the million homes program*. As the first experience of retrofitting with passive house techniques in Sweden, the renovation process started in March 2008 and the whole project is to be completed in 2013 (Morrin 2009).

Energy Efficiency Measures. The main idea behind passive house concept is making heat losses as less as possible. This technology involves sufficient insulation for building envelope as well as making it as air tight as possible. In such a system not much energy is needed for space heating and the air inside the building would be sufficiently warmed by the heat from occupants' body, household appliances etc. In order to provide fresh air in such airtight spaces, the buildings are equipped with heat recovery ventilation systems of high efficiency.

In Brogården the external shell of the buildings was highly insulated. The ground slab was insulated with a total thickness of 200 mm of EPS on both sides. The exterior long side walls which were in a poor condition were replaced with newly built walls with a steel structure and layers of mineral wool and EPS (app. 440 mm) and the insulation layers on the attic floor were replaced with 400-550 mm of loose wool insulation (Morrin 2009). All windows and entrance doors were replaced with xenon gas-filled triple glazed thermo windows and highly insulated doors respectively. The existing recessed balconies which made substantial thermal bridges in the external walls were enclosed as part of the apartment interior space and new balconies were built, standing on a separate structure and mounted on the outside of the façade (Janson 2009). The previous ventilation system was replaced with air-to-air heat exchanger units with 85% efficiency (Janson 2008) installed in each apartment. In very cold days (estimated app. 10 days a year) (Eek 2011), these units can also provide the incoming air with extra heat from the district heating system. The air inlets have been mounted on the living rooms and bedrooms walls and the out lets are in the kitchens and bathrooms. The apartments have been equipped with low-energy household appliances. Almost 60% of the apartments will be accessible by low-energy elevators which store energy from downward motions to be used in upward motions (Morrin 2009).

4 Analysis

In this part the cases are analysed based on the criteria of robust design with regard to changes in some major factors affecting building's energy performance (Table 1). According to our studies robust design deals with reduced vulnerability to any unforeseen situation and thus a thorough robustness analysis entails a comprehensive study of future circumstances. However, among different factors influencing energy efficiency, some of them seem to be more essential and more likely to face uncertainties during building's lifetime including:

- **Household appliances** (using new appliances due to different lifestyles etc.)
- **Occupant behavior** (unexpected patterns of energy consumption)
- **Maintenance support** (Changes in building management etc.)
- **Energy sources** (Introducing different energy supplies due to cost etc.)
- **Technical measures** (issues related to availability of sophisticated systems)
- **Envelope quality** (physical changes due to issues such as aging)
- **Climatic conditions** (issues such as global warming etc.)

Furthermore, among other criteria, adaptability of systems, redundancy of measures, preference for passive techniques and user-oriented design criteria such as users control over IAQ, Transparency of systems to users and facility of maintenance have been chosen as the main robustness criteria for the analytical framework. Since EE buildings are characterized by focusing on three major issues of user comfort, environmental impact and energy cost, the relation between these issues and the aforementioned factors and criteria has been presented in the following diagram (Fig. 1).

While regarding heat loss and energy saving, passive housing seems to be a safer solution due to highly insulated and well air tight envelopes, both cases could be at risk of energy performance reduction in case of unexpected situations in their life time. What seem to be common in both projects are issues related to availability of technical systems, redundancy of systems and feasibility of maintenance. In case of technical solutions such as solar panels, ease and cost of maintenance, as well as availability of the technique and its performance in relation with environmental factors can be questionable whereas energy efficiency in a long-term perspective through passive house method which is quite dependent on building fabric and details, could be vulnerable to issues such as performance loss of thermal insulation materials.

Although the criteria for robust design have been presented with the same level of significance in this table, it is possible to determine more effective factors and criteria by applying methods such as system design to find the leverage points of applied systems according to the specific characteristics of each project.

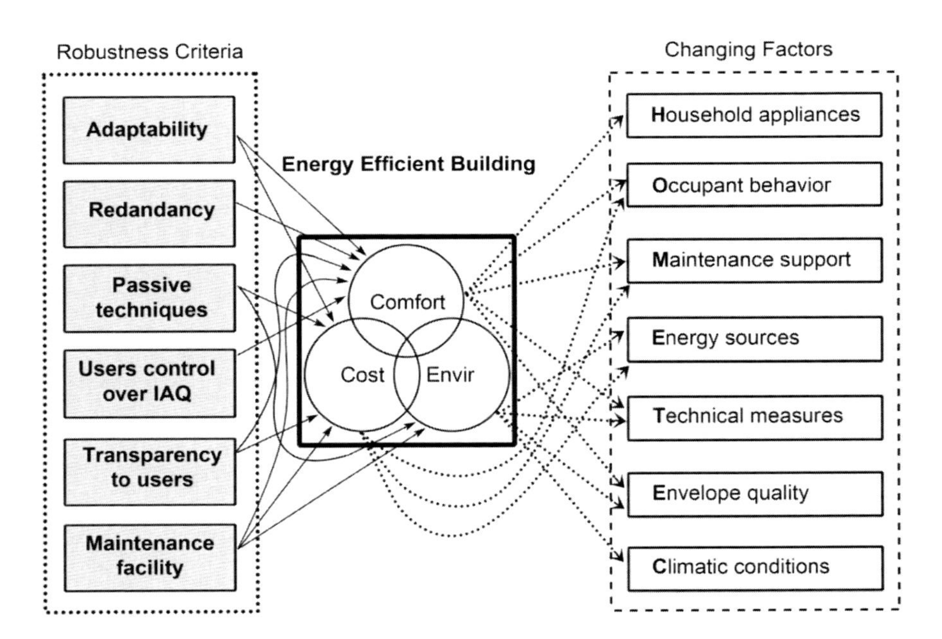

Fig. 1. Relation between main objectives of EE building, robustness criteria and HOMETEC factors

Table 1. Analysis of the cases based on the criteria of robust design and major factors of change[1]

		Gårdsten	Brogården
Robust design criteria	**Adaptability of systems**	+ Glazed balconies adaptable to different outdoor climates (C),(O) - Technical issues of adapting preheating/heating systems to alternative energy sources (ES),(M)	+ Façade material can be easily replaced (C),(O) - Economic issues in case of performance loss in insulation material (EQ),(M)
	Redundancy of measures	+ Possible use of electricity based systems for heating/cooling (C),(O) - Alternative heating/cooling devices not provided in the apartments (O),(H)	+ Possible use of simple sources of energy such as candles for space heating due to highly insulated and air tight envelope (ES) - Lack of fresh air in case of failure in heat exchanger (T),(M)
	Preference for passive techniques	+ Preheating of incoming air through glazed balconies in high-rises (C) + Solar gain through larger windows in living rooms facing south (C) + Sun shading provided both by balconies and operable blinds and curtains (C) - Considerable heat loss through building envelope (C),(T) - Apartments not very well air tight (C),(T) - Heat loss through entrances in the open balconies with no air lock (High-rises) (O),(C)	+ Highly insulated building envelope (C),(T),(EQ) + Xenon gas-filled triple glazed windows (C),(T) + Well air tight apartments (C),(T),(EQ) + Solar gain through larger windows in living rooms facing south, east or west (C) + Sun shading provided both by balconies and operable curtains (C)
	Users control over IAQ	+ Operable windows (C) + Operable glazing panels in balconies (C) + Adaptable indoor temperature (O) + Blind curtains to control daylight (C) - Possible unnecessary use of glazed balconies with extra heating in cold days(O),(H)	+ Operable windows (C) - Integration of heating and ventilation (T)

[1] (H): Household appliances, (O): Occupant behavior, (M): Maintenance support, (ES): Energy Sources, (T): Technical systems, (EQ): Envelope Quality and (C): Climatic conditions.

Table 1. (*continued*)

	Transparency of systems to users	+ Radiating panels used for heating (T),(M) + Glazed balconies to preheat incoming air (T),(M) - The system of solar panels to preheat hot water not easily understandable for layman (T),(M)	+ Highly insulated building envelope + Air tightness of the spaces (T) - Mechanical ventilation and integration with heating (T),(M)
	Facility of maintenance	- Technically sophisticated parts such as solar panels preheating water or heating air in the low-rise building facing south not very easy to maintain (T),(M)	+ The buildings dependent only on one technical system (heat exchanger) which has only a filter to be changed per year (M),(T) - Constant need for technical maintenance (T),(M)

5 Conclusion and Further Remarks

This study aimed at a better understanding of robustness as a building characteristic, especially regarding energy efficiency measures, IEQ and users' comfort. The study indicates that robustness is a qualitative characteristic of systems, specifically buildings in this research, which is generally defined as *the characteristic of measures by which the building or the system lives up to its design purpose in a real life situation*. Consequently, this characteristic is closely related to adaptability of a building and its subsystems. Particularly, for multi-family housing, design for *robustness* seems to be a characteristic which can enhance building *sustainability* from different points of view and support the functional purpose of buildings. Since both notions aim for more durable and reliable systems, design for robustness is quite in sync with sustainable architecture. Therefore, the concept could be applied to assess sustainability of design in a long-term perspective. According to this study robustness of a building and particularly multi-family housing, can be noticeably enhanced through *user-centred* and *climate-oriented* design approaches. These two approaches provide the designers with more comprehensive data to have more realistic predictions and prevent inaccurate assumptions and misestimation of design performance during modelling.

According to the case analysis, there are major factors influencing building's energy performance in a long-term perspective which should be analyzed in the design process in order to assess robustness of design and building's sustainability. These factors which are likely to face unforeseen situations during building's lifetime include: *household appliances*, *occupants behaviour*, *maintenance support*, *energy sources*, *technical measures*, *envelope quality* and *climatic conditions*. On the other hand, aiming for a building with robust energy efficiency measures entails a design process with accurate assumptions in which criteria such as *adaptability*, *redundancy*, *preference for passive techniques*, *users control over IAQ*, *transparency of systems to users* and *facility of maintenance* are taken into account.

An important point of the study to be stressed is that sometimes aiming for a robust design does not necessarily mean to achieve the most efficient performance of the building, especially in a short term perspective. For instance, regarding energy efficiency of a building, some measures seem to save more energy and thus more efficient, but concerning user comfort they are unsatisfactory, likely to cause unexpected behaviours and thus not necessarily robust. Therefore in evaluation of a design or deciding for design characteristics, robustness and efficiency should not be misinterpreted.

References

1. Andersson, P.: Robustness of Technical Systems in Relation to Quality, Reliability and Associated Concepts. Journal of Engineering Design 8(3), 277–288 (1997)
2. Bokalders, V., Block, M.: The whole building handbook, How to Design Healthy, Efficient and Sustainable Buildings. Erthscan, London (2010)
3. Brand, S.: How Buildings Learn: What Happens After They're Built. Viking Press, New York (1994)
4. Dalenbäck, J.O.: Training for Renovated Energy Efficient Social housing (TREES), Section 3 Case studies. Intelligent Energy -Europe programme, contract n° EIE/05/110/SI2.420021, 2 (2007)
5. de Dear, R.J., Brager, G.S., Cooper, D.: Developing an Adaptive Model of Thermal Comfort and Preference. Final report ASHRAE RP-884 (1997), http://sydney.edu.au/architecture/documents/staff/richard_de_dear/RP884_Final_Report.pdf (accessed February 9, 2012)
6. Eek, H.: Architectural qualities of passive houses in Brogården (2011) (interviewed October 7, 2011)
7. Femenías, P.: Demonstration Projects for Sustainable Building: Towards a Strategy for Sustainable Development in the Building Sector based on Swedish and Dutch Experience. Dissertation, Chalmers University of Technology, Gothenburg (2004)
8. Freire-Gonzalez, J.: Methods to empirically estimate direct and indirect rebound effect of energy-saving technological changes in households. Elsevier Sience (2011)
9. Gårdstensbostäder, Solar buildings in Gårdsten (2010), http://www.urbanisztika.bme.hu/segedlet/panel/15-Tanulmanyok-Novak_Agnestol/solar_buildings_Gardsten.pdf (accessed September 8, 2011)
10. Groat, L., Wang, D.: Architectural research methods. Wiley, Chichester (2002)
11. Hass, R., Biermayr, P.: The rebound effect for space heating; Empirical evidence from Austria. Elsevier Sience (2000)
12. Janson, U.: Apartment Building in Brogården, Alingsås SE. IEA SHC Task 37, Advanced Housing Renovation with Solar & Conservation (2008), http://www.iea-shc.org/publications/downloads/task37-Alingsas.pdf (accessed September 19, 2011)
13. Janson, U.: Renovation of the Brogården area to Passive Houses. Passivhus Norden, Gothenburg (2009), http://www.sintef.no/project/eksbo/Janson_Renovation_of_the_ Brogården_area.pdf (accessed September 19, 2011)

14. Juricic, S.: Robustness of a building, Relationship between building characteristics and actual energy consumption and indoor health and comfort perception. Dissertation, Delft University of Technology (2011)
15. Leyten, J.L., Kurvers, S.R.: Robustness of buildings and HVAC systems as a hypothetical construct explaining differences in building related health and comfort symptoms and complaint rates. Elsevier Energy and Buildings (2005)
16. Leyten, J.L., Kurvers, S.R., van den Eijnde, J.: Robustness of office buildings and the environmental Gestalt. Healthy Buildings. Delft University of Technology, Delft (2009)
17. Morrin, N.: Brogården, Sweden. Skanska AB (2009), http://skanska-sustainability-case-studies.com/index.php/sweden?start=10 (accessed September 19, 2011)
18. Nordström, C.: Solar Housing Renovation, Gårdsten, Sweden (2005), http://www.worldhabitatawards.org/winners-and-finalists/project-details.cfm?lang=00&theProjectID=293 (accessed September 12, 2011)
19. Pavard, B., Dugdale, J., Bellamine-Ben Saoud, N., Darcy, S., Salembier, P.: Design of robust socio-technical systems (2006), http://www.resilience-engineering.org/REPapers/Pavard_et_al_R.pdf (accessed October 21, 2011)
20. Roggema, R.: Adaptation to Climate Change: A Spatial Challenge. Springer, New York (2009)
21. Simm, S., Coley, D., de Wilde, P.: Comparing the robustness of building regulation and low energy design philosophies. In: Proceedings of the 12th Conference of International Building Performance Simulation Association, Sydney (2011)
22. Sussman, G.J.: Building Robust Systems an essay. Massachusetts Institute of Technology (2007), http://groups.csail.mit.edu/mac/users/gjs/6.945/readings/robust-systems.pdf (accessed October 21, 2011)
23. Tin, T.: Climate change: faster, stronger, sooner, A European update of climate science (2008)
24. Van der Linden, K.: Robustness of indoor climate & energy concepts. Delft University of Technology (2007), http://virtual.vtt.fi/virtual/fbfworkshop/iea_fbf_kees_van_der_linden.pdf (accessed November 7, 2011)

Chapter 6
Exploring the Courtyard Microclimate through an Example of Anatolian Seljuk Architecture: The Thirteenth-Century Sahabiye Madrassa in Kayseri

Hakan Hisarligil

Meliksah University Department of Architecture Kayseri/TURKEY
hhisarligil@meliksah.edu.tr

Abstract. The aim of the study was to investigate the microscale climatic conditions of courtyard buildings constructed by the Seljuk Turks throughout Anatolia in the thirteenth century. The particular focus was on how semi-open spaces, such as iwan and arcades surrounding the courtyard, are used to control daily and seasonal variation in the harsh semi-arid climate. Sahabiye Madrassa, in Kayseri, was used as a case study. Using ENVI-met 3.1, numerical simulations were run on a three-dimensional microclimate model to observe (a) the variations in microclimatic parameters, such as air temperature, incoming short-wave radiation, outgoing long-wave radiation, wind speed, and mean radiant temperature, and (b) how these variations affect comfort indices, such as the Predicted Mean Vote (PMV) and Predicted Percentage Dissatisfied (PPD). It was found that the madrassa responds dynamically to variation in external parameters and provides its users with increased levels of comfort, which belies its static and massive appearance.

Keywords: Sahibiye Madrassa, Courtyard Microclimate, Thermal Comfort, Seljuk Architecture, Kayseri, ENVI-met.

1 Introduction

The courtyard building is a very old type of architectural construction and has been used by many urban civilisations over the centuries, starting with the Greeks, Romans, and Egyptians. Various examples of such buildings are still seen in many part of the world. Being virtually outdoor rooms, courtyards, whether enclosed or attached, are often referred to as microclimate modifiers in the scientific literature. They are believed to provide better climatic conditions than the surrounding open areas, and are supposed to have a positive effect on the heating and cooling loads of the enclosing building. Martin and March propose that the courtyard is the most effective type of construction for achieving desirable thermal environmental conditions, at both the architectural and urban scale [1]. According to Al-Azzawi, courtyards help to achieve such conditions, not by mechanical devices, but by architectural design in regard to concepts, plans, forms, sections, elevations, and details [2]. Knowles and Koenig present the solar envelope and intersititium as a

A. Håkansson et al. (Eds.): *Sustainability in Energy and Buildings*, SIST 22, pp. 59–69.
DOI: 10.1007/978-3-642-36645-1_6

revisiting of the spatial logic of courtyard, as a remedy for current problems, because they respond to the rhythms of nature by dynamic means that conserve energy and enhance outdoor life [3]. The thermal performance of courtyards has been investigated by such researchers Mohsen [4], Etzion [5], Cadima [6], Muhaisen and Gadi [7], and Muhaisen [8], who paid special attention to the effect of the geometrical and physical parameters of the courtyard. These authors conclude that for the proper protection of the courtyard's surfaces and its surroundings from intense solar radiation and the wind, proper configuration and proportioning of the geometry of courtyard is vital. Failure will result in poor thermal performance, in the form of either too much shadow when solar radiation is needed or too much radiation when it is not desired. It has also been found that the courtyard's orientation is generally secondary in importance to its proportion. Strategies to make courtyards more comfortable to spend time in have been studied experimentally in specific courtyards around the world [9-13]. The results show that the orientation of semi-enclosed open spaces, irrespective of solar angles and wind direction, may create thermal discomfort. Proper orientation, along with dynamic shading (opening it at night and covering it during daytime), can improve their thermal behaviour. The addition of vegetation and the substitution of concrete pavement with soil and grass, and ponds, have similar but milder effects on their immediate surroundings. Further it has been found that sufficient and efficient openings can, if suitably incorporated, improve the thermal conditions of the courtyard's surrounding spaces [14]. Other researchers also studied the effect of courtyard geometry on airflow and temperature stratification. One of these [15], found that some factors, such as the courtyard aspect ratio, the wind speed, and wind direction greatly influence the airflow inside the courtyard. Confined courtyard buildings can have their own thermal environment, due to the minimal mixing of air between the courtyard and the exterior. Almost all studies emphasise that courtyard buildings are the most preferable type, either for reducing the cooling load in hot periods or for heating load in cold period.

This paper is a part of the initial phase of ongoing long-term research on the microscale climatic conditions of courtyard buildings, such as madrassas and caravanserais, the most significant building types of Anatolian Seljuk architectural heritage, appearing from the thirteenth century onwards throughout Anatolian lands. It explores the extent to which the design of courtyards of thirteenth-century buildings meets the criteria suggested for responding to daily and seasonal harsh climatic variations. The key contribution of the study is that it is, to the best of our knowledge, the first to investigate the microclimate and thermal comfort in a courtyard building of thirteenth-century Seljuk architecture using a numerical simulation model, such as ENVI-met in Turkey.

2 A Case Study of a Madrassa Building of the Thirteenth Century

The Seljuks, who established the first Turkish state in Anatolia, built numerous types of structure for various purposes: mosques (cami), schools (madrassa), hospitals (şifahane), tomb towers (kumbet) palaces and pavilions (kosk), roadside inns (caravanserai), baths (hamam), and dervish lodges (tekke). Most of these structures date from the thirteenth century, with a few from the twelfth century. They are to be found throughout the Anatolian peninsula, also called Asia Minor. Among them, the madrassa (a school for higher education in the sciences and religion) is a major type of courtyard building. Constructed in 1267-1268 in Kayseri, Sahibiye Madrassa is a typical example of a Seljuk courtyard building. Its plan follows a traditional four-iwan courtyard madrassa. Typically with a square or rectangular walled exterior and a single portal, and a pond, the courtyard, which is surrounded by chambers, open chambers (iwan-vaulted halls closed on three sides and open at one end) and arcades (covered passageways on the two lateral sides, which can have cells along the sides; or a series of arches supported by columns) is fully open to the sky (Fig. 1).

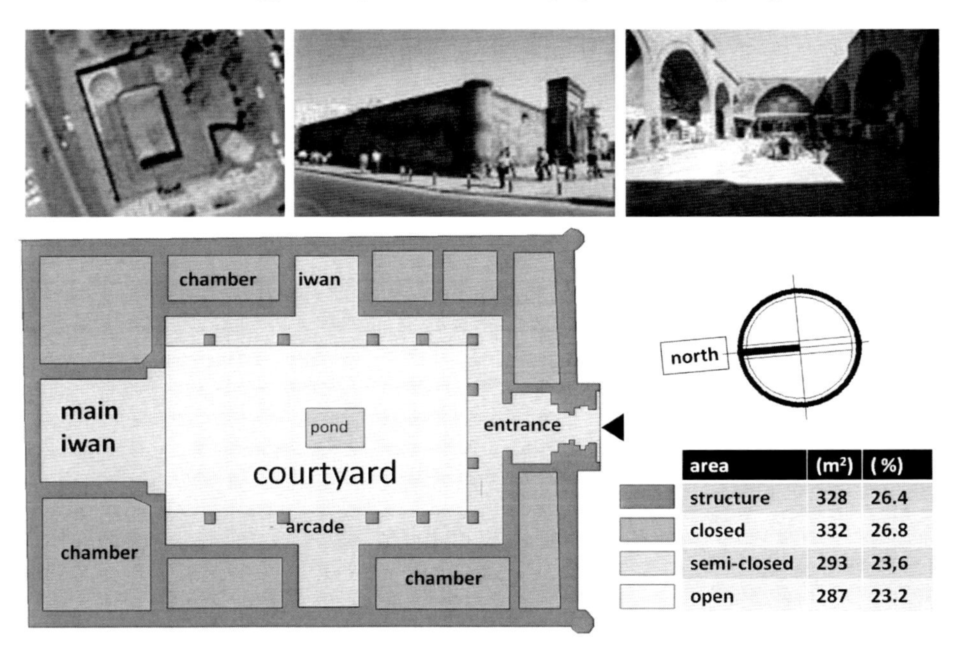

area	(m^2)	(%)
structure	328	26.4
closed	332	26.8
semi-closed	293	23,6
open	287	23.2

Fig. 1. The aerial, outer and inner views of Sahibiye Madrassa in Kayseri with its main portal facing the main iwan, and the percentage of area coverage of spatial components

When the percentage of enclosed, semi-open and open space to total space (1240 m2) is observed, it is found that the percentage of mass constructive walls (327 m2) is 26.3%, the percentage of enclosed space (332 m2) is 26.7%, and the percentage of open and semi-open spaces (626 m2) is 46.8%. There is thus almost twice as much open and semi-open space as enclosed space, which gives the static massive-looking madrassa building the appearance of a dynamic open one.

Sahibiye Madrassa is located in the large, industrialised city of Kayseri in Turkey, close to Cappadocia. Turkey is situated in the large Mediterranean basin, which has a fairly temperate climate in general. However, due to the diverse nature of the landscape and the existence in particular of the mountains that run parallel to the coasts, there are significant differences in climatic conditions from one region to another. While the coastal areas enjoy milder climates, the inland Anatolian plateau experiences extremes of hot summers and cold winters with limited rainfall. Kayseri (38°43' North, 32°29' East) is located at the foot of the extinct volcano Erciyes (3916 m) at an inland Anatolian plateau that has a continental climate (Köppen climate classification *Dsa*) featuring a semi-arid climate (Köppen climate classification *BSk*) [16]. Due to the city's high elevation of 1068 metres, the mean temperature in Kayseri fluctuates between 3°C and 0°C in January (the coldest month) and between 20°C and 22°C in August. The city has a Heating Degree Days (HDD) value of 3174 at a 22°C base temperature and Cooling Degree Days (CDD) value of 76 at a 15°C base temperature (from statistical data for 2011 provided by The Turkish State Meteorological Service). It is thus located in region IV according to the climate classification system of Turkish Standard-825, whereby the country is divided into four climatic regions for insulation purposes. Average annual rainfall varies between 350 mm and 500 mm. Most of the precipitation occurs during the spring and autumn.

3 Microclimate Simulation Model

The simulations reported herein were run using the three-dimensional non-hydrostatic urban microclimate model ENVI-met 3.1 BETA V. This is one of the first models developed to determine the interaction between surface and air that affect the microclimate [17]. Its typical areas of application are Urban Climatology and Planning, Architecture, Building and Environmental Design. ENVI-met has been under constant development since its first appearance in 1998 [18] and was last updated in 2010, which makes it one of the best tools available. It is widely used in the literature and in recent years has been validated for assessing built environments. Providing both lumped and distributed parameter methods [19], the software uses both the characteristics of fluid dynamics, such as air flow and turbulence, and the thermodynamic processes taking place at the ground surface, at walls, at roofs, and at plants. Although it has certain limitations [20-21], this free and user-friendly model calculates the outdoor microclimate and thermal comfort in a systematic way, while consuming relatively few computer resources.

A simulation that uses ENVI-met has three stages. The first stage consists of editing the area input files (.IN), to specify in detail the horizontal and vertical dimensions of the model environment to be analysed. The second stage consists of editing the configuration file (.CF), to define the basic settings, such as location, temperature, wind speed, humidity, PMV parameters, and databases for soil types and vegetation. Using the .IN and .CF files, the model analyses micro-scale thermal interactions within built environments with a typical horizontal resolution from 0.5 to 10 m and a typical time frame of 24 to 48 h, with a maximum time step of 10 sec. To minimize boundary effects that might distort the output data, the model uses an area of nesting grids. In the final stage, the outputs of the first and second stages, which are in the form of binary files (.EDI/.EDT), are imported into a visualisation program LEONARDO 3.75 to map the results.

4 Results and Discussion

4.1 Results for Generic Courtyard Forms

The aim of running the simulations was to explore the microclimate of the enclosed courtyard and its effect on thermal comfort, so the relation between the geometry of an existing building and actual weather condition was observed. For this, we first examined how the proportions and orientation of the courtyard of Sahibiye Madrassa are well-calibrated to achieve thermal comfort during the cold period, because the HDD value of the location of Sahibiye Madrassa is significantly higher than the CDD value. For this, we constructed three generic courtyard forms. We took the height (8 m) of the madrassa as a reference and varied the horizontal dimensions of the courtyard. We used 13 m x 20 m (proportion 1:5) as a base case (B). The other cases were (A) 6 m x 20 m (proportion 3:3), and (C) 20 m x 20 m (proportion 1:1). Simulations were run for 18 hours, from 4 am to 10 pm (Fig. 2). The base parameters in these simulations were set as follows: air temperature, 273 K; wind speed (at 10 m above ground), 2.5 m/sec; wind direction, west; Relative Humidity (RH), 50%; and roughness length in 10 m, 0.1. The results show that the change in ratios has no significant effect on wind speed, which was 0.1 m/sec by the roof level in the courtyard, but reaches 2.8 m/sec around the outer surfaces because of differences in atmospheric pressure around the building. As for the air temperature, the difference between courtyard and outside is around 2 K at 6 am and reaches up to 5 K at noon. Air temperature shows considerable differences among the three types of courtyard: 5 K in type A, 3.8 K in type B and 2.6 K in type C 2.6 K at noon. Thus, the deeper the courtyard, the higher is the air temperature in cold periods.

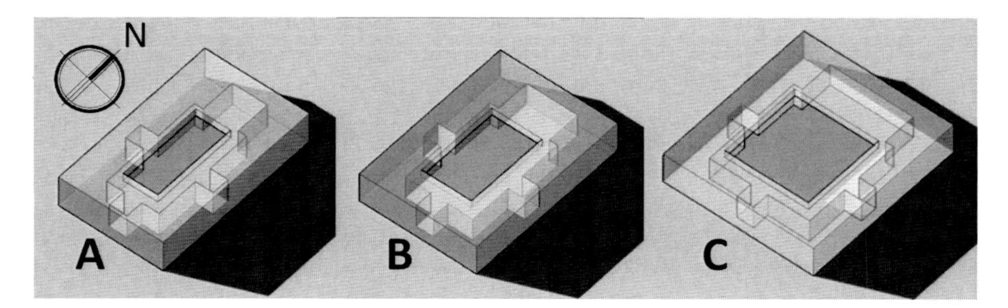

Fig. 2. Generic courtyard forms studied

What is more interesting is that the temperature difference decreases at noon for all types. The highest air temperature measured is in the iwan facing due south, which receives direct solar radiation that is thought to function as thermal mass during the entire day. In all types, the wind speed is less than a light breeze. In sum, the effect of wind on comfort is secondary in comparison with solar radiation. It also relies strongly on the courtyard neighbourhood, so it should be optimized for the specific courtyard environment and cannot be evaluated in a study of a single courtyard.

4.2 Results for Sahibiye Madrassa

After testing the generic courtyards, the madrassa building was analysed for both hot and cold periods. First, for the model simulations, the Sahibiye Madrassa, which is 30 m wide by 42 m long and encloses a courtyard 13 m wide by 20 m long by 8 m high was transformed into a model grid at a spatial resolution of 2 m x 2 m x 2 m. The model area, which comprised a total area of 60 m x 84 m in the horizontal extension and 16 m in the vertical extension, was nested into another model that provided meteorological data at the model borders (Fig. 3). The material assigned to the building and environment was stone; the albedo assumed to be 0.4.

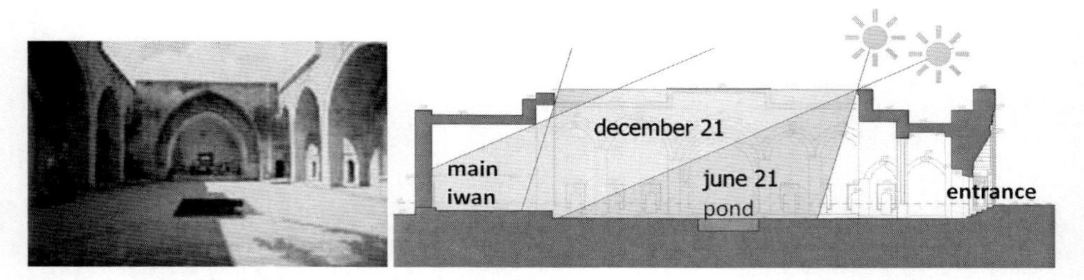

Fig. 3. Plan and perspective view of the Sahibiye Madrassa (left) and section showing solar incident angles for both cold and hot periods (right)

4.2.1 Results for Cold Periods

The base parameters in these simulations were set as follows: air temperature, 273 K; wind speed (at 10 m above ground), 1.7 m/sec; wind direction is south to north, RH: 60%; and roughness length in 10 m: 0.1. The simulation was run for 18 h in total. It was found that the main iwan facing to the south receives incoming solar radiation for almost 4 h between 10:00 and 14:00, with the representative day being 21 December, the shortest day in Kayseri. This finding indicates that the main iwan located at the north end of the courtyard in the courtyard (Fig. 3) is designed to receive as much shortwave radiation as it can during cold periods (Fig. 4).

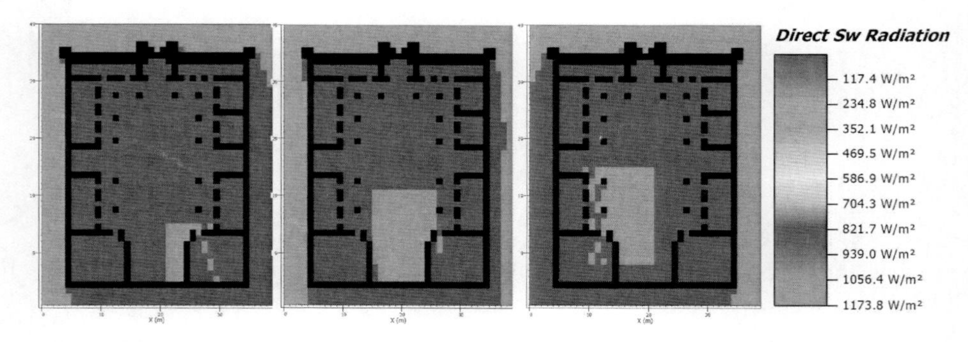

Fig. 4. The time span of exposure of the main iwan to incoming solar radiation

The ambient air temperature measured in the iwan, which is 2 K at noon and at night is 1.5 K higher than the ambient air temperature in other areas in the courtyard, verifies this finding (Fig. 5). It is considered that the iwan helps to improve the thermal environment in the courtyard, providing the dwellers with a pleasant outdoor environment while increasing indoor comfort in cold periods.

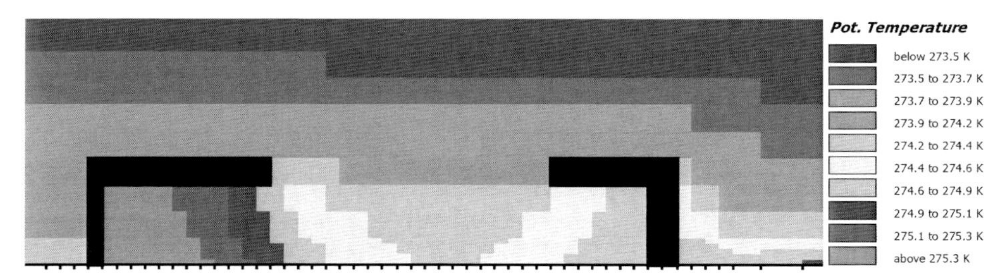

Fig. 5. Air temperature distribution in the courtyard at noon

A wind speed of below 0.1 m/sec in the courtyard, which is less than a light breeze, turns the area into virtually an enclosed space that has constant fresh air, thereby improving the level of comfort for outdoor activities in cold periods (Fig. 6).

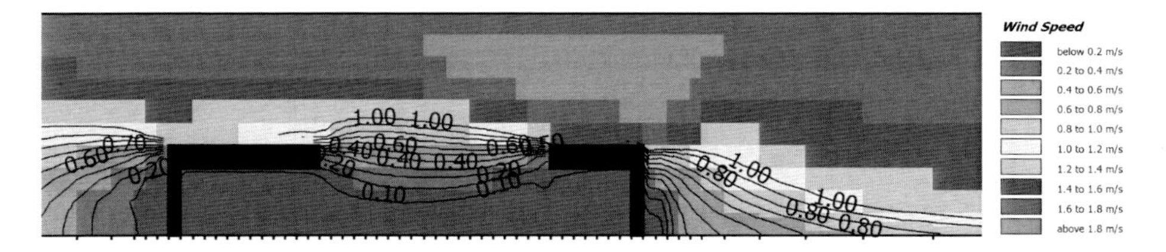

Fig. 6. Section showing the distribution of wind speed in the courtyard

The Predicted Mean Vote (PMV) value, which is regarded as optimum between -0.5 and 0.5, is –0.7 in the courtyard, which is shadowed during the day, and -0.5 around arcades. These findings show clearly why in the past, iwan, where the PMV value is 0, were used as classrooms all day long, even in cold periods.

4.2.2 Results for Hot Periods

The settings for basic parameters that were used in these 16 h of simulations were as follows: initial air temperature, 285 K; wind speed (at 10 m above ground), 2 m/sec; wind direction, south to north; RH, 40%, and roughness length in 10 m, 0.1. Two scenarios in which the main portal door is closed (Scenario 1) or open (Scenario 2) are simulated for hot periods, to compare the effect of ventilation on modification of the courtyard microclimate. As it can be seen from Fig. 7, only the central part of the courtyard, where a pond is located, is exposed to direct shortwave radiation during the daytime. Shortwave radiation affects only in the central part of the courtyard and the Mean Radiant Temperature (MRT) is measured to be below 299.4 K during the daytime in hot periods.

Fig. 7. Exposure to shortwave radiation in the courtyard during the daytime

Accordingly, the air temperature in the courtyard differs slightly in different parts, ranging from 294 K to 296 K. Fig. 8 illustrates the potential temperature patterns in the afternoon. The simulation also shows how the ambient temperature can vary within meters. The surface temperature is lower around the pond, in the main iwan, and around the arcades, which are in shadow throughout the day. In hot periods, the difference in temperature in different parts of the courtyard reaches up to 3 K, ranging between 294.5 K and 297.5 K for Scenario 1. The effect of the pond on air temperature is not known, because ENVI-met cannot simulate water turbulence. The only available strategy for modelling the effects of water is to treat all water as still and treat water bodies as a type of soil when inputting data, and the processes modelled are limited to the transmission and absorption of shortwave radiation [22]. Hence, air temperature fall by 1 K is thought to be an effect of wind reaching into the central part of the courtyard at a speed of 1 m/sec (Fig. 9).

The PMV, which ranges from 0 to 0.6 at noon, reaches an ideal level in the surrounding semi-enclosed parts of the courtyard. According to the results, in the iwan and arcades the MRT is less than 299.4 K at noon, whereas in the central part of the courtyard, which is exposed to direct sunlight, it is around 339.5 K, a difference of almost 40 K (Fig. 10). However, the temperature falls to 314 K around the pond.

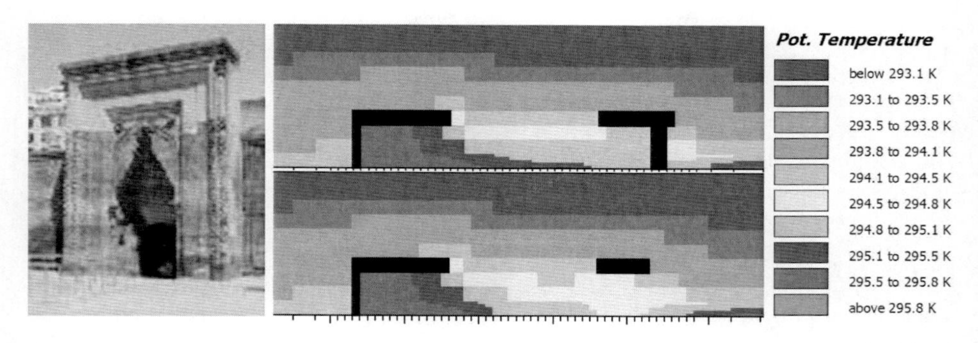

Fig. 8. View from the main portal door and sections, showing the distribution of air temperature in the courtyard when the door is closed (upper) and open (lower)

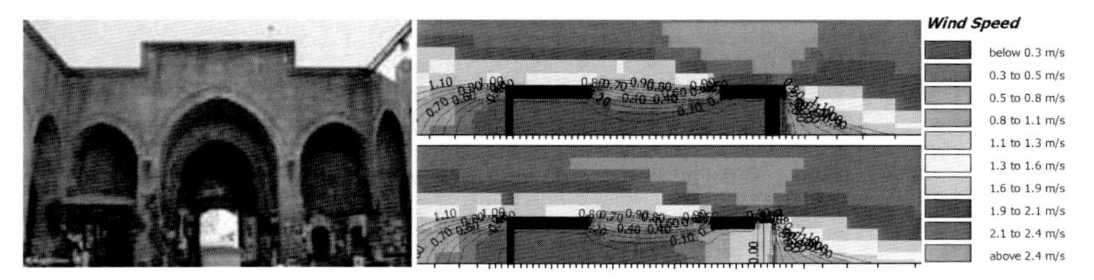

Fig. 9. Distribution of wind speed distribution when the door is closed (upper) and open (lower)

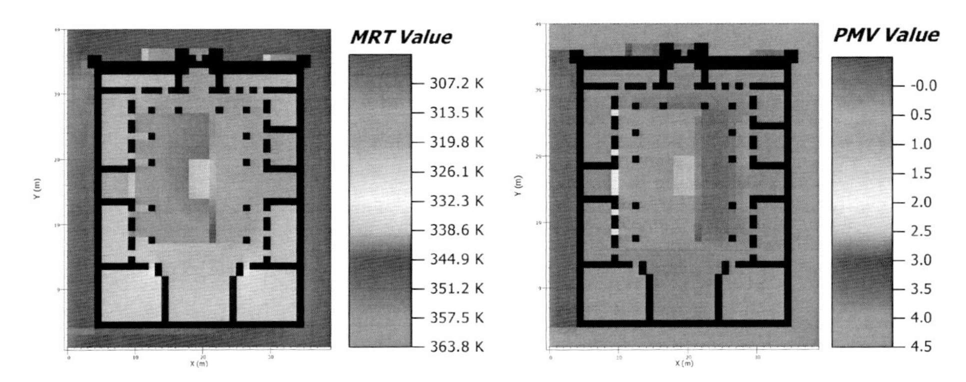

Fig. 10. Horizontal section showing the relation between MRT and PMV at noon

5 Conclusion

The data yielded by the model shows that the predominant factor that affects the thermal performance of the madrassa is the solar radiation that is received by the courtyard surfaces. In addition, the wind speed is much lower in the courtyard than outside; this also has a significant effect on thermal comfort, particularly in cold periods. Such an improvement in the level of comfort would seem to indicate less consumption of energy for heating, cooling and ventilating the madrassa. It is observed that the main iwan receives the most incoming shortwave radiation during daytime and releases it as outgoing longwave radiation at night in cold periods, while the case is the reverse for hot periods. Resembling a breathing open mouth that functions as a thermal regulator for both periods, the main iwan itself exemplifies the design that the builders of the courtyard used to control the microclimate. What is more, there being almost twice as much open and semi-open space as enclosed space has the effect of turning the madrassa building into a virtual open space. It is also worth noting how the enclosed, semi-open and open spaces are organized so as to respond to variations in daily and seasonal temperatures effectively in both cold and hot periods. The madrassa presents a dynamic character contrary to its closed and static appearance, by improving conditions for outdoor activities by offering its users a space that provides thermal comfort all day long. Thus, the design of the madrassa is evidence of a people who were adept at managing the complexity that lies behind overt simplicity. It is likely that other examples of thirteenth-century Seljuk architecture illustrate similar skills. Hence, there is a compelling case for probing further into the physical content of the architectural heritage of the thirteenth century, rather than focusing solely on its historical and symbolic significance.

Acknowledgments. This paper was developed within the research project titled "Microclimate and Thermal Comfort Analysis of Kayseri Sahibiye (Sahabiye) Madrassa" which was supported by Research Fund of the Erciyes University. Project Number: B1027, completed at June 2011.

References

1. Martin, L., March, L.: Urban Space and Structures. Cambridge University Press, Cambridge (1972)
2. Al-Azzawi, S.: Seasonal Impact of Climate on the Pattern of Urban Family Life: Indigenous Courtyard Houses of Baghdad-Regions of the Hot Dry Climates. Renewable Energy 8, 283–294 (1996)
3. Knowles, R.L., Koenig, P.F.: Interstitium: A Dynamic Approach to Solar-Access Zoning (2003), http://www.bcf.usc.edu/~rknowles/interstitium/interstitium.html
4. Mohsen, M.A.: Solar Radiation and Courtyardhouse Forms I- A Mathematical Model. Building and Environment 14(2), 89–106 (1979)
5. Etzion, Y.: The Thermal Behavior of Non-Shaded Closed Courtyards in Hot-Arid Regions. Architecture Science Review 33, 79–83 (1990)
6. Cadima, P.: Solar Radiation of Various Urban Forms for Latitude 38°N. In: PLEA 2000, Cambridge, UK, pp. 448–452 (2000)
7. Muhaisen, A.S., Gadi, M.B.: Effect of Courtyard Proportions on Solar Heat Gain and Energy Requirement in the Temperate Climate. Building and Environment 41, 245–253 (2006)
8. Muhaisen, A.S.: Shading Simulation of the Courtyard Form in Different Climatic Regions. Building and Environment 41, 1731–1741 (2006)
9. Meir, I.A., Pearlmutter, D., Etzion, Y.: On the Microclimatic Behavior of Two Semi-Enclosed Attached Courtyards in a Hot Dry Region. Building and Environment 30, 563–572 (1995)
10. Al-Hemiddi, N.A., Al-Saud, K.A.M.: The Effect of a Ventilated Interior Courtyard on the Thermal Performance of a House in Hot-Arid Region. Renewable Energy 24, 581–595 (2001)
11. Robitu, M., Musy, M., Inard, C., Groleau, D.: Modeling the Influence of Vegetation and Water Pond on Urban Microclimate. Solar Energy 80, 435–447 (2006)
12. Sadafi, N., Salleh, E., Haw, L.C., Jaafar, Z.: Evaluating Thermal Effects of Internal Courtyard in a Tropical Terrace House by Computational Simulation. Energy and Buildings 43(4), 887–893 (2011)
13. Chatzidimitriou, A., Yannas, S.: Microclimatic Studies of Urban Open Spaces in Northern Greece. In: PLEA 2004, Eindhoven, vol. 1, pp. 83–88 (2004)
14. Berkovic, S., Yezioro, A., Bitan, A.: Study of Thermal Comfort in Courtyards in a Hot Arid Climate. Solar Energy 86, 1173–1186 (2012)

15. Alvarez, S., Sanchez, F., Molina, J.L.: Air Flow Pattern at Courtyards. In: PLEA 1998, Lisbon, pp. 503–506 (1998)
16. Sensoy, S., Demircan, M.: Climatological Applications in Turkey (2010), http://emcc.dmi.gov.tr/FILES/ClimateIndices/ClimatologicalApplications.pdf
17. Ali-Toudert, F.: Dependence of Outdoor Thermal Comfort on Street Design in Hot and Dry Climate. Dissertation, Freiburg University (2005)
18. Bruse, M., Fleer, H.: Simulating Surface-Plant-Air Interactions Inside Urban Environments with a Threedimensional Numerical Model. Environmental Software and Modelling 13, 373–384 (1998)
19. Berkovic, S., Yezioro, A., Bitan, A.: Study of Thermal Comfort in Courtyards in a Hot Arid Climate. Solar Energy 86, 1173–1186 (2012)
20. Emmanuel, R., Fernando, H.J.S.: Urban Heat Islands in Humid and Arid Climates: Role of Urban Form and Thermal Properties in Colombo, Sri Lanka and Phoenix. USA. Clim. Res. 34, 241–251 (2007)
21. Nice, K.A., Isaac, P.: The Microclimate of a Mixed Urban Parkland Environment. In: 7th International WSUD Conference, Melbourne, Australia (2012)
22. Bruse, M.: ENVI-met website, http://www.envimet.com

Chapter 7
Analysis of Structural Changes of the Load Profiles of the German Residential Sector due to Decentralized Electricity Generation and e-mobility

Rainer Elsland[1,*], Tobias Boßmann[1], Rupert Hartel[2], Till Gnann[1], Massimo Genoese[2], and Martin Wietschel[1]

[1] Fraunhofer Institute for Systems and Innovation Research ISI, Breslauer Strasse 48, 76139 Karlsruhe, Germany
Rainer.Elsland@isi.fraunhofer.de
[2] Karlsruhe Institute of Technology, Institute for Industrial Production (IIP), Hertzstrasse 16, 76187 Karlsruhe, Germany

Abstract. In this paper, a bottom-up energy demand model is applied to a scenario-based analysis of the load profiles of the German residential sector until the year 2040. This analysis takes into account the increasing diffusion of e-mobility and decentralized electricity generation and addresses questions such as: How much demand has to be met by the electricity supply system and what kind of structural changes in the load profile are to be expected. In order to assess the maximum contribution of decentralized electricity generation, a weekday in summer was chosen for the analysis. Assessing the future residential electricity demand on an hourly basis clearly depicts an increased volatility due to the shift in demand from night-time to daytime hours which is mainly caused by the greater number of ICT appliances. Furthermore, electric vehicles lead to a significant increase in the evening demand peaks. At the same time, electricity generation from photovoltaic sources can entirely compensate this additional demand by e-mobility, if decentralized electricity generation can be matched with the electricity demand via demand-side-management (DSM) or storage devices.

1 Introduction

Mitigation of climate change and the related necessity to transform the energy system are the key challenges for the European energy economy in the upcoming decades. In Germany, the framework for this transformation process is given by the Energy Concept which was published in September 2010 by the Federal Government. The Energy Concept defines goals to reduce greenhouse gas emissions (-80% compared to 1990), to lower primary energy demand (-50% compared to 2008) and to decrease electricity demand (-25% compared to 2008) until 2050 (BMWi and BMU 2010).

* Corresponding author.

A. Håkansson et al. (Eds.): *Sustainability in Energy and Buildings*, SIST 22, pp. 71–84.
DOI: 10.1007/978-3-642-36645-1_7

Thereby, the deployment of energy saving potentials is an inevitable means to cope with these targets.

The residential sector makes a substantial contribution to achieve these goals (BMWi and BMU 2010). Concerning the technological development, the importance of electricity as an energy carrier is increasing. Besides the extensive introduction of heat pumps and electric vehicles, the diffusion of decentralized electricity generation such as photovoltaic systems leads to a structural change of electricity demand from a technological point of view. In this context the question may be raised in what way the patterns of residential electricity demand will change in the upcoming decades, which again has a direct effect on the electricity supply system.

Within this paper a scenario based analysis of the load profiles of the German residential sector is executed until the year 2040 while considering of the future technology specific electricity consumption which includes the increasing diffusion of e-mobility and decentralized electricity generation. In this context, the questions to be answered are which net electricity demand has to be covered by the power plant mix and what kind of load profile changes are to be expected. By applying a technology based analysis of the residential load profiles, a conclusion shall be derived about how the hourly distribution of electricity demand throughout the day could change in the future.

In a first step, the methodological approach to calculate the annually technology based electricity demand is described. Consequently, the annually electricity demand is used as a basis for the scaling up of hourly load profiles. Thereby the structural model framework and the central drivers are described in the first place and hereafter the calculation logic of the modules for residential electricity demand (i.e. electric appliances and heating technologies; RES-module) as well as for e-mobility (EMOB-module) and decentralized electricity generation (DEC-module) are presented. In a second step, three explorative scenarios until 2040 will be developed and parameters will be defined, which are needed for the quantitative analysis. Finally, the results are discussed and conclusions are drawn.

2 Methodological Approach

2.1 Structural Model Framework and Drivers

The energy model is based on a bottom-up approach, while the calculation method is designed as a simulation. The model is divided into three modules, which differ in terms of socio-economic and technological drivers: the RES-module, the EMOB-module and the DEC-module (Fig. 1).[1] Since the decision calculus of the decision-makers for investing in electricity demanding technologies (e.g. refrigerator or electric car) is not essentially based on a homo oeconomicus consideration, but on a variety of non-monetary criteria, e.g. the purchase of a car is based on a lot of

[1] The RES-module and the EMOB-module are combined in one energy demand model called FORECAST-Residential. FORECAST is a modeling platform that contains models for the residential, industrial and the tertiary sector to calculate the electricity demand of the EU27+2 by country.

irrational preferences like the brand or the color of a car. Investments and energy carrier prices are not explicitly part of the calculation. Nevertheless, the development of oil prices as well as the battery prices is used as an indicator for the diffusion of electric vehicles.

The diffusion of decentralized electricity production units is mainly driven by political aspects. The German law for the advancement of renewable energies (EEG 2012) is the main driver for installing decentralized electricity production units. The law influences the economics of the units by tax reductions, bonus payments for the produced electricity and investment benefits. Considering the law incentives the DEC-module therefore analyzes the cost effectiveness of photovoltaic-battery systems and the economic potential of small CHP units based on investment costs of installations, electricity and fuel prices. Due to the individuality of electricity production profiles of CHP units, it is not possible to deflect one general, representative production profile. Therefore, only the photovoltaic simulation and the derivation of the respective electricity generation load profiles are discussed in this article in detail.

The model's algorithm consists of a high amount of equations, which can not be listed in this paper in its entirety. Thus, the following description is confined to the most important algorithms of the respective modules. Besides the determination of the load profiles the calculation method of electricity demand and electricity supply, on which the load profiles are based, will be presented thereby.

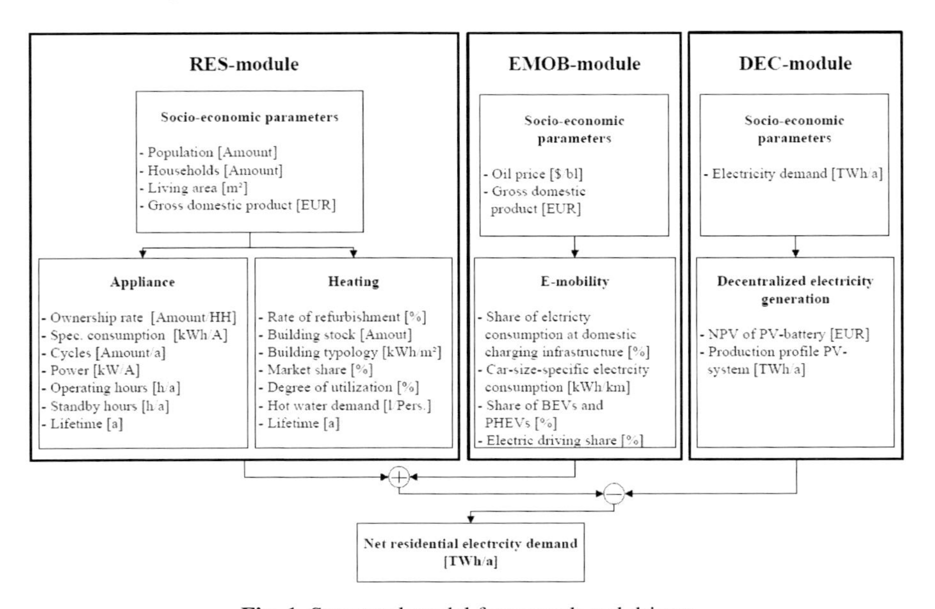

Fig. 1. Structural model framework and drivers

2.2 RES-Module

The electricity demand calculation within the RES-module is carried out separately for appliances and heating technologies. Due to high data availability the appliance based electricity demand is calculated via a vintage stock model.[2] In the projection of the appliance stock a Bass function is applied, which is fitted to the empirical stock development on the basis of the method of least squares. The Bass function has been taken because word-of-mouth is the key element of technology diffusion. The stock turnover is based on an appliance specific lifetime with normal distributed failure probability. The annual electricity demand of all appliances $W_{Appl,s,t}$ is calculated by the product of specific consumption[3] $s_{t,G,T,E,s}$ and average ownership rate $O_{t,G,s}$, multiplied by the number of households HH_t. Besides white appliances, electric stoves and ICT-appliances, the RES-module also compromises lighting as well as air-conditioning. Moreover, all appliances are differentiated by technologies and efficiency classes.

$$W_{Appl,s,t} = \sum_{G=1}^{m}\sum_{T=1}^{l}\sum_{E=1}^{k}(s_{t,G,T,E,s} \cdot O_{t,G,s}) \cdot HH_t \tag{1}$$

Indices:

G : Type of appliance, m = 12
T : Technology, l = 18
s : Scenario
E : Efficiency Class, k = 10
t : Time (year), t = 32

The cumulative electricity demand related to heating technologies $W_{HT,s,t}$ consists of the electricity demand for providing space heating and hot water. The calculation of electricity demand through space heating (first term in formula (2)) is based on the useful heat demand, which is calculated by the product of the number of buildings in stock $HH_{t,B,Y,s}$, the refurbishment rate of the building $R_{t,B,Y,s}$ and the space per building size class $S_{t,B,Y,s}$ (numerator of first term). The electricity based space heating calculation is executed by dividing the useful heat demand by the utilization ratio $U_{t,T,s}$ multiplied by the market share of electricity based heating technologies $M_{t,T,s}$. The driver of electricity demand for hot water supply (second term of formula (2)) is the daily hot water consumption per household. The electricity demand for hot water provision is calculated by the product of hot water demand in liters per household V_s and a constant specific energy consumption for water heating of 21.1 (kWh/l) divided by the utilization ratio of the heating technologies $U_{t,T,s}$ and multiplied by the share of electricity based technologies for hot water supply $M_{t,T,s}$ and the amount of households HH_t.

$$W_{HT,s,t} = \sum_{T=1}^{l}\left(\frac{\sum_{B=1}^{m}\sum_{Y=1}^{l} HH_{t,B,Y,s} \cdot R_{t,B,Y,s} \cdot S_{t,B,Y,s}}{U_{t,T,s}} \cdot M_{t,T,s} + \frac{V_s \cdot 21{,}1\ (kWh/l)}{U_{t,T,s}} \cdot M_{t,T,s} \cdot HH_t \tag{2}$$

Indices:

B : Building age class, m = 3
Y : Type of building size, l = 2
s : Scenario
T : Technology, l = 18
t : Time (year), t = 32

[2] For more detailed information about the calculation approach of the RES-module see (Elsland et al. 2012).

[3] The specific consumption is either based on the operation- and standby-power as well as the operation- and standby-hours (e.g. television) or on the specific consumption per cycle and the amount of cycles per year (e.g. washing machine).

The subsequent step load curves are derived from the electricity demand of appliances and heating technologies. The establishment of a 8760-hours load curve is carried out according to the grid operator's methodology, e.g. (EON 2012), using technology specific 24-hour load profiles $LP_{h,d,T}$, differentiated by typical days. The chronological layout of the year is considered via the number of typical days per year n_d. In combination with the load profiles a synthetic demand structure is generated. Its integral is adjusted by scalar multipliers $m_{t,T,s}$ according to the calculated technology-specific annual electricity demand. There are nine typical days, which differ depending on the day of the week (weekday, Saturday and Sunday) and the season of the year (summer, transition period and winter).[4] The calculation of a load profile for a typical day is based on discrete measurements of household electricity demand (share of a technology of the total electricity demand per hour), which are transferred into hourly load profiles.

$$W_{t,s} = \sum_{T=1}^{l}\sum_{d=1}^{n}\sum_{h=1}^{i} m_{t,T,s} \cdot n_d \cdot LP_{h,d,T} \tag{3}$$

Indices:

d: Typical day, n=9
h : Hour of a typical day, i=24
s : Scenario
T : Technology
t: Time (year), t = 32

2.3 EMOB-Module

The electricity consumption of electric vehicles in households depends on a large variety of factors. Thus, the total annual electricity demand of households from e-mobility $W_{eMob,t,s}$ derives from the number of battery electric vehicles (BEVs) $n_{BEV,i,t,s}$ and plug-in hybrid electric vehicles (PHEV) $n_{PHEV,i,t,s}$, the electric driving share of PHEVs s_E, the vehicle kilometers traveled per year $VKT_{i,t}$, the energy consumption of the vehicles $c_{i,t}$ multiplied by the share of electricity charged domestically $IS_{dom,t,s}$.

$$W_{eMob,t,s} = IS_{dom,t,s} * \sum_{i=1}^{I}(n_{BEV,i,t,s} + n_{PHEV,i,t,s} * s_E) * VKT_{i,t} * c_{i,t} \tag{4}$$

Indices:

t : Time (year), t= 32
s : Scenario
i : Vehicle size
I = {small, medium, large, transportation LDVs}

The load profile of e-mobility derives from a simulation of driving patterns from the German Mobility Panel (MOP 2008) in which we determine the technical feasibility to replace the conventional vehicle by a BEV and electric driving share of PHEVs (Gnann et al. 2012), followed by an economic comparison which vehicles have the lowest total cost of ownership (Kley 2011). Using the driving profiles of the vehicles

[4] The composition of entire annual load curves out of individual load profiles assumes that the middle of January and July represent a typical winter or summer day, respectively.

for which BEVs and PHEVs would be the best options, we simulate the load profiles resulting from the battery profile simulation and determine three typical days *d:* weekday, Saturday and Sunday. Hence, the daily fraction of e-mobility related electricity demand of all households can be written as the power to charge $P_{eMob,t,s,j}$ for all vehicles in time section *t* on the distinct day type *d* divided by the power to charge for the whole week $P_{eMob,t,s,j}$ (see formula (5)).

$$s_{eMob,t,d} = \frac{\sum_{j=1}^{J} P_{eMob,t,d,j}}{\sum_{t_1=1}^{T}\sum_{j=1}^{J} P_{eMob,t_1,d,j}} \tag{5}$$

Indices:

t, t₁ : Time = 15 min — *d* : Typical day ∈ {weekday, Saturday, Sunday}

j: Driving profile — *J:* Total number of driving patterns = 6,629

T= one week in time fractions of 15 mins = 672

We can then calculate the load profile by multiplying the share of power at all households per week $s_{eMob,t,d}$ with the total energy demand of all vehicles $W_{eMob,t,s}$ divided by 52 weeks.

2.4 DEC-Module

The electricity production profile of photovoltaic systems is determined by a weather dataset, called test reference year (TRY) (TRY 2004). The TRY dataset contains hourly measured data for wind speed and wind direction, air temperature and direct and diffuse solar radiation for Germany. The sum of direct $r_{dir,h}$ and diffuse $r_{dif,h}$ solar radiation, the installed capacity Cap_t and the average full load hours f_t are used for the compilation of the hourly electricity production profile $P_{h,t}$ for each considered year *t* according to formula (6).

$$P_{h,t} = \frac{r_{dir,h} + r_{dif,h}}{\sum_{h=1}^{8760} r_{dir,h} + r_{dif,h}} * Cap_t * f_t \tag{6}$$

Indices:

h: Hour — *t* : Time (year)

By dividing the hourly values by the sum of all values of one year the solar radiation profile is normalized. The hourly values of the normalized profiles are then multiplied by the installed photovoltaic capacity and the average full load hours. Using this approach, it is possible to derive the profile of one photovoltaic system or for example of the whole installed capacity in Germany. Since measured values are the basis for the profile, every day has its own production characteristic.

3 Case Study

3.1 Socio-economic Parameters

The calculation of the overall electricity demand and the respective load profiles is conducted with the help of three explorative scenarios: Reference Scenario (RS),

Ambitious Climate Policy Scenario (ACS) and Green Germany scenario (GGS). It is assumed, that in each scenario the yet to be taken energy policy regulations will be implemented. While the technological change in the RS is essentially based on autonomous progress, it is assumed that there is a moderate progression in the ACS and an ambitious progression in the GGS regarding energy policy regulations respectively objectives, which pursuant affects the diffusion dynamics of energy efficient technologies. The base year for the calculation is 2008. In this context it has to be pointed out, that the GDP per capita increases in all scenarios, whereby the GDP of the RS increases more sharply than in the more ambitious scenarios. This development is based on the assumption that ambitious energy policy regulations lead to a moderation of economic growth. A general overview of the socio-economic parameters is given in Table 1.

Table 1. Socio-economic parameters by scenario (StBA 2009, StBA 2011, Prognos et al. 2010, IEA 2011, own assumptions)

Scenario	Parameters	Unit	2008	2020	2030	2040
Reference (RS)	Population	Million	82.0	80.4	79.0	76.8
	Households	Million	39.6	40.7	41.0	41.1
	GDP real (2000)	Bn. EUR	2.277	2.411	2.586	2.773
	Living area	Million m²	3.452	3.723	3.980	3.884
	Oil price	$/bl.	100	106,5	113	120
Ambitious Climate Policy (CS) / Green Germany (GGS)	Population	Million	82.0	79.9	77.4	73.8
	Households	Million	39.6	40.4	40.1	39.5
	GDP real (2000)	Bn. EUR	2.277	2.341	2.436	2.535
	Living area	Million m²	3.452	3.521	3.295	3.042
	Oil price	$/bl.	100	120/130	140/165	160/200

3.2 Technological Parameters

3.2.1 RES-Module

Apart from dish washers and dryers, the ownership rate for large devices is supposed to be close to its saturation level. The ownership rate of ICT-appliances, excluding TVs, will experience a further growth in the upcoming decades. Air-conditioning devices for indoor cooling will feature increasing market diffusion due to rising comfort demands. In terms of lighting technologies, energy saving bulbs and LED lamps will substitute the currently predominant incandescent bulb. The decreasing electricity demand for residential appliances and lighting devices is related to Ecodesign as well as to the Energy Labeling Directive (Hansen et al. 2010, ODYSSEE 2011).

The database for calculating the useful heat demand consists of current studies including information on existing building stock distinguished by construction year, living space and level of refurbishment (Henning et al., 2011; IWU and BEI 2010,

Neuhoff et al. 2011). The specific useful heat demand of single and double family houses ranges from 198 to 7.5 kWh/m² and that of multi family houses from 123 to 6.4 kWh/m². The annual demolition rate was fixed at a level of 0.5% (StBA 2006). The energetic refurbishment rate will increase for the RS equals 1%. In the ACS and GGS, the refurbishment rate will increase to 1.9% and 2.5% respectively, by the year 2030, before dropping back to a level of roughly 1% by 2040. The diffusion of heating technologies is based on technological and regulatory trends such as the phase-out of electric night storage heating systems based on the German Energy Saving Regulation, the declining number of direct electric heating systems, the increasing diffusion of heat pumps, the relatively constant share of district heating, the decreasing number of coal and oil fuelled heating systems and a further extension of solar thermal installations (ODYSSEE 2011).

The database for the generation of aggregated household load profiles for specific typical days consists of data from the Intelliekon project (BMBF 2011). Within this project hourly load profiles were generated based on the electricity consumption monitoring approximately 1,000 German households, covering a representative socio-economic panel of consumers, during the period from May 2009 until May 2010. The determination of technology specific load profiles was carried out by using internal and external studies dealing with the technology based monitoring of residential electricity consumption at a very high time resolution (Klobasa 2006, Seefeldt et al. 2010, SW Mainz 2011).

3.2.2 EMOB-Module

To calculate the annual electricity demand through e-Mobility, we use the same annual driving distance for all light duty vehicles of 16,000 km in 2008 constantly decreasing to 14,300 km in 2040 (BMVBS 2011) and 22,500 km for transportations LDVs (Mock 2010). For vehicle-size-specific energy consumption we use data from

Table 2. Amount of BEVs, PHEVs and share of domestic charging for base years by scenario

Scenario	Parameters	Unit	2008	2020	2030	2040
Reference (RS)	Amount of BEVs	Million	0.003	0.187	0.542	1.332
	Amount of PHEVs	Million	0	0.313	1.459	2.668
	Share of domestic charging	%	100	99%	86,6%	85%
Ambitious Climate Policy (ACS)	Amount of BEVs	Million	0.003	0.381	2.493	7.956
	Amount of PHEVs	Million	0	0.620	3.511	7.047
	Share of domestic charging	%	100	98%	70,8%	65%
Green Germany (GGS)	Amount of BEVs	Million	0.003	0.501	4.651	17.128
	Amount of PHEVs	Million	0	0.799	4.047	7.570
	Share of domestic charging	%	100	99,2%	70%	55%

(Helms et al. 2010). While we keep the annual VKT and the electricity consumption scenario-independent, the numbers of vehicles and their charging infrastructure usage are defined separately for every scenario and every year (Table 2).

The car sizes develop from a ratio of 50% small, 45% medium, 0% large and 5% transportation light duty vehicles in 2008 to 25% small, 55% medium, 15% large and 5% transportation light duty vehicles in 2040, which is the current vehicle size mix in Germany (KBA 2010). In the simulation of driving profiles to determine the load profile, we use battery capacities of 12.5 kWh for PHEVs and 25 kWh for BEVs.

3.2.3 DEC-Module

Since this article deals with the residential sector, it is important to identify the installed photovoltaic capacity belonging to the residential sector. Every renewable unit in Germany is detected by the four network operators and is available to the public via their internet presence. Assuming that photovoltaic applications in the residential sector are only installed on the roof and therefore units are accordingly small, only units smaller than 20 kW_p are considered for the determination of the installed photovoltaic capacity in the residential sector. Summing up the units results in an installed capacity of 4.9 GW_p. This capacity can be converted to an electricity production by multiplying the capacity with the full load hours. In Germany, an average of approximately 950 full load hours can be estimated for a photovoltaic system.

After estimating the installed capacity the development of the capacity has to be forecasted for the three defined scenarios until 2040. In (BDEW 2010) different capacity targets for photovoltaic systems are compared. Thus, the capacity bandwidth results in 20 GW_p to 100 GW_p in 2040. Considering that in 2010 already 17 GW_p of photovoltaic capacity was installed, the capacity target of 20 GW_p seems to be far too low. Based on the assumption that probably half of the maximum installed capacity belongs to the residential sector, we derive a capacity target of 50 GW_p in the GGS which acts as an upper boundary. This reflects an increase by a tenth of the capacity installed in 2010. For the ACS, we assume a capacity of approximately 35 GW_p installed in 2040, which is a seventh of the amount of 2010 and for the RS. It is assumed the installed capacity will quintuple to 25 GW_p.

3.3 Results

The future electricity demand of the German residential sector is expected to continue to grow up to 146.6 TWh by the year 2040, if no political measures are undertaken (cf. RS). This increase of 6.5% compared to the electricity demand in the base year 2008 of 137.7 TWh is mainly related to the diffusion of new ICT appliances. The intensification of energy efficiency, as assumed under the ACS and GGS, can overcompensate the growing electricity demand despite an extensive commissioning of heat pumps leading to a short term increase. By 2040, overall electricity demand will drop to 121.6 TWh (ACS) and 97.9 TWh (GGS), respectively. An increasing number of plug-in electric vehicles, which mainly charge at home may compensate the savings through more energy efficient products in the long term. The electric car fleet assumed in the ACS and GGS would add 18.8 TWh and 27.1 TWh on top of the electricity demand by the year 2040. However, this additional demand can be entirely covered by the intensified installation of photovoltaic cells on residential buildings. Their contribution to the electricity supply ranges from 25.3 TWh in the RS up to 44.6 TWh in the GGS. Thus, by 2040 decentralized electricity generation leads to a net electricity demand reduction of 7.5% in the RS, 25.9% in the ACS and 41.6% in the GGS compared to today's level despite the intensified use of electric vehicles.

In order to assess the impact of the changes in electricity consumption and decentralized electricity generation on the load profile, the technology specific load profiles of a summer weekday in 2008 and 2040 are compared. The summer season is chosen due to the maximum contribution from photovoltaic installations and hence from decentralized generation to cover electricity demand. Given the limited availability of technology specific load profiles, ICT appliances are displayed as an aggregate featuring only one characteristic load profile. With regard to the structural change of the load profile, the share of electricity consumption during noon and midnight increases compared to the consumption in the morning and at night for all scenarios (Fig. 2). On the one hand, this effect can be explained by the phase-out of night storage heating systems and their substitution through heat pumps, which, however can only be observed to a limited extent during the summer season. On the other hand, the growing electricity consumption of ICT appliances mainly takes place in the afternoon and evening hours, which implies a further increase of the existing consumption peak.

Under the RS, the load profile in 2040 exceeds the one from the base year given the strong increase in electricity demand through ICT appliances, air-conditioning devices and the block of 'new and other appliances that cannot be distinguished in a more detailed manner. The latter are supposed to experience a further growth in electricity demand due to the continuous diffusion of small devices, mainly ICT devices, that was observed in the past. The peak of the summer weekday load profile occurs at 8pm (20th hour) at a level of 23 GW. The ACS features a load profile beneath the base year as a consequence of the increased energy efficiency of white appliances and lighting devices and despite the increased electricity demand that is triggered through ICT devices as well as 'new and other' appliances. The consumption peak of the summer weekday load profile in 2040 occurs also at 8pm but only at a level of 18.5 GW. The GGS features the lowest load profile given the electricity demand reduction of all appliances apart from the 'new and other' appliances. The consumption peak arises at 8pm at a level of 15 GW.

As previously shown, the summer weekday load profile in 2040 features a relative increase of the consumption peak during day hours. In the following step the consumption load profile is completed by the load profile for e-mobility and decentralized electricity generation in order to get a holistic picture of the electricity consumption pattern that actually needs to be covered through the public grid. Fig. 3 depicts on the left hand side the load profile of the electric appliances and heating technologies en bloc in blue and of e-mobility in red, underlining that the electricity consumption through electric vehicles is comparably low in the base year. The right hand side of Fig. 3 displays the cumulative consumption load profile in total diminished by the load profile for decentralized electricity generation. The demand being covered by photovoltaic systems is shown in green. Hence, the brown area depicts the residual load that needs to be covered through the public grid. It becomes obvious that the decentralized electricity generation mainly contributes to cover the electricity demand around midday.

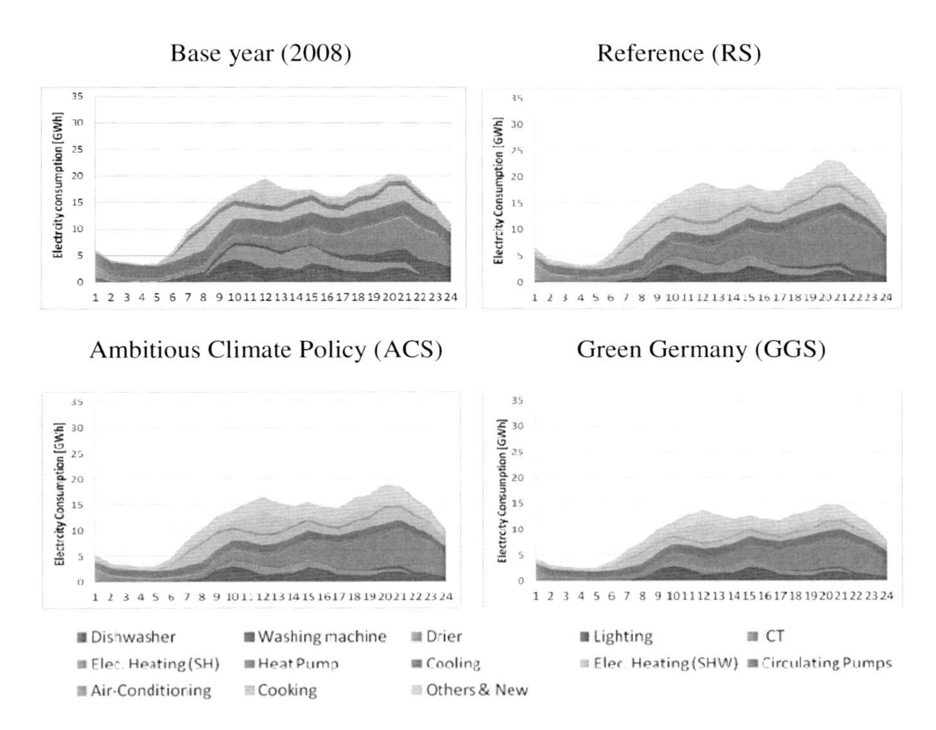

Fig. 2. Load profile on a summer-weekday in 2008 and by scenario in 2040

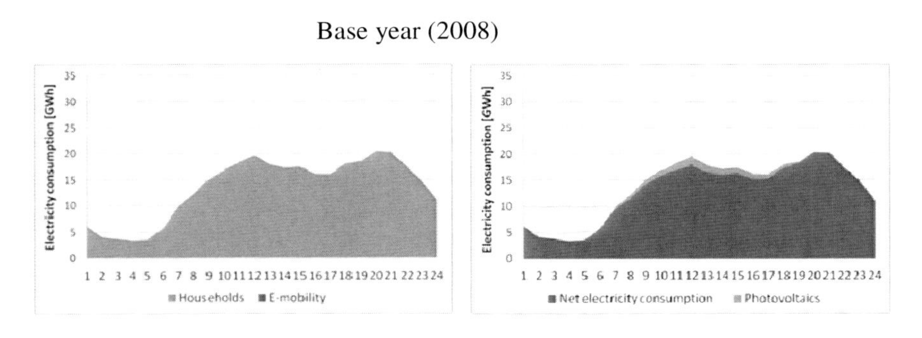

Fig. 3. Load profile of household and e-mobility electricity demand (left) and of net electricity consumption (right) on a summer workday in 2008

In Fig. 4 we observe that the consideration of e-mobility with 'uncontrolled charging after the last trip' leads to a significant increase in the evening demand peaks. We find the highest increase in the evening peaks of 24.7 GW at 8pm (20th hour) in the RS-scenario. By comparing the PV electricity generation profile with the demand profile of the residential sector and the demand profile of the electric mobility one may observe that in 2040 during midday hours the full electric demand can be covered by the decentralized electricity production (cf. GGS).

Reference (RS)

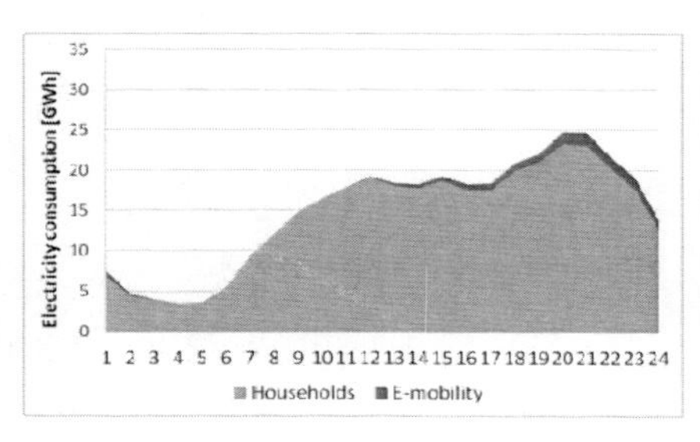

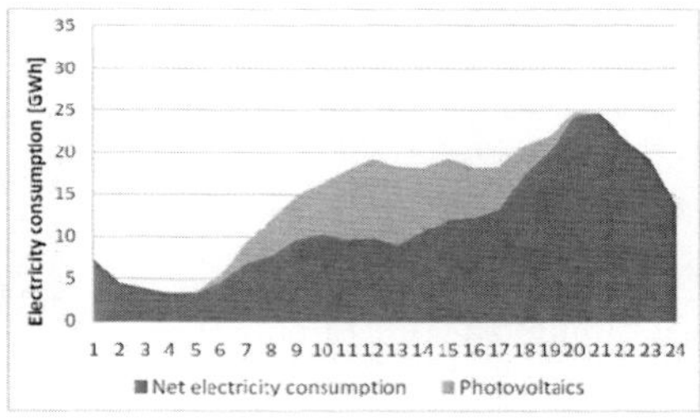

Ambitious Climate Policy (ACS)

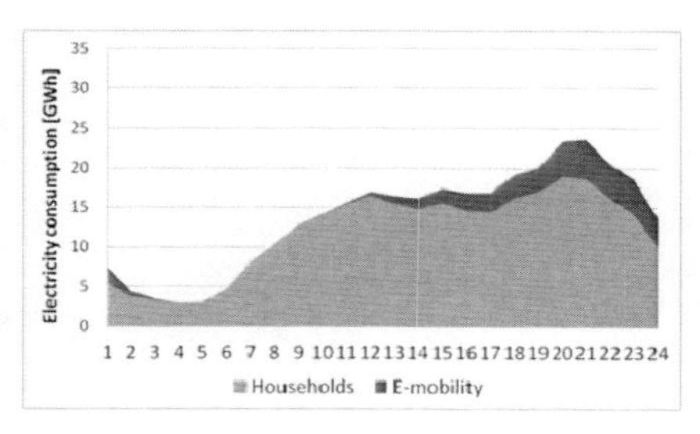

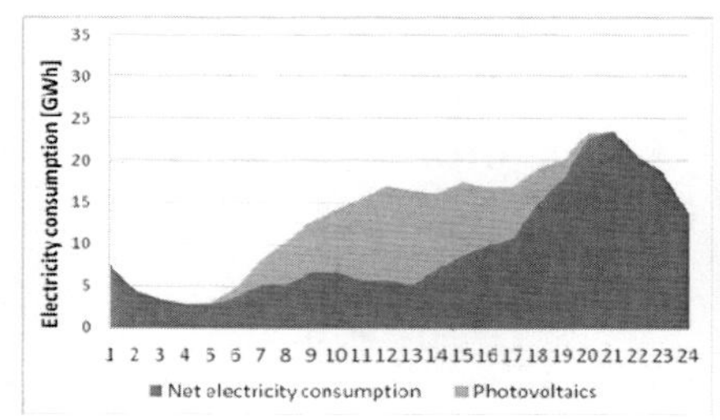

Green Germany (GGS)

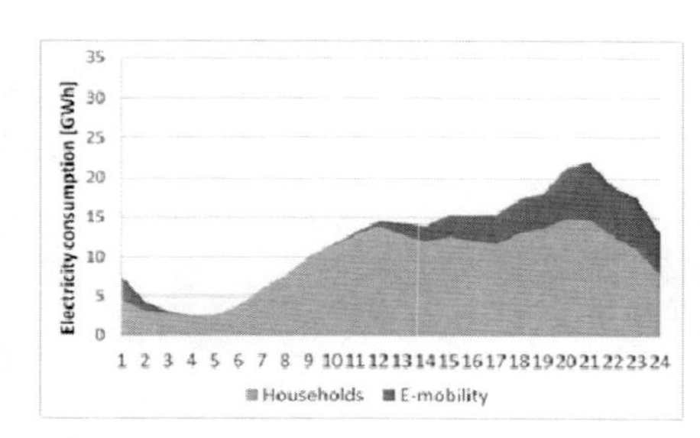

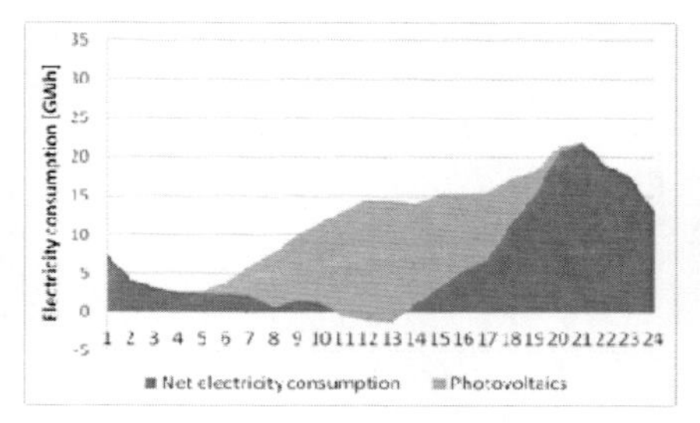

Fig. 4 Load profile of household and e-mobility electricity demand (left) and of net electricity consumption (right) on a summer workday in 2040

4 Conclusion and Critical Appraisal

The analysis of the future electricity demand of German households by the year 2040 permits the conclusion that electricity demand will experience further growth up to a level of 146.6 TWh under 'business-as-usual' conditions. However, assuming that the implementation of additional efficiency measures can potentially reduce the demand to 97.9 TWh a broad market introduction of plug-in vehicles would trigger a further demand increase of up to 27.1 TWh. On the other hand this effect can largely be overcompensated by the installation of roughly 50 GW_p of photovoltaic systems, generating up to 44.6 TWh per year.

Assessing the future residential electricity demand on an hourly basis clearly depicts an increased volatility, given the demand shift from night to day hours due to new ICT appliances and the intensified use of heat pumps. Hence, the need for flexible electricity generation capacities, storage devices or demand-side-management strategies is becoming increasingly significant despite an overall reduction of the total demand under the Ambitious Climate Policy Scenario (ACS) and Green Germany Scenario (GGS).

The strong advantage of the present work consists in the combination of a technology based demand forecast approach for the residential electricity demand with a simulation of upcoming electricity demand triggered by electric vehicles as well as the potential supply through local photovoltaic installations. Among others, this approach permits a holistic assessment of the future need for electricity supply through the central power plant mix.

References

BDEW: Studiensynopse Energieprognosen: Prognosen zur Entwicklung der Stromversorgung und Einordnung der Energieszenarien für ein Energiekonzept der Bundesregierung, Berlin (2010)

BMBF: Nachhaltiger Energiekonsum durch intelligente Zähler-, Kommunikations- und Tarifsysteme, Berlin (2011)

BMVBS: Verkehr in Zahlen 2010/2011, DVV Media Group, Hamburg (2011)

BMWi and BMU 2010: Energiekonzept für eine umweltschonende, zuverlässige und bezahlbare Energieversorgung, Berlin (2010)

EEG: Gesetz für den Vorrang Erneuerbarer Energien (Erneuerbare-Energien-Gesetz – EEG), Saarbruecken (2012)

Elsland, R., Boßmann, T., Wietschel, M.: Technologiebasierte Analyse der Stromnachfrage im deutschen Haushaltssektor bis 2050. In: Conference Proceedings, 12. Symposium Energieinnovation. Graz University of Technology, Graz (2012)

E.ON Bayern AG: Standardlastprofile E.ON Bayern AG (2012), `http://www.eon-bayern.com/pages/eby_de/Netz/Stromnetz/Netzzugang/Lastprofilverfahren/Standard_Lasprofile/Verfahrensbeschreibung_SLP.pdf` (accessed June 01, 2012)

Gnann, T., Plötz, P., Kley, F.: Vehicle charging infrastructure demand for the introduction of plug-in electric vehicles in Germany and the US. In: Conference Proceedings, Electric Vehicle Symposium 26, Los Angeles (2012)

Hansen, P., Matthes, F.C.: Politikszenarien V – auf dem Weg zum Strukturwandel, Treibhausgas-Emissionsszenarien bis zum Jahr 2030, Schriften des Forschungszentrums Jülich Reihe Energie & Umwelt, Band 62, Jülich (2010)

Helms, H., Hanusch, J.: Energieverbrauch von Elektrofahrzeugen, ifeu - Institut für Energie- und Umweltforschung Heidelberg GmbH, Heidelberg (2010)

Henning, M., Ragwitz, M., Bürger, V., Jochem, E., Kranzl, L., Schulz, W.: Erarbeitung einer Integrierten Wärme- und Kältestrategie – Bestandsaufnahme und Strukturierung des Wärme- und Kältebereichs, Freiburg (2011)

IEA: Oil Market Report - Annual Statistical Supplement - 2011 Edition, Paris (2011)

IWU, BEI: Datenbasis Gebäudebestand Datenerhebung zur energetischen Qualität und zu den Modernisierungstrends im deutschen Wohngebäudebestand, Darmstadt (2010)

KBA: Fahrzeugzulassungen (FZ) - Bestand and Kraftfahrzeugen und Kraftfahrzeuganhängern nach Haltern (2010), http://www.kbashop.de/webapp/wcs/stores/servlet/ProductDisplay?langId=-3&storeId=10001&catalogId=10051&productId=22503 (accessed May 17, 2011)

Kley, F.: Ladeinfrastrukturen für Elektrofahrzeuge - Analyse und Bewertung einer Aufbaustrategie auf Basis des Fahrverhaltens, Fraunhofer Verlag, Karlsruhe (2011)

Klobasa, M.: Fraunhofer-ISI-interne Datenbank zu technologiespezifischen Lastprofilen für einzelne Typtage in Sommer und Winter basierend auf Auskünften von verschiedenen Unternehmen und Verbänden zu Lastganglinienverläufen, Karlsruhe (2006)

Mock, P.: Entwicklung eines Szenariomodells zur Simulation der zukünftigen Marktanteile und CO2-Emissionen von Kraftfahrzeugen (VECTOR21), Stuttgart (2010)

MOP: Mobilitätspanel Deutschland 1994-2008, Clearingstelle Verkehr des DLR-Instituts für Verkehrsforschung (2008), http://www.clearingstelle-verkehr.de (accessed November 29, 2008)

Neuhoff, K., Amecke, H., Novikova, A., Stelmakh, K.: Thermal Efficiency Retrofit of Residential Buildings: The German Experience, Climate Policy Initiative, Berlin (2011)

ODYSSEE: Odyssee – Energy Efficiency Indicators in Europe (2012), http://www.odyssee-indicators.org (accessed January12, 2012)

Prognos AG, EWI, GDS: Energieszenarien für ein Energiekonzept der Bundesregierung, Basel, Köln, Osnabrück (2010)

Seefeldt, F., Wünsch, M., Schlomann, B., Fleiter, T., Gerspacher, A., Geiger, B., Kleeberger, H., Ziesing, H.-J.: Interne Studie zur Bewertung von Energieeffizienzmaßnahmen, Karlsruhe (2010)

StBA: Bevölkerung Deutschlands bis 2060 – 12. koordinierte Bevölkerungsvorausberechnung, Wiesbaden (2009)

StBA: Bruttoinlandsprodukt, Bruttonationaleinkommen, Volkseinkommen – Lange Reihen ab 1950, Wiesbaden (2011)

SW Mainz: Synthetisches Lastprofil für Elektrospeicherheizungen, Wärmepumpen und Direktheizungen, Mainz (2011)

TRY: Testreferenzjahre von Deutschland für mittlere und extreme Witterungsverhältnisse TRY, Offenbach a. Main (2004)

Chapter 8
The Impact of Hedonism on Domestic Hot Water Energy Demand for Showering – The Case of the *Schanzenfest*, Hamburg

Stephen Lorimer[1], Marianne Jang[2], and Korinna Thielen[3]

[1] UCL Energy Institute, University College London, 14 Upper Woburn Place, London WC1H 0NN, United Kingdom
s.lorimer@ucl.ac.uk
[2] LSE Cities, London School of Economics, Houghton Street, London WC2A 2AE, United Kingdom
[3] Institut für Entwerfen, Stadt und Landschaft, Technische Universität München, Raum 2237, Arcisstrasse 21, 80333 München, Germany

Abstract. The causes of variation in energy demand for hot water in showering or bathing within the same dwelling is often difficult to understand. This study followed the activities of a study group living as a large household and working on projects in architecture studios. Consumption diaries for lifestyle choices and showering times was triangulated with electric meter data to examine energy use behaviours and explore changes in hot water demand. This occurred over a two week period that included a street festival, the *Schanzenfest,* on the weekend. The study found that total energy demand amongst the same group could double depending on the amount of hedonism, or the seeking of fun and self-gratification, on display the previous evening. Therefore, this paper proposes that the increase of opportunities to have fun and the lack of structure on the weekend significantly increase domestic energy demand compared to a more structured weekday and should be the subject of further research.

1 Introduction

The aim of the research presented here is to assess the impact of lifestyle choice on hot water demand from a sociological perspective. This paper considers how household actions are patterned, shared amongst and between groups, and shaped by larger trends [1,2]. This view of energy is different from the prevailing physical-technical-economic model or building science-economics model [2-4]. These models assumes that energy efficient choices emerge out of the improved awareness of technologically superior alternatives, and that the cost of these alternatives comes down to a level that enables people to make the 'right' and rational decision to pursue these alternatives.

This kind of survey is noteworthy in the following respects: it measured actual energy consumption through electricity meters as well as self-reported environmentally-significant behaviours [5] at the group level alongside extensive consumption diaries.

A. Håkansson et al. (Eds.): *Sustainability in Energy and Buildings*, SIST 22, pp. 85–94.
DOI: 10.1007/978-3-642-36645-1_8

These diaries asked the participants to describe their socialising as well as quantifying both their total and "wasteful" showering time over an 11-day period instead of a single weekday and weekend day in standard surveys and allow analysis between and within individual participants.

Hedonism is an abstract concept that cannot be directly measured. The concept is described by the European Social Survey as "pleasure and sensuous gratification for oneself" [6]. The particular prompts used in this large-scale survey were around "seeks every chance to have fun" and "likes to spoil him/herself" [7]. Alcohol use is often used as a predictor of hedonism [8,9], and mentions of alcohol use are surveyed as predictors of water demand in the participants.

Although this study makes correlations between individuals' self-reported behaviour, actual energy demand was also measured for the group. Behaviour that is observed is preferable to asking people to self-report their behaviour. The strength of associations between observed and self-reported behaviour can be unpredictable [10]. The reasons for this can include survey design [11], social desirability bias [12], and imperfect short-term memory. The lack of memory can often a problem in research into habitual environmentally-significant behaviours such as showering [5], leading to estimations of these activities instead of recollections. However, without individualised data, one cannot get to the bottom of which attitudinal factors are the most important in these behaviours as different people respond to different stimuli for behaviour change [13,14]. Therefore, because of understandable privacy concerns, self-reporting of water use was the primary way of assessing the effect of behaviour change of individual participants.

As of 2007, the built environment currently contributes 17% of all carbon emissions in Germany [15], two-thirds of which are from the residential sector. In Germany, space heating and water heating still account for approximately 78% of all domestic energy demand.

The European Union as constituted at the time of the Kyoto Protocol (EU-15) collectively committed to reduce 2012 greenhouse gas emissions to 12.5% below 1990 levels [16]. Hot water heating makes up around 12% of electricity consumption in the EU-27 [17]. In this study, the heating of water was fuelled by electricity. Tonnes of carbon dioxide generated per kilowatt-hour of electricity use (0.655) calculated by the energy supplier for this building is more than three times the tonnes of CO_2 generated per kilowatt-hour of natural gas use (0.191).

1.1 Aims, Objectives and Research Questions

The study's focus was to explore the potential for reducing energy demand for hot water by raising awareness around the connection between displaying hedonism during the evening and then the human exception paradigm the following morning. To meet this aim, three research questions were developed:

- How much energy do the participants use for hot water for showering? And how much time was spent in the shower?
- Did hedonism use make an impact on hot water demand?
- Did the level of hedonism have an impact on hot water demand?

In addition to answering these questions, the study sought to develop a methodology for connecting consumption diaries with qualitative data that would go some way to explain poor self-reporting of habitual behaviour versus one-off occurrences.

2 Method

A group of built environment, environmental policy, and environmental engineering students (N=16) attended the University of the Neighbourhoods (UdN) Summer School, organised by HafenCity University Hamburg, for two weeks in August 2011. *The UdN was* located in a disused health centre in the Wilhelmsburg district of Hamburg. During the International Summer School, the participants were housed in dormitory rooms in one wing of the building and worked in the opposite wing. The result is that the building was a hybrid of residential and office uses with a spatial division between the two uses in the wings of the building with the participants' "commute" merely a walk across the kitchen to the workspaces of the building.

The students (N=16) completed a consumption survey that asked them about their consumption of food and drink and their showering habits over ten days. The daily journals were designed to help participants take stock and become aware of the volume, rate and diversity of their consumption of materials - from food, to water to the disposal of cotton buds. This was complemented by a parallel survey of the use of electricity, water, food and production of waste from the Summer School entity itself.

Alcohol consumption was analysed using the question "What did you eat and drink today outside of mealtimes?" in the consumption survey. The data was analysed for key words that described alcoholic drinks. The number of mentions of alcohol consumption by each person per day was summed up.

The actual number of alcoholic drinks was not a question in the survey, but the number of mentions is still a valid assessment of use. A recent review of measuring alcohol consumption in epidemiological studies [18] questioned the use of surveys that tried to obtain exact amounts of alcohol by trying to obtain both drink volume and strength. In other cases that only asked for the total number of drinks, researchers have long noted that individuals have a downward bias in the total number of drinks reported [19].

Both electricity use and temperature were measured over the course of these two days of detailed measurement. Overall electricity use was collected by a smart meter in increments of one hour (see http://www.vattenfall.de/de/smartmeter.htm). Plug monitors were then used to measure the energy use in the workspaces, the lecture room, and in the kitchen. Each plug monitor was checked every half-hour, and a description of the activities and the intensity of use of the building was included.

In order to estimate the amount of electricity used by non-monitored end-uses such as hot water for showering, assumptions needed to be made about the amount of electricity used for lighting, major cooking appliances, boiler operation, and hot water heating. A separate day, evening, and night "baseline" and "peak" amount of energy was calculated from the data and different end-uses assigned to them. The baseline was the median value for total electricity for different times of day. The day was divided for operational purposes into three sections:

Table 1. Baseline and peak electricity use

	Day (0800 – 1800)	Evening (1800 – 2200)	Night (2200 – 0800)
Baseline End-uses	Monitored plugs Boiler operation Lighting	Monitored plugs Boiler operation Lighting	Monitored plugs Boiler operation
Peak end-uses	Hot water Lecture (monitored plug)	Cooking Lecture (monitored plug)	Hot Water

The activities for electricity monitoring from the plug meters were grouped into the categories. This enabled a breakdown of domestic (food and socialising) and non-domestic (lectures and work) uses in the building. An example of this summing of plug monitors is found in Figure 1.

Table 2. Peak electricity uses, 18-19 August 2011

Time	0800 - 0859	0900 – 0959	1000 – 1059	0000 – 0059	0100 – 0159
Peak Activity	Showering	Showering	Lecture	Showering	Showering
Baseline	Day	Day	Day	Night	Night
Electricity Use (kWh)	4.66	12.26	3.92	4.46	2.8

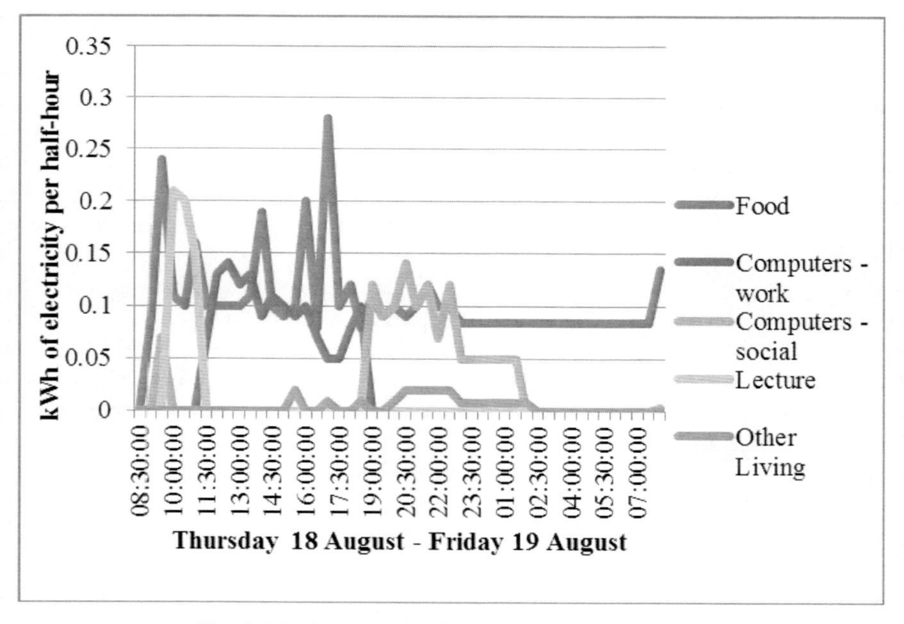

Fig. 1. Monitored electricity use, 18-19 August 2011

For each day, hours of electricity use were selected that displayed "peaks" of electricity use and if these were associated with showering (hot water), lecture (projector), cooking (oven and hob). The activities were recorded by researchers living and working within the same accommodation as the participants. An example from 18-19 August is in Table 2.

3 Results

3.1 Amount of Time and Electricity Used for Hot Water

Figure 2 summarises participants' showering duration and electricity consumption for hot water along with the number of mentions of alcohol use in the group. At first glance, there appears to be a much stronger relationship between the total number of mentions and total energy consumption that between the total number of mentions and the average duration of showers by the participants. Mentions of alcohol use by the group ranged from 2 (Tuesday 16 August) to 19 (Saturday 20 August), with a maximum number of 3 mentions by 3 different people in that evening. The average duration of showering ranged from 5.92 (Tuesday 23 August) to 9.7 (Sunday 21 August) minutes, and the total energy consumption from showering ranged from 10.56 (Wednesday 17 August) to 25.68 (Sunday 21 August) kilowatt-hours. The mean number of mentions of alcohol use was 7.9, the mean number of self-reported minutes of showering was 7.10 minutes, and the mean energy consumption for showering was 16.34 kilowatt-hours. The percentage of electricity that was used for heating hot water for showering in the building ranged from 17% (Wednesday 17 August) to 42% (Saturday 20 August) with a mean of 29%.

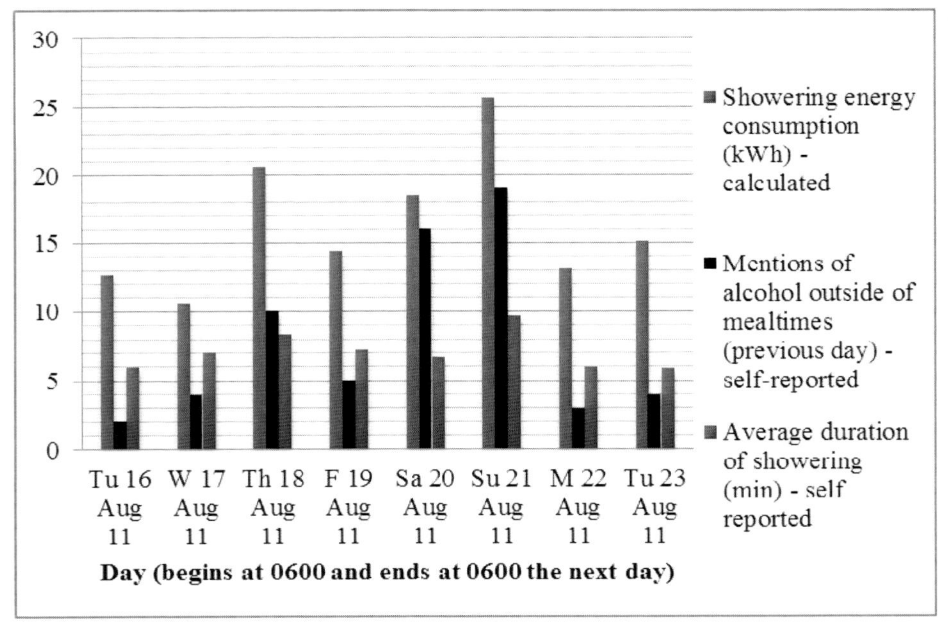

Fig. 2. Total measured electricity consumption and mean self-reported hot water demand compared to total mentions of alcohol use per day

The increase in alcohol use can be understood through setting and through changes in the price of alcohol during the week. A supplementary question in the consumption survey was “where did you get this food and drink?” On Wednesday 17 August, the participants left the premises to visit bars in central Hamburg. In the diaries, there are mentions of areas known for numerous drinking establishments, “St. Pauli”, “Schanze”, or simply “outside”. On days where there were low mentions of alcohol use, references were made to the residence of the participants or other domestic activities, such as “home”, “UdN” *(Universität der Nachbarschaften)*, or “supermarket”. During the weekend, mentions of locations such as “Reeperbahn” reappeared, before the greatest change in the location of obtaining food and drink on Saturday 20 August.

On 20 August, the locations of consumption centred on the *Schanzenfest* in the Schanze district of Hamburg. The *Schanzenfest* is a street festival held annually in the summer in the Sternschanze (Schanze) district. Schanze one of the "alternative" quarters of Hamburg and renowned for its riots: police clashes twice a year with locals: once on May 1st, and once at the *Schanzenfest*, an unannounced alternative street party. Starting peacefully in the afternoon, with music, flea markets, picnics, international food stalls, and lots of beer, it turns with predictable regularity into a street battle between left activists and the police at night. Around dusk fires are started, grafitti is sprayed, bottles are thrown and at some point in the early morning, the water cannons are called in and the area is cleared.

Notably, alcohol is sold cheaply throughout the area in an informal fashion. The mode price for a standard 330ml bottle of beer was €1 in the market stalls; for a basic cocktail with 25ml of spirit the mode was €1.50. This was considerably cheaper than the mode for a similar beer and cocktail was €3 and €5 respectively. Recent work on the connections between a sudden drop in the price of alcohol [20,21] support a connection to be made between across-the-board discounting and increased drinking.

3.2 Impact of Hedonism on Hot Water Demand

The Krushal-Wallis distribution-free test was conducted on the reported mentions of alcohol use and self-reported minutes of hot water demand. There were eleven days of data collected for 16 individuals (N=133). Hot water demand was significantly affected by mentions of alcohol use $H(4) = 9.89, p < .05$.

The Krushal-Wallis test cannot state what level of hedonism causes hot water demand to go up, and further post-hoc tests were necessary. Wilcoxon two-sample tests were used to follow up this finding by testing the effect of each number of mentions of alcohol use compared to no mentions at all. The participants mentioned alcohol use up to four times in one day. However, three (1) and four (2) were seldom the number of mentions and they were discarded from the analysis. A Bonferroni correction was applied so all effects are reported as a .025 level of significance. It emerged that self-reporting showering time was no different when one mention was made of alcohol use *(Mean=61.0, U=2010, p=.922)* compared to none *(Mean=60.3)*. However, when two mentions of alcohol were made, the difference was significant *(Mean=68.1, U=681, p=.0201)* compared to no mentions *(Mean=46.8)*. Thus, if the participants drank to a level outside of mealtimes that they mentioned alcohol use only once, hot water demand is not expected to increase. However, if alcohol use is mentioned twice, hot water demand is expected to go up, and go up by a small to medium amount *(r=.236)*.

A further question in the survey asked if the participants could have used less water in their shower. Quantitatively, no conclusions could be reached, but two participants answered "too tired" in their survey on the morning of Sunday 21 August, the day with the highest reported water consumption.

4 Discussion

This study set out to answer three research questions concerning the increase for hot water demand amongst participants that displayed hedonism the previous evening and night. Although the sample size was small, intensive consumption surveys and monitoring of electricity use provided valuable insights which can be explored in further research, and could ultimately inform policies and interventions that could reduce future domestic electricity consumption.

The mean electricity consumption for hot water for all of the participants during the week of monitoring was 16.34 kWh per day with a range of 10.56 to 25.68 kWh per day. This can be compared with typical hot water demand in residential buildings devised by the Association of German Engineers (VDI) in 1982 and cited in Quaschning [22] of an expected heat content of between 0.94 and 1.4 kWh/person/day. Of the 8 days covered by the electricity monitoring survey, three of the days actually fell below this range, four days fell within it, and the day after the *Schanzefest*, Sunday 21 August, was higher (1.7 kWh/person/day).

The design of the experiment, if there is additional research in this area of sociability, hedonism, and energy use, could be more focused in its recruitment of participants, observation of activities, measuring of hedonism by alcohol use, monitoring of water use, and monitoring of energy usage. There a range of options and opportunities that are outlined below.

The recruitment of participants in a future study could either focus an age group where alcohol use and binge drinking are more likely (18-25 year olds, for instance) or form a more representative sample of the population. Any future study should cover people as they are living in a more typical residential setting.The measuring of hedonism in this study was effective and repeatable, but comparable studies in the field of epidemiology have had more sophisticated ways of measuring alcohol consumption in their survey designs [18]. Consumption surveys could encourage participants to tally the number of drinks or units of alcohol that they consumed during an evening. The study was also limited to a predictor variable that only had three categories: none, one, or two mentions of alcohol use. If the same study was repeated using the same consumption survey, it should take place over a longer time interval than a day – a full week or even a weekend may be more appropriate.

There is a serious policy consequence to this study that reaches in to the policy realms of climate change, sustainable lifestyles, and sustainable events. There are many who argue that "big" events that involve huge amounts of sociability and hedonism disproportionately inflict environmental damage [23]. Events are one-off occasions, breaking us out of our normal habits, including ones that are built up to limit the environmental consequences of our actions.

The nighttime and event economy is, of course, not going to and will not be allowed to disappear in developed nations to satisfy environmental concerns. The Enterprise and Industry Directorate of the European Commission estimates that the EU tourism industry generates more than 5% of the EU GDP, with 1.8 million business employing 5.2% of the European labour force [24]. The environmental impact of having large scale events that attract visitors, whether they have emissions from international travel, is significant when they are there at the event, and remains significant when they arrive back at their own homes or hotel rooms that evening. This study can inform new policy directions and awareness campaigns around "letting your hair down, and letting your guard down" around environmental issues – and a temporary re-emergence of the human exception paradigm in highly sociable contexts.

5 Conclusions

Although the study was small, non-representative of the general population, and the participants were decidedly pro-environmental in their attitudes, a number of conclusions can be taken forward for further work. Firstly, while there was largely average hot water use during most of the week, an event that triggered hedonism also triggered higher than average hot water use in the participants. When surveyed about their water use, some even expressed that they could not limit their water consumption because of the impact of the previous night's hedonism on their well-being the next morning. If hedonism did not cause more use of hot water the next morning, these otherwise environmentally-aware and motivated participants would have been using significantly less than average energy consumption for hot water in a household. Instead, their average energy consumption rose to around average for a person in Germany. The finding does require testing with larger samples. If this result proves to generalise to the entire population, which would be expected, it indicates a greater need to educate and make aware to the public the consequences of hedonism in domestic energy and water use the following morning. Until near-zero carbon homes become standard, there may be no such thing as a sustainable event or night out because of significantly increased carbon emissions from hot water for showering (or bathing) created in the domestic sector the following day.

References

1. O'Neill, B.C., Chen, B.S.: Demographic determinants of household energy use in the United States. Popul. Dev. Rev. 28, 53–88 (2002)
2. Lutzenhiser, L.: A Cultural Model of Household Energy-Consumption. Energy 17(1), 47–60 (1992)

3. Guy, S.: Designing urban knowledge: competing perspectives on energy and buildings. Environ. Plann C 24(5), 645–659 (2006), doi:10.1068/C0607j
4. Sullivan, M.: Behavioral Assumptions Underlying Energy Efficiency Programs for Businesses. CIEE Behavior and Energy Program. California Institute for Energy and Environment, Oakland, Calif, USA (2009)
5. Stern, P.C.: Toward a coherent theory of environmentally significant behavior. J. Soc. Issues 56(3), 407–424 (2000)
6. European Social Survey Education Net (2012) Introduction to the Values Theory, http://essedunet.nsd.uib.no/cms/topics/1/1/1.html (accessed February 9, 2012)
7. European Social Survey, ESS Round 5 - 2010 (2010), http://ess.nsd.uib.no/ess/round5/ (accessed February 9, 2012)
8. Henson, J.M., Carey, M.P., Carey, K.B., Maisto, S.A.: Associations among health behaviors and time perspective in young adults: Model testing with boot-strapping replication. Journal of Behavioral Medicine 29(2), 127–137 (2006), doi:10.1007/s10865-005-9027-2
9. Keough, K.A., Zimbardo, P.G., Boyd, J.N.: Who's smoking, drinking, and using drugs? Time perspective as a predictor of substance use. Basic Appl. Soc. Psych. 21(2), 149–164 (1999)
10. Tourangeau, R., Rips, L.J., Rasinski, K.A.: The psychology of survey response. Cambridge University Press, Cambridge (2000)
11. Steg, L., Dreijerink, L., Abrahamse, W.: Factors influencing the acceptability of energy policies: A test of VBN theory. J. Environ. Psychol. 25(4), 415–425 (2005), doi:10.1016/j.jenvp.2005.08.003
12. Fisher, R.J.: Social Desirability Bias and the Validity of Indirect Questioning. J. Consum Res. 20(2), 303–315 (1993)
13. de Groot, J.I.M., Steg, L.: Mean or green: which values can promote stable pro-environmental behavior? Conserv. Lett. 2(2), 61–66 (2009), doi:10.1111/j.1755-263X.2009.00048.x
14. Lindenberg, S., Steg, L.: Normative, gain and hedonic goal frames guiding environmental behavior. J. Soc. Issues 63(1), 117–137 (2007)
15. European Commission, EU Energy In Figures 2010: CO2 Emissions by Sector. European Commission, Brussels (2010)
16. European Commission, Kyoto Protocol on Climate Change (2010), http://europa.eu/legislation_summaries/environment/tackling_climate_change/l28060_en.htm (accessed November 16, 2011)
17. Atanasiu, B., Bertoldi, P.: Latest Assessment of Residential Electricity Consumption and Efficiency Trends in the European Union. Int. J. Green Energy 7(5), 552–575 (2010), doi: 10.1080/15435075.2010.515449
18. Greenfield, T.K., Kerr, W.C.: Alcohol measurement methodology in epidemiology: recent advances and opportunities. Addiction 103(7), 1082–1099 (2008), doi:10.1111/j.1360-0443.2008.02197.x
19. Rehm, J., Greenfield, T.K., Walsh, G., Xie, X.D., Robson, L., Single, E.: Assessment methods for alcohol consumption, prevalence of high risk drinking and harm: a sensitivity analysis. Int. J. Epidemiol. 28(2), 219–224 (1999)
20. Makela, P., Osterberg, E.: Weakening of one more alcohol control pillar: a review of the effects of the alcohol tax cuts in Finland in 2004. Addiction 104(4), 554–563 (2009), doi:10.1111/j.1360-0443.2009.02517.x

21. Heeb, J.L., Gmel, G., Zurbrugg, C., Kuo, M., Rehm, J.: Changes in alcohol consumption following a reduction in the price of spirits: a natural experiment in Switzerland. Addiction 98(10), 1433–1446 (2003)
22. Quaschning, V.: Understanding renewable energy systems. Earthscan, London (2005)
23. Collins, A., Flynn, A., Munday, M., Roberts, A.: Assessing the environmental consequences of major sporting events: the 2003/04 FA Cup Final. Urban Stud 44(3), 457 (2007)
24. European Commission Supporting European Tourism (2012), `http://ec.europa.eu/enterprise/sectors/tourism/index_en.htm` (accessed February 14, 2012)

Chapter 9
The Process of Delivery – A Case Study Evaluation of Residential Handover Procedures in Sustainable Housing

David Bailey, Mark Gillott, and Robin Wilson

Energy and Sustainability Group,
Built Environment,
Nottingham University,
Nottingham NG7 2RD, UK

Abstract. At present research groups are developing a growing body of evidence quantitatively demonstrating through post occupancy evaluation, a significant gap between the actual physical performance characteristics and the design predictions of sustainable dwellings. In examining this documented performance variability this paper argues that a substantial proportion of this gap may be the result of mismanagement and misuse of sustainable systems by the occupants who have received little to no training in the specialised equipment and design techniques regularly employed in modern sustainable housing. Specifically this paper looks into the training and guidance given to new house owners during the critical handover phase. The research adopts a direct observational methodology in conjunction with a suitable housing case study and the associated handover process. By recording and analysing the handover procedures of a representative housing developer the study hopes to gain valuable insight into the current technological training and guidance provided to new tenants of modern ecologically certified housing. The study finds occupants are not receiving adequate training and guidance with regard to the sustainable measures employed in their housing. In addition the survey suggests that residents struggle to absorb the information provided in the current format. Ultimately the study proposes a complete reform of the handover process, based on existing commercial precedence and focusing on both the accessibility and content of the handover procedure.

1 Introduction

This paper has been compiled as part of a larger body of work seeking to address the widely accepted, yet little documented phenomenon of performance variability in domestic applications. There is a growing body of evidence (Herring & Roy, 2007) (Johnston, 2010) (Taylor *et al.*, 2010) (Wingfield, 2011) quantitatively showing a significant gap between actual physical performance characteristics and design predictions. As the housing industry comes under greater pressure to develop more ecological products, a technology oriented approach has dominated the search for

A. Håkansson et al. (Eds.): *Sustainability in Energy and Buildings*, SIST 22, pp. 95–105.
DOI: 10.1007/978-3-642-36645-1_9

sustainability. However the following work examines the importance of the often marginalised variable of socio-technological interaction and occupant behaviour within the role of housing performance. (Marsh, 2010) (Herring & Roy, 2007)

Modern sustainable housing requires a significant amount of technology oriented solutions which in turn require some level of occupant interaction, ranging from changing a filter every 5 years to daily operational contact. This paper asks the question: have handover and training procedures targeting new house owners evolved and become more formalised with the advent of more complex sustainable technologies and design concepts in the ecologically oriented housing sector?

2 Background and Context

2.1 Government Policy

In 2007 the UK Government introduced a new housing policy objective aimed at reducing the residential sector's estimated 26% share in green house gas emissions (DECC, 2011). This widely debated legislation charges the housing industry with the goal of producing fully zero carbon homes by the year 2016 (DCLG, 2007). While subject to much controversy the policy is supported by the overarching mandate of the Climate Change act that commits the UK to legally-binding targets for emissions reductions of 80% by 2050 and at least 34% by 2020, against a 1990 baseline (OPSI, 2008). In response to the Government policies, research organisations and housing developers have focused on developing technologies that both reduce overall energy consumption and produce renewable energy to supplement and replace that drawn from the national grid (Marsh, 2010). This focus and emphasis on technology and good building practice is prevalent throughout the industry but often comes at the expense of social considerations. "There is growing recognition that building performance studies should take more account of occupant behaviour and needs. In the past there has been over reliance on, for example, predictions from design models and estimations." (HCA, 2010, p. 3) Understanding how occupants interact with a building and the subsequent variations that may cause in the building performance is vital as occupant behaviours vary widely and can impact energy consumption by as much as 100% for a given dwelling (Dutil et al. 2011) Note that here occupant interaction and participation refers to activities that have a direct or indirect impact upon building energy consumption. With the advent of sustainable construction and the associated user participation, it is important to understand the interaction between occupant and building (DEFRA, 2008). The findings of a report by the NHBC Foundation (2011, p. 6) indicate further research is required to examine both occupant behaviour and the "best ways to inform users how to make the most efficient use of their homes and the systems in them…Understanding what information should be provided in user guides and what level of detail and in what format should this information be provided."

2.2 The Significance of Handover Procedures

Given the importance and concurrent difficulties associated with Post Occupancy Evaluation (POE) as a relatively new practice in domestic applications (Stevenson, 2009), it is important that this study targets the principal instance of occupant education. The construction of a new housing project necessitates 5 key phases.

1. Design
2. Planning
3. Specification
4. Construction
5. Handover

(BSRIA, 2009)

Within a mainstream sustainable housing development the purchaser of the house would only become part of the development process during the final stages of construction or even post-construction. Thus the first contact that many people have with their houses is during that initial handover stage. Currently housing handovers are a non-regulated, informal procedure. The handover stage generally comprises of an instructional tour around the house and usually some form of literary backup such as instruction manuals for the more complicated equipment (DCLG, 2010). It is during this brief period of contact that the developer or relevant sales person, must convey the entirety of the design concepts and technological installations that make the house sustainable. Alternatively the homeowner is expected to absorb all the information given to them and then apply it throughout their ownership of the house. With the exception of the handover procedure there is little formal contact between occupant and developer, that allows for in-depth guidance and instruction on the systems and technologies employed in the property under current standard practice.

3 Methodology

3.1 Strategy

The study adopts a direct observational protocol in combination with a specialised case study. The direct observational methodology is a branch of qualitative field research that simply aims to gather data from specific sampled situations within the natural setting of a case study. In order to reinforce the data gathered through the direct observation techniques, a suitable case study with relevant parameters is used as the test bench. Relevant parameters in this context refer primarily to the ecological certification of the sample dwellings. The dwelling's ecological rating needs to be such that it conforms to the environmental protocols required by the government and in conforming to these protocols, it utilizes representative sustainable technologies and energy saving techniques characteristic throughout industry. The handover of such a property should, in theory, reflect the level of training and guidance required to introduce a layperson to the subtleties and functionality of the sustainable technology that they will be required to use on a daily basis. A critical analysis of these sample handovers then gives an indication of how the industry is adapting to accommodate new technologies and ideas.

3.2 Case Study Outline

The case study development is located in the Meadows, Nottingham. The Green Street development is a newly constructed housing scheme accredited at CSH Level 4. The 6 houses under investigation incorporate numerous technologies and sustainable design techniques making them an ideal testing ground for the handover procedure. The observed handover procedure in the 7 houses of phase 2 took place over a nine day period. Each handover took between 30-60min. During the handover, the homeowners were guided around the property by the a member of the development agency in charge of the Green Street Project, starting outside then moving upwards through the floors of the property. Each handover was recorded electronically with additional handwritten notes from two observers. It is important to note that the sample size of 6 houses means conclusions from this site are by no means definitive, rather they are seen as indicative of what may be widespread practice in the industry.

3.3 Design of Data Analysis

At present there is no formal protocol governing the handover of a building and its documentation (Graves et al., 2002) It is therefore necessary to establish a benchmark of expectations based on the handover's position within the delivery process, introduced in section 2, case study literature (DCLG, 2010) and best practice recommendations from similar studies (Graves et al., 2002) (BSRIA, 2009) (Stevenson & Leaman, 2010). Based on these sources, a best practice handover should provide a thorough and accurate introduction to all occupant driven attributes of the dwelling such that the occupant is confident in their functionality and operation. With regard to sustainable properties, this introduction should include any and all sustainable attributes or systems that may be affected by occupant behaviour. The methodological protocol therefore dictates a critical analysis of the handover procedure that compares these best practice expectations with the case study evidence to determine the proficiency of the procedure. The following pilot study is derived from the stipulations detailed in the previous section, the results being a compilation of the sustainable design concepts from the case study homes which should be covered in a best practice handover. The information is gathered from a variety of sources including marketing information, the architects specification, Standard Assessment Procedure (SAP) analysis and first hand observation. These concepts and technologies are then categorised with respect to the intentions of the study and the feasible scope of a typical handover procedure to produce a list of elements that should, in best practice, be introduced to the occupants. The intentions of the research dictate a focus on the behavioural determinants which can be defined as the energy saving characteristics which relate directly to occupant management of the building and occupant control of systems and components. (Yao & Steemers, 2005) The behavioural determinants are further subdivided into active and passive components (in table 1) where:

- Active components – Occupants must be actively taught how to use the particular technology or design element.

- Passive components – Occupants intuitively know how to use these elements, but need to be encouraged and taught best practice so as to avoid problems such as the rebound effect (Nassen & Holmberg, 2009) (C. J. M. van den Bergh, 2011).

Table 1. Breakdown of behavioural determinants into active and passive components

Status	Identifier*	Behavioural Determinant
Active	A	Photo-voltaic panels
	B	Mechanically ventilated heat recovery system
	C	Stack Ventilation
	D	Radiators with timed and temperature zone controls
Passive	E	90% Condensing Combination Gas Boiler
	F	100% Low Energy dedicated internal lighting
	G	Washing Lines
	H	Bike Sheds
	I	Low flow toilets and taps and showers
	J	'A' rated Kitchen appliances
	K	Recycling Bins

*** Throughout this paper, this unique identifier will be used to label quotations and excerpts drawn from the handover observations.**

Using this matrix the core content required of the handover, from a sustainable perspective, is narrowed to 4 vital elements, supported by 7 sub-factors. The training and support inherently involved with the introduction of these behavioural determinants is seen as part of the fundamental requirements of a handover and deemed feasible within the scope of the handover process. While these particular elements are case study specific, the pilot study protocol can be applied by any developer or relevant handover personnel, to any new environmentally focused development. This facilitates the prioritisation of those technologies and design attributes that require a more in-depth and instructional introduction.

4 Results

This section into divided into 2 subsections focusing, respectively, on an analysis of the results as defined by the pilot study protocol and the substantiation of these results with extracts from the audio transcripts and handwritten notes.

4.1 Analysis

The grade based scale introduced in tables 2&3 is founded on extensive knowledge concerning each of the elements in table 1. An understanding of each of these elements allows the development of a hypothetical model of what constitutes a good explanation by the demonstrator. An ideal description represents a 3 on the scale, the element is then downgraded if the demonstrator is seen to omit vital bits of information or mislead the occupier. In table 2, functionality is defined as the demonstrator's ability to practically explain how an element works. The bike shed or bin storage for example is fairly intuitive, requires little training and will generally score highly, thus indicating no need to change current methods of handover procedure. However mechanically ventilated heat recovery (MVHR) is far more complex and would therefore require much more explanation.

Table 2. Functionality Scale

Functionality (F)	
Grade	Definition
0	No mention of item at all or mentioned, but explained falsely.
1	Brief explanation of functionality
2	Adequate explanation of functionality
3	Thorough explanation of functionality

Table 3. Sustainability Scale

Sustainability (S)	
Grade	Definition
0	No mention of item at all or mentioned, but explained falsely.
1	Brief explanation of sustainability
2	Adequate explanation of sustainability
3	Thorough explanation of sustainability

In table 3 "S" denotes Sustainability. This is the level to that the demonstrator has addressed the sustainable ethos of the element in question. Examples of this would be the inclusion of a map of cycle routes in the surrounding area when introducing the bike shed, or a breakdown of how much money and CO_2 can be saved when using a washing line as compared to a conventional dryer thereby encouraging a occupant to actually use a component that they might otherwise ignore. The purpose of the dual rating system is to show, particularly in the case of the "passive" items that functionality or "how" can be covered without actually dealing with the sustainable ethos behind the item or the "why." As industry looks into a more holistic solution to sustainability the why becomes

Table 4. Analysis of individual handover processes

Plot ID*	**Behavioural Determinant ID**																							
	A		**B**		**C**		**D**		**E**		**F**		**G**		**H**		**I**		**J**		**K**		**L**	
	F	**S**	**F**	**S**	**F**	**S**	**F**	**S**	**F**	**S**	**F**	**F**	**S**	**S**	**F**	**S**	**F**	**S**	**F**	**S**	**F**	**S**	**F**	**S**
P1	1	1	1	1	0	0	3	2	2	0	0	0	0	0	3	0	2	0	2	0	2	0	2	0
P2	2	0	0	0	0	0	3	1	2	0	0	0	3	0	3	0	2	1	2	1	2	0	2	0
P3	1	0	1	0	0	0	3	3	2	0	0	0	0	0	3	0	2	0	2	0	2	0	2	0
P4	1	0	1	0	0	0	0	0	2	0	0	0	0	0	3	0	1	1	2	0	2	0	2	0
P5	N/A	N/A	0	0	0	0	2	0	2	0	0	0	0	0	3	0	2	1	2	0	3	0	2	0
P6	1	0	1	0	0	0	0	0	2	0	0	0	0	0	3	0	2	0	2	0	3	0	2	0

*** Throughout this paper, this unique identifier will be used to label quotations and excerpts drawn from the handover observations in combination with the statement's position on the audio recording eg. (P1, 12:45). Plot handovers are listed in chronological order.**

just as important as the how. Addressing the motivations behind using a particular technology can encourage an individual to use an element that they would otherwise remain reluctant to interact with (Hargreaves et al., 2010). In best practice the handover should adequately explain both the functionality of an element and the sustainable ethos behind it or else risk the inevitable performance gap between design and reality when occupants fail to use their homes as they were intended to be used.

Table 4, Analysis of individual handover processes, clearly shows that sustainability is a marginalised variable within the handover process, across both active and passive determinants. This is due to 3 key reasons:

1. Regarding the active behavioural determinants such as MVHR, and PV the failure is primarily the result of inadequate knowledge on the part of the demonstrator supported by statements by the demonstrator revealing that they were unfamiliar with the equipment, particularly the MVHR, during the initial handovers. (P1, 10:30)
2. Passive determinants are affected by the endemic problem of insufficient previous experience with sustainable housing. There is no knowledge infrastructure or even aspiration to support the handover of sustainable housing, due paradoxically to the fact that housing has never before included this level of sustainable design and technology. The ever growing field of sustainable construction however provokes an integral shift in the responsibilities of the modern handover with a far greater emphasis on a holistic process (BSRIA, 2009).
3. Finally there was the assumption by the demonstrator that it (the handover) isn't much different from a normal house. (D1) Changing this attitude requires collaboration from the designers and architects who initially conceive the various sustainable elements and the developer who is contractually obligated to perform the handover.

The figures also support the notion that practical explanation is not necessarily indicative of sustainable instruction. Behavioural determinants H,I,J,K,L are all addressed adequately in terms of functionality but the demonstrator pays little attention to the sustainability as evidenced by only 4 points across all 4 determinants.

4.2 Result Validation and Basic Interpretation

This section goes through the thought process used to downgrade the determinant values (see tables 3&4) in table 5, and substantiates these values using direct quotes and excerpts from the observational stage of the study in line with observational data analysis protocol. The following is a typical example of the level of emphasis placed on the bike sheds and recycling facilities provided:

These are your bike sheds and these are your bins. (P3, 2:20)

The bike sheds are a vital element in achieving Code for Sustainable Homes (CSH) Level 4 accreditation (BRE, 2010). It would seem prudent therefore to expand on their utility beyond simply one sentence and perhaps include a map of local bike paths in the handover pack and discuss the protection and security offered by the shed.

Simple, additions that cost little to nothing to the developer, but may encourage an occupant to use the house as it was intended. Kitchen appliances have all been specially selected to conform to CSH standards and yet throughout the study there is only 1 mention (P2: 12:30) of an "A rated" appliance in the form of the dishwasher:

Bog standard microwave, bog standard oven, and bog standard fridge. (P4, 6:55)
"Again, this dishwasher, no different from any other dishwasher..." (P6, 12:20)

The study records no mention of the low power lighting installed throughout the housing and the integral stack ventilation provided by the stairwell and window design. The washing line, a simple but significant energy saving device during the summer time is only functionally introduced once, with no mention of energy savings. In these instances there may be a reluctance to state the obvious in regards to functionality, however the integral sustainability and motivations behind the inclusion of these elements becomes important when they are viewed as a holistic package which reduces the ecological footprint of a property. The photovoltaic (PV) array on the roof is functionally considered in all handovers, with emphasis on the isolation switches for safety purposes and the meter box. However, basic knowledge such as the number of panels on the roof and the purpose of one of the primary components (the inverter) is significantly lacking:

Confusing and uninformed explanation of PV panels (P1, 11:10-13:00)
"I don't actually know how many PV panels are on the roof." (P3, 3:40)

This confusion is coupled with a lack of sustainable emphasis (see section 4.1). The MVHR unit, is probably the most technically challenging piece of equipment in the house, but is realistically an indicative example of the technology employed in modern sustainable housing. This case study exhibits a worrying lack of knowledge when faced with one of the key sustainable features of the house:

"It is really just to remove odours, smells and condensation..." (P3, 19:00)
On opening a window the occupant asks "Will this affect the MVHR?" Reply from demonstrator "It doesn't, it doesn't, no." (P2, 7:00)
"To be completely honest with you I would be lying if I said I knew how it (MVHR) worked because I don't." (P6, 43:10)

It should be noted that the case study evidence on that this report is based is not intended to reflect on the proficiency or professional bearing of the individual or company that was examined. Its purpose it to provide an evidence founded platform on that to address the growing issue of handover procedures in sustainable housing.

5 Discussion

The most striking issue raised by this study is the lack of emphasis and time dedicated to the actual handover of a sustainable property. These dwellings are laboriously designed and constructed with often complex and occupant driven characteristics and yet standard handover practice (as derived from the case study) is roughly an hour long tour supported by a collection of technical manuals referring to individual pieces of equipment. In addition there is no specialised technical training for the demonstrator, as evidenced by quotes in section 4.2. These findings are supported by research that suggests that the physical implementation of sustainable technologies in homes often fails to incorporate social considerations (Herring & Roy, 2007). A literature review by Marsh (2010, p. 5) examines this "techno-rational construct" concluding that: "Whilst many housing developments look to achieve sustainability by incorporating technological indicators, the result has been shown to be inefficient if they lack social awareness or understanding of the potential occupants." What the data and literature analysis from this study suggests, is that by not considering the socio-technical interaction of the occupiers, particularly during the handover stages, there will invariably be shortcomings in the specified efficiency (HCA, 2010) (Dutil et al., 2011). The case study results clearly show the representative handovers lack sufficient technical and sustainable content. Subsequently it is reasonable to assume that numerous other sustainable housing handovers are similarly deficient. The analysis goes on to show not simply a need for greater technical detail, but a far more fundamental obligation to include comprehensive social considerations throughout the handover process, thus breaking with the traditional techno-centric mindset. Throughout the results analysis and review of the audio transcripts a key theme has emerged that does not fall within the remit of the functional/sustainable methodology. However it is seen to significantly impact the fundamental research objective. As the handovers progress it soon becomes evident through both the tone and language of the occupiers, that they are having a hard time absorbing all the information. Input from the second handover observer (D1) talks of the systems in the house being explained extremely briefly and at a rapid pace. This is supported by comments from the homeowners:

Tenant post boiler explanation – "Ohhh I'm lost now, start again!" P1 (8:50)
Overwhelmed occupiers – "No no no too much information!" P1 (15:45) (17:50)
After seeing the confused look on the occupiers face: Demo – "Is everything ok?" Tenant – "Yes, we are just trying to take it all in." P5 (16:30)

This inability to absorb complex information quickly and efficiently is by no means an isolated or unusual occurrence (DCLG, 2010), and may become especially acute when dealing with the elderly or individuals for whom English is a second language. This key observation calls into question the format and protocol currently used to deliver the information within the domestic handover process. It implies the answer goes far beyond simply adding more information to the handover itself, suggesting an examination of both the content and propriety of the process as a whole (Stevenson & Leaman, 2010) (DCLG, 2007). The denotive results and key themes raised in sections 5.1 & 5.2 of the study show:

1. Demonstrators are not aware of the significance of sustainability and the impact that occupant behaviour has on the performance of a house.
2. Demonstrators struggle to understand many sustainable technologies and concepts which are taken for granted within the modern sustainable housing industry.
3. Demonstrators struggle to appropriately communicate worthwhile information in the standard 1 hour tour and handover manual format.

Given these challenges, how can the industry impart sufficient knowledge required for the correct and efficient use of the dwelling, while ensuring that this knowledge is passed on in an appropriate manner? Throughout this study a process known as the Soft Landings Framework (BSRIA, 2009), has been instrumental in providing precedence for a solution. While based on the construction and commissioning of non-domestic property, the framework and ideology mirrors the problems facing the domestic building sector. Originally conceived by Mark Way in 2004 and formalised by BSRIA in 2009 the Soft Landings Framework has become the de-facto approach in ensuring non-domestic buildings achieve specified performance values. The process is based on 5 steps and involves a graduated handover period, that is predicated by consultation with the design and construction team to ensure that all performance related elements are first understood and then explain and implemented properly to a buildings new occupants. The adaptation and analysis of this existing framework can serve as a foundation as the industry looks to enact the fundamental changes required in domestic handovers for sustainable construction.

6 Conclusions

The direct observational methodology gives an unambiguous picture of representative handover procedures performed in the modern UK housing market. It is clear from the results of the study that these occupants are not receiving appropriate guidance and encouragement with respect to the sustainable measures employed in the housing and therefore it is reasonable to assume that much of industry faces similar challenges. A review of both the content and delivery methods employed suggests that struggle, even under current handover procedures, to absorb the information provided.

Thus a tenable solution must focus initially on the accessibility of the information during the handover procedure itself, presenting the information in an appropriate manner. (DCLG 2010) This appropriate manner is a subject for further investigation and involves understanding how people efficiently assimilate information and the most effective methods to get occupiers engaging with the sustainable design of their houses. Ultimately the inherent level of technology and environmental design concepts required to reach Government backed ecological standards necessitates a complete reform of the handover process. The precedence exists in the form of the Soft Landings Framework (BSRIA, 2009). Further study is required in order to understand how to adapt the concepts and practices from this predominantly commercial procedure, to a domestic perspective, particularly when looking at aftercare and the responsibilities of individual project stakeholders.

Subject to this review the developer in question has already set in place measures to modify and improve their practice for phase 3 of the housing project, placing them at the forefront of a new and growing evolutionary process in handover procedure.

Acknowledgments. The research was supported by input from Jennifer White, Nottingham University (D1) and funded by the Technology Strategy Board under the 'Performance Evaluation of Buildings Programme' and with the permission and ongoing support of Blueprint (Developer) and Lovell (Contractor).

References

1. (HCA) Homes & Communities Agency, Monitoring Guide for Carbon Emissions, Energy and Water Use - The Carbon Challenge: developing an environmental evaluation of housing performance in new communities. Homes & Communities Agency, Warrington, UK (2010)
2. BRE, Code for Sustainable Homes - Technical Guide. Department for Communities and Local Government, London, UK (2010)
3. BSRIA, The Soft Landings Framework. Building Services Research and Information Association, Bracknell, UK (2009)
4. van den Bergh, J.C.J.M.: Energy Conservation More Effective With Rebound Policy. Environ Resource Econ. 48, 43–58 (2011)
5. DCLG (Department for Communities and Local Government), Building a Greener Future: policy statement. The Crown, London, UK (2007)
6. DCLG (Department for Communities and Local Government), The Code for Sustainable Homes: Case Studies vol. 2. The Crown, London, UK (2010)
7. DECC. 2009 Final UK Figures - Climate Change Data (March 31, 2011), http://www.decc.gov.uk, http://www.decc.gov.uk/en/content/cms/statistics/climate_stats/data/data.aspx (retrieved January 6, 2012)
8. DEFRA. A Framework for Pro-Environmental Behaviours. Colgate, Dept. for Environmental Food and Rural Affairs, UK (2008)
9. Dutil, Y., Rousse, D., Quesada, G., Quesada, G.: Sustainable Buildings: An Ever Evolving Target. Sustainability (3), 443–464 (2011)
10. Graves, H., Jaggs, M., Watson, M.: HOBO protocol - Handover of Office Building Opperations. BRE Digest 474, pp. 1–12 (December 13, 2002)
11. Hargreaves, T., Nye, M., Burgess, J.: Making Energy Visible: A qualitative field study of how householders interact with feedback from smart energy monitors. Energy Policy 38(10), 6111–6119 (2010)
12. Herring, H., Roy, R.: Technological innovation, energy efficient design and the rebound effect. Technovation 27(4), 194–203 (2007)
13. Johnston, D.: Fabric testing: Technical approaches and processes. Ecobuild - Earls Court. London: Leeds Metropolitan University (2010)
14. Marsh, P.: Sustaining Technical Efficiency and the Socialised Home: Examining the Social Dimension within Sustainable Architecture and the Home. The Int. Journal of Interdisciplinary Social Sciences 5(5), 287–298 (2010)
15. Nassen, J., Holmberg, J.: Quantifying the rebound effects of energy efficiency improvements and energy conserving behaviour in Sweden. Energy Efficiency (2), 221–231 (2009)
16. NHBC Foundation. How occupants behave and interact with their homes - The impact on energy use, comfort, control and satisfaction. IHS BRE Press, Watford, UK (2011)
17. OPSI (Office of Public Sector Information). The Climate Change Act 2008. The Crown, London, UK (2008)
18. Stevenson, F., Leaman, A.: Evaluating housing performance in relation to human behaviour: new challenges. Building Research & Information 38(5), 437–441 (2010)
19. Taylor, T., Littlewood, J., Geens, A., Counsell, J., Pettifor, G.: Development of a strategy for monitoring the environmental performance of apartment buildings in Swansea: measuring progress towards zero carbon homes in Wales. COBRA/RICS, London, UK (2010)
20. Wingfield, J.: Fabric Testing: Whole House Heat Loss, London, UK (April 14, 2011)
21. Yao, R., Steemers, K.: A method of formulating energy load profile for domestic buildings in the UK. Energy and Buildings (37), 663–667 (2005)

Chapter 10
Sustainable Renovation and Operation of Family Houses for Improved Climate Efficiency

Ricardo Ramírez Villegas and Björn Frostell

Division of Industrial Ecology, The Royal Institute of Technology, Stockholm
{rrv,frostell}@kth.se

Abstract. In the developed world the existing stock of houses will provide shelter to the majority of population in the upcoming years. Houses are physical objects that consume material and energy and need to be maintained, repaired and restructured from time to time. In order to fulfill the requirements of the Kyoto Protocol and be comfortable for their inhabitants, the existing stock needs to be renovated. Strong disagreements between different parts of the scientific community and overlapping and contradictory concepts make the definition of sustainable renovation confusing. In this study, therefore, an approach of renovation and operation for higher energy efficiency and lower climate impact has been the main focus. Based on a systems analysis approach, the aim of this work is to evaluate cost and benefits of possible actions and choosing the most energy and cost effective approach of a series of alternatives. With the result of this analysis, a sustainable renovation and operation staircase is proposed. The work found that it is possible to develop a staircase manual for sustainable renovation and operation of family houses that follows a logical step-by-step approach and could result in considerable life cycle reductions in both costs and climate impact. The work also suggests that it is possible for academic experts to develop material in a simpler form and language to reach the public in a more understandable form.

Keywords: sustainable renovation, energy savings, CBA, LCCA, sustainable staircase.

1 Introduction

Since the 1990's there is a tendency in North America and Western Europe where the number of buildings to renovates is higher than the number of buildings to be built (Botta 2005). According to the European Union (EU) the existing stock will provide shelter for the majority of population of Europe, with no more than a 15% of new buildings until the year 2020. This existing stock needs to be renovated in order to fulfill the requirements established in the Kyoto Protocol.

Renovation can be defined as the *"improvements in the quality of the built environment, which are closely linked with needs expressed by the actors concerned*

A. Håkansson et al. (Eds.): *Sustainability in Energy and Buildings*, SIST 22, pp. 107–117.
DOI: 10.1007/978-3-642-36645-1_10

(users), especially improvements in comfort and reductions in the cost-in-use and maintenance of residential and non-residential buildings" (Mørck et al. 2003).

The scope of sustainable renovation is to make a house healthier and energy and resources efficient, but sustainable renovation is understood and implemented differently in each country (Kaklausas et al. 2008).

Strong disagreements among the scientific community and conflictive information make the concept of sustainable renovation confusing. The debate is developed in academic environments while in the field there is a lack of practical knowledge (Botta 2005).

Organizations around the world provide guidelines and support to house owners and general public, like the *"Energy Star upgrade manual";* an initiative of the U.S. Department of Energy (DOE). Guidelines are also available on the web, helping house owners to implement sustainable measures. Renowned universities publish their own sustainability guidelines on the internet, and organizations around the globe developed Carbon Footprint Calculators.

2 Methodology

Environmental, economic and social impacts need to be considered when discussing sustainable renovation. In this study, the analysis was limited to economic and environmental aspects, covered by Life Cycle Costing (LCC), energy use and Cost Benefit Analysis (CBA).

2.1 Life Cycle Costs (LCC)

LCC is an economical approach that sums all the costs of a product process (Ness et al. 2007) helping to choose the most cost effective approach from a series of alternatives (Barringer 2003). Operational costs are usually higher than purchasing costs; LCC shows if an investment is justifiable for a long term decision (Barringer 2003). According to Ellis (2007), LCC is a technique for evaluating total costs of mutually excluding alternatives. LCC helps to calculate long term costs (State of Alaska – Department of Education and Early Development 1999). Ellis (2007) states that realistic assumptions can be made evaluating the performance over time of different options whit a well-established time period. Then, LCC results must be presented as Net Present Value (NPV) using carefully a discount rate driving to a cradle to grave NPV analysis (Barringer 2003).

2.2 Energy Use

Energy use may be regarded as an important indicator of climate impact since approximately 80% of the average global energy use is related to energy use. Energy savings therefore are related to environmental performance improvements. The difference is great between different countries, however, and e.g. Sweden has a comparatively low climate impact from energy use thanks to a high proportion of electricity supply is based on nuclear power and hydropower.

2.3 Cost-Benefit Analysis (CBA)

CBA is used to evaluate public and private investments by weighting costs and expected benefits, selecting a Policy, Program or Project (PPP) which is efficient in their use or resources.

CBA is performed by compiling costs and benefits of a PPP and translating them into monetary values (Wierenga 2003). In connection with sustainability, CBA is an effective tool to measure social and economic costs and benefits in connection with accountable impacts e.g. energy and transportation (Ness et al. 2007). To analyze a PPP all the aspects must be expressed in a common unit, money being the most favored (Watkins 2004). CBA can account impacts in a time scale, a clear advantage over other Systems Analysis Tools (Wrisberg et al. 1999).

2.4 Description of the Selected Tool

The Energy Star program of the U.S. Environmental Protection Agency (EPA) and DOE, (2009), developed tools for *"Purchasing and procurement"* available for free use on the program website. Based on the tools developed by EPA and DOE, it is possible to create a modified version. To perform the analysis in the Swedish market, some changes are proposed:

- Compare between multiple options instead of two
- Check for the Swedish labor costs and discount rate

The analysis starts with a calculation of the annual energy consumption of the product

$$\begin{aligned}Energy\ Costs\ per\ year\,(\$) = Energy\ Consuption\ per\ year\,(kWh) * \\ Energy\ Rate\,(\$/kWh)\end{aligned} \tag{1}$$

Then the maintenance costs are calculated by:

$$\begin{aligned}Maintenanc\ e\ Costs\ \ per\ \ year\ (\$) = \\ \big(Labor\ \ Cost(\$) * Labor\ \ time\ (hours)\big) + Purchase\ \ Cost(\$)\end{aligned} \tag{2}$$

The LCC for this purpose are the energy, purchasing and maintenance costs for the lifetime which are calculated by:

$$PV\,(\$) = \left(PMT \big/ 1 + r_1 \right) + \left(PMT \big/ 1 + r_2 \right) + \ldots + \left(PMT \big/ 1 + r_n \right) \tag{3}$$

Where:

PV: Present Value of Money ($)

PMT: Payment (Energy cost and maintenance respectively) ($)

r: Central Bank discount rate

Finally, indicators are developed to compare the costs and benefits of the different measures.

$$Initial\ Cost\ Difference(\$) = Initial\ Cost_1 - Initial\ Cost_2 \tag{4}$$

$$Life\ Cycle\ Savings(\$) = (Energy\ Cost_1 + Maintenance\ Cost_1) - (Energy\ Cost_2 + Maintenance\ Cost_2) \tag{5}$$

$$Net\ Life\ Cycle\ Savings\ (\$) = LCC_1 - LCC_2 \tag{6}$$

$$Simple\ Payback\ of\ additional\ Costs\ (years) = \left(\frac{Initial\ cost\ difference}{Annual\ operation\ costs_2} \right) \tag{7}$$

$$Life\ Cycle\ Energy\ Saved(kWh) = Energy\ Cosumption\ Lifetime_1 - Energy\ Consumption\ Lifetime_2 \tag{8}$$

Where 1 denotes a traditional product and 2 an alternative solution.

In some steps of the sustainable renovation, a simple LCC-CBA analysis cannot be performed due to the complexity of the decision. When the proposed methodology cannot be applied, the information will be taken from reliable sources available in the literature.

3 Results

It is difficult to find an established Sustainable Renovation Staircase. EPA and DOE (2004) through the Energy Star program states that it is necessary to consider all the aspects of heat flow in buildings as a system to upgrade it maximizing the energy and cost reductions. In this study various measures are considered to improve the climate indoor efficiency.

3.1 Lighting

Lighting devices producers have a wide range of products available. This analysis is made for Philips products available in the Swedish market for the year 2009. For the LCC analysis a discount rate of 2% was assumed according with the Central Bank of Sweden (Riksbank). The price of electricity used for this analysis is 1.11 Swedish Krona (SEK)/kWh (Eurostat 2007).

Table 1. CBA-LCC and energy results for lighting

-	Type of lamp			
CBA indicators (compared with the highest LCC)	Incandescent	Compact Fluorescent (CFL)	Halogen	Light Emitting Diode (LED)
Initial Cost Difference (SEK	0,00	124,00	26,00	377,00
Net Life Cycle Savings (SEK)	0,00	3.909,00	1.476,00	5.089,00
Life Cycle Energy saved (kWh)	0,00	2.160,00	810,00	2.385,00

The result showed that the prices of the devices increase with the lifetime and the energy consumption decreases. In a lifetime the energy saving devices can save up to 2000 kWh.

3.2 Electric Equipment

Cooking Devices. For stoves an analysis was made for Electrolux products available in the Swedish market. The energy prices and the discount rate are the same as used for lighting and will be the same through the whole study. An assumption of four hours per day of use for a resistance range and two hours per use for induction range was taken.

The energy consumption calculations assumed that induction devices need half of the time used by resistance appliances to transfer heat to the food cooked and the heat transfer efficiency is 65% for resistance ranges and 86% for the induction ones, This shows that despite the higher power needed to run induction devices the high efficiency make them cook food faster due that they do not radiate heat but induce an electromagnetic field.

Table 2. CBA-LCC and energy results for stoves

-	Type of range	-
CBA indicators (compared with the highest LCC)	Resistance	Induction
Initial Cost Difference (SEK	0,00	3.090,00
Net Life Cycle Savings (SEK)	0,00	43.151,00
Life Cycle Energy saved (kWh)	0,00	33.327,00

The technology used in heating reduces time and energy consumption, despite of the power. Induction ranges have similar efficiency as gas ranges transferring heat.

Analysis for ovens is developed for Electrolux products available in the Swedish market. All the options have similar dimensions. A daily use was assumed to calculate the energy consumption per year.

Table 3. CBA-LCC and energy results for ovens

-	Type of oven	-	-
CBA indicators (compared with the highest LCC)	Vapor	Pyrolitic	Warm air
Initial Cost Difference (SEK	0,00	-2,125,00	-9.525,00
Net Life Cycle Savings (SEK)	0,00	4.961,00	13.577,00
Life Cycle Energy saved (kWh)	0,00	2.044,00	2.920,00

In this particular case, a radiative oven which heats the air is the most energy efficient. The thermodynamics involved in the different processes can change the way the ovens cooks and operates.

Refrigerators. Three options were analyzed, side-to-side, combi and 2 doors. All devices are from Electrolux and are available at the Swedish market.

The results of the analysis are:

Table 4. CBA-LCC and energy results for refrigerators

-	Type of refrigerator	-	-
CBA indicators (compared with the highest LCC)	Side-by-side	Combi	2 door
Initial Cost Difference (SEK	0,00	-5.860,00	-9.460,00
Net Life Cycle Savings (SEK)	0,00	9.258,00	14.079,00
Life Cycle Energy saved (kWh)	0,00	1.225,00	1.664,00

The thermodynamics involved in a side-by-side refrigerator increase the losses leading to more operations of the compressor to take the heat out of the refrigerator. Two door refrigerators have a configuration that minimizes the energy consumption.

Washing Machines. Assumptions of 6 uses per week at full load were taken for the energy consumption.

Table 5. CBA-LCC and energy results for washing machines

-	Type of washing machine	-
CBA indicators (compared with the highest LCC)	Top loaded	Front loaded
Initial Cost Difference (SEK	0,00	-1.047,00
Net Life Cycle Savings (SEK)	0,00	2.093,00
Life Cycle Energy saved (kWh)	0,00	754,00

Front loaded machines spend less energy than their top loaded counterparts and are cheaper in the market.

Dryers and Dishwashers. It is not possible to perform an analysis for dryers and dishwashers because most models use the same technology and energy consumptions rates. Some operational options can be performed to save energy.

3.3 House Division

For this measure it was not possible to provide an analysis due the complexity of the system. However, some recommendations can be made. Dividing the different spaces and isolate them with doors and walls is a common practice in modern architecture. Keeping doors closed and separate thermal zones can be energy efficient and save large amounts of money per year.

3.4 The Building Envelope

Insulation. The *K* value assumed is 350 kJ*mm/h*m^{2}*$^{\circ}$C, an energy consumption of 2,7x10^{-4} kWh/kJ (Çengel 2003) and a lifetime of 15 years.

Notice that the analysis were performed just to give a picture of the performance of one square meter and does not consider the wall as a system, also the maintenance costs assumed the price of the insulation.

Table 6. CBA LCC and energy results for insulation

-	Thickness of insulation (mm)	-	-	-	-	-	-
CBA indicators (compared with the highest LCC)	45	70	95	120	145	170	195
Initial Cost Difference (SEK	0,00	9,00	19,00	29,00	41,00	51,00	60,00
Net Life Cycle Savings (SEK)	0,00	1.818,00	2.679,00	3.182,00	3.511,00	3.743,00	3.916,00
Life Cycle Energy saved (kWh)	0,00	1.966,00	2.897,00	3.440,00	3.796,00	4.047,00	4.234,00

The economic thickness can be noticed in this analysis, as the life cycle savings increases with the thickness the price also increases. The indicators show that high savings can achieve by increasing the thickness of insulation, but relative savings decrease as the insulation gets thicker.

Windows. Two types of windows were analyzed; 2 glass panes and 3 glass panes. The *U* factors for these windows were taken from DOE (1997) and the windows are available in the Swedish market. The space between the panes is filled with air.

Energy losses were calculated for a temperature difference of 20ºC between the sides of the window. The analysis results are:

Table 7. CBA LCC and energy result for windows

-	Glazing type	-
CBA indicators (compared with the highest LCC)	2 glass	3 glass
Initial Cost Difference (SEK	0,00	1,097,00
Net Life Cycle Savings (SEK)	0,00	2,248,00
Life Cycle Energy saved (kWh)	0,00	6,027,00

The results show that adding a single glass pane can decrease the energy losses in more than 6000 kWh in a lifetime.

3.5 Ventilation

Two ventilation options are considered: forced ventilation and Heat Recovery Ventilation (HRV). Analysis was made for devices and not for the whole system; besides, as input energy needed to run the systems is the same; the analysis was performed considering heat wasted. The calculations refer to devices manufactured by Fläkt Woods available at the Swedish market. A lifetime of 30 years was considered and the costs of installation were added to the maintenance costs.

Assumptions of energy of 24,22 kJ/m^2, a flow of 100 l/s and a 60% rate of heat recovered were made.

The analysis results are presented in Table 8.

Table 8. CBA-LCC and energy results for ventilation

-	Ventilation system	-
CBA indicators (compared with the highest LCC)	Forced Ventilation	Heat-Recovery Ventilation
Initial Cost Difference (SEK	0,00	16.220,00
Net Life Cycle Savings (SEK)	0,00	327.139,00
Life Cycle Energy saved (kWh)	0,00	123.733,00

The introduction of Heat Recovery Ventilation to a system can lead to high energy reductions in the lifetime. Heat Recovery Ventilation saves up to 66% on heat losses.

3.6 Heating System

Heating is a complex system of a house. It is closely linked to the building envelope performance, the ventilation system, the household equipment loads, the size of the living space, the occupation and other loads. The amount of data needed for the measure to be analyzed leaves this chapter of the work out of the boundaries of the model used. However, recommendations based on available literature can be made. Also, the complexity of the analysis of this measurement helps to position it in the sustainable staircase.

3.7 The Sustainable Behavior

All the measures described were analyzed based on the assumption of good household management. There are non-cost measures to save energy but they depend on individual behavior. The sustainable behavior is the way that each individual uses their resources with responsibility saving energy and carbon emissions.

3.8 Defining the Sustainable Staircase

As stated before, it is important to renew and operate a house to improve energy efficiency sequentially. The analysis of the proposed different part of a staircase was made, and the impact of different measures in energy and money savings was determined. This work is based on the concept that reducing energy consumption reduces also resources consumption to produce energy, cutting greenhouse gases emissions (DOE 2008) aiming to three main components of sustainability, economic, social and environmental improvement. In the following analysis for the sustainable staircase, a qualitative analysis is made based on the results of the LCC-CBA analysis performed.

Based on a qualitative analysis, a staircase is proposed:

Table 9. Suggested Sustainable Renovation Staircase

Measure	Result
Lighting	Low cost and high energy savings
-	Easy installation
-	Not dependent of other measures
Electric Equipment	Low to medium costs
-	Medium to high energy savings
-	Easy installation
House division	Is the base to plan how the building envelope, ventilation and heating system will be develop
-	High complexity of implementation. Professional work will be needed
Building envelope	High energy savings
-	Have to be planned after the house division
-	Complexity of installation
Ventilation	High energy savings
-	Interdependent with Heating system
-	Dependent of the building envelope
Heating system	Can lead to high energy savings
-	High complexity of implementation
-	High dependability of other measures

For better results, this sustainable renovation staircase must be accompanied by a sustainable behavior. A good household management can lead to high energy savings without spending capital.

4 Discussion

In elaborating a sustainable renovation staircase, important considerations influenced on the priorities of the different measures. They will shortly be discussed here.

For lighting, it is a common behavior among users to leave lights on when leaving a room rather than turning on and off several times. It is true that a small peak of current is used in the starting event, but this is negligible compared with the energy consumption of a light turned on. In the analysis it was shown that CFLs and LEDs can save up to 2300 kWh of energy in a lifetime. These savings are for one lamp, multiple devices at home can lead to higher energy reductions and money savings. Light bulbs can be changed without other investments than lamp purchase and thus were set as the first option for energy and climate improvement in renovation and operation.

Electrical appliances are present in almost every home in developed countries; washing machines, cooking devices, refrigerators and other devices are used every day. At the end of the lifetime, they do not fail to work, but efficiency decreases, the results therefore give a picture of how to renew when needed. For example, a refrigerator must be replaced every 15 to 20 years and probably most of the people have one at work. To chance it for a more energy efficient one before time will not lead to energy savings, but having different options when buying a new one can be really valuable. These considerations apply to heat, ventilation and air conditioning systems also. It is therefore important for the house owner to perform an analysis of every system change to choose which is the most beneficial for the budget and expectations.

Dividing a house is a hard job, but most people already use spaces differently during seasons. Dividing a house considering the use of the spaces can lead to savings in energy and money, because the occupation defines the amount of heat and insulation needed for one space.

The building envelope is an important component of a shelter system. Windows and insulation play a major role in how and when spaces need to be heated. The appearance of new technologies for windows and the good use of old ones reduce the needs of heating. Windows are important to take advantage of radiative heating. As insulation becomes thicker, energy savings increase, but not at a constant rate; insulation has an optimal point and it is necessary for each house owner to determine it. When money is not a constraint, super-insulation can be considered. There are examples in the literature of houses that can have a human thermal comfort without heating. Low energy houses that use less than 50kWh/m^2 of heating can be found in Northern Germany and Southern Sweden, Like the Lindås project and Värnamo.

Ventilation can vary from an open window to a complex system. In the analysis it was shown that, despite the high prices, HRV leads to substantial savings. The system recycles close to 66% of the wasted heat and uses it to heat the income air saving money and energy.

The heating system must be designed according to the needs of the building, the materials and orientation, the building envelope and the heat flux. Passive houses were built in cold climates, leaving mechanical heating out of the equation. However, it was possible to notice differences between systems when heating is needed, but, the size and components must be defined by the building characteristics.

Behavioral tips can be given by thousands, however, the aware user takes time to read the manufacturer considerations and understand that common sense can be applied to household management.

5 Conclusion

There are many guidelines to sustainable renovation. Most of them, however, deal with specific and often debated parts of sustainable renovation. The analysis made here, showed that a staircase guideline was possible to established, where measures are proposed from the cheapest and easier to implement to the most expensive and complex.

Other perspectives could be considered for the establishment of similar staircases using other tools, or addressing other objectives like improving social communication or integrating neighborhoods. The achievements of this study were to provide a guideline to a sustainable renovation based on energy and monetary savings. A simple communication is an important primary aim of the study, the deepness can be discussed, but a need of a simple language in the academy to reach the stakeholders is needed. By providing simple recommendations of renovation, retrofitting and operating can induce an interest in the reader, leading to a deeper investigation of literature and search for an own, tailored analysis system.

References

1. Barringer, H.P.: A Life Cycle Cost Summary. In: International Conference of Maintenance Societies (ICOMS 2003), Perth (2003)
2. Botta, M.: Toward Sustainable Renovation. Three Research Projects. Dissertation. The Royal Institute of Technology (2005)
3. Çengel, Y.A.: Heat Transfer: a practical approach. McGraw Hill, Blacklick (2003)
4. Ellis, B.A.: Life Cycle Cost. Jethroproject (2007), `http://www.jethroproject.com/Life%20Cycle%20Cost1.pdf` (accessed February 26, 2010)
5. Eurostat Electricity –market prices – half year prices – data until 2007. EUROSTAT (2009), `http://epp.eurostat.ec.europa.eu/portal/page/portal/product_details/dataset?p_product_code=NRG_PC_206_H` (accessed April 07, 2009)
6. Kaklauskas, A., Naimariciene, J., Tupenaile, L., Kanapeckiere, L.: Knowledge base model for sustainable renovation. Modern Buildings Materials, Structures and Techniques. Vilnius Gediminas Technical University. Vilnius (2008)
7. Mørck, O.C., Charlot-Valdieu, C., Nagy, L.: Methods and tools for Sustainable Renovation of Urban Neighbourhoods in the HQE2R Project. The International Solar Energy Society (ISES), Freiburg (2003)
8. Ness, B., Urbel-Piirsalu, E., Anderberg, S., Olsson, L.: Categorising tools for Sustainable Assessment. Ecol. Econ. 60, 498–598 (2006), doi:10.1016/j.ecolecon.2006.07.023
9. State of Alaska – Department of Education and Early Development. Life Cycle Cost Analysis Handbook, Juneau (1999)
10. U.S. Department of Energy (DOE). Insulation Fact Sheet, Washington, D.C. (2008)
11. U.S. Department of Energy (DOE). Energy Savers Booklet, Washington, D.C. (2008)
12. U.S. Environmental Protection Agency (EPA), U.S. Department of Energy (DOE). Energy Star Building Upgrade Manual, Washington D.C. (2004)
13. Watkings, T.: An introduction to Cost Benefit Analysis. San Jose State University (2004), `http://www.sjsu.edu/faculty/watkins/cba.htm` (accessed February 26, 2010)
14. Wierenga, M.: A brief introduction to Environmental Economics. Environmental Law Alliance Worlwide, Eugene (2003)
15. Wrisberg, H.A., de Haes, U., Triebswetter, U., Eder, P., Cliff, R.: Analytical Tools for Environmental Design and Management in a System Perspective. The combined use of analytical tools. Academic Publishers, Dordrecht (1999)

Chapter 11
Solar Collector Based on Heat Pipes for Building Façades

Rassamakin Boris[1], Khairnasov Sergii[1], Musiy Rostyslav[2], Alforova Olga[1], and Rassamakin Andrii[1]

[1] National Technical University of Ukraine "Kiev Polytechnic Institute", Kyiv, 6 Polytechnichna Str., of. 709, Kyiv-56, 03056, Ukraine
Sergey.Khairnasov@gmail.com
[2] Department of Physical Chemistry of Combustible Minerals L. M. Lytvynenko Institute of Physical-organic Chemistry and Coal Chemistry NAS of Ukraine 3a Naukova Str. 79053, Lviv, Ukraine

Abstract. A variety of liquid thermal solar collectors designs used for water heating have been developed by the previous researchers. But the majority of them do not meet the requirements on small weight, easy assembling and installing, versatility, scalability, and adaptability of the design, which are particularly important when they are façade integrated. In order to avoid the above mentioned drawbacks of the liquid thermal collectors the article authors propose to apply to them extruded aluminum alloy made heat pipes of originally designed cross-sectional profile with wide fins and longitudinal grooves. Such solar collectors could be a good solution for building façade and roof integration, because they are assembled of several standard and independent, hermetically sealed and light-weight modules, easy mounted and "dry" connected to the main pipeline. At that, their thermal per-formances are not worse than of the other known ones made of heavier and more expensive copper with higher thermal conductance, or having entire rigid designs. Some variants of the developed solar collectors shaping of the assembled modules for building façade or roof integration are proposed. Variously coloured coatings to the absorbers are developed and made of carbon–siliceous nano-composites by means of sol-gel method. Their optical performances were compared with "anodized black". It is stated that coloured coatings have a good prospects in thermal SC adaptation to building facades decoration, but the works on study and upgrade of their performances should be continued.

Keywords: solar collector, heat pipe, colored coating, façade integration.

1 Introduction

Solar thermal collectors are more and more used as sustainable energy devices, but their design constructions are far from being perfect. Therefore, a lot of attention is dedicated to developments of solar collectors (SC) with Heat Pipes (HPs) as their main parts. HPs application to solar collectors' designs as highly efficient heat absorbing and transferring components makes possible to avoid several disadvantages of

A. Håkansson et al. (Eds.): *Sustainability in Energy and Buildings*, SIST 22, pp. 119–126.
DOI: 10.1007/978-3-642-36645-1_11

conventionally used ones. Thus, solar collectors with HPs as a core [1, 2] ensure very low hydraulic resistance, constant liquid flow, isothermal heat absorbing surface. Nevertheless, ensuring of high quality contact of HP's external surface with heat absorbing surface is one of the main problems for copper made heat pipes applications to solar thermal collectors. For example, in the known evacuated tube solar collectors copper HPs are soldered, welded or pressed to the heat absorbing surface. The most important task is effective heat transfer from the HPs to the working liquid of the solar system loop at minimal hydraulic resistance. Besides that, the majority of traditionally designed SC doesn't meet the requirements on small weight, easy assembling and installing, versatility, scalability, and adaptability of the design. The latter features are particularly important when SCs are façade integrated.

In order to solve the above mentioned problems, the authors propose to apply extruded aluminum HPs to the constructions of solar thermal collectors [3, 4]. The other reasons for using aluminum alloy made HPs in SCs are the following: they are cheaper, lighter, and stronger than copper ones. This kind HPs technology is successfully applied to space engineering and satellites, and, moreover, it is perfectly worked out. Due to the proposed production method cylindrical heat pipe and flat heat absorbing surface are obtained as a single unit. So, it's not necessary to think about the HP's contact with heat absorbing surface that means the design of heat exchanger could be improved fundamentally. To the authors' opinion, such solution can reduce the cost of solar collectors and improve their hydraulic and thermal performance. Beside that, SCs based on aluminum HPs could be a good solution for building façade and roof integration.

2 Solar Collectors Based on Aluminum Heat Pipes

It was proved by the study, that the most optimal design of aluminum HP applied to solar collectors is, so-called, finned HP, where HP's container is made as a single unit with its heat absorbing - releasing surface. There is no thermal contact resistance between them. Such HPs can function within the span of tilt angle values from 0 C deg (horizontal collector attitude) up to 90 C deg (vertical collector attitude).

Special flat heat exchanger is used for heat removal from the HP. Heat is transferred from the condensation zones of the HPs to the circulating coolant by means of contact method.

Experimental flat solar collector prototype (fig. 1) with the following dimensions: 2.13 m - length, 1.0 m - width, 0.085m - height, 1.98 m^2 – area of anodized (black) aperture heat absorbing surface has been manufactured in order to prove the expediency of the HPs application to solar collectors.

Prototype of SC's panel for absorbing heat from solar irradiation consists of eight finned aluminum heat pipes, made of aluminum alloy by means of extrusion. Solar heat is collected by flat longitudinal fins of the HPs and then immediately transferred to the condensation zones of HPs, which contact with small flat heat exchanger. Such effective heat transfer line ensures very low hydraulic resistance of solar collector. Thus highly effective operation of solar heating plants, where many SCs are connected in serious, is possible, as well as electric power consumption of circulation pumps could be lessened.

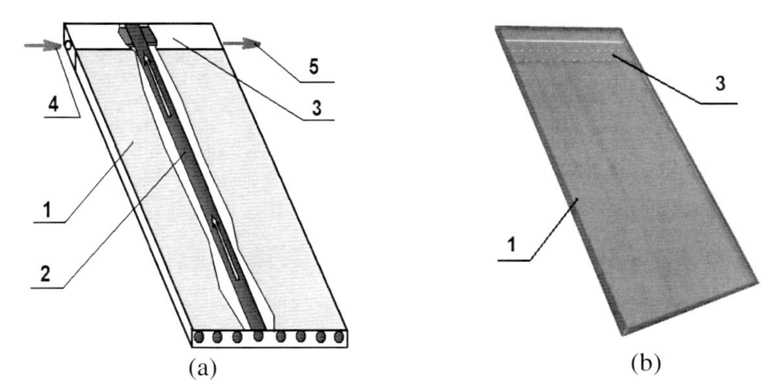

Fig. 1. A flat plate collector based on the aluminum HPs: a - SC structrure, b - general view; 1 – heat absorber made of HPs' fins, 2 – HP container, 3 - heat exchanger, 4 – inlet of coolant,5 – outlet of coolant heated

The merits of this kind SCs are of great importance for photovoltaic powered autonomous solar systems, where low power consumption circulation pumps are used. The result of such technical solution is shorter payback period of solar heating plants.

Besides that, modular solar systems of various shapes and appearance could be created by means of aluminum HPs. Varying numbers and sizes of the HPs like a building kit one can create various kinds of modular systems. These make easy not only façade layout of solar heating systems, but also using them as building construction elements (fig. 2). Solar heating systems with HPs could be upgraded easily by removing or adding HPs, and there is no need to discharge the working fluid from the loop, to stop it, and to recharge.

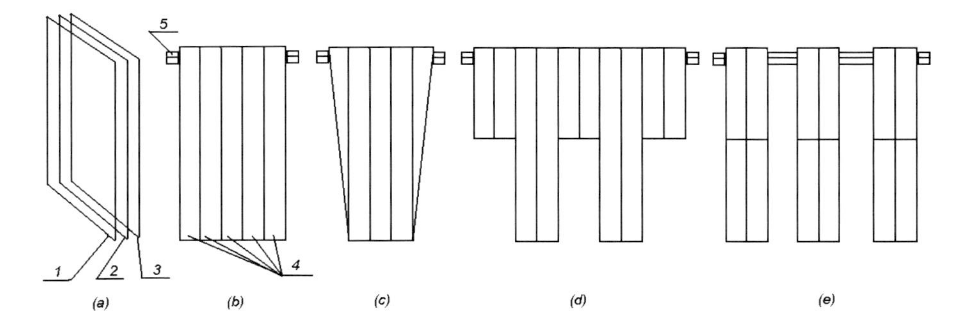

Fig. 2. Variants of flat SCs made of aluminum profiled HPs as an element of building façade shell: 1 – glazing; 2 – absorber, made of HPs ; 3 – insulation; 4 – finned HPs; 5- main pipeline; a - structure; b – full size SC; c – SC of trapezoidal shape; d – SC with HPs of different length; e – several SC modules

In the authors' opinion the cost of 1 m^2 of the developed flat SC can be 1.2...1.4 times less than the cost of the other known flat SC, considering their use as a part of the facade of the building.

Moreover this kind of thermal SCs could be easily installed to the building façade or roof. They might be adapted to the façade outer view and construction by means of their various shaping and assembling of the similar light weight modules (Fig.2). Each aluminum alloy made HP is autonomous hermetically sealed device which might be connected "in dry" to the main pipeline. They need no service and, if it becomes necessary can be simply dismounted without recharging of cooling agent.

3 Efficiency Study of Solar Thermal Collectors

Efficiency study of solar collectors, having aluminum heat pipes as a core, was carried out at two independently functioning full-scale solar heating plants in Ukraine.

The calculated hydraulic resistance of SCs with aluminum HPs is not higher than 70 Pa at coolant flow rate of about 130 liters per hour.

Solar thermal efficiency is defined as a ratio of useful solar heat

$$Q_{use} = G_B \, C_p \, (t_{out} - t_{in}) \tag{1}$$

to total solar heat, which falls to the collector

$$Q_{total} = E_{irr.} \, F_{h.a.}, \tag{2}$$

where, C_p, G_B - heat capacity and flow rate of a coolant in the circulation loop; t_{in}, t_{out} – temperature values of the coolant at the inlet and outlet of the collector manifold; $F_{h.a.}$ – heat absorbing area.

Thermal efficiency of various solar collectors designs versus $x = \Delta t / E_{irr}$ are given in Figure 3, where Δt - is temperature difference between heat absorbing surface and ambient air, E_{irr} - irradiative solar flux value:

1. Calculated values for E=800 W/m^2 for flat SC with non-selective coating. Data were provided by SolarTek.
2. Calculated values for E = 800 W/m^2 were obtained using data from German certification center «DIN CERTCO» (Solar KEYMARK certification, reg. No 011-7S329-F) for flat SC Vitosol 100-F prototype (Co = 0.776, C1 = 4.14, C2 = 0.0145). The prototype has an absorber, covered with selective coating layer.
3. The calculated values were obtained from the empirical formula given in reference [5] for flat SCs with non-selective coating.
4. Experimental data SC based on the aluminum HPs were obtained in the range of total solar irradiation values of 750 W/m^2 to 950 W/m^2.

Analyzing the experimental data, we could say that efficiency of flat solar collector with aluminum HPs is not worse than of well known evacuated tube solar collector, and for small **x** values (for low temperature values of heat absorbing surface and water in the storage tank) are even better. But that is true only for the summer period of the year.

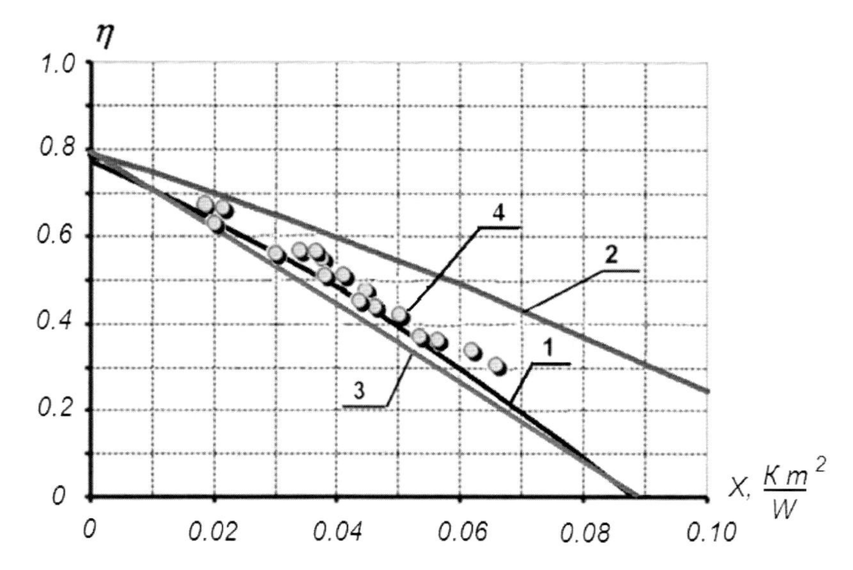

Fig. 3. The efficiency of flat SCs: 1 - calculated values for the SC with heat-absorber covered with non-selective coating, 2 - calculated values for the Vitosol 100-F; 3 - calculated values obtained according to the formula given in reference [5] for flat SC with non-selective coating, 4 - experimental values for the SC based on aluminum HPs

Efficiency of the presented SC model is not worse than of well-known flat SCs designs with non-selective coating as it's evident from the test results, depicted in Figure 3.

Although, the use of selective coating on the heat absorber surface of the SC is important for the temperature level higher than 55 ... 60 ^{0}C and low values of solar irradiation. So, as a rule, seasonal solar plants operate when X values are less than 0.05. In this case, the designed SC based on HPs (see Fig. 3) is not much inferior to the collectors with selective coating, and the difference in efficiency is below 10% within the range of X = 0… 0.05.

Anyway, heat absorbing surface is one of the main constructional element, which can influence greatly on the optical efficiency of SC. Better performance of the absorber means the increased share of absorbed thermal energy. In order to improve the absorber performance, the authors are creating a new kind of selective coating to the absorber. Such coatings consists of the carbon–siliceous nano-composite, which are obtained by means of sol–gel technology [6]. We are developing spectral–selective composite coatings made of carbon *nano*particles dispersed in a dielectric matrix of SiO_2 and NiO.

The main feature of such coating is, the possibility of variously colored selective layers (Fig. 4), and due to that, the use of SCs in building constructions might be expanded. One have a choice in color of the flat SC with higher performance in order to match or decorate the building façade.

Fig. 4. Coatings of various colors

Preliminary results on the optical performances of the developed coatings with the following colors (Fig. 4): black (sample number 1), brown (sample number 2), purple (sample number 3) were obtained. The results (Table 1) are given in comparison with coated "anodized (black)" (number 4 in the Table 1). The samples were made of aluminum alloy 6060 plates sized as 50x50x3 mm and then covered with variously colored experimental selective coatings.

The absorption factor As was determined with the help of FM-59 photometer (Russia) in accordance with GOST 92-0909-69 within the wavelength band of 0.3 ÷ 2.4 μm. The emissivity ε was determined by means of special device - "AE1" emissometer (USA) within the wavelength band of (3 ÷ 30 μm).

Table 1. Measurement results on the optical performances of the experimental selective coatings

No	Absorption factor As	Emissivity ε
1	0.89	0.52
2	0.86	0.58
3	0.76	0.72
4	0.94	0.95

It was revealed by the test results, that the coatings have a relatively high absorption factor As. But the emissivity ε for colored coatings ## 1-3 is also very high, which does not meet the requirement to the high selectivity. For the selective coating emissivity should be minimal.

On the other hand, if we compare the optical coefficients of the colored coating with "anodized (black)" coating, which is used on the experimental sample of the SC (Fig. 1), their values do not differ significantly. Therefore, it can be stated, that efficiency value of SC with applied colored coating would be near to the one of the developed specimen (Fig. 3).

4 Conclusions

In the research the successful application of aluminum HPs to various kinds of solar collectors was proved. Aluminum HPs could be used as elements of solar collectors and serve for efficient heat absorption and transfer to the coolant, which circulates in the solar heating system.

Expediency of aluminum alloy HPs usage in SC's design was verified by operation and tests on the produced prototype of flat SC and full-scale solar heating plant.

Low hydraulic resistance of solar heat collectors with aluminum HPs allows to use successfully many solar collectors in one system or PV controlled circulation pumps. Also, high modularity of SCs is obtained due to using of separate HPs sections. Designs of various shape and size could be assembled from the finned aluminum alloy made HPs. The described solar heating plants could be integrated into building constructions. Moreover, repairing or upgrading of the whole system, doesn't require draining the coolant out of the solar system loop, stopping it or refilling.

Works on the efficiency increase of colored coatings for solar SCs' absorbers are being continued.

Abbreviations:
SC –solar collector
HP - heat pipe

Nomenclature:
t – temperature, Celsius degrees
Q_{use} - useful solar heat, W
Q_{total} - total solar heat , W
$E_{irr.}$ – irradiative solar flux, W/m^2
$F_{h.a.}$ – heat absorbing area, m^2
η - efficiency factor of solar collector
As - absorption factor;
ε - emissivity

Subscripts:
out – outlet
in – inlet

References

[1] Burch, J.: An Overview of One-Sun Solar Thermal Technology. SEET Solar Thermal Seminar, 96 p. (July 26, 2007)

[2] Facão, J., Oliveira, A.C.: Analysis of a Plate Heat Pipe Solar Collector. In: SET 2004 - International Conference on Sustainable Energy Technologies. Nottingham, UK, June 28-30 (2004)

[3] Rassamakin, B., Khairnasov, S., Zaripov, V., Rassamakin, A.: Solar water-heating installation with high-performance types of collectors on the basis of aluminum heat pipes. Nova Tema 3(26), 27–29 (2010)
[4] Baturkin, V., Rassamakin, B., Khayrnasov, S., Shevel, E.: Grooved heat pipes with porous deposit to enhance heat transfer in the evaporator. In: International Conference "Heat Pipes for Space application, Moscow, Russia, September 15-18, p. 11 (2009); Laplace, S., Pascal, W.: Effects of magnetic field to the water droplets. J. Physics 20 (3), 395–399 (2005)
[5] Puhovoy, I.: Experimental researches steam generator flat solar collector with double and triple glass. Renewable Energy, № 2, 16–20 (2005)
[6] Musiy, R., Khairnasov, S.: EMLG-JMLG Annual Meeting. Complex Liquids "Modern trends in exploration, understanding and application", Lviv, September 5-9, c. 121 (2010)

Chapter 12
ICT Applications to Lower Energy Usage in the Already Built Environment

Anna Kramers

KTH Royal Institute of Technology, Department of Urban Planning and Environment and Centre for Sustainable Communications
kramers@kth.se

Abstract. ICT could play a role as a key enabler for decreasing energy usage in buildings. This study identifies, list and describe ICT applications that can reduce energy use in buildings without the need for refurbishment or extensive change. For each area of application, there is a study from the actor perspective to understand who can make use of the different ICT applications to influence energy usage.

Keywords: ICT, buildings, energy efficiency.

1 Introduction

There is a great inertia against change in dense urban structures that are established where the built environment has cultural values and provides historical continuity. Conservation of the already built environment is a potential conflict with measures to drastically reduce energy use and reducing climate impacts. Energy is a limited resource that we need to use more conservatively than we do today. The residential and service sector used 40 per cent of total energy production in Sweden 2010 (Swedish Energy Agency 2011). Residential houses, holiday homes and commercial premises apart from industrial facilities accounted for almost 90 per cent of the sector's energy use. Throughout the life cycle of a building, most energy (approximately 80 per cent) is used during the operational stage (REEB 2009a). Almost 60 per cent of the sector's energy use consists of heating and hot water, the rest is electricity use divided between operating electricity, household electricity and electricity for heating (Swedish Energy Agency 2011).

Information and Communication Technologies (ICT) could play a role as a key enabler for decreasing energy usage in existing buildings. Aside from greenhouse gas (GHG) emissions associated with deforestation, the largest contribution to man-made GHG emissions comes from power generation and fuel used for transportation (GeSI 2008). According to the Smart 2020 report (GeSI 2008) ICT can lead to reductions on a global basis five times the size of the ICT sector's own greenhouse gas emissions, which is 15 per cent of total global emissions in a 'business as usual' scenario for 2020. The greatest role that ICT could play according to the Smart 2020 report is to support improved

A. Håkansson et al. (Eds.): *Sustainability in Energy and Buildings*, SIST 22, pp. 127–135.
DOI: 10.1007/978-3-642-36645-1_12

energy efficiency in buildings and factories that demand power, in power transmission and distribution and in the use of transportation to deliver goods. Global emissions from buildings, including the energy used to run buildings, are in a 'business as usual' scenario estimated to be 11.7 GtCO2e by 2020. It has been estimated that ICT has the potential to reduce these emissions by 15 per cent in 2020 (GeSI 2008).

2 Aim, Objectives and Research Question

The aim of this paper was to explore how Information and Communication Technologies (ICT) solutions can contribute to lower energy usage and green house gas emissions in buildings without the need for refurbishment or extensive change. The first objective was to identify, list and describe ICT applications that can reduce energy use in existing buildings. The second objective was to analyse which actors can make use of the different ICT applications to influence energy usage. The specific research question was "Which ICT applications can reduce energy use in buildings without the need of refurbishment and by whom?"

The focus of this article has been on ICT software, services and hardware that already exist, are pilot products or future visions. It is the usage, management and operation of buildings that were in focus, thus, production technology was not included. The stakeholders that were selected and analysed in this study were the resident, the building owner and the energy provider. A stakeholder in this area that was omitted for practical reasons were district-heating provider, as we only considered electricity providers in this study. However, the ICT solutions presented can be used both to reduce the usage of electricity as well as district-heating and district-cooling. The results refer in the first instance to dense parts of Swedish cities with multi-family houses, but in the second instance to all of Sweden. Much of the analysis is also applicable to other high-income countries, especially in Europe.

3 Methods

A literature review of current ICT solutions, visions and ideas that can reduce energy use in buildings was made. The Web of Science (ISI) and Google Scholar were used to find relevant articles. The following search words were combined: Buildings, Facility, House*, Apartment or Flat; Manag*, Control, Use; ICT, Information Technology, Telecommunication, Application, System, Program, Software; Energy efficiency, Intelligent, Smart; Sustainable, Environmental, Low Carbon, Green. After thorough analysis and comparison of approximately 70 abstracts, some 20 articles were selected for further study. Early in the literature search The European Strategic Research Roadmap to ICT enabled Energy Efficiency in Buildings and Construction (REEB) project result was found and identified as interesting and relevant (REEB 2010b). The REEB project developed a vision for energy efficiency in construction (REEB 2009a), a best practice guide to existing projects (REEB 2010a) and an overview of ongoing research projects (REEB 2009b). The project resulted in a strategic research roadmap for ICT supported energy efficiency in constructions (REEB 2009c).

In their study, the REEB project clustered ICT applications into five areas. These five areas were used as a basis to category ICT applications in this article. The different application areas were 'tools for energy-efficient design and production management', 'intelligent and integrated control', 'user awareness and decision support', 'energy management and trading' and 'integration technologies'. The first area was left out of this study since it was concerned with the design and productions of buildings. Five additional application areas and/or concepts with potential to reduce energy were found in the literature review. The additional application areas/concepts are 'participatory sensing', 'social media technologies', 'persuasive technologies', 'cloud computing' and 'energy management without smart meters'. These areas were integrated into the REEB application areas and shaped five new areas and developed to be: 'Monitor and manage energy use of a building', 'User awareness and influence of behavioural change', 'Energy management and trading on the energy market', 'Collaboration, social media and knowledge sharing' and 'Interoperability, Standards and Cloud computing'.

For each area of application in this study the question of energy decrease *by whom* is further elaborated. Which of the different actors, the resident, the building owner and energy provider can influence energy use by the different ICT applications due to their sphere of influence?

4 Findings

The main findings of this study is a compilation and identification of five clusters of ICT application that can reduce energy use in buildings and a study of the actors perspective on who can influence energy usage by using the different ICT solutions during the building's usage phase.

4.1 Monitor and Manage Energy Use of a Building

ICT will be able to monitor and manage energy use of a whole building and provide better quality of service for all energy usage devices. Information about energy use can be gathered from all energy-using applications such as *heating, ventilation and air-conditioning* (HVAC), air-cooling, lighting, hot water, laundry, dishwashing, cooking and consumer electronics.

Smart meters can record and report energy use information automatically. The information can then be sent on a daily, hourly or real-time basis to all actors involved, so that they can analyse it and take appropriate measures (REEB 2009a). It is also possible to monitor and control residential energy use without the use of smart meters. A network architecture using existing nodes in the home such as internet modem or home gateway is an alternative and cost-effective solution that can be used for the same purpose as a smart meter network architecture. The energy monitoring and control functions are realised over the Internet Protocol (IP) and can therefore be exploited in user applications running on any sort of device (mobile phone, PC, PDA, etc.). Through this technique it is possible to switch on or off the actual residential appliance such as TV, dishwasher, etc. and also get alerts about abnormal energy usage. The benefits from this would be a cheaper and more flexible system, since the smart metering devices are expensive and are proprietary in two ways: The energy provider owns them and they are not built on open standard platforms (Tompros et al. 2009).

Another way of collecting data is through the use of mobile phones. Since mobile phones are capable of sending data (such as images, sounds, location and other information) interactively or autonomously, they could be used as a wireless sensor network if the right architecture is provided (Burke et al. 2006). A participatory network can enable public and professional users to gather, analyse and share local knowledge of the performance of a building. The technologies in the mobile phone platform that can be used are microphones and imagers that can record environmental data, get location and time synchronisation of data and also interact with data from local or remote servers or processors (Burke et al., 2006).

ICT systems are also able to manage local energy production and usage in buildings based on information inside and outside the house. Wireless sensors can measure or detect light, temperature, pressure, noise, humidity, air quality, presence, activity, etc. and communicate information via a wireless network. Intelligent HVAC systems can for example use data from sensors (temperature, occupancy, light, etc.), weather forecast and user behaviour information to optimise processes. Smart lighting gets information from occupancy sensors, daylight and ambient light sensors to turn on lighting in rooms when people are present. Ideally, sensors will need no external energy supply, instead relying on energy harvesting technologies from vibrations, temperature gradients, electromagnetic waves or light via photovoltaic cells.

Through ICT it will be possible to ensure better quality of service, predictive control and maintenance will be possible. Diagnostics will be improved since systems will be able to detect malfunctions in all connected devices. One example is a system that receives information when a light bulb has to be replaced.

4.2 User Awareness and Influence on Behavioural Change

ICT systems can play an important role in making users aware of how much energy they are using and as decision support tools. Visualisation of energy use by interactive interfaces means that energy use can be analysed in real-time. Future display technologies can move away from display screens such as LCD to very thin, flexible, paper-like display materials that can be added for example to walls or furniture's (REEB 2009a). A real-world object could incorporate the interactive interface. An example of a real world object is a flower lamp that changes shape according to the level of energy use (Interactive Institute, 2012). In order to make the lamp more beautiful, the energy use must go down. Real-time pricing systems can be tailored to stimulate behavioural change. By using smart meters it will be possible to send price signals by communicating instantaneous kWh pricing and voluntary load reduction programmes to residents or building owners who can decide how much money they want to spend on energy (REEB 2009a).

Computers As Persuasive Technologies (Captology) is the exploration of computing technology as persuasive, its theory and design intended to influence and change people's attitudes and behaviour through persuasion or social influence, but without coercion (Stanford 2010). Fogg (1999) distinguishes three levels of interaction between computer and user, seeing the computer as *tool*, *medium* and *social actor*, which show how computers can be used in different ways to influence people's attitudes and behaviours. As a *tool*, the computer can provide humans with new abilities, for example allow them to do things more easily. As a *medium*, the computer can convey symbolic content such as text and icons. As a *social actor*, the computer can invoke social responses from users. Mass interpersonal persuasion is a new phenomenon that brings together the power of interpersonal persuasion, where the persuasive experience is distributed from one friend to another with the reach of mass media, where millions of connected people are reached very quickly (Fogg, 2008). Using persuasive ICT services could be a way to achieve step-wise change in energy usage behaviour. The user follows the argumentation and can agree or not agree to the proposals.

4.3 Energy Management and Trading on the Energy Market

A 'smart grid' is a solution with both software and hardware tools that can route electricity more efficiently. The smart grid allows the electricity to go in a two-way direction, which allows real-time two-way information exchange with customers for real-time demand side management (GeSI 2008). The building will be able to act as a key node in the electric grid and its owner/manager will become a participant on the energy market, becoming a 'prosumer' that both produces energy and uses energy (REEB, 2009a). In order to achieve this vision, the four ICT systems described below are needed according to REEB 2009a.

An *Energy Management System (EMS)* handles information about the usage of energy and stores information of production of energy from a certain building. The EMS communicates with all devices and equipment installed within the building that can be remotely monitored by the energy provider. The information generated about the usage and production of energy can be used for district energy management and optimisation and/or trading between buildings.

An *Advanced Meter Infrastructure (AMI) system* is needed that provides two-way communication between the measuring devices in the building and the systems used by the energy suppliers, building owner and residents. The energy provider will send price and reliability signals to the buildings that it serves via the AMI system. The building owner's *Building Automation System (BAS)* will then optimise its loads and energy resources generation based on the pricing and reliability signals it receives. Real time pricing and dynamic energy prices makes it possible for the energy provider to steer energy usage to times when the load is lower or to be able to decrease energy usage from buildings at certain times when the demand for energy is higher from critical facilities in society.

A *Building Information Model (BIM)* will play a major role in storing information about the building in shareable data repositories such as building usage, component installation, assembly and operating rules, maintenance activities and history during the buildings whole life cycle.

4.4 Collaboration, Social Media and Knowledge Sharing

ICT can be used to enable different actors to collaborate and share knowledge, Meetings between different actors can be made more effective using virtual meeting techniques such as multiple collaboration tools, audio, chat screen, whiteboards, etc., with less resource and energy usage as a result. Social media technology is a new phenomenon that has emerged during the last years where the main function is to find and manage contacts with which an individual wants to communicate. Once the connection is established, there are further opportunities to build groups or communities for special interest areas. Groups can be formed around different topics for knowledge sharing and learning. Social media focusing on energy efficiency could be used both for professionals and residents. One example is the service Personal Energy Efficiency Rewards (Efficiency 2.0 2011). It provides a service for energy providers, which are used for the sharing of knowledge and encouragement of energy efficiency. Functions available are such as personalised recommendations and insights, display of the customer's energy savings and comparison of neighbours, goal setting and personalised e-mail tracking customers' progress in lowering their energy bills and where they are using the most energy, issue reward points to customers based on how much energy they save etc.

4.5 Interoperability, Standards and Cloud Computing

In the building and construction sector there are many actors involved, each with different ICT tools and systems for a variety of applications that need to share information. The vision is that different actors will be able to collaborate across ubiquitous and multi-platform ICT tools as if there were no geographical or organisational boundaries (REEB 2009a). Another vision is that each new ICT component in a building should be recognised automatically, which means that each new component can be easily connected and unnecessary components removed from the network (plug-and-play). A similar concept for easy system integration is Service Orientated Architecture (SOA), which makes it easier to integrate or remove different services from time to time (REEB 2009a). A common platform for the 'building operation system' rather than separate hardware as a host for the different software systems is preferable (REEB 2009a). Setting up energy-efficient ICT applications such as smart grids and building management systems for the purpose of lowering energy usage will require servers, computing resources and software applications, all using energy to function. Cloud computing could reduce energy use through more efficient use of hardware and storage of data. Cloud computing is an emerging model for enabling on-demand network access to a shared pool of configurable computing resources. (e.g. networks, servers, storage, applications, and services) that can be rapidly provisioned. This cloud model promotes availability and is composed of resource pooling where the providers computing resources are dynamically assigned according to consumer demand and also by rapid elasticity to quickly scale out and scale in (Mell and Grance 2009).

4.6 Actors Who Can Influence Energy Usage by ICT Applications

There are three main actors involved that can influence energy usage and are able to introduce ICT solutions for energy efficiency during the buildings usage phase. It is the resident, the building owner and the energy supply company. In the following there is an elaboration of what ICT technology, software and services that can be introduced by the three actors due to their sphere of influence.

There are wireless sensor systems that can be installed by *residents* themselves to get informed about the energy usage of different appliances. By using the internet modem or home gateway the information can be sent to an energy usage display in a central place in the apartment or via the telecom network to any device anywhere. Residents can choose displaying device after their taste and needs. There are different devices ranging from real world objects as the flower lamp (Interactive Institute, 2012) to mobile phone applications.

Social media technologies can be used by *residents* to create groups of people with interest in energy reduction among friends and neighbours. A group of residents in a building can build their own groups to encourage each other to find new ways of reducing energy usage via knowledge sharing, games or competitions.

The *building owner* could install and operate a Building Information Model (BIM) system that will be a good support to keep track of all information about the building. Performance management tools could be installed to measure physical parameters such as temperature, light level, air quality etc. or social parameters such as desirable temperature, light level etc. Combining the performance management with a participatory sensing system where residents are able to send data (images, sounds, location etc.) from their mobile phones could lead to enhanced knowledge of the buildings performance.

Wireless sensor systems could be installed by the *building owner* for all electric appliances both within each apartment as well as for the common of the building, such as lighting, washing machines and HVAC (heating, ventilation, air-conditioning and cooling). To encourage the residents to save more energy the building owner can install displays that show the whole energy usage for the building with statistics on if the usage is higher or lower than another day or period. It would also be possible to show the usage of each apartment in the entrance to make it possible for residents to compare.

By subscribing to most of the ICT applications and to store the data centrally using cloud technique the *building owner* can save energy by sharing resources with others and the systems can be optimized more easily.

Energy providers are responsible by law to install smart meters in Sweden to measure energy usage each month since 2009. The smart meters that are installed today are mainly used for the billing system and use one-way communication. To enable the smart grid to work the next generation smart meters systems AMI that provides two-way communication needs to be installed by the electricity providers. AMI makes it possible to connect the energy-measuring device in the building with the energy providers different ICT systems and make them communicate with each other. AMI is a requirement to get demand management to work and to make micro generation (production of energy) from the building possible.

Energy providers could also be the driver of employing different collaboration platform for making collaboration between different actors more efficient. The energy provider could offer knowledge sharing systems for the building owner and for the residents to present solutions to decrease energy usage. There are different social media platforms and Internet knowledge platforms that can be used.

5 Concluding Discussion

There are a number of technical barriers, economic challenges and policy and regulatory restrictions that need to be overcome before many of the ICT applications discussed in this study are common in anyone's home or building. Better knowledge needs to be developed to understand which ICT solutions for energy efficiency in buildings really can make a change in energy use. On the other hand, increasing energy costs and shortage of energy will make the users of energy aware of the need for more efficient solutions.

The lack of standards and insufficient inter-operability is another barrier. If the smart grid is to include real-time electricity pricing, infrastructure needs to be put in place. Development of standards and policies is needed and should be driven by firms that develop the technology and solutions. To justify the usage and investment of different energy saving technologies 'easy to use' and 'plug-and-play' are concepts that must be developed.

New business models need to be developed, as do financial mechanisms to support investment in energy efficiency. Policies needs to be developed and local and national authorities need to be in place in order to understand who could benefit and make a business out of this new and emerging area.

Systems for collecting data on users' energy performance in a way interfere with privacy. In many countries the laws are not keeping pace with technology development, which leaves the user of the new technology without protection. Therefore, the actor that collects the data about energy usage needs to be a trustworthy company that follows rules and regulations and makes agreements with the user of the system who provides personal data.

One of the main sources used to understand ICT's potential to reduce energy usage in buildings is the Smart 2020 report that was commissioned by the Global eSustainability Initiative (GeSI), which represents the IT and Telecom industry. If ICT solutions contribute to a decrease in GHG emissions, this will lead to more demand for these types of solutions and hopefully profitable business for the firms within this sector. Therefore, we can assume that it lies within the ICT sector's interest for the report to present positive figures about the decrease in GHG emissions.

ICT for energy efficiency in buildings is currently mainly focusing on the electricity grid. However, since Sweden has a large district-heating grid, it would also be possible to make energy savings in that system. There is thus huge untapped potential for ICT to support these systems in the best possible way.

References

Burke, J., Hansen, M., Parker, A., Ramananathan, N., Reddy, S., Srivastava, M.B.: Participatory Sensing (2006)

Efficiency 2.0. (2011), http://efficiency20.com (June 20, 2011)

Fogg, B.J.: Persuasive Technologies - Now is your chance to decide what they will persuade us to do—and how they'll do it. Communications of the ACM 42(5) (May 1999)

Fogg, B.J.: Mass Interpersonal Persuasion: An Early View of a New Phenomenon. In: Oinas-Kukkonen, H., Hasle, P., Harjumaa, M., Segerståhl, K., Øhrstrøm, P. (eds.) PERSUASIVE 2008. LNCS, vol. 5033, pp. 23–34. Springer, Heidelberg (2008)

Ge, S.I.: SMART2020: Enabling the low carbon economy in the information age. A report by the Climate Group on behalf of the GeSI-Global e-Sustainability Initiative (2008)

Interactive Institute (2012), http://tii.se/node/5988(June 21, 2012)

Mell, P., Grance, T.: The NIST definition of Cloud Computing, Version 15, 090710 (2009)

REEB. D4.1 Vision for ICT supported Energy Efficiency in Construction, REEB European strategic research Roadmap to ICT enabled Energy Efficiency in Buildings and constructions (2009a)

REEB. REEB Map of European Research Projects and International Initiatives on Energy Efficiency in Constructions (2009b)

REEB. D4.2 Strategic Research Roadmap for ICT supported Energy Efficiency in Construction (2009c)

REEB. ICT-based Energy efficiency in Construction – Best Practices Guide, REEB European strategic research Roadmap to ICT enabled Energy Efficiency in Buildings and constructions (2010a)

REEB. ICT for Sustainable Growth: The European strategic reserach Roadmap to ICT enabled Energy-Efficiency in Buildings and constructions (2010b), ftp://ftp.cordis.europa.eu/pub/fp7/ict/docs/sustainable-growth/fp7-reeb_en.pdf (retrieved April 24, 2012)

Stanford. Stanford University Persuasive Technology Lab (2010), http://captology.stanford.edu/ (September 14, 2010)

Swedish Energy Agency. Energiläget 2010 (2011)

Tompros, S., Mouratidis, N., Draaijer, M., Foglar, A., Hrasnica, H.: Enabling Applicability of Energy Saving Applications on the Appliances of the Home environment 0890-8044/09 2009 IEEE (2009)

Chapter 13
Using Dynamic Programming Optimization to Maintain Comfort in Building during Summer Periods

Bérenger Favre and Bruno Peuportier

Centre Energétique et Procédés, Mines-ParisTech
5 rue Léon Blum, 91120 Palaiseau France
berenger.favre@mines-paristech.fr

Abstract. Being increasingly insulated, new buildings are more and more sensitive to variations of solar and internal gains. Controlling solar protections and ventilation is therefore becoming essential. In this publication, we study the possibility to maintain comfort in the building by controlling either mechanical ventilation for night cooling or solar protections or both of them during hot periods. The proposed energy management is a predictive set of optimal commands issued from a dynamic programming optimization knowing in advance the weather, occupation and internal gains for the next 24 hours. This method is tested on a bioclimatic house situated in Chambery, France with an annual heating demand of 26 kWh/m².

Keywords: Dynamic programming, comfort, mechanical ventilation, shutters, building.

1 Introduction

The main objective for control systems in buildings during summer is to reduce the energy consumption of air conditioning or to maintain comfort using passive cooling. Previous studies concerned the control of solar protections, e.g. [1], [2], ventilation [3], and active cooling [4], [5]. Night ventilation can be used to cool the building structure and a high thermal mass reduces the temperature elevation during the day corresponding to a passive storage [6].The stocking and destocking of heat at the right time requires a predictive controller able to anticipate the variation of ambient temperature, solar irradiance and internal loads. Many advanced control systems are reviewed in [7]. For predictive controllers, a thermal model of the building is required [8], [9], [10]. Due to the time step of this model, a combinatorial optimization is required. Among these methods, the A* [11] and the Branch and Bound algorithms [12] need an assumption of the lower or upper bound not available here. Dynamic programming is then chosen because of its exact optimization character. It has served in a building context mainly for winter operation of the heating system [9],[13]. In this publication, a dynamic programming optimization is used to set up a predictive controller knowing in advance ambient temperature, solar gains and internal loads. This

A. Håkansson et al. (Eds.): *Sustainability in Energy and Buildings*, SIST 22, pp. 137–146.
DOI: 10.1007/978-3-642-36645-1_13

controller serves to maintain comfort in the building by controlling mechanical ventilation during nighttime and solar protection during daytime.

2 Methodology

The main objective of this study is to maintain comfort in the building even in a worst case scenario with an important heat wave. We have first to define what kind of comfort is considered and then to present the thermal model of the building and the optimization method.

2.1 Adaptive Comfort

Comfort is difficult to define. It depends on the direct thermal environment of the inhabitants but also on their bodies' metabolism. It is usually defined as the state of mind which expresses satisfaction with a given thermal environment. Among the many parameters influencing thermal comfort, the adaptive approach states that the indoor comfort temperature depends on the ambient temperature T_C (°C) [14] or its variation over a week [15]:

$$T_C = a\,T_{RM} + b \qquad (1)$$

with T_{RM} the running mean temperature over a week (°C) and a, b are constants determined experimentally in the Smart Controls and Thermal Comfort project [15]. For France, the relation becomes:

$$T_C = 0{,}049\,T_{RM} + 22{,}58 \qquad \text{if } T_{RM} \leq 10°C$$

$$T_C = 0{,}206\,T_{RM} + 21{,}42 \qquad \text{if } T_{RM} > 10°C \qquad (2)$$

with $T_{RMn} = 0{,}8\,T_{RMn-1} + 0{,}2\,T_{MOYn-1}$, T_{MOYn-1} being the daily mean temperature of day n-1 (°C). This is only a thermal comfort without any consideration for air velocity or humidity level. This indoor temperature cannot be maintained at this exact value at all time. The Predicted Mean Vote (PMV) [16] approach is partially used, and we consider that the comfort is maintained if:

$$T_C - 2°C < T_C < T_C + 2°C \qquad (3)$$

T_C corresponds to an operative temperature, accounting for air but also wall surfaces because comfort is influenced by convective and radiative transfer.

2.2 Thermal Model of the Building

The building is modeled as zones of homogenous temperature. For each zone, each wall is divided in meshes small enough to also have a homogeneous temperature.

There is one more mesh for the air and furniture of the zone. Eventually, a thermal balance is done on each mesh within the building:

$$C_{mesh}\dot{T}_{mesh} = Gains - Losses \tag{4}$$

C_{mesh} being the thermal capacity of the mesh, T_{mesh} its temperature, Gains and Losses including heat transfer by conduction, radiation and convection but also possible internal heating and cooling from equipment and/or appliances.
For each zone, repeating equation (4) for each mesh and adding an output equation leads to the following continuous linear time-invariant system [17]:

$$\begin{Bmatrix} C\dot{T}(t) = AT(t) + EU(t) \\ Y(t) = JT(t) + GU(t) \end{Bmatrix} \tag{5}$$

with

- ✓ T mesh temperature vector
- ✓ U driving forces vector (climate parameters, heating, etc)
- ✓ Y outputs vector (indoor temperatures accounting for air and wall surfaces)
- ✓ C thermal capacity diagonal matrix
- ✓ A, E, J, G matrices relating the temperature and driving forces vectors

In order to simulate such a model, it is important to know the occupancy of the building, which defines the emission of heat by inhabitants and appliances, the thermostat set point influencing the heating/cooling equipment, and possible actions regarding ventilation and solar protections. Another important aspect is the weather model, influencing the loss due to heat transfer with the ambient temperature and the gain with solar irradiance. All the data of the occupancy and weather models are contained in the driving forces vector *U*.

A high order linear model is now available. Its state dimension is too large to allow a fast convergence of an optimization algorithm. A reduction method is applied to lower the state dimension and thus to make the algorithm faster

2.3 Optimization Algorithm

The dynamic programming algorithm is a sequential optimization method which gives the optimal set of commands over a period. A state variable describing as well as possible the system is discretised temporally:

$$x(t) = x_t \in X_t, X_t \subset R^{Ne} \tag{6}$$

with *Xt* the set of possible states, *Ne* the dimension of X_t. There is also a control vector with *Nc* dimension:

$$u(t) = u_t \in U_t, U_t \subset R^{Nc} \tag{7}$$

with *Ut* the set of possible control. The state equation at each time step *t* is then:

$$x(t) = x_t, x(t+1) = f(x(t), u(t), t) \tag{8}$$

We now define a value function v_t which is the cost to go from *x(t)* to *x(t+1)* :

$$v_t(x_t, x_{t+1}), x_{t+1} \in \Gamma_t(x_t) \tag{9}$$

Γt being the set of possible state variable at time *t*. The cost function is then the sum of all the value functions at each time step:

$$V_0^t = \sum_{j=0}^{t-1} v_j(x_j, x_{j+1}) \tag{10}$$

This equation gives us a set of control to go from x_0 to x_t. The optimization seeks to maximize or minimize the following objective function over *N* time steps:

$$J = Max[V_0^{N-1}] \tag{11}$$

Bellman's principle of optimality is applied to accelerate this optimization by breaking this decision problem into smaller sub-problems:

> An optimal policy has the property that whatever the initial state and initial decision are, the remaining decisions must constitute an optimal policy with regard to the state resulting from the first decision [18]

Then (11) becomes:

$$J = Max[V_0^{N-1}] = Max(v_0(x_0, x_1)) + Max(V_1^{N-1}) \tag{12}$$

To resume, we have to find a set of command $U_N = (u_0, u_1, \ldots, u_N)$ maximizing (12) from a system described in (8) with constraints on the state variable (6) and on the controls (7).

3 Application on a Case Study

3.1 Building Description

The building under study is a French single-family house. The actual building is an experimental passive house part of INCAS platform built in Bourget du Lac, France. The house has two floors for a total living floor area of 89 m². 34% of its south facade surface is glazed while the north facade has only two small windows. All the windows are double glazed except for the north façade with triple glazed windows. The south facade is also equipped with solar protections for the summer period. The external walls are made with a 30 cm-thick layer of concrete blocks and the floor is composed of 20 cm reinforced concrete. The insulation is composed of 30 cm of glass-wool in the attic, 15 cm in external walls and 20 cm of polystyrene in the floor. According to thermal simulation results using Pléiades+COMFIE [17], the heating load is 26 kWh/(m².year) which is typical for such type of house.

3.2 Optimization Parameters

The chosen state variable is the total energy of the building. This energy is calculated as follows:

$$E = \sum_{i=1}^{nbr\_meshes} E_i = \sum_{i=1}^{nbr\_meshes} C_i T_i \tag{13}$$

with E the total energy of the building, C_i the thermal capacity of the mesh i, and T_i the temperature of the mesh i. The model of the building is mono-zonal, there is only one control for the whole building.
The optimization is done over 14 days, a very hot week for a worst case scenario and a normal summer week after (Fig.1), the simulation includes also a week initialization period. The occupancy of the building is a typical four people family. The building is non-occupied only during the working days from 8.00 a.m. to 17.00 p.m.. Each occupant emits 80 W due to his metabolism, there are also small internal loads from appliances during occupied hours.

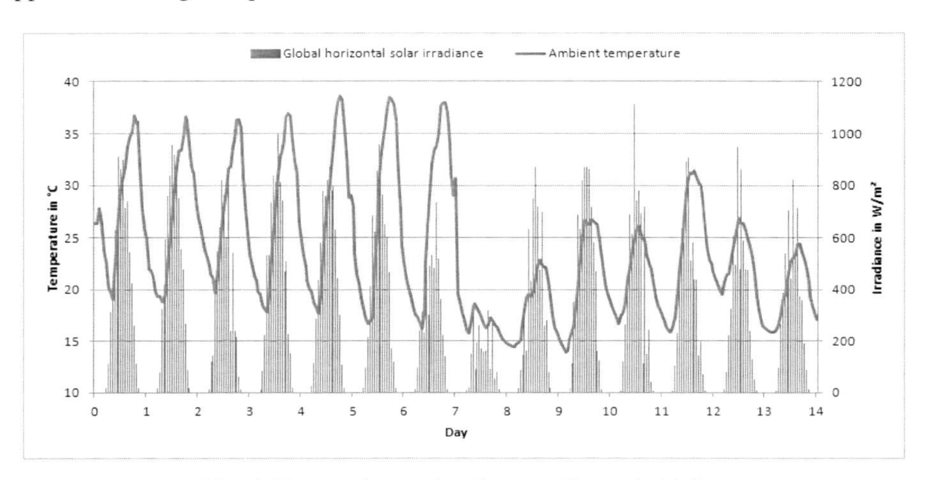

Fig. 1. Two weeks weather data used for optimization

4 Results

4.1 Mechanical Ventilation Controller

The mechanical ventilation controller is first optimized, the roller blinds being open at all time during the two weeks. The air flow rate can vary between 0.6 and 6 ach (air change per hour) with no heat recovery in summer. The mechanical ventilation consuming electricity, the objective is to maintain comfort while minimizing its use, the value function is then:

$$v_t(E_t, E_{t+1}) = abs(T_{\text{in}} - T_c) + \cos t * \frac{vent}{100} \tag{14}$$

with *vent* the control in percentage of the maximum ventilation, T_c the comfort temperature and T_{in} the indoor temperature. The results for *cost* = *1* are presented in Fig.2.

At the beginning of the very warm week, the indoor temperature is under the value of the comfort temperature, then the mechanical ventilation is operating during the night. The comfort condition (3) is always maintained during this very warm week. During the second week, the mechanical ventilation is more often used but at lower value. During the two first days normal ventilation is sufficient to follow the decrease of the comfort temperature. Then night ventilation allows cooling the thermal mass of the building in order to maintain comfort during daytime. Without a regulation, the night cooling is very limited because of the constant air flow rate value (0.6 ach), and the comfort condition (3) is maintained but with a high temperature.

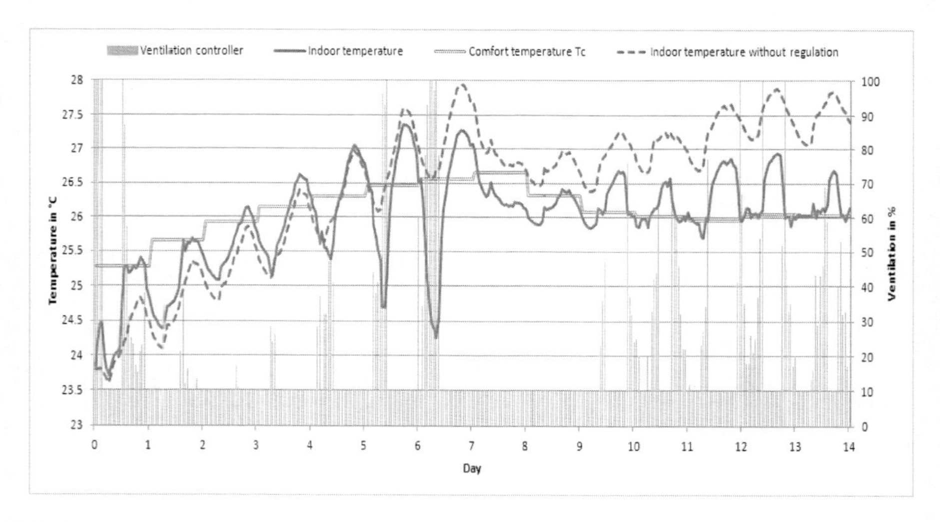

Fig. 2. Variation of indoor temperature and of the mechanical ventilation controller over the two considered weeks

The electricity consumption is reasonable because the average flow rate over the period is 1.2 ach. If the objective function isn't minimizing the utilization of mechanical ventilation (cost = 0), the air flow rate over the period is 1.9 ach (Fig. 3.).

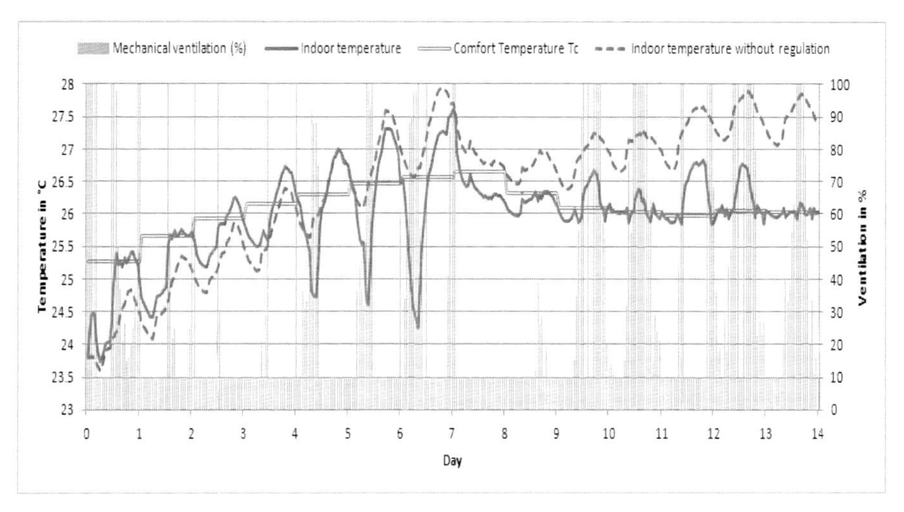

Fig. 3. Variation of indoor temperature and of the mechanical ventilation controller with no cost of use of ventilation

Figure 4 presents the results relating comfort and ventilation depending on the cost of use of ventilation. The more the weight is put on minimizing the use of ventilation, the bigger is the thermal discomfort. Further studies will concern Pareto frontiers and natural ventilation.

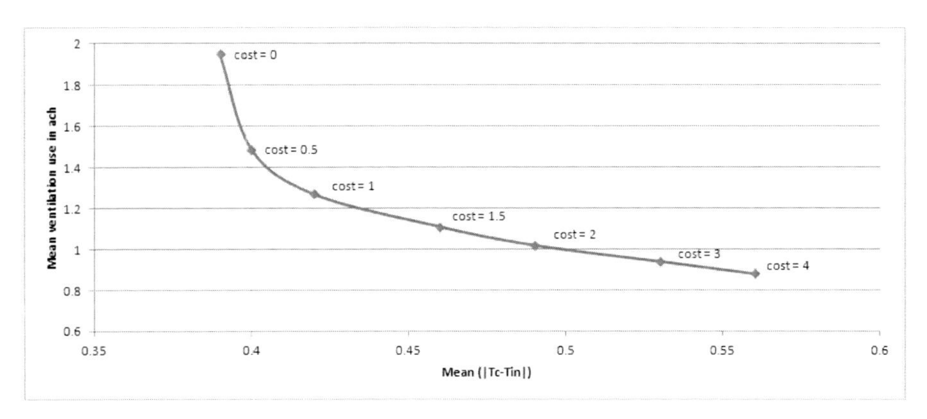

Fig. 4. Balance between ventilation use and thermal comfort depending on the cost of use of ventilation

4.2 Solar Protection Controller

The roller blind control is now studied, considering a constant 0.6 ach mechanical ventilation. The opening interval is from 0% to 100%. In the value function the electricity consumed for opening or closing the roller blades is supposed negligible.

$$v_t(E_t, E_{t+1}) = abs(T_{\text{in}} - T_c) \quad (15)$$

The results of this optimization are presented in Fig. 5. :

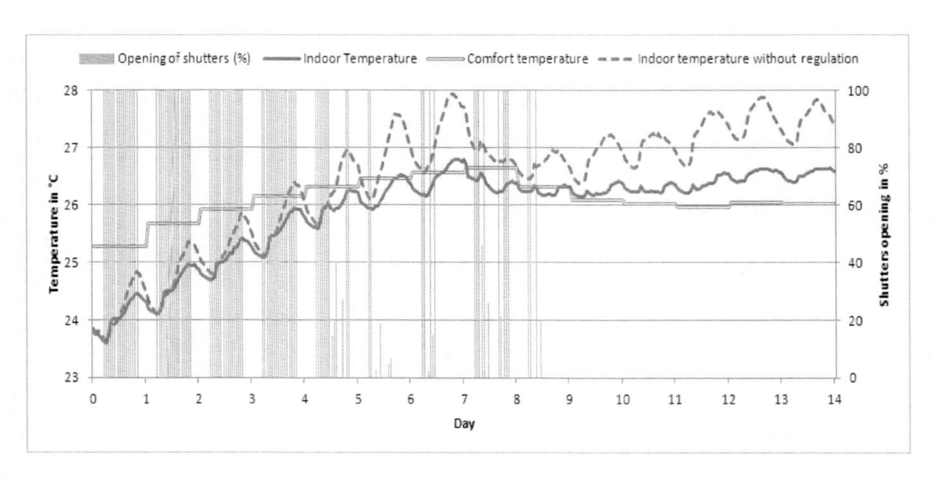

Fig. 5. Variation of indoor temperature and of the shutters controller over the two considered weeks

The roller blinds reduce efficiently the solar gains; therefore the temperature variation is reduced during the day compared to the ventilation control. But during the second week, even if the roller blinds are always closed, the indoor temperature is always higher than the comfort temperature because no important night cooling is possible. This controller allows reducing the amount of gains but can't clear it off once in the building. Still, the comfort condition (3) is always maintained.

4.3 Controlling Solar Protection during the Day and Mechanical Ventilation during the Night

The mechanical ventilation is 0.6 ach during the day and it is controlled as soon as the global solar irradiance is under 200 W/m², globally at night. Solar protection is controlled during the day and closed at night. The optimization is done using the value function described in (14). The main goal is to increase the comfort condition even further while decreasing the use of mechanical ventilation (Fig.6).

Combining of the two controllers is very effective. Except for the first day the difference between the indoor temperature and the target comfort temperature is under 1°C. During the second week this difference is even under 0.5°C. Solar protection control is the most used because there is no cost for operating it. The operating of mechanical ventilation is minimized, with a mean value of 0.72 ach.

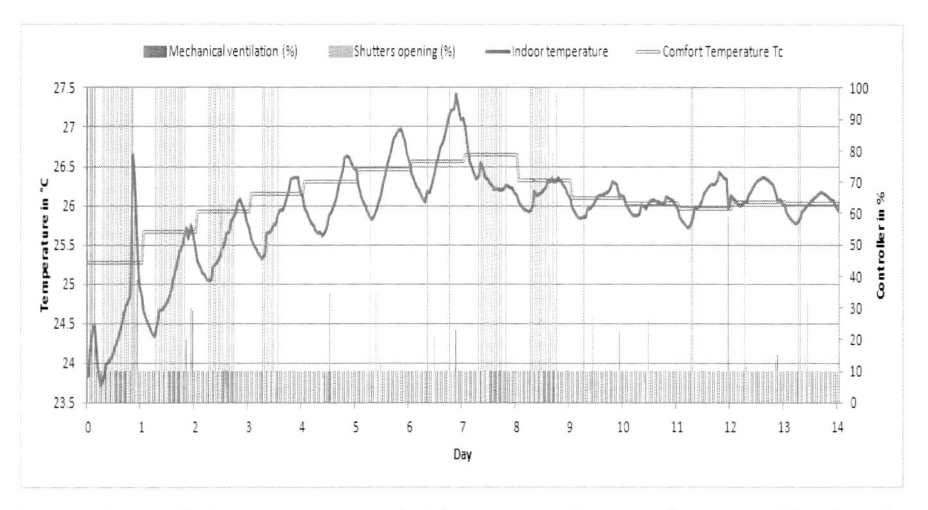

Fig. 6. Variation of indoor temperature and of the two controllers over the two considered weeks

5 Conclusion

Dynamic programming optimization has been used to study the control of ventilation and solar protections in a low energy building. A control strategy can be identified to optimize comfort and minimizing energy consummation. Further studies will address natural ventilation.

Acknowledgments. This work has been supported by the chair ParisTech-VINCI "Ecodesign of building and infrastructure" and by the ANR project SIMINTHEC.

References

1. Nielsen, M.V., Svendsen, S., Jensen, L.B.: Quantifying the potential of automated dynamic solar shading in office buildings through integrated simulations of energy daylight. Solar Energy 85(2010), 757–768 (2011); American Control Conference (2010)
2. Lollini, R., Danza, L., Meroni, M.: Energy efficiency of a dynamic glazing system. Solar Energy 84, 526–537 (2010)
3. Chahwane, L., Stephan, L., Wurtz, E., Zuber, B.: On a novel approach to control natural and mechanical night ventilation. In: 12th Conference of International Building Performance Simulation Association (2011)
4. Mathews, E.H., Arndt, D.C., Piani, C.B., van Heerden, E.: Developing cost efficient control strategies to ensure optimal energy use and sufficient indoor comfort. Applied Energy 66, 135–159 (2000)
5. Le, K., Tran-Quoc, T., Sabonnadière, J.C., Kieny, C., Hadjsaid, N.: Peak load reduction by using air-conditioning regulators. In: Electrotechnical Conference, the 14th IEEE Mediterranean (2008), doi:10.1109/MELCON.2008.4618519

6. Braun, J.E., Montgomery, K.W., Chaturvedi, N.: Evaluating the Performance of Building Thermal Mass Control Strategies. HVAC&R Research 7, 403–428 (2001)
7. Dounis, A.I., Caraiscos, C.: Advanced control systems engineering for energy management in a building environnement – A review. Renewable and Sustainable Energy Reviews 13, 1246–1261 (2009)
8. Oldewurtel, F., et al.: Energy Efficient Building Climate Control using Stochastic Model Predictive Control and Weather Predictions
9. Morel, N., Bauer, M., El-Khoury, M., Krauss, J.: Neurobat, a predictive and adaptive heating control system using artificial networks. International Journal of Solar Energy 21, 161–201 (2001)
10. Freire, R.Z., Oliveira, G.H.C., Mendes, N.: Predictive controllers for thermal comfort optimization and energy savings. Energy and Buildings 40, 1353–1365 (2008)
11. Hart, P.E., Nilsson, N.J., Raphael, B.: A Formal Basis for the Heuristic Determination of Minimum Cost Paths. IEEE Transactions of Systems Science and Cybernetics 4, 100–107 (1968)
12. Narendra, P.M., Fukunaga, K.: A Branch and Bound Algorithm for Feature Subset Selection. IEEE Transactions on Computers 6, 917–922 (1977)
13. Nygard Ferguson, A.M.: Predictive thermal control of building systems. Thesis from Ecole Polytechnique Federale de Lausanne (1990)
14. Humphreys, M.A.: Standards for thermal comfort: indoor air temperature standards for the 21st century (1995)
15. Mc Cartney, K.J., Nicol, J.F.: Developing an adaptive control algorithm for Europe. Energy and Buildings 34, 623–635 (2002)
16. Fanger, P.O.: Calculation of thermal comfort: Introduction of a basic comfort equation. ASHRAE Transactions (1967)
17. Peuportier, B., Sommereux, I.B.: Simulation tool with its expert interface for the thermal design of multizone buildings. International Journal of Sustainable Energy 8, 109–120 (1990)
18. Bellman, R.: Dynamic Programming. Princeton University Press (1957)

Chapter 14
Assisting Inhabitants of Residential Homes with Management of Their Energy Consumption

Michael Kugler[1], Elisabeth André[1], Masood Masoodian[2], Florian Reinhart[3], Bill Rogers[2], and Kevin Schlieper[3]

[1] Human Centered Multimedia
University of Augsburg
Augsburg, Germany
mail@michaelkugler.de, andre@informatik.uni-augsburg.de
[2] Department of Computer Science
The University of Waikato
Hamilton, New Zealand
{masood,coms0108}@cs.waikato.ac.nz
[3] Bottled Software GmbH
St. Leon-Rot, Germany
{florian,kevin}@bottledsoftware.de

Abstract. Although there are already a range of energy monitoring and automation systems available in the market that target residential homes, mostly with the aim of reducing their total energy consumption, very few of these systems are directly concerned with how those energy savings are actually made. As such, these systems do not provide tools that would allow users to make intelligent decisions about their energy usage strategies, and encourage them to change their energy use behaviour. In this paper we describe a system designed to facilitate planning and control of energy usage activities in residential homes. We also report on a user study of this system which demonstrates its potential for making energy savings possible.

Keywords: Residential energy consumption, efficient energy usage, energy usage management, smart homes.

1 Introduction

Energy consumed in private homes, particularly in developed countries, can constitute a large percentage of total energy consumption (e.g. 29% in Europe [6]). This, combined with the fact that the worldwide energy consumption is increasing, has led to development of a range of technologies in recent years directed at reducing energy use in private households. These technologies aim to assist people by providing them with information about their energy usage, with the hope that this information will lead to them reducing the amount of energy they consume. However, the problem of saving energy use in residential homes

A. Håkansson et al. (Eds.): *Sustainability in Energy and Buildings*, SIST 22, pp. 147–156.
DOI: 10.1007/978-3-642-36645-1_14

is not just about reducing the total amount of energy consumed, but it is also about the better management of energy usage particularly at peak times (see [3]), which require a short-term energy production much higher than the daily average, as well as providing larger grid capacity by energy suppliers .

Furthermore, in an attempt to reduce reliance on fossil-fuel, more emphasis is currently being put on generation of electricity through renewable resources such as hydro, wind, and solar energy. Unlike traditional fossil-fuel or nuclear power plants, most renewable energy generation options are less predictable, and require more careful management.

Fortunately, although there are many tasks in a typical household which require use of energy, not all these tasks need to be carried out at specific times, but rather during certain time frames. For example, an electric car can generally be charged any time between the afternoon when the user arrives home and the next morning when the user leaves for work. Better management of these types of tasks could lead to use of energy when it is more readily available (e.g. off peek or when renewable energy is available).

To support the process of improving the management of energy use in private homes, it is important to identify where and how energy is being used by different devices in a household, and then provide tools which assist individual residents with changing their energy use behaviour.

Although there are now some home automation and energy monitoring devices in the market, most of these technologies only aim to allow users to reduce their total (e.g. daily, weekly, or monthly) energy use. Using these systems it is often very difficult to find out when and how much energy each device in a household is using, or if there are more than one person living in a house, how much energy each person consumes.

In this paper we briefly introduce a system we have developed, called USEM (Ubiquitous Smart Energy Management), which allows monitoring energy consumption by individual devices and inhabitants of private homes [8]. Here we focus more specifically on those mechanisms provided by USEM for controlling devices and scheduling tasks that consume energy, which would allow making energy savings in residential homes in an intelligent manner. We also briefly discuss a user study of USEM along with its related findings.

2 Energy Monitoring and Home Automation Systems

Existing technologies designed to assist users with saving energy in residential homes can be divided into two categories, those for monitoring energy consumption, and those for providing home automation capabilities.

Energy monitoring systems such as Current Cost[1], the Energy Detective[2] and Wattson[3] generate energy consumption statistics, which users can then analyse to see where and how energy could be saved in the future. Although these systems

[1] http://www.currentcost.com
[2] http://www.theenergydetective.com
[3] http://www.diykyoto.com

are helpful to some extent, they often only give total energy use measurements for the entire household, rather than providing details at the individual appliance level. Furthermore, these systems do not actively control appliances to save energy, neither do they make any suggestions to the user as to how energy savings could be made.

Home automation systems, on the other hand, are designed to actively control household appliances, as well as any other possible sources of energy use. Examples of such systems are HomeMatic[4], Gira[5] and Intellihome[6]. Home automation systems are usually equipped with sensors (e.g. for temperature, motion detection, etc.), which are monitored by the system to allow it to react to environmental changes (e.g. someone enters a room, a window is left open, etc.) by controlling various actuators (e.g. for opening and closing doors and windows, turning lights on and off, etc.).

Unfortunately, despite their potential benefits for saving energy, home automation systems are not widely used in private homes. One of the main reasons for this is because home automation systems need to be able to control off-the-shelf devices. However, at present there are no widely adopted home automation communication standards (e.g. KNX[7]). Due to this lack of communication standards most appliances cannot be effectively controlled by home automations systems, other than perhaps just turning them on or off. Although this level of control is sufficient for simple devices (e.g. lights), it is less than ideal for most devices which have programmable functionality (e.g. washing machines, dishwashers, ovens, etc.).

The other reason why home automation systems are not widely used is probably due to the complex configuration of such systems. Most people do not want to, or simply cannot deal with complex system set-up and operation required by home automation technology.

Several recent attempts have been made to overcome the shortcomings of energy monitoring and home automation systems by combining these two types of technologies to allow more intelligent monitoring and management of energy use in residential homes. These types of combined systems are viewed as being particularly useful for off-the-grid households where better management of energy use is much needed [2].

One such system is AIM [4] which combines measurement and automation approaches to reduce energy consumption. However, AIM requires the users to install specific hardware which can only control specific types of supported devices, and therefore, cannot be used to control existing off-the-shelf appliances.

A system that aims to support off-the-shelf devices is the Energy Aware Smart Home [7], designed to make energy usage more transparent by providing real-time energy consumption information at the device level. To measure energy consumption for each device, "Ploggs" socket adapters are used to transmit

[4] `http://www.homematic.com`
[5] `http://www.gira.de`
[6] `http://www.intellihome.com`
[7] `http://www.knx.org`

energy consumption data wirelessly via Bluetooth or ZigBee[8] to a server. The consumption data is displayed to the user through a stationary interface, as well as a "UbiLense" augmented reality technology. The system focuses more on providing energy consumption data rather than allowing interaction with devices.

A similar system has been developed by Intel® as part of their Intelligent Home Energy[9] management platform. This proof-of-concept system uses the ZigBee technology to communicate wirelessly with home appliances that are plugged into remote controllable power sockets. Although the system allows monitoring energy consumption, and some device control, it does not provide scheduling functionality to automatically manage execution of energy consuming household tasks.

Yet another example of a system that combines energy measurement and control hardware is the Home Energy Saving System (HESS) [5]. However, HESS only aims to reduce standby power consumption by switching off devices when they are not in use.

In summary, most of the above mentioned approaches either require the installation of special hardware, or provide very little control over energy using devices, other than perhaps turning them on and off, with no task scheduling functionality to allow automatic management of their use in an energy efficient manner.

3 Requirements of an Automatic Control and Scheduling System

As mentioned earlier, most existing residential home technology for energy usage monitoring and automation aim to assist users with reducing their total energy consumption, without caring too much about how those savings are made at the individual appliances level. This is because in most countries electricity is supplied to domestic users at a single rate, independent of when the electricity is used, or how it is generated. Although some differential rates (e.g. day and night time rates) may exist, these apply to only a certain range of energy using tasks (e.g. heating, hot water). This is however changing, as electricity suppliers aim to reduce peak time usage which is costly in terms of generation and the grid capacity they have to support. With the introduction of "smart metering" in an increasing number of residential homes in developed countries, users are given the option of changing their usage patterns to consume electricity when it is cheaper to supply.

Furthermore, as our reliance on renewable sources of energy, with their inherent variability, increases it becomes crucial to manage the use of energy intelligently so that energy is used when it is more readily available, and savings are made when it is not. This is particularly important for houses that are off-grid

[8] http://www.zigbee.org

[9] http://www.intel.com/p/en_US/embedded/applications/energy/energy-management

and/or are reliant on their own renewable small-scale electricity generation (e.g. wind turbine, solar panel, etc.).

In this paper we propose an intelligent system that allows users to specify their preferences, in the form of *rules*, *tasks* and *levels*, which can then be used to automatically control how and when home appliances are used by scheduling household energy usage activities. To do this, we identify two categorises of appliances:

- **Regular devices** are used to perform specific individual tasks that have a set duration. Most home appliances (e.g. dish washer, TV, oven, etc.) fall into this category, and can have two or more modes of operation. Simple devices are either on or off (e.g. lights), while more advanced devices have many different modes of operation (e.g. a washing machine can be on, off, on standby, in colour wash or warm wash mode, etc.).
- **Continuous devices** operate more automatically without much manual control by the user. Devices such as a hot water heater, air conditioner, refrigerator, etc. fall into this category. These devices do not generally have an operating mode like regular devices do, but rather try to automatically maintain a value (e.g. water or air temperature) within a user specified range.

For regular devices we define *rules* and *tasks* as:

- **A rule** has a set of *conditions*, that once met, allow the rule to be executed to cause some *effect*. There are a range of conditions that the system can check (e.g. weather conditions, presence of people, energy prices, etc.). The effect of a rule being executed is usually to switch one or more regular devices to a specified operating mode. Rules allow users to configure the system to automatically perform *reoccurring tasks* (e.g. turn the lights off if nobody is at home).
- **A task** is a *one off activity* that the user gets the system to *schedule* and *perform*. Tasks have a number of variables which the user can specify. For example, the operating mode in which the device should run, for how long, etc. As with rules, execution of tasks can also be dependant on certain conditions being met (e.g. the washing is done only if there is someone at home).

For continuous devices, which automatically maintain a value within a user specified range, rather than defining rules and tasks we define *levels*.

- **A level** is similar to a rule, in that it allows setting conditions for continuous devices. In addition to all the conditions available for rules, a time condition can also be set for a level. This makes it possible, for instance, to set a level which tells the system to reduce heating to a lower temperature to if nobody is at home on weekdays between 8a.m and 4p.m.

4 Controlling Appliances and Scheduling Tasks in USEM

We have developed a prototype system called USEM [8] to allow monitoring and control of energy usage in residential homes. The system provides interfaces for smart phones and tablets which can be used to control and monitor energy use while mobile. The system also provides a web interface for setting rules, tasks, and levels for controlling regular and continuous devices. In this section we describe the web interface of USEM.

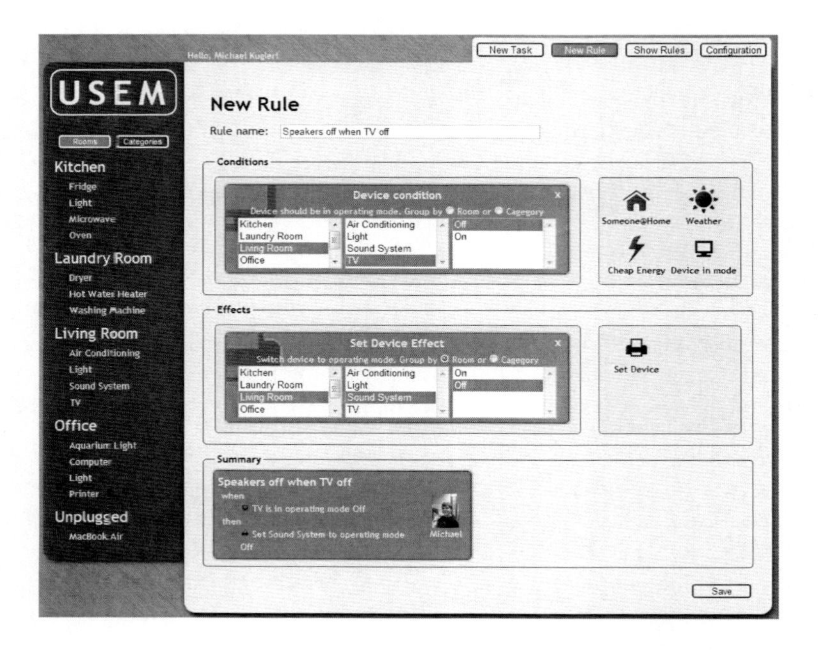

Fig. 1. The Rule Editor of the web interface component of USEM

Figure 1 shows the Rule Editor component of USEM which can be used to create new rules, or edit the existing ones. The interface allows the user to choose conditions and effects from a set of templates (shown on the right), for devices that are connected to the system. After adding conditions and effects the user can then set further parameters (e.g. a temperature range for the weather condition). The interface displays a summary of the rule which will be created at the bottom. Users can also view all the rules they have created. To make finding existing rules easier, the rules are grouped by their conditions (e.g. Weather, Someone@Home, etc.), and the parameters for these conditions (e.g. rainy, cloudy, etc. for the Weather condition).

The Task Editor component of USEM, shown in Figure 2, enables users to create new tasks, or edit the existing ones. Users can choose a device, and an affiliated operating mode, for which they want to create a task. They can then specify the time boundaries for the task (e.g. when the tasks should be finished by, a time before which the task should not start, or a timespan in which the task should be executed). Users can also add conditions to the task in a manner

Fig. 2. The Task Editor of the web interface component of USEM

similar to that of the Rule Editor. The web interface provides a summary of the task which is going to be created at the bottom.

Once a task has been created, USEM attempts to schedule the task, based on the conditions set by the user, including the time boundaries, etc. The user can view all the scheduled tasks (see Figure 3) using a visualization based on glyphs [1]. The user defined time boundaries of a task are shown in a gray box in the background, and the actual planned execution time by the system is shown as a colored box in the foreground. A color range of green to red is used to represent the amount of renewable energy which is likely to be available when the task starts, based on the estimation provided by USEM. The amount of energy required for the task is shown by a speedometer icons (e.g. less than 0.5kWh). These are obtained from a device profile database maintained by the system.

USEM uses the JBoss Drools Expert[10] framework to manage the execution of all its rules. The task scheduling and optimization is done by using the JBoss Drools Planner[11] component. Rules and tasks are directly converted into the Drools Rule Language (DRL) format, readable by these components.

---

[10] `http://www.jboss.org/drools/drools-expert`

[11] `http://www.jboss.org/drools/drools-planner`

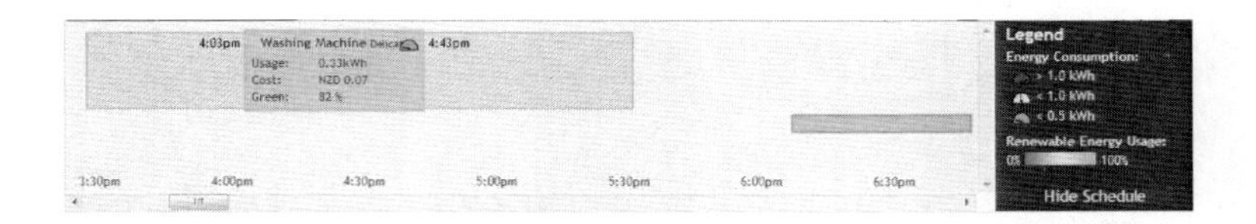

Fig. 3. An example schedule with detailed information shown for a selected task

5 User Evaluation

We conducted a user study to evaluate the effectiveness of the various interfaces provided by USEM. In particular we aimed to see how easy the users would find the concepts of rules and tasks, and the process of setting up rules and scheduling tasks in USEM.

Twenty people participated in this study. They were 15 males, 5 females; 11 were students, 2 academics, 1 teacher and 6 others; aged 20 to 62 years old with an average age of 35. All of the participants used computers on a daily basis.

5.1 Methodology

Each study session started with a short tutorial describing the functionality of the interface to be used. The participants were then given sufficient time to familiarise themselves with the system, before performing the actual study tasks. Each task was described in a couple of sentences, followed by several steps which the participants had to perform using the system.

The participants carried out four tasks using the web interface. Task 1 was to create a new rule; Task 2 was to set up a level for a continuous device; Task 3 was to create a new task for a regular device; and Task 4 was to interpret a schedule with several upcoming tasks.

After performing each task, the participants were required to answer two questions:

1. How easy was it to perform this task?
2. How useful do you find the functionality?

At the end of the session the participants were asked to complete a final questionnaire, with the following questions:

1. How easy would it be for you to adapt to using USEM for tasks, where you do not have to change your daily routine very much? (e.g. create tasks for doing the laundry, instead of just switching the washing machine on manually?)
2. Would you adapt your daily routine in order to use more renewable energy? (e.g. start cooking dinner an hour later?)
3. How useful do you find the overall system with regard to efficient energy usage?

A Likert scale of 1 (not easy/not useful) to 7 (very easy/very useful) was used for each of the questions used after the tasks and in the final questionnaires.

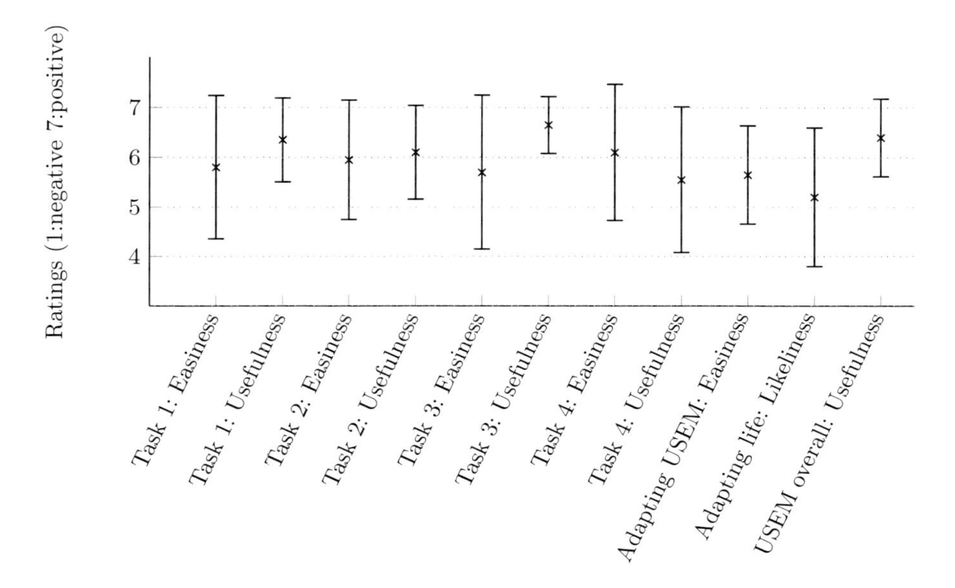

Fig. 4. A summary of the results of the questionnaires used in the study

5.2 Results and Discussion

A summary of the results of the questionnaires used in the study is shown in Figure 4. Generally the participants rated all the four tasks as being easy to perform, with an average rating between 5.7 (for Task 3) and 6.1 (for Task 4). In terms of the usefulness of the functionality provided by USEM, the ratings were once again very positive, ranging on average between 5.7 (for Task 4) and 6.7 (for Task 3).

In response to the first question of the final questionnaire the participants gave an average rating of 5.7. This means that they believe it would be easy to adapt to using USEM for scheduling their tasks, rather than just doing them when they want a task to be done (e.g. asking USEM to schedule and do the laundry when it is cheaper rather than doing it right now). Of course unless a long-term evaluation of USEM is carried out, one cannot be sure that this is indeed what users will adapt to do.

Even in response to the question of whether the participants would adapt their daily routine in order to use more renewable energy, they gave an average rating of 5.2. This means that our participants were in favour of consuming renewable energy, and would schedule their tasks to be done when there is more renewable energy available. However, once again this is only their opinion, and may or may not lead to actual change of behaviour.

Overall, the usefulness of USEM with regard to efficient energy usage was rated 6.4. This indicates that our participants believe USEM could assists them in using energy more efficiently.

6 Conclusions

This paper has described some of the components of USEM, which have been designed to facilitate scheduling and control of energy consuming tasks in residential homes. The results of a user study conducted to evaluate USEM demonstrate its potential benefits for assisting users in making energy saving tasks easy and useful. However, as we have pointed out, this study has only obtained users' opinions on the perceived usefulness of USEM. Therefore, a long-term study of USEM in real-life is still needed to confirm whether these results translate into actual practice, and if USEM would help users change their behaviour to save energy more intelligently.

Acknowledgements. We would like to thank the participants of our user study. This research is related to the IT4SE project, funded by BMBF. For more information visit http://www.it4se.net.

References

1. Aigner, W., Miksch, S., Thurnher, B., Biffl, S.: PlanningLines: novel glyphs for representing temporal uncertainties and their evaluation. In: Proceedings of the 9th International Conference on Information Visualisation, pp. 457–463 IV 2005 (2005)
2. Banerjee, N., Rollins, S., Moran, K.: Automating energy management in green homes. In: Proceedings of the 2nd ACM SIGCOMM Workshop on Home Networks, HomeNets 2011, pp. 19–24. ACM (2011)
3. Brown, R., Koomey, J.: Electricity use in California: past trends and present usage patterns. Energy Policy 31(9), 849–864 (2003), http://linkinghub.elsevier.com/retrieve/pii/S0301421502001295
4. Capone, A., Barros, M., Hrasnica, H., Tompros, S.: A New Architecture for Reduction of Energy Consumption of Home Appliances. In: TOWARDS eENVIRONMENT, European Conference of the Czech Presidency of the Council of the EU (2009), http://www.e-envi2009.org/presentations/S3/Barros.pdf
5. Choi, K., Ahn, Y., Park, Y., Park, W., Seo, H., Jung, K., Seo, K.: Architectural design of home energy saving system based on realtime energy-awareness. In: Proceedings of the 4th International Conference on Ubiquitous Information Technologies & Applications, ICUT 2009, pp. 1–5. IEEE (2009)
6. European Union: Energy - Yearly statistics 2008. Yearly statistics (2010), http://epp.eurostat.ec.europa.eu/cache/ITY_OFFPUB/KS-PC-10-001/EN/KS-PC-10-001-EN.PDF
7. Jahn, M., Jentsch, M., Prause, C., Pramudianto, F., Al-Akkad, A., Reiners, R.: The energy aware smart home. In: Proceedings of the 5th International Conference on Future Information Technology, FutureTech 2010, pp. 1–8 (May 2010)
8. Kugler, M., Reinhart, F., Schlieper, K., Masoodian, M., Rogers, B., André, E., Rist, T.: Architecture of a ubiquitous smart energy management system for residential homes. In: Proceedings of the 12th Annual Conference of the New Zealand Chapter of the ACM SIGCHI, CHINZ 2011, pp. 101–104 (2011)

Chapter 15
Raising High Energy Performance Glass Block from Waste Glasses with Cavity and Interlayer

Floriberta Binarti[1], Agustinus D. Istiadji[1], Prasasto Satwiko[1], and Priyo T. Iswanto[2]

[1] Architecture Department, University of Atma Jaya Yogyakarta, Indonesia
flo.binarti@gmail.com
[2] Mechanical Engineering Department, Gadjah Mada University, Indonesia

Abstract. The main glazing energy performance measure in warm humid climates is light-to-solar-gain ratio (LSG), which denotes the ratio of the visible light transmittance (VT) and its solar heat gain coefficient (SHGC). In laminated glazing the LSG depends on the design of the cavity and (inter)layers. This study explored the contribution of cavity and interlayer in raising high energy performance glass block from laminated waste glasses. Analytical method and computational simulations using comparative method and heat balance model were employed to obtain glass block model with the most optimum combination of the VT, the SHGC and its thermal transmittance (U). The effect of cavity on increasing the VT was showed by simulation and laboratory test results. Based on SHGC laboratory tests, the presence of interlayer declined 69-89% of the simulated SHGC. Laminated glass block with certain number of closed cavity and interlayer can raise 4.35 of the LSG.

Keywords: cavity, glass block, interlayer, light-to-solar-gain ratio, solar heat gain coefficient, visible light transmittance.

1 Introduction

Building energy consumption can be reduced by adopting high energy performance glazing. In warm humid climates high energy performance glazing should have high ratio of the visible light transmittance (VT) and its solar heat gain coefficient (SHGC), which is called as Light-to-Solar-Gain Ratio (LSG). Thermal transmittance (U) is important for air conditioned buildings. Low-U building envelope can reduce the conductive heat transfer rate, which further cuts down the building cooling load. According to Energy Conservation Code 2006 vertical fenestration in warm humid climates is recommended to have 0.25 for the maximum SHGC and 3.177 W/m^2.K for the maximum U with 0.27 of the minimum VT for small fenestration area [1].

Lamination is the selected method to produce new glass block from waste glasses. This low-technology method potentially creates low U and low SHGC material. The SHGC, the VT, the U and the mechanical strength of the layers bonding depend on the layer number, the interlayer and lamination technique. Chen and Meng [2] studied

A. Håkansson et al. (Eds.): *Sustainability in Energy and Buildings*, SIST 22, pp. 157–165.
DOI: 10.1007/978-3-642-36645-1_15

the contribution of interlayer by examining the effect of polyvinyl butyral (PVB) laminated glass application on the building cooling load. The simulation results showed that application of 7 mm PVB laminated glass created lowest cooling load compared to the application 12 mm clear glass and 6 mm low-e coated glass.

Cavity was introduced in glass block as thermal resistance. Heat transfer across the cavity depends on the cavity number, dimension, the optical and the thermal properties of the material [3-6]. Cavity avoids significant reduction of the VT due to the transparency for visible light. Material with higher refractive index (RI) is less transparent. Air has 1 for the RI, whereas ordinary clear glass has 1.52.

This study explored the contribution of cavity and interlayer in raising high energy performance by obtaining and testing optimum combination of the layer number, the cavity type, number, width and position. Analytical and computational simulation approaches were used to design glass blocks with proper cavities. Contribution of the interlayer would be examined in laboratory tests.

2 Methods

This study employed several methods that will be explained chronologically. The first step is interlayer selection. Some criteria in selecting interlayer are transparency, emissivity, thermal conductivity, compressive and tensile strength, durability, curing time and price. Clear epoxy resin was selected as the interlayer material. Epoxy resin can form extremely strong durable bonds with glass (50 MPa). Generally epoxy resin has 0.02–0.1 W/m.K for the thermal conductivity and 0.8 for the emissivity. The maximum RI is 1.57.

Cavity inside glass block can be designed as open cavity and closed cavity. In this study cavity type was examined as 1 m^2 vertical fenestration in a 3 m x 3 m x 3 m adiabatic building system using a Computational Fluid Dynamic (CFD) – ACE software package. The accuracy of simulation results of CFD - ACE has been remarkable. Validation conducted by Satwiko et al. [7] described that the air flow analysis are close to the field measurements with deviation from 0.003 until 0.027 for three dimensions with standard k-ε turbulent model. In CFD - ACE geometry and mesh were created in CFD-GEOM. Models were constructed from 285,345 unstructured cell number. Simulations were conducted with steady state laminar model with low air velocity (0.2 m/s). This condition was reached after 200 iterations and 0.0001 for the convergence. Heat flux was set to 540 W/m^2 (at the peak local condition). The exterior surface temperature was set to 50 ^{0}C, whereas the interior surface temperature was set to 26.85 ^{0}C, which represents the lowest average temperature to describe significant effect of the convective heat transfer across each model.

The next step is development of glass block models with selected cavity type. Formula (1) was employed to estimate the U of each model.

$$U=1/[(1/f_0)+(b_1/k_1)+(b_2/k_2)+(b_1/k_1)+R+(b_1/k_1)+(b_2/k_2)+(b_1/k_1)+(1/f_i)] \quad (1)$$

The accuracy of the formula depends on the determination of the external surface conductance in W/m^2.K (f_0), the internal surface conductance in W/m^2.K (f_i), the thermal resistance of cavity in $m^2.K/W$ (R), the glass conductivity in W/m.K (k_1) and the interlayer conductivity in W/m.K (k_2). b denotes the layer thickness in m'. Variation in R depends on the cavity width. Only interlayers among the layers were calculated.

The effects of heat transfer through each model on the airflow rate inside the cavity, the external and the internal surface temperature were simulated by CFD individually. To lighten the central processing unit (CPU) burden, interlayer in each model was neglected. This is also valid for other simulations.

The VT of glass block models was estimated using comparative method of illumination levels simulated by Radiance (plug in Ecotect). Accuracy of the simulation results rely on the models, ray tracing method [8, 9], and simulation setting. Models were constructed to simulate field measurement of the VT with reference to the simulation procedures developed by Laouadi and Arsenault [10]. Each glass block in the VT simulation was installed as a top lighting of a black box with zero reflectance (Fig. 1).

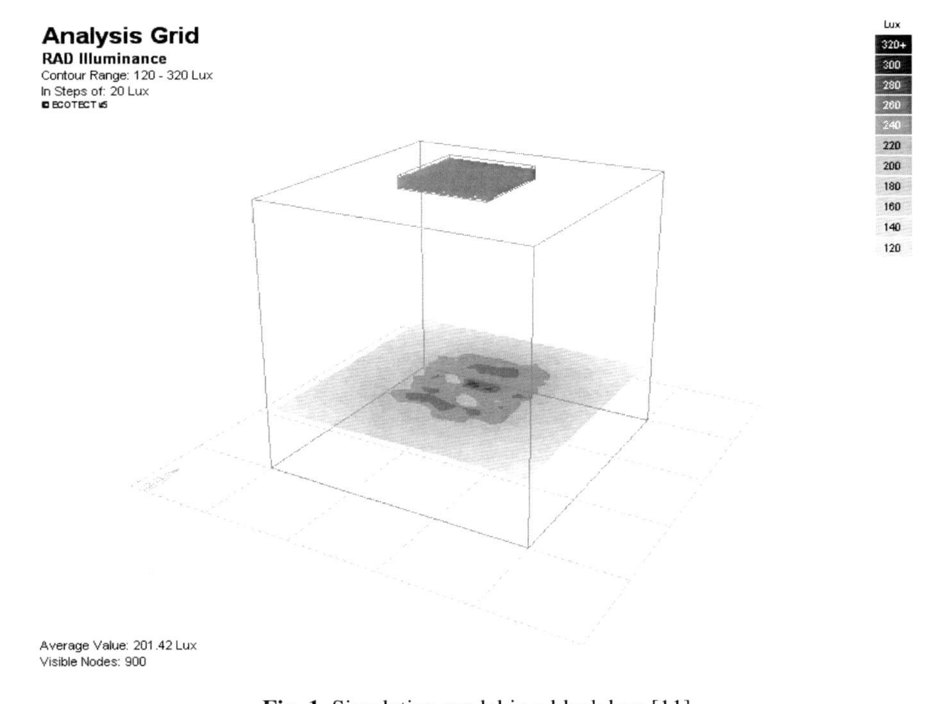

Fig. 1. Simulation model in a black box [11]

The VT of a glass block was obtained from the ratio of the illumination levels transmitted through each glass block to the illumination level captured by a light sensor in the center of black box without glass block. Sky illuminance set up for all simulations is 9897 lux. In this condition (normal incidence angle) the VT is maximum. Validation was conducted using VT field measurement results of single and multiple-glasses with and without cavities. The correction factor is 12% higher than simulation results, which is obtained from the deviation value between the simulation results and the field measurement results.

The SHGC values were obtained by comparing the simulated direct solar gains (in W) of each glass block model ($Q_{g\text{-}glassblock}$) to the one of 3 mm standard glass ($Q_{g\text{-}3mm}$), which were calculated by Ecotect (2).

$$SHGC_{glassblock} = (Q_{g-glassblock} / Q_{g-3mm}) * 0.87 \tag{2}$$

In Ecotect solar heat gain is calculated by Losses and Gains. This facility can analyze heat transfer with admittance method, which works based on cyclic variation concept and is valid under steady state condition. The simulation models were constructed as horizontal fenestration on a roof plane of a zero U and painted black zone. Since the simulation date was set on the hottest day, the results describe the maximum SHGC.

Simulation of heat balance analyzes the quantity of the conductive heat gain (Q_c) and Q_g transferred through each glass block model and the internal heat gain produced by lamp to subtitute the lack of daylighting levels (Q_{lamp}). Models were built as 1 m^2 fenestration on 3 m x 3 m x 3m adiabatic room. The internal heat gain was the heat released by a lamp (Q_{lamp}), which was supplemented to reach the same illuminance level as the illuminance level created by 3 mm standard glass application. One wattage fluorescent lamp power was assummed to produce 11 lux of illumination level and the 20% of the energy was released as heat. The total heat gain was compared to that of the 3 mm glass. In an adiabatic and air-tight room the indirect solar gains, the inter-zonal gains, and the ventilation and infiltration gains are zero.

Q_c refers to conductive heat gains (in W) through a surface area in m^2 (A) due to the air temperature differential (in K) between inside (T_i in ^0C) and outside (T_0 in ^0C) the space and the thermal transmittance (U) of the surface (3).

$$Q_c = U * (T_0 - T_i) * A \tag{3}$$

Whereas, Q_g is the solar radiation (in W) transmitted through a transparent/translucent surface. It depends on the SHGC of the transparent/translucent surface, the total incident solar radiation on the transparent/translucent surface (E in W/m^2) and the glazing area in m^2 (4).

$$Q_g = SHGC * E * A \tag{4}$$

The last step is VT and SHGC laboratory tests to obtain the real LSG. Measurements of the VT referred to the experimental method developed by Wasley and Utzinger [12] with average relative error less than 5% compared to the manufacturer's data. Artificial lighting (Spotone PAR 80 W) replaced the sun to provide weather-independent measurement with less shading and reflection effects from the surrounding environment. Luxmeter Lutron LX-101 with 5% of accuracy deviation was used to measure the illuminance level (Fig. 2). The laboratory VT was obtained from

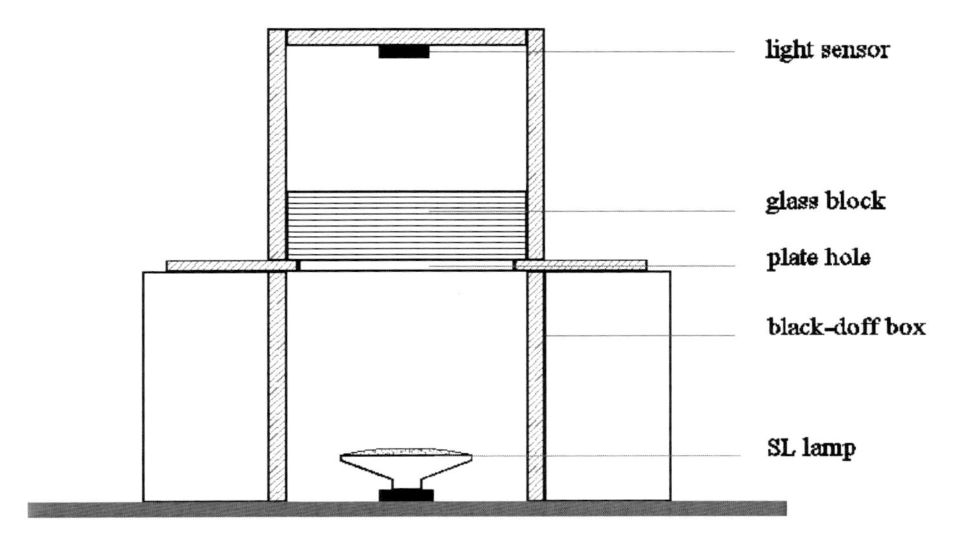

Fig. 2. Schematic apparatus of VT laboratory test

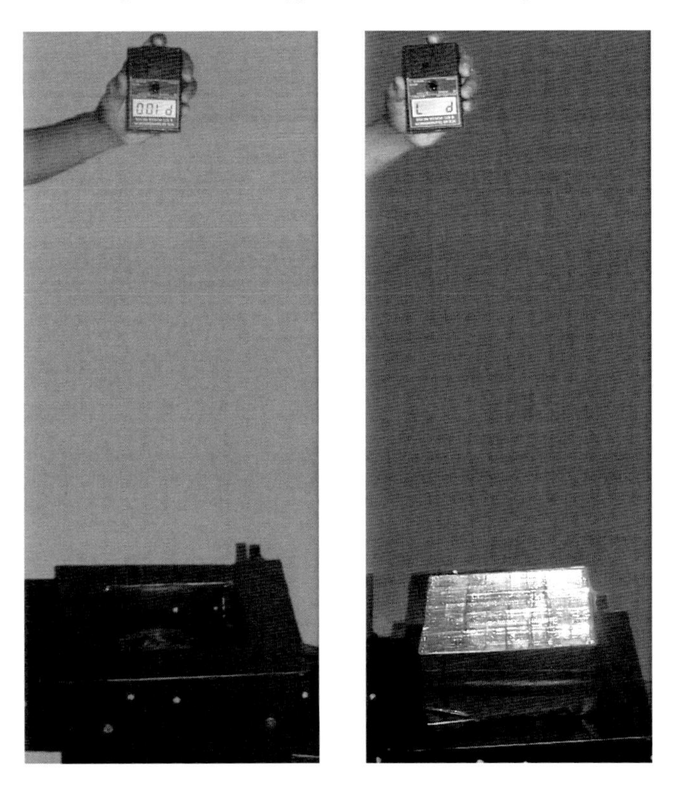

Fig. 3. SHGC measurement using Power meter SP2065: defining the apparatus position (left) and measurement of model l18_l3x2_r2_30 (right)

the ratio of the glass block illuminance level to the illuminance level when no glass was installed. Validation was conducted by comparing the VT measurement of 5 mm clear glass to the standard VT of 5 mm clear glass.

SHGC measurements used digital Solar Transmission and Power Meter model SP2065, which has been factory calibrated to a National Institute of Standards and Technology (NIST) traceable thermopile and requires no field adjustment. An infra red heat lamp (PAR38 150 W) was used as heat radiation source. Self-calibration was done by pressing the power mode. When the display was read P100, the power meter is ready to measure the SHGC of the specimen (Fig. 3). The accuracy of self calibration, therefore, depends on the performance of control microprocessor and the apparatus position consistency. Comparative result of the laboratory SHGC of 5 mm clear glass to the standard value was used to validate the results.

3 Results and Discussions

When glass block models installed as vertical fenestration of 9 m^3 building, lowest indoor temperature was achieved by model with closed cavity (Fig. 4). Closed cavity inside glass block truly functions as thermal insulator. Twelve models with closed cavity, then, were developed with various thickness, cavity width, cavity number, which were selected based on the mechanical strength, the effectiveness of the cavity width and production cost. Table 1 shows that all models have lower U compared to the standard U established by the Conservation Energy Code 2006. Models with more than 30 mm in width cavity have relative high SHGC. None raised 1 for the LSG. Only 4 models with $\geq$ 10 cm thickness reached low indoor surface temperature (27 ^{0}C).

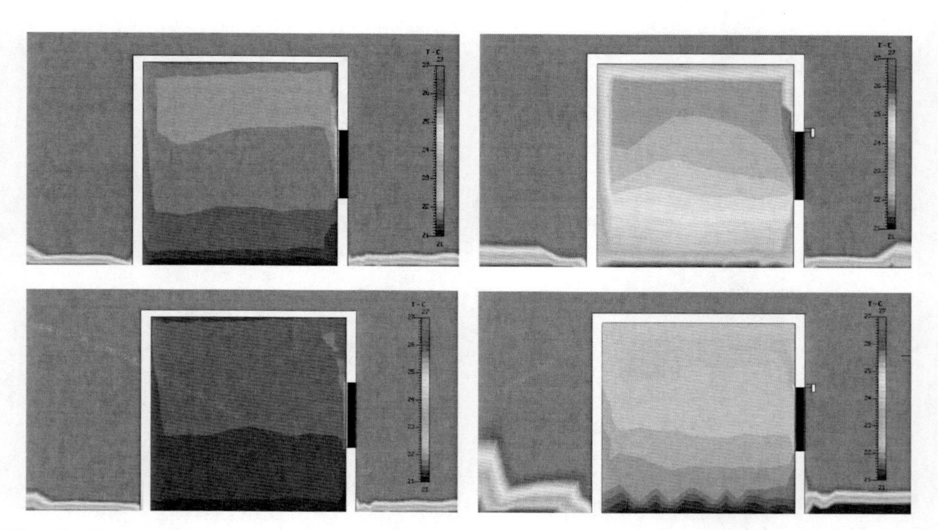

Fig. 4. Temperature profile of application of glass block without cavity (top-left), with open cavity (top-right), with closed cavity (bottom-left), and with open cavity in cooler environment (bottom-right) [11]

Table 1. Energy Performance of Glass Block Models Based on Analytical and Simulation Approach

Model codes[a]	U (W/m^2K)	T_{s0}[b] (^0C)	T_{si}[c] (^0C)	VT	SHGC	LSG
l10_l2x2_r1_30	2.60	77.5	31.5	0.52	0.65	0.80
l11_l2x3_r1_25	2.54	65.5	30	0.41	0.65	0.63
l12_l2x2_r1_40	3.24	78	30	0.52	0.74	0.70
l13_l2x3_r1_35	3.17	67	29	0.40	0.73	0.55
l14_l4x2_r3_10	2.55	75	28	0.40	0.71	0.56
l14_l3x2_r2_20	2.56	77.5	28	0.40	0.72	0.56
l15_l3x3_r2_15	2.50	67	28	0.28	0.70	0.40
l15_l2x3_r1_45	3.06	67	28	0.40	0.71	0.56
l18_l3x2_r2_30	2.60	77	27	0.31	0.57	0.54
l19_l3x3_r2_25	2.54	67	27	0.27	0.59	0.54
l20_l4x2_r3_20	2.30	78	27	0.31	0.67	0.46
l21_l4x3_r3_15	2.28	65	27	0.19	0.66	0.29

[a] Models are coded using lA_lBxC_rD_E formula, which means that A is the total layer number, B is the glass layer number per group, C is the group number, D is the cavity number, and E is the thickness of each cavity in mm.
[b]T_{s0} = outdoor surface temperature
[c]T_{si} = indoor surface temperature

Model l10_l2x2_r1_30 was selected to develop, since it has highest LSG, whereas model l18_l3x2_r2_30 was selected due to its combination of the lowest SHGC and the medium VT (Fig. 5).

Fig. 5. Prototype of model l10_l2x2_r1_30 (left) and model l18_l3x2_r2_30 (right)

Application of best models, i.e. l10_l2x2_r1_30 and l18_l3x2_r2_30 with closed cavity, in 9 m^3 building model produced 60% to 80% lower heat gain compared to the application of 3 mm clear glass. Table 2 presents the simulation results of the maximum and the minimum heat gains of the best models compared to 3 mm clear glass. Glass block with lower SHGC is more efficient than the one with higher VT.

The SHGC and the VT of each prototype were measured 3-5 times. Table 3 shows that laboratory tests of two prototypes resulted in much lower SHGC compared to the

Table 2. Heat Balance of Best Models

Models	Q_c (W)	Q_g (W)	Q_c+ Q_g (W)	Q_{lamp} (W)	Q_{total} (W)	Efficiency (%)
			Oriented to East			
110_12x2_r1_30	142	328	470	5.8	476	60%
118_13x2_r2_30	142	151	293	6.2	299	80%
3 mm clear glass	322	1007	1329	0.0	1329	0%
			Oriented to South			
110_12x2_r1_30	142	106	248	5.8	254	60%
118_13x2_r2_30	142	49	191	6.2	197	70%
3 mm clear glass	322	327	649	0.0	649	0%

simulation results. Low standard deviation in laboratory SHGC, i.e. 1.2%, proved that the results are valid and reliable. Small reductions occurred in the VT with acceptable standard deviation (3.3-4.3%). The LSG of real glass blocks increases due to the lower laboratory SHGC than the simulated SHGC. The real glass blocks consist of interlayer, which contributes more significant in decreasing the SHGC than in decreasing the VT.

A big difference between the percentage difference of simulated SHGC and laboratory SHGC shows that adding glazing interlayer reduced the SHGC more than adding glazing layer. The lower emissivity of the interlayer (0.8) compared to the clear glass emissivity (0.9-0.95) made the glass block emit less heat to the interior. The less transparent (slight higher RI) interlayer compared to the clear glass might create small (percentage) difference of simulated VT and laboratory VT.

Table 3. Laboratory Test Results of the VT and the SHGC

Properties	110_12x2_r1_30	118_13x2_r2_30	Percentage Difference
Laboratory VT	0.47	0.30	36%
Simulated VT	0.52	0.31	40%
Laboratory SHGC	0.18	0.06	67%
Simulated SHGC	0.65	0.57	12%
Laboratory LSG	2.50	4.35	(-) 74%
Simulated LSG	0.80	0.54	32%

The wide discrepancy values of the SHGC were probably caused by the accuracy of the simulation program. In simulated SHGC Ecotect did not calculate the absorption and the back transmission of solar radiation occurring among the glazing layers. Whereas, Radiance proved its accuracy in calculating inter-reflections among glass layers described in the simulated VT.

4 Conclusions

Closed cavity with medium width admits optimum visible light and low solar radiation transmitted across the glass block. Cavity width should be no more than 30 mm to avoid high SHGC. Glass block's thickness is another factor determining the SHGC and the indoor surface temperature. The presence of interlayer (epoxy resin) in laminated glass block reduces the SHGC significantly with small reduction in the VT.

Contribution of interlayer in reduction of the SHGC depends on the emissivity. Certain combination of closed cavity and interlayer number can help the glass block to raise high energy performance, i.e. LSG. New interlayer with lower emissivity and lower refractive index can effectively create a higher energy performance laminated glass block.

Acknowledgments. Authors are grateful to Directorate General of Higher Education (Dirjen DIKTI), Ministry of National Education for funding this research in the scheme of "National Competitive Grant" (*Hibah Bersaing*) under the contract no. 055/SP2H/PP/DP2M/IV/2009 on April 6, 2009 and no. 301/SP2H/P/DP2M/IV/2010 on April 12, 2010. Thank is also addressed to Mr. Widijanto (laboratory staff) for preparing the specimens.

References

1. International Institute of Energy Conservation and USAID. Energy Conservation Building Code 2006. Bureau of Energy Efficiency, New Delhi (2006)
2. Chen, Z., Meng, Q.: Analysis and research on the thermal properties of energy-efficient building glass: a case study on PVB laminated glass. In: Proc. Sixth International Conference for Enhanced Building Operations, Shenzhen (2006)
3. Evoy, M., Southall, R.G.: Validation of a Computational Fluid Dynamics Simulation of a Supply Air Ventilated Window. In: Proc. 20/20 Vision CIBSE/ASHRAE conference, Dublin, Ireland (2000)
4. Gosselin, J.R., Chen, Q.: A dual airflow window for indoor air quality improvement and energy conservation in buildings. HVAC & R Research 14(3), 359–372 (2008)
5. Powles, R., Curcija, D., Kohler, C.: Solar absorption in thick and multilayer glazings. In: Proc. the World Renewable Energy Congress VII, Cologne (2002)
6. Xaman, J., Alvarez, G., Hinojosa, J., Flores, J.: Conjugate turbulent heat transfer in a square cavity with a solar control coating deposited to a vertical semitransparent wall. Heat and Fluid Flow 30(2), 237–248 (2009)
7. Satwiko, P., Locke, N., Donn, M.: Reproducing the real pressure coefficient using a computational fluid dynamic program: how close is close enough? In: Proc. the 32nd Annual Conference of the Australia and New Zealand Architectural Science Association, Wellington (1998)
8. Mardaljevic, J.: Verification of program accuracy for illuminance modeling: assumptions, methodology and an examination of conflicting findings. Lighting Research & Technology 36(3), 218–238 (2004)
9. Reinhart, C.F., Andersen, M.: Development and validation of a Radiance model for a translucent panel. Energy and Buildings 38(7), 890–904 (2006)
10. Laouadi, A., Arsenault, C.: Validation of skylight performance Assessment Software. ASHRAE Transactions 112(2), 1–13 (2006)
11. Binarti, F., Istiadji, A.D., Satwiko, P., Iswanto, P.T.: Analytical and Computational Simulation Approaches to Design Low Energy Glass Block (in Ind). Makara seri Teknologi 15(2), 115–122 (2011)
12. Wasley, J.H., Utzinger, M.: Vital Signs 2. Johnson Controls Institute for Environmental Quality in Architecture, School of Architecture and Urban Planning, University of Wisconsin- Milwaukee, Milwaukee (1996)

Chapter 16
A New Model for Appropriate Selection of Window

Abdolsalam Ebrahimpour[1] and Yousef Karimi Vahed[2]

[1] Department of Mechanical Engineering, Tabriz Branch, Islamic Azad University, Tabriz, Iran
Salam_ebr@yahoo.com
[2] Department of Structure and Material, University Technology Malaysia

Abstract. Appropriate selection of window is important task to reduce the building energy consumption. Calculating of window energy transfer is difficult and must be calculate with the computer simulation softwares. Therefore, simple software is necessary to estimate the window energy transfer and to compare the different window types without complex computer simulation. In this study, a new software (named as Panjare) has been prepared to calculate hourly window energy transfer and to select appropriate window in the building. Using this software, the window can be designed based on the minimum building energy consumption.

Keywords: building, window, Simulation, Energy Transfer, Overhang.

1 Introduction

The window is an external envelope of the building that has more effect on the building energy consumption. The rate of the energy transfer from the window is depends on several parameters such as window type, overhang, etc [1]. Heat transfer in the windows includes conduction, convection and radiation, simultaneously. So, calculating window energy transfer is difficult. Also, testing and measuring of the window heat transfer is very complicated, in the last 20 years computer building simulation have been used to evaluate and to compare the building systems [2].

Many investigations presented dealing with the performance of the window systems (with or without overhangs). Some research works can be found about derived equation or method to calculate and to evaluate the window heat transfer. Karlsson [3] in a research paper evaluated and compared different window. The comparisons were made for different European climates, types of buildings and orientations. Fang [4] in an experimental research with a HotBox, calculated the window heat transfer coefficient with or without inside cloth curtain in china. He did some experimental tests about the two types of the window systems; the double-glazing with or without a cloth curtain, respectively.

In this study, new software has been prepared to calculate hourly window energy transfer and to appropriate selecting of window. Using this software, the window can be designed based on the minimum window energy transfer.

A. Håkansson et al. (Eds.): *Sustainability in Energy and Buildings*, SIST 22, pp. 167–178.
DOI: 10.1007/978-3-642-36645-1_16

2 Simulation Arrangement

Panjare software has been prepared based on building energy simulation result of Energyplus™ software. First Energyplus™ software used to calculate the window energy transfer in the buildings and then the Panjare software has been prepared of the Energyplus results. Energyplus is made available by the Lawrence Berkley Laboratory in USA. Energyplus calculates thermal loads of the buildings by the heat balance method [5, 6].

In Energyplus, the heat transfer by radiation, convection and conduction is calculated at each time step. The U-values are not constant through the simulation because the irradiative and convective heat transfer is calculated by algorithms that take into account parameters such as temperature difference between the surface and the air [7].

3 The Procedure

First, a single and a double pane-glazing window have been selected separately as reference model. Then with changing each parameter of the reference window (effectual on the window heat transfer), the relationship between the variations of these parameters and the window heat transfer has been derived using simulation in the Energyplus software. The derived equations are as follow:

- The variation formula of the thickness of the glass (outside or inside layer)
- The variation formula of the heat conduction coefficient of the glass (outside or inside layer)
- The variation formula of the area of the window
- The variation formula of the depth of the horizontal overhang
- The variation formula of the width of the horizontal overhang
- The variation formula of the depth of the vertical overhangs (side fins)
- The variation formula of the width of the vertical overhangs (side fins)
- The variation formula of the orientation of the vertical overhang (eastern, western)
- The variation formula of the height of the earth's surface
- The variation formula of the optical property of the glass
- The variation formula of the gap thickness between window layers
- The variation formula of the type of gas in the gap space

After deriving the formulas that show separately the effects of changing in each parameter of the window in comparison with the reference model, the general equation has been derived to calculate the window heat transfer. General equation has been derived with the combination of all mentioned formulas and then the Panjare software have been prepared based on the achieved general equation. The algorithm has following procedure:

a) *Selecting the reference model (for double and single pane glazing window separately)*

Changing each parameter of the reference window at the state that other parameters are constant

b) Deriving the relationship between the variations of these parameters and the window heat transfer using simulation in the Energyplus software
c) Deriving general equation by combination of the all mentioned formulas and by fitting the Energyplus results
d) Evaluating the accuracy of general equation and Energyplus results (with selecting randomly window with different optical and physical properties, with varying the height of the earth's surface, with varying the inside temperature and with varying the horizontal and vertical overhangs)
e) Preparing the Panjare software based on the achieved general equation

4 The Formulas of the Single Pane-Glazing Window

A single clear pane glazing, without overhang, southern orientation, without frame, minimum height of the earth's surface (floor 1, 1.5 m distance center of glass from the earth's surface), 3mm thickness of glass layer, 1m^2 area of the window, 23 ^{o}C inside temperature, 10 ^{o}C outside temperature and without solar radiation has been selected as the reference model that it's characteristics has been displayed in Table 1. The reference model is simplest window that has the maximum heat exchange. The outside and inside conditions and characteristics of the reference model for single and double pane have been selected based on the default model in the Window software. [8]

After selecting the characteristics of the reference model, with changing each parameter of the reference window (at the state that other parameters are constant) the relationship between variations of these parameters and the window heat transfer coefficient (u_m) has been derived using simulation in Energyplus software as follow:

1) variation in the thickness of glass

$$u_m = 6.872 - 12.984 x_1 \quad (1)$$

2) variation in the heat conduction coefficient of the glass

$$u_m = 6.873 - \frac{0.039}{x_2} \quad (2)$$

3) variation in the area of the window

$$u_m = 12.886 - 3.065 (x_{3a} \times x_{3b})^{0.5} - \frac{3.023}{(x_{3a} \times x_{3b})} \quad (3)$$

4) variation in the area of the horizontal overhang

$$u_m = 6.8485 - 3.059 x_4 + 1.224 x_4^2 - 0.4646 x_5 \quad (4)$$

5) variation in the area of the vertical overhang(The right side of the window)

$$u_m = 7.248 + 0.4107 x_6^2 - 1.711 x_6^{0.5} - 0.355 (x_7 \times x_6) \quad (5)$$

variation in the area of the vertical overhang(The left side of the window)

$$u_m = 6.92 - 1.791x_8 + 1.1789x_8^{1.5} - 0.273(x_8 \times x_9) \quad (6)$$

6) variation in the height of the earth's surface

$$u_m = 6.921 - \frac{0.118}{x_{10}} \quad (7)$$

7) variation in the optical property of window glass

$$u_m = \frac{1}{0.318 - 0.189\tau^{0.5} + (\ln(1 - \rho_1\varepsilon_1 + \rho_2\varepsilon_2))^2} \quad (8)$$

Where, x_1 is the thickness of glass layer vs. m. x_2 is the glass layer heat conduction coefficient vs. *W/m.k.* x_{3a} is the length of the window and x_{3b} is the width of the window vs. *m*. x_4 is the depth of the horizontal overhang and x_5 is the width of the horizontal overhang vs. *m*. x_6 is the depth of the vertical overhang(right side) and x_7 is the width of the vertical overhang(right side) vs. *m*. x_8 is the depth of the vertical overhang(left side) and x_9 is the width of the vertical overhang(left side) vs. *m*. x_{10} is the height center of the window from the earth's surface vs. *m*. τ : solar transmittance at normal incidence, ρ_2 : solar reflectance at normal incidence: back side, ε_1 : IR hemispherical emissivity: front side , ε_2 : IR hemispherical emissivity: back side.

Table 1. Characteristics of the reference window

Name of Glass	Solar transmittance at Normal Incidence	Solar reflectance at Normal Incidence(Front)	Solar reflectance at Normal Incidence(Back)	Visible Transmittance at Normal Incidence	Visible reflectance at Normal Incidence(Front)
Ref	0.834	0.075	0.075	0.899	0.083
Visible reflectance at Normal Incidence(Back)	**IR Transmittance at Normal Incidence**	**IR Hemispherical Emissivity Front Side**	**IR Hemispherical Emissivity Back Side**	**Conductivity W/mK**	
0.083	0	0.840	0.840	1	

5 The Formulas of the Double Pane-Glazing Window

For the double pane glazing window like single pane, a double clear pane glazing without overhang, southern orientation, without frame, minimum height of the earth's surface (floor 1, 1.5 m distance center of glass from the earth's surface), 4mm thickness of glass (outside or inside layer), 1m^2 area of the window, 12mm thickness of gap between glass layers that has been filled by air, 23 °C inside temperature, 10 °C outside temperature and without solar radiation has been selected as the reference model that it's characteristics has been displayed in Table 1.

After selecting the characteristics of the reference model, with changing each parameter of the reference window (at the state that other parameters are constant) the relationship between variations of these parameters and the window heat transfer coefficient (u_m) has been derived using simulation in Energyplus software as follow:

1) variation in the thickness of glass

$$u_m = 4.594 - 5.727x_{1a} - 7.416x_{1b} \tag{9}$$

2) variation in the heat conduction coefficient of the glass

$$u_m = 4.587 - \frac{0.021}{x_{2a}} - \frac{0.0281}{x_{2b}} \tag{10}$$

3) variation in the area of the window

$$u_m = \frac{1}{0.1322 + 0.00425(x_{3a} \times x_{3b})^{1.5} + 0.0837/(x_{3a} \times x_{3b})} \tag{11}$$

4) variation in the area of the horizontal overhang

$$u_m = 6.85 - 3.062x_4 + 1.227x_4^2 - 0.465(x_4 \times x_5) \tag{12}$$

5) variation in the area of the vertical overhang(The right side of the window)

$$u_m = 4.785 + 0.268x_6^2 - 1.107x_6^{0.5} - 0.243(x_6 \times x_7) \tag{13}$$

variation in the area of the vertical overhang(The left side of the window)

$$u_m = 4.562 - 1.106x_8 + 0.721x_8^{1.5} - 0.177(x_8 \times x_9) \tag{14}$$

6) variation in the height of the earth's surface

$$u_m = 4.5561 - \frac{0.0464}{x_{10}^2} \tag{15}$$

7) variation in the optical property of the glass

$$u_m = 0.778 - 0.015u_a + 0.082u_a u_b \tag{16}$$

$$u_a = \frac{1}{0.318 - 0.189\tau_a^{0.5} + (\ln(1 - \rho_{1a}\varepsilon_{1a} + \rho_{2a}\varepsilon_{2a}))^2}$$

$$u_b = \frac{1}{0.318 - 0.189\tau_b^{0.5} + (\ln(1 - \rho_{1b}\varepsilon_{1b} + \rho_{2b}\varepsilon_{2b})^2}$$

8) variation of the gap thickness between window layers

$$u_m = \frac{1}{0.1418 - 3.784x_{11} + 1.129x_{11}^{0.5}} \tag{17}$$

9) variation the type of gas in the gap space

$$u_m = \frac{1}{0.2351 - 0.0105 x_{12}^2}$$

$$x_{12} = (\%gas_1)\frac{\alpha_1}{\mu_1} + (\%gas_2)\frac{\alpha_2}{\mu_2} + (\%gas_3)\frac{\alpha_3}{\mu_3} \quad (18)$$

Where, x_{1a} is the thickness of outside glass layer and x_{1b} is the thickness of inside glass layer vs. m. X_{2a} is the outside glass layer heat conduction coefficient and X_{2b} is the inside glass layer heat. X_{3a} is the length of the window and X_{3b} is the width of the window vs. *m*. x_4 is the depth of the horizontal overhang and X_5 is the width of the horizontal overhang vs. *m*. X_6 is the depth of the vertical overhang and X_7 is the width of the vertical overhang vs. *m*. X_8 is the depth of the vertical overhang and X_9 is the width of the vertical overhang vs. *m*. X_{10} is the height center of the window from the earth's surface vs. *m*. τ : solar transmittance at normal incidence, ρ_2 : solar reflectance at normal incidence: back side, ε_1 : IR hemispherical emissivity: front side, ε_2 : IR hemispherical emissivity: back side And *a* index indicate the outside glass layer and *b* index indicate the inside glass layer. X_{11} is the thickness of the gap space vs. *m*. α : thermal expansions of the gas, μ : viscosity coefficient of the gas, *%gas*: the ratio of the gas in the gap space vs. % And the index 1 to 3 indicate the gas type in the gap space.

6 Combination the Formulas

General equation has been derived with the combination of the all mentioned formulas and by fitting of Energyplus results. The general equation must estimate the window heat transfer with varying value of each parameter of window or with the simultaneous varying of several parameters of window in different climates and orientations and with or without overhangs at the total hours of the year. Before presenting the general equation, a coefficient as efficiency coefficient (Ec) has been defined that it equals to the heat transfer coefficient with variation in the value of one parameter divided by the reference window heat transfer coefficient:

$$Ec = \frac{u_m}{u_{ref}} \quad (19)$$

u_m is heat transfer coefficient with variation in the value of one parameter of the reference model and u_{ref} is the reference window heat transfer coefficient and equals to 4.5419 $W/m^2.K$ for double pane glazing reference window and equals to 6.8358 $W/m^2.K$ for single pane glazing reference window that they are constant value. Also, the *generic efficiency coefficient* has been calculated as the follow:

a) For double pane glazing window

$$Ec_T = Ec_1 \times Ec_2 \times Ec_3 \times Ec_4 \times Ec_5 \times Ec_6 \times Ec_7 \times Ec_8 \times Ec_9 \quad (20)$$

b) For single pane glazing window

$$Ec_T = Ec_1 \times Ec_2 \times Ec_3 \times Ec_4 \times Ec_5 \times Ec_6 \times Ec_7 \tag{20}$$

Ec_T is *generic efficiency coefficient* and also, Ec_1: is *Ec* of variation in the value of the optical property of the window , Ec_2: is *Ec* of variation in the value of the thickness of the glass, Ec_3: is *Ec* of variation in the value of the heat conduction coefficient of the glass, Ec_4: is *Ec* of variation in the value of the area of the window, Ec_5: is *Ec* of variation in the value of the area of horizontal overhang, Ec_6: is *Ec* of variation in the value of the area of vertical overhang, Ec_7: is *Ec* of variation in the value of the height of the earth's surface , Ec_8: is *Ec* of variation in the value of gap thickness between window layers, Ec_9: is *Ec* of variation in the value of the gas property in the gap space. Also, *overhang efficiency coefficient* (Ec_s) has been defined as follow:

$$Ec_s = Ec_5 \times Ec_6 \tag{21}$$

From Eq.20, it can be concluded that the amount of the Ec_t equals to 1 while the all parameters of the window set to be the same with the reference model values. The value of Ec_t varies for the different windows and it's value changes from 0.1 to 1.5. Also it can be seen that the different combinations of the parameters of the window may have the similar values of Ec_t. In the other words, by separately varying of the two parameters of the reference window, the amount of Ec_t may be acquired equality.

6.1 The Sol-Air Temperature on External Surface TSol-Air

The sol–air temperature, $T_{Sol\text{-}Air}$, includes the effects of the solar radiation and convection heat transfer. To calculate $T_{Sol\text{-}Air}$, the heat convection transfer coefficient on external surface (*h*) must be calculated. The heat convection transfer coefficient on external surface can be calculated using wind velocity on external surface. The velocity variation can be calculated using Eq.(22) which has been presented by reference [7, 9] in different height of the earth's surface.

$$V_z = V_{met}\left(\frac{\delta_{met}}{z_{met}}\right)^{\alpha_{met}}\left(\frac{z}{\delta}\right)^{\alpha} \tag{22}$$

z = altitude, height above ground, V_z = wind speed at altitude z, α = wind speed profile exponent at the site, δ = wind speed profile boundary layer thickness at the site, z_{met} = height above ground of the wind speed sensor at the meteorological station, V_{met} = wind speed measured at the meteorological station (from weather data file), α_{met} = wind speed profile exponent at the meteorological station, δ_{met} = wind speed profile boundary layer thickness at the meteorological station.

The wind speed profile coefficients α, δ, α_{met}, and δ_{met} are variables that depend on the roughness characteristics of the surrounding terrain. In this study, α and δ values assumed $\alpha=0.33$ and $\delta=460\ m$ to be according to reference [7]. So, the default value for z_{met} wind speed measurement is 10 *m* above the ground and the default values for α_{met} and δ_{met} are 0.14 and 270 m, respectively. Also, these values have been assumed from reference [7]. So, the following simple algorithm has been used to calculate the exterior convection heat transfer coefficient:

$$h = D + EV_z + FV_z^2 \tag{23}$$

h = convection heat transfer coefficient, V_z = local wind speed calculated at the height above ground of the surface, *D, E, F* = material roughness coefficients. This coefficients for a glass are *D=8.23, E=3.33, F=-0.036* that have been used from reference [7]. In this study, after calculating convection heat transfer coefficient on the external surface and using of dry bulb temperature T_o (from weather data file), solar radiation on the window surface q''_{solar} and inside temperature T_{in}, the sol–air temperature, $T_{Sol\text{-}Air}$, has been defined as the follow:

$$T_{Sol-Air} = \frac{h_o T_o + q''_{Solar} + u_i T_{in}}{h_o + u_i} \quad and \quad u_i = u_{ref} \times Ec_T \tag{24}$$

u_{ref} is the reference window heat transfer coefficient and equals to 4.5419 $W/m^2.K$ for double pane glazing reference window and equals to 6.8358 $W/m^2.K$ for single pane glazing reference window.

6.2 The General Equation

The hourly window heat transfer, Q $[W/m^2]$ can be calculated from the following equations during the year. But the calculated window heat transfer from these equations is just for the center of the glass and the heat transfer from the frame must be calculated separately.

a) For double pane glazing window

$$Q = 4.5419(T_{Sol-Air} - T_{in}) \times (1 + 0.15\, Ec_T \times \tau_{ave})\left(\frac{f}{Ec_s} + Ec_T\right) \Big/ m \tag{25}$$

$$f = 0.27 - 0.61 Ec_T + 3.989\, Ec_T^2 - 2.06 Ec_T^3 \tag{26}$$

And τ_{ave} is the mean solar transmittance at normal incidence of inside and outside glass layers (Eq. (20)) and the value of coefficient m can be selected from Table 2.

$$\tau_{ave} = \frac{\tau_a + \tau_b}{2} \tag{27}$$

τ is the solar transmittance at normal incidence and a index indicate the outside glass layer and b index indicate the inside glass layer.

Table 2. m values for equation 25

m values	*South*	*East*	*West*	*North*
With overhang	2.043	1.988	2.087	2.555
Without overhang	2.653	2.386	2.528	2.705

b) For single pane glazing window(south window)

$$Q = 6.8358\left(T_{Sol-Air} - T_{in}\right) \times f \tag{28}$$

$$f = 0.326 + 1.19Ec_T^2 \tag{29}$$

c) For single pane glazing window(other side window)

$$Q = 6.8358\left(T_{Sol-Air} - T_{in}\right) \times \left(\frac{f}{Ec_s} + Ec_T\right) \Big/ m \tag{30}$$

$$f = 0.326 + 1.19Ec_T^2 \tag{31}$$

The value of coefficient m is 1.9 for north, 1.8 for west and 1.7 for east window.

7 The Results

In this study, using the computer simulation with Energyplus software, a new method has been presented to calculate window energy transfer. In this part, the accuracy of this method has been evaluated. For that, several windows with different optical properties, different area and direction and in different floors of the building have been selected randomly and their window heat transfer has been calculated using simulation in Energyplus software. Then, the window heat transfer has been calculated using the presented equation (Equation 25 for double pane glazing and Equations 28 & 30 for single pane glazing) for these mentioned windows. The results of the Energyplus software and the presented equation has been compared based on yearly average and yearly sum of the window heat transfer in Tehran, Tabriz, Ahwaz, Yazd, Rasht and Bandarabass. These cities have been selected based on yearly cooling and heating energy demand and type of the metrological weather [10].

The results have been presented based on the average of errors of all windows for predicting yearly average window energy transfer and average of errors of all windows for predicting yearly sum window energy transfer in Tables 3 and 4 for different cities and in cases with or without overhangs. It can be seen that the error of presented equation is 10-15% against the result of Energyplus software. The error in different metrological weathers varies but it is about 10-15%. The main results have been summarized as follows:

The results of the presented equation have about 10-15% error against the results of the Energyplus software for different windows types (single pane or double pane, with or without overhang). This error is negligible because the energy simulation programs (Energyplus, Doe, Blast, …) have different result and K.J.Lomas [11] showed that the energy simulation programs have error against the empirical results. So, the results of the presented equation have good agreements with the Energyplus software results. Also, the results shows that the window heat transfer increases with the increasing of Ec_t but this point is not correct for close values of Ec_t, In the case, with the close values of Ect, the windows heat transfer do not has significantly change.

After selecting the characteristic of the window, the Ec_t number can be calculated. Using it, the optimum window can be selected. Therefore, the Ec_t number can be found equally with different combination of the window parameters, so the different window designing can be selected in a building.

Table 3. Average errors of all windows for predicting yearly sum and yearly average window energy transfer (double pane glazing with and without overhang)

City	Ahwaz		Bandar Abass		Rasht		Tabriz		Tehran		Yazd	
Without Overhang / Direction	Yearly Average Error%	Yearly Sum Error%	Yearly Average Error%	Yearly Sum Error%	Yearly Average Error%	Yearly Sum Error%	Yearly Average Error%	Yearly Sum Error%	Yearly Average Error%	Yearly Sum Error%	Yearly Average Error%	Yearly Sum Error%
East	13.3	13.3	14.7	14.6	14.5	14.5	13.6	13.7	14.8	14.8	15.0	15.0
North	12.4	12.4	17.6	17.6	12.6	12.5	14.7	14.6	11.4	11.4	11.4	11.4
South	13.5	13.4	13.8	13.8	14.3	14.3	14.6	14.6	16.0	15.9	15.0	15.0
West	13.0	13.0	14.0	14.0	14.2	14.2	12.3	12.3	12.9	13.0	12.9	12.9

City	Ahwaz		Bandar Abass		Rasht		Tabriz		Tehran		Yazd	
With Overhang / Direction	Yearly Average Error%	Yearly Sum Error%	Yearly Average Error%	Yearly Sum Error%	Yearly Average Error%	Yearly Sum Error%	Yearly Average Error%	Yearly Sum Error%	Yearly Average Error%	Yearly Sum Error%	Yearly Average Error%	Yearly Sum Error%
East	15.2	15.2	15.7	15.7	16.3	16.3	14.7	14.8	16.3	16.3	16.4	16.4
North	12.7	12.7	14.7	14.6	14.0	14.0	14.6	14.5	12.4	12.4	12.2	12.2
South	16.4	16.4	15.6	15.5	16.5	16.6	16.5	16.5	18.2	18.2	17.5	17.5
West	14.6	14.6	15.3	15.3	15.3	15.3	13.8	13.8	14.7	14.7	14.6	14.6

Table 4. Average errors of all windows for predicting yearly sum and yearly average window energy transfer (single pane glazing with and without overhang)

City	Ahwaz		Bandar Abass		Rasht		Tabriz		Tehran		Yazd	
Without Overhang / Direction	Yearly Average Error%	Yearly Sum Error%	Yearly Average Error%	Yearly Sum Error%	Yearly Average Error%	Yearly Sum Error%	Yearly Average Error%	Yearly Sum Error%	Yearly Average Error%	Yearly Sum Error%	Yearly Average Error%	Yearly Sum Error%
East	9.4	9.4	10.9	10.9	12.6	12.5	9.1	9.2	10.9	10.9	11.0	11.0
North	8.9	8.8	14.1	14.1	7.7	7.6	8.7	8.7	6.7	6.7	6.6	6.6
South	9.0	9.0	9.2	9.2	9.8	9.8	9.3	9.3	11.1	11.1	10.2	10.2
West	9.3	9.3	10.5	10.5	11.2	11.2	7.0	7.1	8.9	9.0	8.9	8.9

City	Ahwaz		Bandar Abass		Rasht		Tabriz		Tehran		Yazd	
With Overhang / Direction	Yearly Average Error%	Yearly Sum Error%	Yearly Average Error%	Yearly Sum Error%	Yearly Average Error%	Yearly Sum Error%	Yearly Average Error%	Yearly Sum Error%	Yearly Average Error%	Yearly Sum Error%	Yearly Average Error%	Yearly Sum Error%
East	13.0	13.0	13.5	13.4	12.9	12.9	10.8	10.8	12.4	12.5	12.4	12.5
North	8.7	8.7	11.1	11.1	8.9	8.9	7.7	7.7	6.8	6.8	7.0	7.0
South	12.3	12.2	11.5	11.5	10.1	10.2	9.9	9.9	11.6	11.6	11.5	11.5
West	12.0	11.9	12.6	12.5	11.5	11.5	9.6	9.6	10.9	10.9	11.0	11.0

8 The Panjare Software

The panjare software has been prepared based on the achieved general equation. The Fig 1. shows main interface of the panjare software. This software has been created using Visual C++ programming language. The following hints show some capabilities of this software.

1- It has a database for selecting windows
2- It can be use in different climates

3- It can be compare the several different windows in one time run and compare up to 10000 type of windows in 1 minutes(with different type, different overhang and side fin and different parameters)
4- It can be compare a reference window with changing the different parameters
5- It calculates the hourly windows heat transfer and Ec_t parameter

Using this software is very simple. The user only has to input some simple data and the software automatically will create the different available type of the windows.

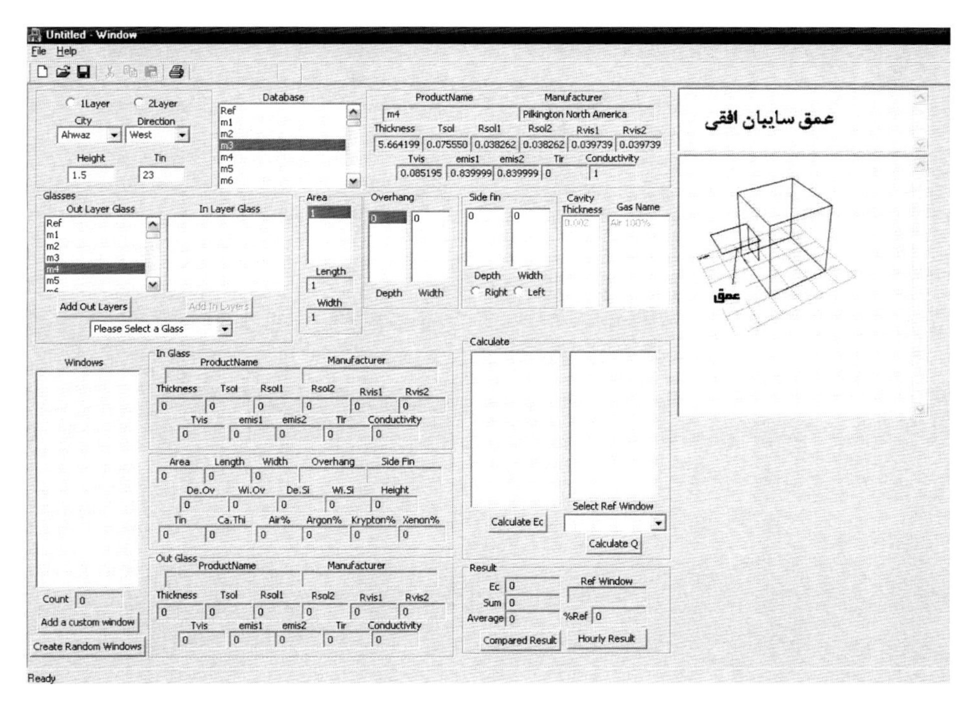

Fig. 1. The main interface of Panjare software

9 Conclusion

In this paper, the Panjare software has been prepared based on a mathematical method. The result shows that this software has good agreement with the Energyplus software results. A building engineer must simulates the different windows types (with or without overhangs) to select the optimum window using computer simulation software. It takes a long time and needs advanced knowledge about the building simulation software. The main application of the result of this paper is that building engineer can uses of the presented software to predict the window heat transfer and to select the optimum window without using of the simulation software. Using the Panjare software is simple and a building engineer can be comparing the several window types in fewer minutes.

References

[1] Ballocco, C., Forastriere, M.A., Grazzini, G., Righini, G.C.: Experimental result of transparent, reflective and absorbing properties of some building material. Energy and Building 32, 315–321 (2000)
[2] Witte, J.M., Henninger, H.R.: Testing and validation of new the building energy simulation program. In: Seventh International IBPSA Conference Rio de Janeiro, Brazil, August 13-15 (2001)
[3] Karlsson, J., Roos, A.: Evaluation of window energy rating models for different houses and European climates. Solar Energy 76, 71–77 (2004)
[4] Xiande, F.: A study of the U-factor of a window with a cloth curtain. Applied Thermal Engineering 21, 549–558 (2001)
[5] Henninger, R.H., Witte, M.J.: Energyplus Testing with ASHRAE 1052-RP Toolkit–building Fabric Analytical Tests. U.S. Department of Energy Office of Energy Efficiency and Renewable Energy building Technology, State and Community Programs (2006)
[6] Witte, M.J., Henninger, R.H., Glazer, J., Crawley, D.B.: Testing and validation of a new building energy simulation program. In: Proceedings of The building Simulation, IBPSA (2001)
[7] Energyplus Engineering Document. The US Department of Energy (2006), http://www.energyplus.com (accessed June 6, 2012)
[8] The window 5 Software, www.the windows.lbl.gov (accessed June 6, 2012)
[9] ASHRAE, Handbook, heating, ventilating, and air-conditioning applications (2005)
[10] Iranian standards for buildings, ministry office of buildings and constructions (2001)
[11] Lomas, K.J., Eppel, H., Martin, C.G., Bloomfield, D.P.: Empirical validation of building energy simulation programs. Energy and Building 26, 253–275 (1997)

Chapter 17
Improved Real Time Amorphous PV Model for Fault Diagnostic Usage

Mehrdad Davarifar, Abdelhamid Rabhi, Ahmed EL Hajjaji, Jerome Bosche, and Xavier Pierre

University of Picardie "Jules Verne", Laboratory of Modeling, Information and Systems (M.I.S), Amiens, France
{mehrdad.davarifar,abdelhamid.rabhi,ahmed.hajjaji,
jerome.bosche,xavier.pierre}@u-picardie.fr

Abstract. Amorphous PV panel is modeled in this paper to improve electrical characteristic and curve fitting in real time data processing such as fault diagnostic and Maximum Power Point Tracking (MPPT). The proposed model uses the basic circuit model of PV solar cell by manipulating component parameters, and also changing online shunt resistance by considering solar irradiation and temperature variation effects. Irradiation and temperature data of the PV panel are captured by National Instrument data acquisition system (NI DAQ USB-6212) and applied to the simulation in Matlab software to calculate the I-V curve of PV panel in real time. Then simulation outputs are compared with measured voltage and current for fault diagnostic deliberation. This model is done for triple layers Amorphous PV panel (Unit-solar ES-62T), which is installed in MIS laboratory energy renewable platform.

Keywords: Amorphous model, shunt resistance, real time, Matlab Simulink, fault diagnostic.

1 Introduction

Now a days photovoltaic (PV) panels are used in several sector of industry. They could be found not only on the main power production plants but also on the hand-held calculator, on top of illuminated highway signs, on building roofs and much more markets. Photovoltaic markets are growing fast because of their advantages such as: pollution free, safety, noiseless, easy installation, and short construction period.[1]

Amorphous thin-film photovoltaic cells are encapsulated in UV-stabilized polymer and so they are light in weight. Because of their flexibility and weight amorphous thin-film PV cells do not require mounting racks for fixing on to building structure. Thus, the overall installation cost of the amorphous thin-film PV cells is usually much less than the crystalline silicon modules, which are embedded in glass layers. Another advantage of amorphous thin-film PV laminates is that they can be installed on the roof structure easily by "peel-and-stick" process, by using a series of "clamping batten system"[2].

A. Håkansson et al. (Eds.): *Sustainability in Energy and Buildings*, SIST 22, pp. 179–188.
DOI: 10.1007/978-3-642-36645-1_17

For building-integrated applications the material can be deposited on glass or flexible substrates, which allows for products like roofing shingles and integrated PV/building glass. The material also has a uniform surface, which is ideal for many architectural applications. Amorphous silicon modules perform well in warm weather and have a small temperature coefficient for power. Depending on the building load, this may be beneficial when compared to crystalline systems [3]. For mentioned reasons, the amorphous silicon technology is deliberated.

Fault detection and monitoring in solar photovoltaic (PV) arrays is a fundamental task to increase reliability, efficiency and safety in PV systems. Without proper fault detection, unclear faults in PV arrays not only causes power losses, but also might lead to safety issues and fire hazards in new building [3], as PV panel are normally installed in invisible aria in the building.

Numerous fault diagnostic methods for PV modules/arrays have been proposed in literatures. PV fault detection models based on long-term energy yield and power losses have been proposed in [4]. An extension fault detection method based on the extended correlation function and the matter-element model is proposed to identify specific fault types of a PV system [5]. The study in [6] uses the discrepancy between simulated and real I-V curve of PV systems to detect and identify the faults. To prevent PV components from fire hazards, DC arc detection and protection methods for PV arrays have been studied in [7]. At PV-string level, PV string monitoring has been proposed for real-time fault detection in [8] and decision tree-based fault detection and classification method [4].

However, all this methods need very particular real time model with good performance in variable condition characteristic. In this work a triple layers Amorphous cells/module model is proposed based on a combination of mathematical and electronic components-based modeling [9]. The hybrid proposed model is implemented in Matlab Simulink/Simscape library. The five-parameter model of fundamental photovoltaic solar cells, which is an equivalent electrical circuit, consist of: diode, resistance and dependent current supply with solar irradiance and temperature dependent component, is considered [10]. For accurate modeling of amorphous, the R_{sh} should be corrected according to the irradiance. Also Amorphous junctions differ from other junctions by the presence of an "intrinsic" layer (p-i-n junction). In [11] it is proposed to take the recombination losses in this i^{th} layer into account, by adding a term in the general I/V equation. This term is equivalent

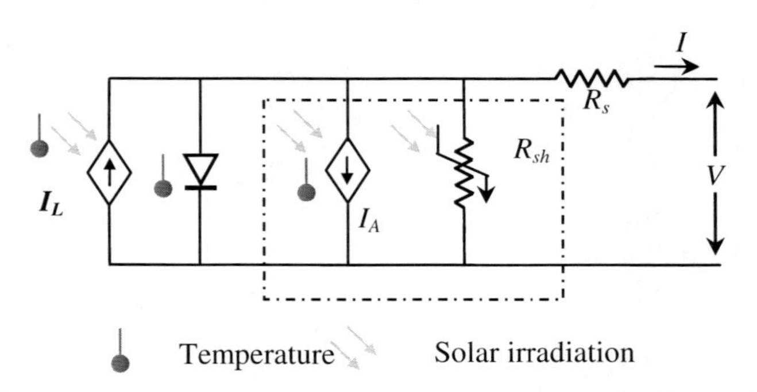

Fig. 1. Improved PV standard five-parameters model for real time amorphous modeling

to adding an element to the equivalent circuit, representing a current leak depending on the photo-current and the voltage, Fig. 1. The model clearly separates effects related to the technology of the device (series and parallel resistance) and effects related to the physics of the pin-junction (recombination losses). It should be noted that the recombination loss in the i-layer I_A is negligible, because I_A is function of V and needs some iterative solution, however this new term doesn't modify significantly the procedure used for getting the model and so it is neglected in this work.

2 Determination of PV Solar Model Parameters

Manufacturers of photovoltaic panels normally provide simply a few experimental data about electrical and thermal characteristics [14].Typically, flowing information could be finding in PV panel datasheets:

V_{ocn} Nominal open-circuit voltage;
I_{scn} Nominal short-circuits current;
V_{mp} Voltage at the maximum power point;
I_{mp} Current at the maximum power point;
K_v Open-circuit voltage/temperature coefficient;
K_i Short-circuit current/temperature coefficient;
$P_{max,e}$ Maximum experimental peak output power;

This information is always provided with reference to the nominal or Standard Test Conditions (STC) of temperature and solar irradiation according to IEC 61646 standards. Though, some manufacturers provide *I-V* curves for several irradiation and temperature conditions, which make the adjustment and the validation of the desired mathematical *I-V* equation easier. Basically, this is all the information one can get from datasheets of photovoltaic panel. However, some of essential parameters for adjusting hybrid photovoltaic panels models cannot be found in the manufacturer's data sheets, such as the light generated current, I_L, the diode reverse saturation current, I_{sat}, the diode ideality constant, a, the band-gap energy of the semiconductor, E_g, the series and shunt resistances, R_s and R_{sh} respectively [12]. Great numbers of publication tried to find parameters of the five-parameters model. Most of these methods are evaluated in standard test conditions and even large numbers of them are done only for crystalline solar cells. Among these methods, empirical or mathematical, one could be chosen and developed for the real time Amorphus PV panel modeling.

A new current loss term I_A , which explicitly takes into account the recombination losses in the i-layer of the device could be added into the equivalent circuit (represented by the dashed section in Fig. 1). This current I_A is a function of V and I_L,which could be calculated as [11]:

$$I_A = I_L.\left(di^2 / [\mu_t.(V_{bi} - (V + I.R_s)]\right) \tag{1}$$

$$\mu_{\tau eff} = 2.\ \mu_n\ .\mu_p\ / \left(\mu_n + \mu_p\right) \tag{2}$$

where *di* is the thickness of the intrinsic i-layer (of the order of 0.3 m), μ_t is the total diffusion length of the charge carriers of p and n layers. V_{bi} is the intrinsic voltage (built-in voltage) of the junction.

Regarding equation (1) to calculate I_A, some extra information from datasheet and manufacturer are needed, which are not normally available in commercial PV panel catalogs. For instance, empirical researches consider the di^2/μ_t quantity as one only parameter, that optimized the V_{co} response of the model [13]. As results, initiation of di^2/μ_t around 1.4 V gives excellent results and corrects quite well the V_{co} distribution. Besides, the value of the intrinsic voltage of the junction, V_{bi}, could be considered constant, and is about 0.9V for an Amorphous junction.

Considering this new term, the general one-diode model *I-V* expression could be written as:

$$I = \underbrace{I_L - I_L.\left(di^2/\left[\mu_t.(V_{bi} - (V + I.R_s))\right]\right)}_{I_A} - \underbrace{I_{sat}\left[exp\left(\frac{V + R_S I}{a.V_{tn}}\right) - 1\right]}_{I_D} - \underbrace{\frac{V + R_S I}{R_{sh}}}_{I_{Rsh}} \qquad (3)$$

3 Real Time Simulation of Multi-layer Amorphous PV Panels

Photovoltaic cells are connected in series and parallel to form a PV module. For triple layer Amorphous panel, based on single cell circuit module, three sub cells without bypass diode in series combine together to make each triple layer cell, the cells are assembled in the form of series-parallel configuration in PV panel, Fig.2.

In this work the PV model is simulated in real time without solving equations (1) and (3). Solving these equations needs some iterative methods, while real-time simulation does not support models containing algebraic loops. However, the result

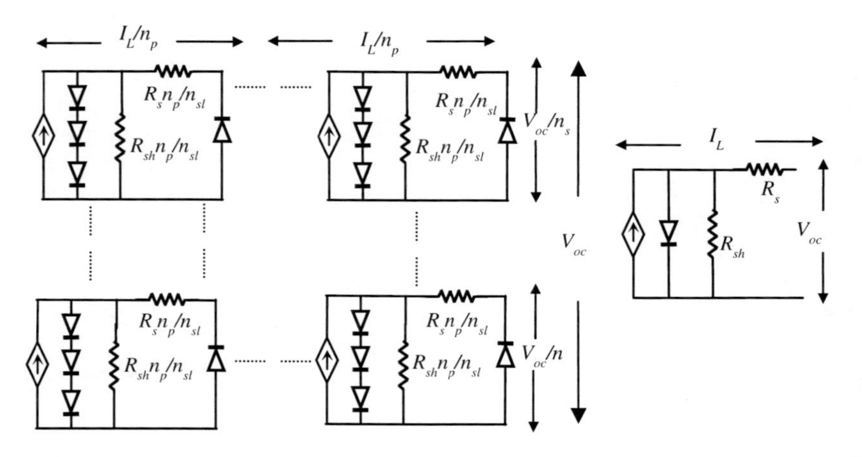

Fig. 2. Triple Amorphous PV panel model circuit with a controlled current source and adapted diode

has good accuracy because temperature effect on diode exponential equation for Amorphous triple layer diode could be passed up [10]. Also simplified proposed model for amorphous PV panel is adapted for any environment conditions (PV panel temperature and solar irradiation) in this paper. By this improvement simulated current and voltage curve is beter fitted to real measured value.

3.1 Degree of Ideality of the Diode

The light generated current of the photovoltaic cell depends linearly on the solar irradiation and is also influenced by the temperature according to the following equation [7]:

$$I_L = \left(I_{LN} + K_i . \Delta T\right) . G / G_n \tag{4}$$

where I_{Ln} is the light-generated current at the nominal condition (usually 25°C and 1000-W/m2) and ΔT is the difference of nominal temperature and actual temperature, I_{ln} could be find from equation (3)

$$I_{LN} = I_{scn} . [(R_{sn} + R_{shn}) / R_{shn}] \tag{5}$$

3.2 Parallel Exponential Resistance Correction versus Variable Irradiance

The shunt resistance R_{sh} corresponding to the inverse of the slope of the *I-V* curve around short circuit area is considered as a constant parameter in the standard one-diode model and in many previous work considered fixed value in STC. Nevertheless, in [15] and [14] is demonstrated that on amorphous *I-V* curve families this slope decreases with the irradiance.

This distribution is approximated by the following exponential expression [13]:

$$R_{sh} = R_{shn} [1 + 10 \ exp(\frac{-5.5.G}{G_n})] \tag{6}$$

It has been noted that this empirical equation is validated for several amorphous PV family by [13].

3.3 Manipulating Band-Gap Energy in Matlab Diode Block

It is possible to obtain the value of band-gap energy for a PV panel regarding the information of its datasheet and substitute in the required diode parameter in the simulation [12]. For that, open-circuit voltage of the model is matched with the open-circuit voltage of the real panel regarding its temperature $T_n < T < T_{max}$ as well as short-circuit current (Noted that all PV product is tested according IEC 61646 standard). Considering $T = T_{max} = 85°C$ for number of large data manufacture, diode ideality factor could be estimated according equation (8) where:

$$I_{scTmax} = I_{scn} + K_i.\Delta T \tag{7}$$

$$V_{oc,Tmax} = V_{oc,n} + K_v.\Delta T \tag{8}$$

$$V_{Tn} = n_{sl}.k.T_n / q \tag{9}$$

$$V_{Tm} = n_{sl}.k.T_{max} / q \tag{10}$$

$$E_g = -Ln\left[\frac{\left(\frac{I_{scTmax}}{I_{satn}}\right)\left(\frac{T_n}{T_{max}}\right)^{\frac{3}{a}}}{exp\left(\frac{V_{ocTmax}}{a.V_{Tm}}\right)-1}\right].\frac{a_n.k.T_n.T_{max}}{q.(T_n - T_{max})} \tag{11}$$

3.4 Adopted Diode Saturation Current under Standard Test Condition

The diode saturation current I_{sat} is depends on the temperature which could be expressed for a PV panel with n_s cells by [12]:

$$I_{satn} = \frac{I_{scn} + K_i.\Delta T}{exp\left(\frac{V_{ocn} + K_v.\Delta T}{a.V_{Tn}}\right)-1} \tag{12}$$

We can manipulate this parameter in Matlab environment diode block to specify according manufacture datasheet information such as K_i, K_v, V_{oc}, … .

4 Simulation and Monitoring of Simplified Real Time Model for Multi-layer Amorphous PV Panel

Triple layer simplified real time model consist of three sub cells which are connected without bypass diode together to form one triple cell and each triple layer cell combine together in series or parallel in the panel Fig. 2. For each cell model, the value of resistance inherit from on diode standard model, noted that this value is calculated in standard test condition, parallel exponential resistance is corrected in variable irradiance (section 3. 1) and band-gap energy is manipulated in Matlab diode block (section 3. 2). Diode saturation current is adapted according to average temperature which is mentioned in section 3.3. It is important to consider the improved average value of the diode parameters such as: saturation current I_{sat} , band gap E_g and diode ideality factor a_n, because it is not possible to change diode block parameter in real time.

In PV panel each cell voltage would be equal to the diode voltage, $V_{oc}/n_{sl} \approx Vd$ for the open circuit condition. Besides the portion of each cell, for series R_{scell} and shunt loss resistances R_{shcell}, would be as:

$$R_{scell} = R_{sn}.n_p / n_{sl} \tag{13}$$

$$R_{shcell} = R_{shn}.n_p / n_{sl} \tag{14}$$

with

$$n_{sl} = n_s.n_L$$

where n_L, n_s, n_p are the number of layers, series and parallel cells respectively.

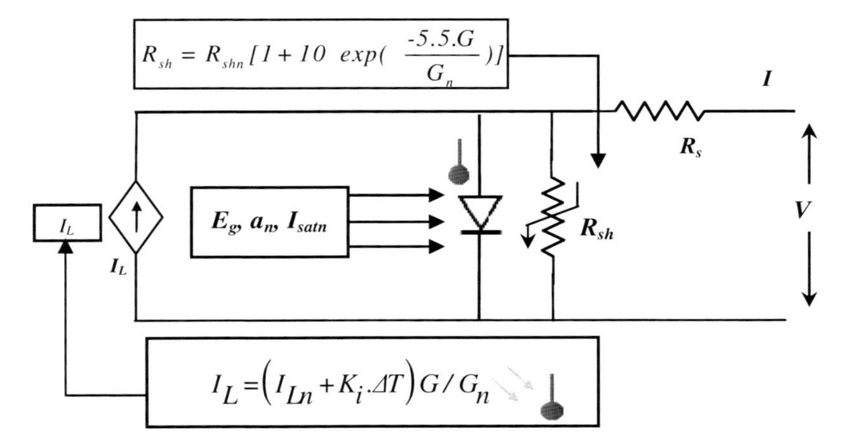

Fig. 3. Triple Amorphous PV panel model circuit with a controlled current source, adapted diode parameters and corrected shunt resistance

It should be noted that in simplified model, recombination loss in the i-layer I_A is negligible, because I_A is a function of V which needs iterative solution, however this new term doesn't modify significantly the procedure used for getting the model. For instance, for UNI-SOLAR Solar Shingles SHR-17 tripple-junction, the V_{oc} drops only about 3.3% to 0.7%.

4.1 Simulink under Uniform Shading Conditions (USC) to Find

Fig. 4. is illustrated *I-V* curve of proposed simplified model for triple layers Amorphous PV panel, [Uni-solar Es-62] solar panel that is installed in MIS laboratory energy renewable platform (from table 1, Amorphuse model's parameter could be extracted [13]). The correction applied in the proposed method results in better fitting characteristic. Another advantage of using this model appears when partial shading condition simulation for the PV panel is required. For instance, if a hot spot or unclear cell fault appear in the panel, *I-V* curve of the panel would change. The *I-V* curve a for sample Amorphous Panel is obtained from proposed model's Simulink under partial shading conditions (PSC) in Fig. 4.

Table 1. Characteristics of Uni-solar ES-62T Amorphous Panel

V_m	15 V	Ns	10	R_s	0.32Ω
I_{mp}	4.10 A	Number of layer	3	R_{sh}	27 Ω
V_{oc}	21 V	T_{max}	85°C	a_n	4.41
I_{sc}	5.10 A	T_n	25°C	$di^2 / \mu t$	1.38
K_v	--81 mV/K	K_i	5.1 mA/K	V_{bi}	0.9

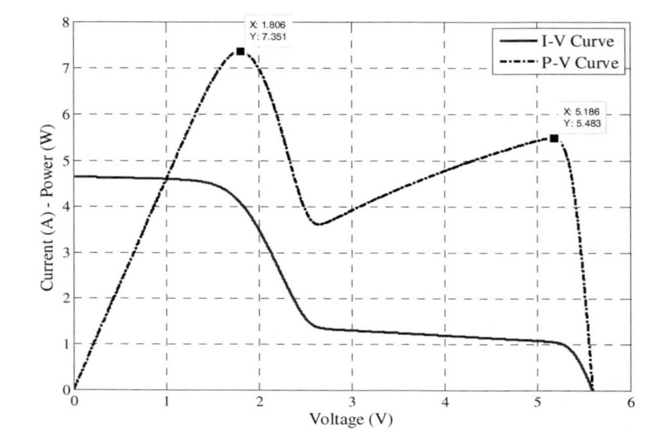

Fig. 4. I-V curves for PV panel combined in one string in series form, under PSC condition

4.2 Real Time Monitoring and Fault Diagnostic Application

Real time modeling is necessary for fault diagnostic application. Using proposed simulation model, current and voltage could be calculated in real time by computer which is so closed to the measured data. In this system, solar irradiation data is captured by pyrometer CS300 and PV temperature is sensed with a thermocouple of type K, also RMS voltage and current value are captured by national instrument data acquisition system (NI DAQ 6212 USB), Fig 5.

In Fig. 6. and Fig. 7. voltage and current of the simulation are compared with measured voltage and current for two series UNI-SOLAR ES-62T Amorphous panels.

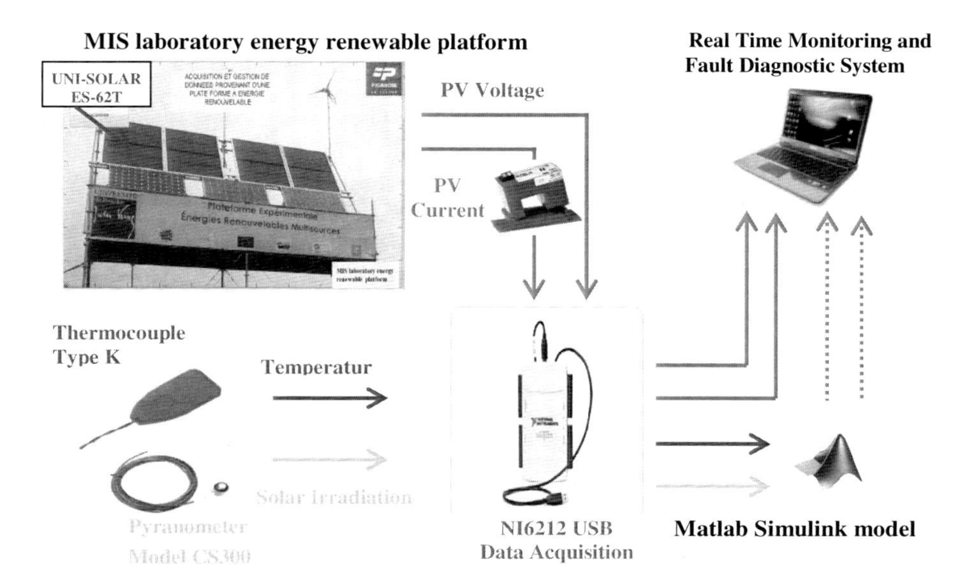

Fig. 5. Evaluated simulink model in MIS laboratory energy renewable platform

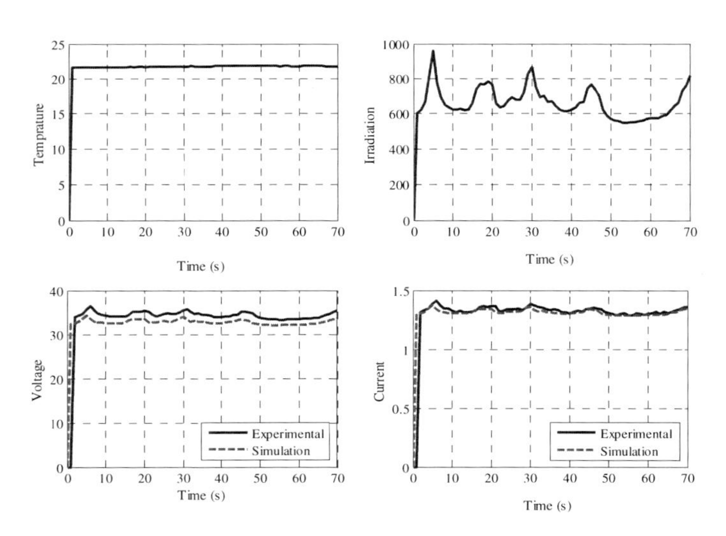

Fig. 6. Data capture and simulation results of the UNI-SOLAR ES-62T PV panel in 60seconds, (a) Temperature, (b) Irradiation, (c) Voltage, (d) Current

5 Conclusions

Real time PV model, particularly Amorphous family model is needed for online monitoring and fault diagnostic application. In this work, a real time triple layers amorphous photovoltaic solar cells/module model is proposed. The hybrid model is implemented in MatlabSimulink/ Simscape/ SimElectronics / Pspice library. In proposed model, algebraic loop for finding *I-V* curve is not needed. Thereby the code could be generated to use this simulation in real time, also partial shading and normal conditions are simulated. This model could be easily interfaced to the electronic devices and power converters for maximum power point tracking studies.

References

1. Housing market still a stabilizing force for Austin's economy, `http://austin-green-builder.com/`
2. Wu, Eddie W.K., LAU, Iris P.L.: The Potential Application of Amorphous Silicon Photovoltaic Technology in Hong Kong. In: The HKIE (Electrical Division) 25th Annual Symposium on Sustainability and Environmental Challenges in Electrical Engineering, ASHRAE Hong Kong (2007)
3. Kroposki, B., Hansen, R.: Performance and Modeling of Amorphous Silicon Photovoltaics for Building- Integrated Applications. In: Solar 1999 Growing the Market, Portland, Maine, June 12-17 (1999)
4. Ye, Z., Ling, Y., Lehman, B., de Palma, J., Mosesian, J., Lyons, R.: Decision tree-based fault detection and classification in solar photovoltaic arrays. In: 2012 Twenty-Seventh Annual IEEE Applied Power Electronics Conference and Exposition (APEC), February 5-9, pp. 93–99 (2012)
5. Chao, K.-H., Ho., S.-H., Wang, M.-H.: Modeling and fault diagnosis of a photovoltaic system. Electric Power Systems Research (2008); Smith, T.F., Waterman, M.S.: Identification of Common Molecular Subsequences. J. Mol. Biol. 147, 195–197 (1981)
6. Stellbogen, D.: Use of PV circuit simulation for fault detection in PV array fields. In: 1993 Conference Record of the Twenty Third IEEE Photovoltaic Specialists Conference, pp. 1302–1307 (1993)
7. Haeberlin, H.: Arc Detector for Remote Detection of Dangerous Arcs on the DC Side of PV Plants. In: 22nd European Photovoltaic Solar Energy Conference, Milano, Italy, 200
8. Takehara, N., Fukae, K.: Abnormality detection method, abnormality detection apparatus, and solar cell power generating system using the same. 5669987 (1997)
9. Davarifar, M., Rabhi, A., El Hajjaji, A., Bosche, J.: Real Time Modeling of Triple Layers Amorphous Photovoltaic Panels. In: IEEE-ICREGA 2010, Al-Ain, UAE (March 2012)
10. Davarifar, M., Rabhi, A., El Hajjaji, A.: Modeling of Solar Photovoltaic Panels in Matlab/Simscape Environment. In: 2nd International Conference on Systems and Control, Marrakech, Morocco, June 20-22 (2012)
11. Merten, J., Asensi, J.M., Voz, C., Shah, A.V., Platz, R., Andreu, J.: Improved equivalent circuit and analytical model for amorphous silicon solar cells and modules. IEEE Transactions on Electron Devices 45(2), 423–429 (1998)
12. Villalva, M.G., Gazoli, J.R., Filho, E.R.: Comprehensive Approach to Modeling and Simulation of Photovoltaic Array. IEEE Trans on Power Electronics 24(5), 1198–1208 (2009)
13. Geneva University: User's Guide, `http://files.pvsyst.com/pvsyst5.pdf`
14. Ramaprabha, R., Mathur, B.L.: MATLAB Based Modelling to Study the Influence of Shading on Series Connected SPVA. In: ICETET, December 16-18, pp. 30–34 (2009)
15. Gottschalg, R., Betts, T.R., Infield, D.G., Kearney, M.J.: The effect of spectral variations on the performance parameters of single and double junction amorphous silicon solar cells. Solar Energy Materials and Solar Cells 85(3), 415–428 (2005)

Chapter 18
An Investigation of Energy Efficient and Sustainable Heating Systems for Buildings: Combining Photovoltaics with Heat Pump

Arefeh Hesaraki and Sture Holmberg

Division of Fluid and Climate Technology, School of Civil and Architectural Engineering, KTH Royal Institute of Technology, Handen-Stockholm, Sweden
Arefeh.Hesaraki@byv.kth.se

Abstract. Renewable energy sources contribute considerable amounts of energy when natural phenomena are converted into useful forms of energy. Solar energy, i.e. renewable energy, is converted to electricity by photovoltaic systems (PV). This study was aimed at investigating the possibility of combining PV with Heat Pump (HP) (PV-HP system). HP uses direct electricity to produce heat. In order to increase the sustainability and efficiency of the system, the required electricity for the HP was supposed to be produced by solar energy via PV. For this purpose a newly-built semi-detached building equipped with exhaust air heat pump and low temperature-heating system was chosen in Stockholm, Sweden. The heat pump provides heat for Domestic Hot Water (DHW) consumption and space heating. Since selling the overproduction of PV to the grid is not yet an option in Sweden, the PV should be designed to avoid overproduction. During the summer, the HP uses electricity only to supply DHW. Hence, the PV should be designed to balance the production and consumption during the summer months. In this study two simulation programs were used: IDA Indoor Climate and Energy (ICE) as a building energy simulation tool to calculate the energy consumption of the building, and the simulation program WINSUN to estimate the output of the PV. Simulation showed that a 7.3 m^2 PV area with 15 % efficiency produces nearly the whole electricity demand of the HP for DHW during summer time. As a result, the contribution of free solar energy in producing heat through 7.3 m^2 fixed PV with 23° tilt is 17 % of the annual heat pump consumption. This energy supports 51 % of the total DHW demand.

Keywords: Sustainable development, Solar power, PV-HP system,Domestic hot water.

1 Introduction

Due to the scarcity of fossil fuel sources and their environmental impact, renewable energy has become an increasingly important topic over the past decades. On a global scale renewable energy sources only contribute less than 15 % (Lund 2007) to the

A. Håkansson et al. (Eds.): *Sustainability in Energy and Buildings*, SIST 22, pp. 189–197.
DOI: 10.1007/978-3-642-36645-1_18

primary energy supply; however, during the last few decades this percentage has increased considerably in some countries.

Energy from solar radiation can be obtained in two ways, passively and actively. Passive solar design is based on the optimal design of a building's shape leading to the capture of as much solar radiation as possible for space heating. Active solar design is based on converting solar radiation into energy by using solar thermal collectors or photovoltaics. Photovoltaics (PV) convert sunlight into electric power by a solid-state device called solar cells. The common PV module converts 15-20 % (Tyagia et al. 2012) of the incoming solar radiation into electric energy, depending on the type of solar cells. Our total present energy demand can be supplied if 0.1 % of the earth's surface were covered with solar cells with 10 % efficiency (Tyagia et al. 2012). Solar thermal collectors convert solar radiation into thermal energy through a transport medium, liquid or gaseous. Solar cells are more efficient than solar thermal collectors, since they are performing during the winter months at low irradiation with constant efficiency while the solar collector has very low efficiency during hours of low intensity due to a high heat loss (Gajbert 2008).

The electricity generated by PV could be utilised by a heat pump to produce heat. Depending on the Coefficient of Performance (COP) of the heat pump, the energy required for space heating and Domestic Hot Water (DHW) may be decreased by a factor of the COP. The outputs of 1 m^2 of plane solar collector delivering hot water at 50 °C and 1 m^2 of PV modules of 15 % efficiency in combination with a heat pump with COP 3.2 in Stockholm were compared using the WINSUN program, see Figure 1. It can be seen that the combination of the PV module and the heat-pump (PV-HP system) has a higher annual output than the solar thermal collector. During winter time (low temperature and low irradiance) the solar collector has zero efficiency due to high heat loss. The heat loss of the solar collector is dependent on the collector efficiency factor (F´), heat loss coefficient from absorber to ambient (U_L ,W/m^2K) and temperature difference between ambient temperature (T_a) and collector temperature (T_c), Equation 1 (Duffie and Beckman 1991). PV is usable year round performing even at low intensity since it works on light not heat. In the summer time both the solar collector and PV have good characteristics. As a result, a PV-HP system is more efficient since it provides a higher annual solar fraction in comparison with a solar thermal collector system. Solar fraction (SF) gives the fraction of energy provided by solar energy to an annual heating demand. SF varies between 0 when no energy is supplied by solar energy to 1 when all required energy is supplied by solar technology.

$$P_{loss\_collector} = F^{'} * U_l * (T_c - T_a) \quad (1)$$

The present paper points to the possibility of combining PV with heat pump in an actual building equipped with an exhaust air heat pump in Stockholm. For this purpose a Building Energy Simulation (BES) program, IDA Indoor Climate and Energy (ICE) was used to calculate the energy demand in the building. Then, using the System Simulation Program (SSP) WINSUN the output of PV with 15 % efficiency was calculated to produce the electricity for the HP. Since there is yet no regulation in

Sweden concerning selling excess solar power to the grid, overproduction of electricity was avoided. Hence, the area of the PV modules required should be calculated to balance energy consumption and production when the production is at the highest point and the consumption is at the lowest value (during the summer season). However; in the future there might be a policy for selling extra electricity to the grid. It might then be profitable to produce more electricity than needed and export it to the grid.

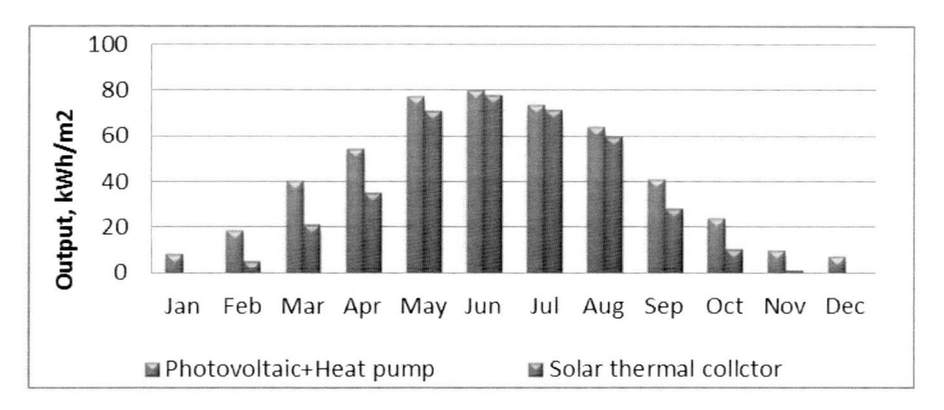

Fig. 1. Output comparison between photovoltaic + heat pump and solar thermal collector

2 Method

In this study two simulation programs were used, one for finding the energy requirement of the building and the other one for calculating the output of appropriate PV to partly meet this demand.

2.1 Building Energy Simulation (IDA ICE)

IDA Indoor Climate and Energy (ICE) is a Building Energy Simulation (BES) tool to calculate thermal comfort, indoor air quality and energy consumption in buildings. This program was partly developed at KTH (Jokisalo et al. 2008). An early validation of the IDA ICE program was conducted by comparing the results predicted by simulation with measurements; they showed good agreement (Travesi et al. 2001).

The purpose of using the IDA ICE program in this study was to calculate the monthly energy consumption for a whole year in order to find the electricity consumption of the heat pump. The electricity consumption for providing domestic hot water and space heating by the heat pump was calculated by dividing energy consumption by the Coefficient of Performance (COP) of the heat pump.

When performing IDA ICE analyses the input data include building location, geometry, construction type, HVAC system, internal heat gain (number of people, light and equipment) and DHW consumption. Referring to the Swedish building regulations (Boverkets byggregler, BBR), the average value for DHW usage is 30 kWh/m^2

of floor area. The area of the studied building is 160 m^2 leading to 4800 kWh/year of DHW. Through creating a mathematical model and solving heat balance equations, the energy consumption in the building was estimated by IDA ICE. Details of the simulation and validation of results for the studied building may be found in a previous study (Hesaraki and Holmberg 2012).

2.2 System Simulation Program (WINSUN)

Simulation tools are preferred for analysing a system rather than implementing an actual device on site, since studying a model is essentially easy and inexpensive. WINSUN is a system simulation program developed at Lund University for designing solar collectors and PV (Hatwaambo et al. 2008). WINSUN is an abbreviation of Windows version of MINSUN usable in DOS (Boström et al. 2003). WINSUN is based and developed completely on PRESIM, TRNSYS and TRNSED version 14.2 (Boström et al. 2003). PRESIM is a graphical modelling program used for producing input data for TRNSYS, developed by the Solar Energy Research Center in Sweden (Beckman et al. 1994). In TRNSYS (TRaNsient Systems Simulation) thermal energy equations are solved based on a modular approach depending on the input data (Beckman et al. 1994). TRNSED is text editor program to create a user-friendly TRNSYS interface of a solar energy system, and convert the input file to a TRNSED-formatted document to be usable for other users, developed at the University of Wisconsin, Madison (Beckman et al. 1994, Hatwaambo et al. 2008). WINSUN aims to provide an output of a solar thermal collector and PV in kWh/m2 depending on the location of system, azimuth, tilt, tracking mode and efficiency of PV. The program uses the weather data during 1983-1992 including diffuse and beam radiation collected by the Swedish Meteorological and Hydrological Institute (SMHI) (Boström et al. 2003). Comparing the solar radiation intensity during the long term at all weather stations, the divergence between measured solar radiations is less than 2 % (Hatwaambo et al. 2008). Hence, the results of the program though using old weather data appear reliable. The validation of the WINSUN program was conducted by comparing the simulation results with site measurements, which showed good agreement (Gajbert 2008). The results of WINSUN may be found in two ways: one as a table and plot file, and the other as an online plot where several variable changes may be watched and analysed at the same time as TRYNSY calculations are running to solve thermal energy equations. In this study, WINSUN simulation was conducted to evaluate the performance and output of fixed PV with 15 % efficiency at 23° tilt towards the south (according to the plan of the chosen building). Simulation was based on monthly net metering. Export of overproduction to the grid was avoided. Input data to the simulation include starting day of simulation, month and length of simulation, site of place, climate, tracking mode (1 for fixed, 2 for turning around vertical axes, 3 for turning around an axis in the plane of the glass, 4 for 2 axes tracking), ground reflectance (typical value 0.2-0.3), slope of surface from horizontal plane and azimuth of surface (azimuth angle from the south i.e. -90 for east, 0 for south and 90 for west).

3 Results and Discussion

3.1 IDA ICE Simulation Results

Using the IDA ICE program, the results for electricity usage in the heat pump to provide heat for space heating and DHW are shown in Figure 2. The DHW consumption was assumed to be equal for all months (Gajbert 2008) since it may not be dependent of weather condition. Considering the COP of the heat pump, which is 3.2 according to production data from the manufacturer, heating demand was decreased by a factor of 3.2. The electricity consumption by the HP for space heating depends on the weather conditions, heat losses i.e. ventilation loss, transmission loss and leakage loss, and passive heating (contribution of free heating source such as people, equipment, solar energy and lights in heating the house).HP consumption for space heating varies for different months, it is at its highest level in winter and zero in summer; however, the electricity consumption by the HP for DHW was assumed to be constant over the year (1500 kWh/year or 125 kWh/month).

The efficiency of an electrical pump in converting input energy into useful energy is usually 80 % (Bhargava et al. 1991) due to losses in the compressor. Thus, for running the pump, the solar cells must generate electricity equal to P/0.8 = 1.25 P, where P is the heat pump consumption. So, the required electricity for the heat pump increases by a factor of 1.25, i.e. for DHW consumption the required electricity to be generated by PV is 125*1.25 = 156.25 kWh/month

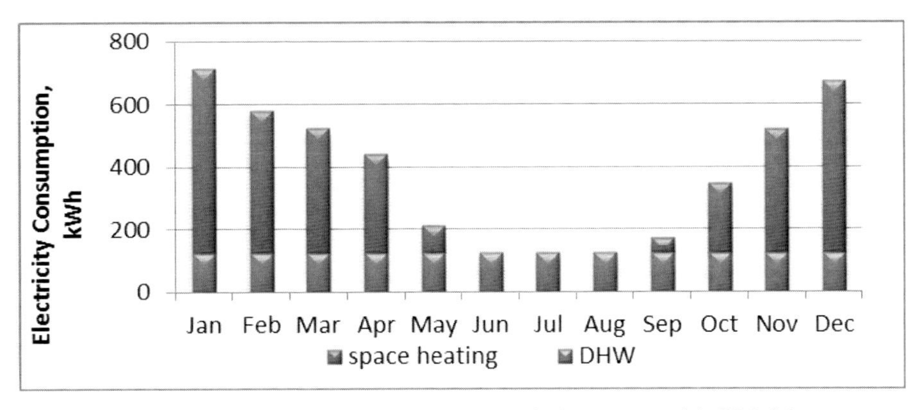

Fig. 2. Monthly electricity consumption of a heat pump with COP 3.2

3.2 WINSUN Results

Using the WINSUN program, the output for a fixed photovoltaic system is given in Figure 3. This is for one year with 15 % efficiency and southward orientation with a roof slope of 23° in Stockholm. As shown, the electricity produced by PV varies during the year depending on the solar irradiation. PV can work with acceptable output even during the cold and mostly dark winter time since they can work even on slight light. The output of the PV drops considerably during the winter months due to the low solar intensity in comparison with the summer months when the output has its maximum value. The annual output for the chosen PV is 156 kWh/m^2 year independently of the cell temperature.

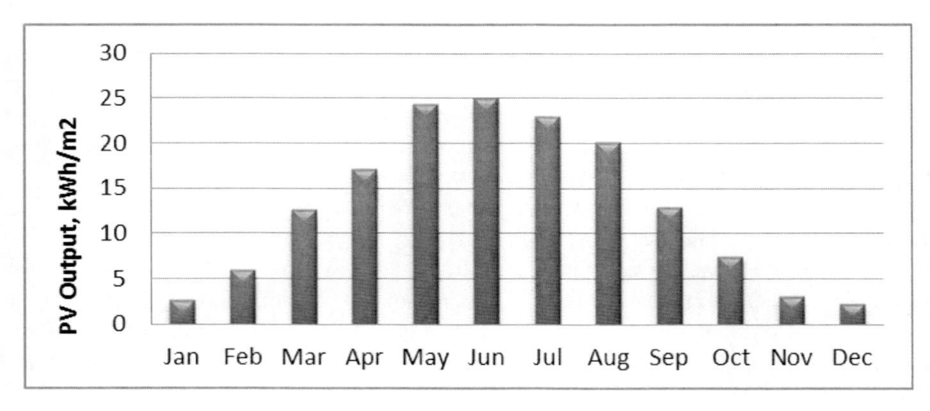

Fig. 3. Photovoltaic output at 15 % efficiency in Stockholm

The efficiency of the PV decreases by increasing the cell temperature, there are many factors that determine the operating temperature of PV such as ambient air temperature, type of module, intensity of sunlight and wind velocity.

The efficiency of a solar cell is usually determined under standard test conditions, i.e. cell temperature is 25 °C and normal incidence is 1000 W/m^2. Solar cells are generally exposed to temperatures ranging from 15 °C to 50 °C (Singh and Ravindra 2012). Mainly, the operating temperature of the module is higher than 25 °C and the angle of incidence is larger than 0° which is not considered in the WINSUN program. So, the output of PV should be multiplied by a correction factor φ. The correction factor varies for different months depending on the cell temperature, solar intensity and solar angle. Measurements conducted in Switzerland from 1992 to 1996 monitored the monthly correction factor for a PV with 45 ° tilt, see Figure 4 (Häberlin 2012).

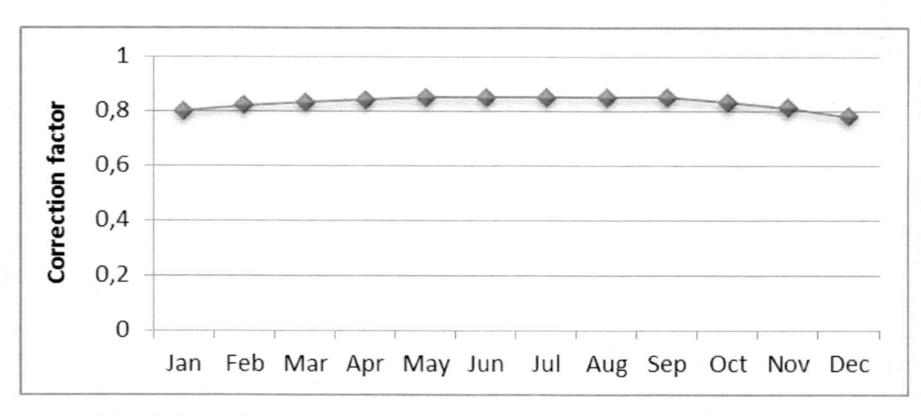

Fig. 4. Correction factor of PV output for different months (Häberlin 2012)

As mentioned before, to avoid overloading in the summer the PV should be designed to supply the electricity for DHW production. By monthly net metering of the simulation, to determine the module area (Equation 2), the month which gave the maximum output of PV was June. Thus, the production and load are in balance in June. Hence, to find the required area for the PV the energy demand (kWh) for this month was divided by the PV output (kWh/m^2) for the same month. So, by dividing 156.2 kWh by 21.3 kWh/m^2 the area demand of the PV was found to be 7.3 m2.

$$Area(m^2) = \frac{\text{energy demand in Jun(kWh)}}{\text{energy produced by PV in Jun(kWh/m2)}} \qquad (2)$$

To find the annual solar fraction (SF, Equation 3) the electricity generated by 7.3 m^2 PV was divided by the electricity consumption by the HP during the whole year. Figure 5 gives the comparison between production and consumption of electricity for each month. Simulation results showed that the solar fraction is 17 %. However, if export to the grid were allowed the solar fraction might be improved by increasing the area of the PV and then ignoring the overloading problem.

$$SF = \frac{\text{Electricity generated by PV (kWh/year)}}{\text{Electricity consumed by HP (kWh/year)}} * 100 \qquad (3)$$

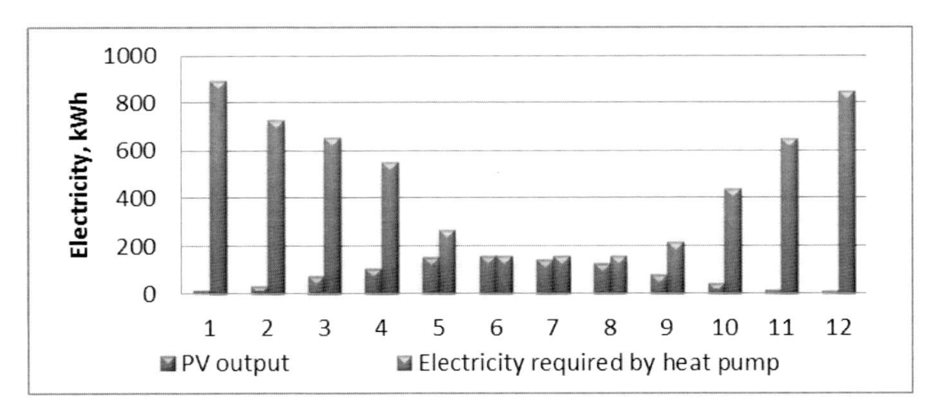

Fig. 5. Annual solar fraction calculation for total heating demand (space heating + domestic hot water)

4 Conclusion

The objective of this study was to investigate and evaluate the performance of the PV-HP system for the building in Stockholm. In addition, the solar fraction was calculated. The method included two simulation programs, one for calculating the energy demand of the building and the other for designing a PV system. Investigation of energy performance was conducted using the IDA ICE program. Simulation showed that the total energy consumption in the building is 14876 kWh/year including 10076 kWh for space heating and 4800 kWh for DHW. To avoid overproduction with monthly net metering, the PV-HP system should be designed to create a balance between production and demand during summer. WINSUN, a system simulation program, was used to design appropriate PV. Simulation showed that using 7.3 m^2 PV with 15 % efficiency would create a good balance during the summer season in generating and consuming electricity by PV and HP, respectively. In other words, the electricity consumed by HP is more or less totally supplied by PV during summer. The annual solar fraction (SF) in a designed HP-PV system was 17 %. If electricity generated by PV is used only for DHW the SF value is 51 % (Equation 4), i.e. more than half of the DHW need would be covered by implementing only 7.3 m^2 PV, Figure 6.

$$SF_{DHW} = \frac{\text{Electricity generated by PV (kWh/year)}}{\text{Electricity consumed by DHW (kWh/year)}} * 100 \tag{4}$$

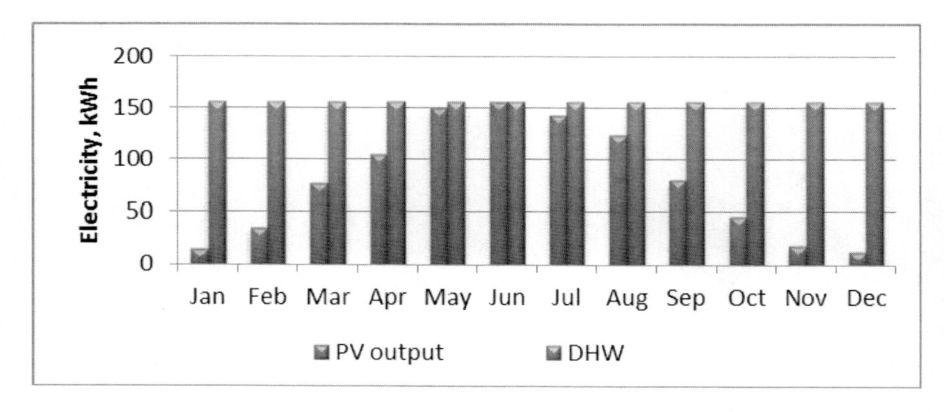

Fig. 6. Annual solar fraction calculation for domestic hot water consumption

It can be concluded that by implementing the HP-PV system with monthly net metering a relatively high solar fraction is achieved by the small PV area. The solar fraction may be increased when the heat loss is decreased which can be reached by minimising transmission, ventilation and leakage losses, i.e. using low U-value materials with high air tightness in the building envelope or using energy-saving equipment such as a heat pump or heat exchanger ventilation. Also, the solar fraction may be increased by increasing the PV area if the monthly net metering is disregarded.

In this project, combining the heat pump with PV system (PV-HP) was introduced as an energy-efficient and sustainable solution. The need to supply 20 % of energy demand by renewable energy by 2020, as a European Union target (European Environment Agency Report 2006), will lead to the design of more sustainable and energy-efficient systems. Using renewable and sustainable systems particularly in the building sector will reduce environmental problems such as carbon dioxide concentration and the global warming effect.

References

Beckman, W.A., Broman, L., Fiksel, A., Klein, S.A., Lindberg, E., Schuler, M., Thornton, J.: TRNSYS, The most complete solar energy system modeling and simulation software. Renewable Energy 5, 486–488 (1994)

Bhargava, A.K., Garg, H.P., Agarwal, R.K.: Study of a hybrid solar system-solar air heater combined with solar cells. Energy Convers 31, 471–479 (1991)

Boström, T.K., Wäckelgård, E., Karlsson, B.: Design of solar system with high solar fraction in an extremely well insulated house. In: Proceeding In Solar Energy Conference in Gothenburg, Sweden (2003), http://www.lu.se/o.o.i.s?id=12722&postid=1040078

Boverkets byggregler, BBR, Swedish Building Regulation (2002)

Duffie, J.A., Beckman, W.A.: Solar engineering of thermal processes, 2nd edn. Wiley-interscience publication, New Jersey (1991)

European Environment Agency Report No 8/2006: Energy and environment in the European Union - Tracking progress towards integration, ISBN 92-9167-877-5

Gajbert, H.: Solar thermal energy systems for building integration, Licentiate Dissertation, University of Lund, Sweden (2008)

Häberlin, H.: Introduction in Photovoltaics: System Design and Practice. John Wiley & Sons, Ltd., Chichester (2012), doi:10.1002/9781119976998

Hatwaambo, S., Jain, P., Perers, B., Karlsson, B.: Projected beam irradiation at low latitudes using Meteonorm database. Renewable Energy 34, 1394–1398 (2008)

Hesaraki, A., Holmberg, S.: Energy Performance Evaluation of New Residential Buildings with a Low-Temperature Heating System. In: Proceedings of the Second International Conference on Building Energy and Environment, Colorado, USA (2012)

Jokisalo, J.: On design principles and calculation methods related to energy performance of building in Finland. Doctoral Dissertation, Helsinki University of Technology, Finland (2008)

Lund, H.: Renewable energy strategies for sustainable development. Energy 32, 912–919 (2007)

Singh, P., Ravindra, N.M.: Temperature dependence of solar cell performance analysis. Solar Energy Materials and Solar Cells 101, 36–45 (2012)

Travesi, J., Maxwell, G., Klaassen, C., Holtz, M.: Empirical validation of Iowa energy resource station building energy analysis simulation models. IEA Task 22. Subtask A (2001)

Tyagia, W., Kaushika, S.C., Tyagib, S.K.: Advancement in solar photovoltaic/thermal (PV/T) hybrid collector technology. Renewable and Sustainable Energy Reviews 16, 1383–1398 (2012)

Chapter 19
Assessment of Solar Radiation Potential for Different Cities in Iran Using a Temperature-Based Method

Farivar Fazelpour[1], Majid Vafaeipour[1,*], Omid Rahbari[1], and Mohammad H. Valizadeh[2]

[1] Azad University, Tehran South Branch, Tehran, Iran
{F_Fazelpour,St_m_vafaeipour}@azad.ac.ir, Rahbari.omid@gmail.com
[2] Shiraz University of Technology, Shiraz, Iran
Valizadeh.mh@gmail.com

Abstract. The amount of solar irradiation in any region, is the most important required parameter for sizing and installing solar systems. Unavailability of this data has led to presenting different models for estimating its value. Using temperature based models is one of the most considered methods due to its simplicity and validity. Introducing various temperature based methods, Hargreaves and Samani's model has picked out to evaluate solar radiation potential in 4 different cities with various climate conditions and latitudes in Iran. Solar Radiation has been estimated in each city and the results are discussed. This investigation shows high solar radiation potential of Iran specially in Shiraz city.

Keywords: Energy Resources, Solar Energy, Solar Radiation Potential, Temperature Based Method.

1 Introduction

Due to environmental awareness, limitation of fossil resources, global warming and increasing energy demands, the importance of renewable energies is obvious to everyone. Implementing of renewable energy systems is considerably expensive. According to load profile, size determination is necessary for such systems which is called sizing. Cost effectiveness and economic issues plays an important role in sizing and planning for renewable energies. Detection of high potential regions is a significant preliminary step to have a Low-priced, optimized utilization and development in these systems which requires a vast study. For this, the hourly and annually collected long-term data must be analyzed in detail. Solar energy is one of the most considered renewable energies nowadays. Observing data to reach the accessibility and variability of solar radiation needs meteorological stations and devices (Pyranometers or Actinometers), but unavailability of these equipment for many regions is a problem of measuring Global Solar Radiation (GSR) and its density. In USA and Britain less than 1% of stations are capable of measuring solar radiation, so it can be guessed less than

* Corresponding author.

A. Håkansson et al. (Eds.): *Sustainability in Energy and Buildings*, SIST 22, pp. 199–208.
DOI: 10.1007/978-3-642-36645-1_19

that globally [1]. Therefore developing various empirical, multi parameter, temperature-based, neural networks, sunshine-based, physical and etc. techniques to estimate global solar radiation [2] has been the interested field of many researchers. The distribution of global solar radiation on horizontal surface depends on factors such as extraterrestrial radiation, atmospheric transmittance, latitude, sunset hour angle, duration of sunshine and cloudiness of the location [3]. The amount of global solar radiation on earth surface changes site to site due to these reasons. Many studies have been done on Angstrom empirical method to calculate GSR on horizontal surface using measured sunshine duration [4-7]. The original Angstrom-type equation related monthly average daily radiation to clear day radiation at the considered location and average fraction of possible sunshine hours. Later on, J.K Page represented the equation based on extraterrestrial radiation on horizontal surface (H_0), instead of clear day radiation, to estimate monthly average daily global solar radiation on horizontal surface (H).

The modified form of Angstrom equation is ($\frac{H}{H_0} = a + b\frac{S}{S_0}$), where S and S_0 are the monthly average daily hours of bright sunshine and monthly average day length in hours respectively where a and b are empirical coefficients [8].

Researchers have developed the modified type of Angstrom's empirical relation based on sunshine duration for specified locations in Turkey, Greece, Spain and China [9-12]. The others have made efforts to use multi parameter models to estimate the GSR based on longitude, latitude, altitude and available meteorological parameters data for considered locations in Turkey, Egypt, Nigeria [13-15]. In Iran, a comparison between calculated amount and measured data using R^2, RMSE and MBE for 7 models has been done to recommend the most appropriate model in a semi-arid climate [16]. From the literature, it can be concluded that for different locations, the most fitted model to real amount of GSR is not always similar and using some models need various meteorological measured data which their absence leads to inefficiency of the method. For this, employing models which use the most available data and minimum parameters are of the considerable methods, such as sunshine-based models and temperature-based models. Temperature is one of the most hand in hand parameters with GSR. In addition, the average, maximum, and minimum temperature are the most readily and measurable data which can be employed to estimate GSR. Using different temperature-based models, researchers have estimated GSR for different locations [17-21].

Table 1 focuses on some presented models that utilize temperature data, their developed date and their required parameters for predicting GSR. The difference between the models with the same required parameters is driven from number of empirical coefficients used in models, the calculation method of parameters and the fundamentals of developing each one.

The mentioned models have been developed by modifying their coefficients. The base for most of the modified methods is Hargreaves and Samani's model [22]. At the current study, Hargreaves and Samani's model is employed to estimate GSR in different climates with various latitudes in Iran to get a step closer to evaluate the potential of solar energy for the selected city.

Table 1. Temperature-Based Models for estimation of Solar Radiation

Model (Authors et al.)	Year	Model Requirements
Almorox	2011	H_0, T_{max}, T_{min}
Duat	2011	H_0, T_{max}, T_{min}
Mahmood	2002	$H_0, T_{max}, T_{min}, \phi$,DOY,LDY
Annandale	2002	H_0, T_{max}, T_{min}, Z
Winslow	2001	$H_0, T_{max}, T_{min}, T_{mean}$,Hday, ϕ
Goodin	1999	H_0, T_{max}, T_{min}
Donatelli	1998	H_0, T_{max}, T_{min}
Allen	1997	H_0, T_{max}, T_{min}
Bristow	1984	H_0, T_{max}, T_{min}
Hargreaves	1982	H_0, T_{max}, T_{min}

H_0 extraterrestrial solar radiation, T_{max} daily maximum air temperature, T_{min} daily minimum air temperature, T_{mean} mean annual temperature, ϕ latitude, DOY day of year, LDY longest DOY, Hday half-day length.

2 Model and Methodology

Hargreaves and Samani, presented Eq 1. To estimate R_s (Solar Radiation), employing the difference between maximum and minimum temperatures.

$$R_s = K_r (T_{max} - T_{min})^{0.5} R_a \tag{1}$$

Where K_r is an empirical coefficient, R_s is solar radiation and R_a is extraterrestrial radiation. The unitless coefficient K_r which varies for different atmospheric conditions can be taken equal to 0.17 for arid and semi- arid climates. It also presented by Hargreaves equal to 0.16 and 0.19 for interior and coastal regions respectively [23]. Later on, Allen [24] and Annandale et al. [25] introduced correction factors for K_r which can be seen on equations (2) and (3).

$$K_r = K_{ra} (\frac{p}{p_0})^{0.5} \tag{2}$$

$$K_r' = (1 + 2.7 \times 10^{-5} Z) K_r \tag{3}$$

Where p (kPa), p_0 (kPa) and Z are the mean atmospheric pressure of the site, the mean atmospheric pressure at sea level (101.3 kPa) and elevation in meters respectively. Mean pressure of the site (p), can be calculated using measured data on the site or can be estimated using the elevation of the site according to Burman equation as it follows [26]. Measured mean pressure data has employed at the current study for each location.

$$P = P_o (\frac{293 - 0.0065Z}{293})^{5.26} \quad (4)$$

At the Eq 2., the value of empirical coefficient K_{ra} which should be calculated first to be applied in Eq 1. is equal to 0.17 and 0.2 and has been suggested by Allen for interior and coastal regions respectively [24]. At the present study Bushehr where is located beside Persian Gulf is a coastal region and the other cities are interior regions. It can be noted that the represented correction coefficient in Eq 2. is on the basis of the relativity of elevation and volumetric heat capacity of the atmosphere and Eq 3. takes the effects of reduced atmospheric thickness on R_s into account, also it is showed that Eq. 2 performs inefficient for elevations more than 1500 meters [17] which is not in contrast with the elevation of chosen sites of this study.

Mentioned in the literature review, Samani suggested the modified form of the empirical coefficient K_r. Using maximum and minimum temperature difference, Eq 5 is applicable for latitudes $7^{\circ}N$ to $55^{\circ}N$.

$$K_r = 0.00185(T_{max} - T_{min})^2 - 0.0433(T_{max} - T_{min}) + 0.4023 \quad (5)$$

To calculate the extraterrestrial radiation (R_a), for a given latitude equations 6 to 10 are conducted. Declination angle (δ), is the angle between the joining line of the centers of the sun and the earth and its projection on the equatorial plane , depends on the nth day of the year and can be obtained using Eq. 6.

$$\delta = 23.45 Sin[\frac{360}{365}(284 + n)] \quad (6)$$

Introducing the hour angle (ω) which the earth must rotate to take meridian plane under the sun, the sunset hour angle (ω_s) will be as follows in Eq. 7.

$$\omega_s = Cos^{-1}(-tan\phi \, tan\delta) \quad (7)$$

Where ϕ is the latitude angle of the location. The Eq 8 defines the angle between the sun's ray and perpendicular line to the horizontal plane which is called Zenith (Polar) angle.

$$Cos\theta_z = Cos(\phi)Cos(\delta)Sin(\omega_s) + \omega_s Sin(\phi)Sin(\delta) \quad (8)$$

The solar radiation outside the atmosphere on a horizontal plane (I_o) for nth day of the year, considering I_{SC} (solar constant) which is defined equal to 1367 $\frac{W}{m^2}$ by World Meteorological Organization (WMO) standard, is given in Eq 9.

$$I_0 = I_{SC}[1 + 0.033\,Cos(\frac{360n}{365})]\,Cos\theta_z \tag{9}$$

To calculate the integrated daily extraterrestrial radiation on a horizontal surface (R_a), Eq 10 can be employed.

$$R_a = \frac{24\times3600}{\pi} I_0 \tag{10}$$

3 Area and Data

For, Iran is a vast country with a large climate variety, 4 cities with different latitudes and different meteorological conditions is chosen to investigate the solar radiation using the introduced temperature-based model. The properties of the selected cities to study (Tehran, Shiraz, Yazd as interior regions and Bushehr as coastal region) are presented in Table 2.

Long-term measured maximum, minimum and average data of the temperature (°C) and pressure (kPa) for 2011 have obtained from the daily recorded meteorological data of Wunderground for each city [28]. For the year 2011, monthly maximum and minimum temperature as well as average atmospheric pressure data is given in Table 3 from measured data. Observing the data for the last 6 years, it concluded that the variety of temperature in each selected city was not too large, so the most updated data (2011) is used for this study. Using data, Figures 1 to 4 illustrate the monthly temperature changes for each selected city in 2011.

Table 2. Considered cities and their properties

Site	Elevation(m)	Latitude(°N)	Longitude(°E)
Tehran	1190.8	35.41	51.19
Yazd	1230.2	31.54	54.24
Shiraz	1488.0	29.36	59.32
Bushehr	19.6	28.59	50.50

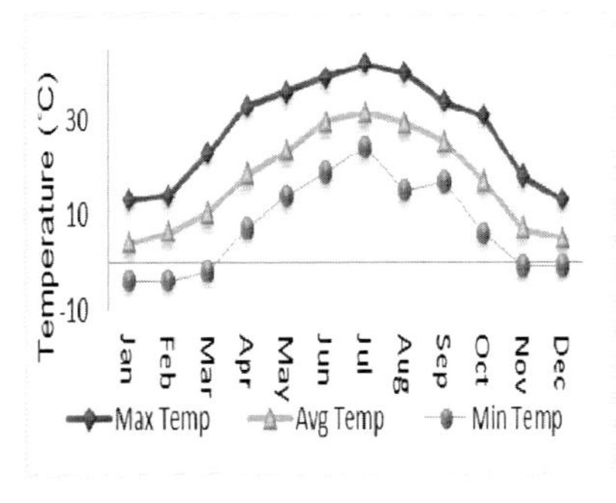

Fig. 1. Temperature, Tehran, 2011

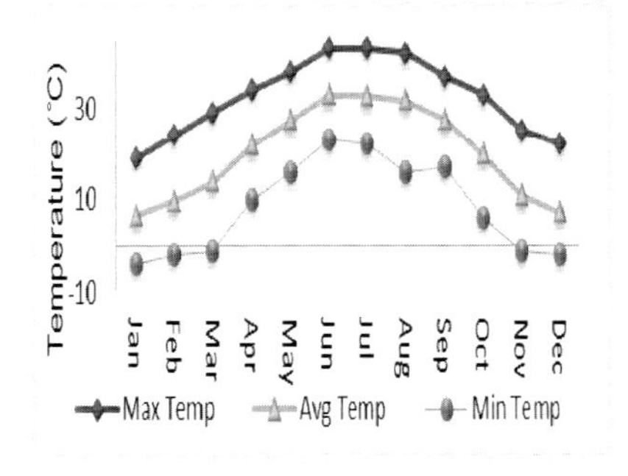

Fig. 2. Temperature, Yazd, 2011

Table 3. Maximum, minimum temperature and average atmospheric pressure of selected cities

	Tehran			Yazd			Shiraz			Bushehr		
	Max Temp	Min Temp	AVG P(kPa)	Max Temp	Min Temp	AVG P(kPa)	Max Temp	Min Temp	AVG P(kPa)	Max Temp	Min Temp	AVG P(kPa)
Jan	13	-4	101.85	19	-4	101.97	26	-6	102	24	8	101.67
Feb	14	-4	101.25	24	-2	101.45	21	-3	102	25	8	101.37
Mar	23	-2	101.73	29	-1	101.83	26	-2	102	32	12	101.41
Apr	23	7	101.43	34	10	101.58	30	4	102	37	14	100.91
May	36	14	101.45	38	16	101.51	37	7	102	43	22	100.53
Jun	39	19	100.94	43	23	100.94	40	11	101	39	25	99.81
Jul	42	24	101.06	43	22	101.04	40	17	101	45	26	99.66
Aug	40	15	101.1	42	16	101.03	40	5	101	40	25	99.7
Sep	34	17	101.43	37	17	101.46	36	12	102	39	23	100.27
Oct	31	6	101.81	33	6	101.91	32	3	102	36	16	101.07
Nov	18	-1	101.9	25	-1	102	25	-2	102	32	11	101.51
Dec	13	-1	102.28	22	-2	102.41	18	-5	102	24	7	101.92

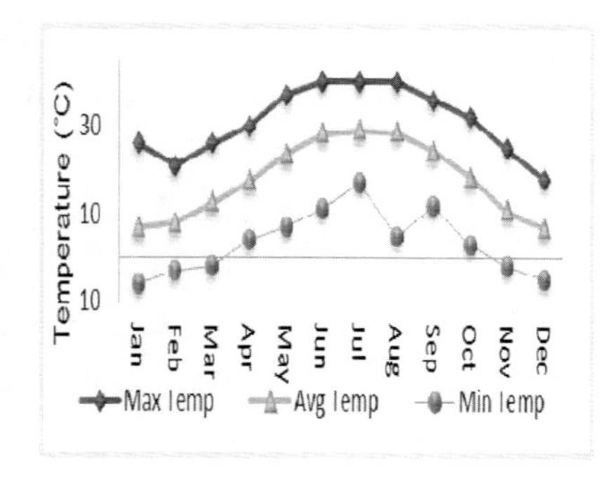

Fig. 3. Temperature, Shiraz, 2011

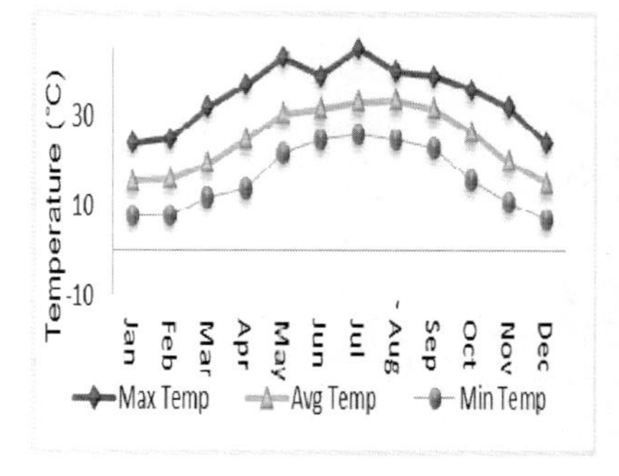

Fig. 4. Temperature, Bushehr, 2011

4 Discussion and Results

Employing the mentioned equations of previous section, Fig 9. And figures 5 to 8 illustrate the received monthly and daily R_s in 2011 for the studied cities respectively. It can be seen in figures 1 to 4, Shiraz and Yazd due to their desert characteristic have more temperature differences (ΔT), in comparison to Tehran and Bushehr. Considering the effect of climatal parameters as well as the locations, the illustrated graphs of R_s are logical in comparison (Fig 9.). Shiraz and Yazd which are located in arid climate have higher solar radiation potential than Tehran and Bushehr in all seasons of the year. The received solar radiation graph for Bushehr which is a southern humid coastal city and the nearest city to the equator, locates above Tehran's graph in cold seasons (Bushehr experiences more hot days than Tehran in a whole year). Althogh Bushehr is located at a lowest latitude and is nearer to the equator, its lower total R_s in comparison to other cities can be explained by its climate and coastal location (higher humidity).

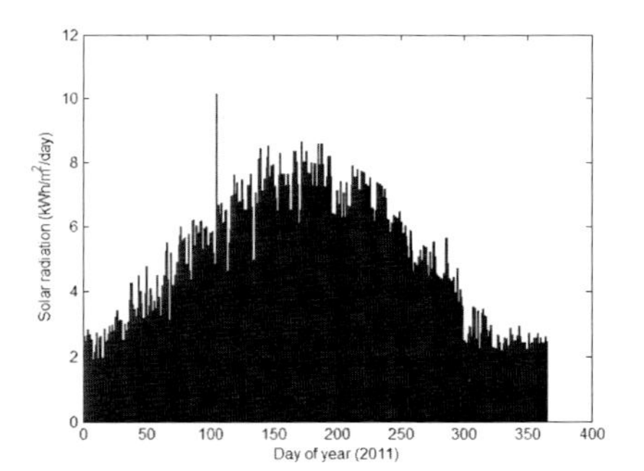

Fig. 5. Daily solar radiation (Tehran-2011)

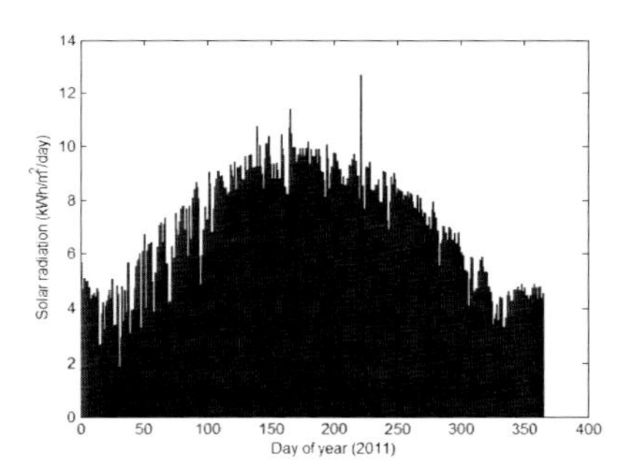

Fig. 6. Daily solar radiation (Shiraz-2011)

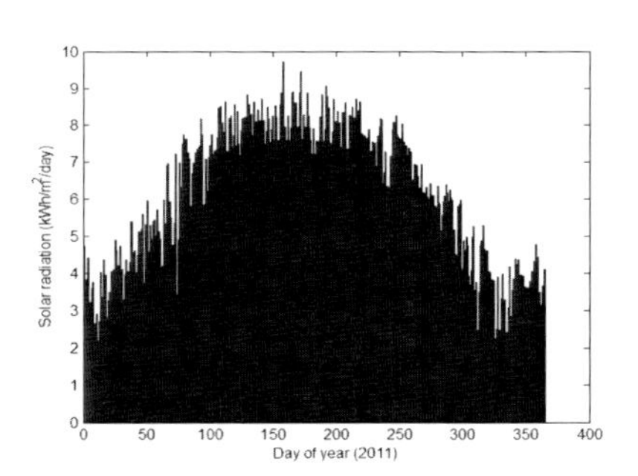

Fig. 7. Daily solar radiation (Yazd-2011)

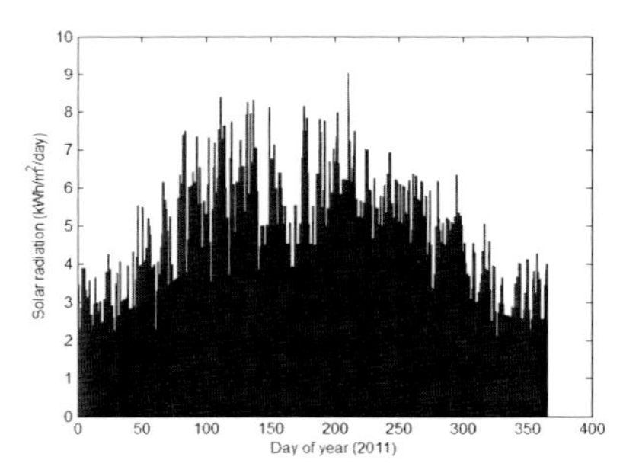

Fig. 8. Daily solar radiation (Bushehr-2011)

Table 4. Daily average of K_r for each city

	Tehran	Yazd	Shiraz	Bushehr
Jan	0.2005	0.2007	0.2006	0.2004
Feb	0.2	0.2001	0.2002	0.2001
Mar	0.2004	0.2005	0.2005	0.2001
Apr	0.2001	0.2003	0.2005	0.1996
May	0.2001	0.2002	0.2003	0.1992
Jun	0.1996	0.1996	0.1998	0.1985
Jul	0.1997	0.1997	0.1998	0.1984
Aug	0.1998	0.1997	0.1998	0.1984
Sep	0.2001	0.2002	0.2002	0.199
Oct	0.2004	0.2006	0.2006	0.1998
Nov	0.2006	0.2007	0.2006	0.2002
Dec	0.201	0.2011	0.2009	0.2006

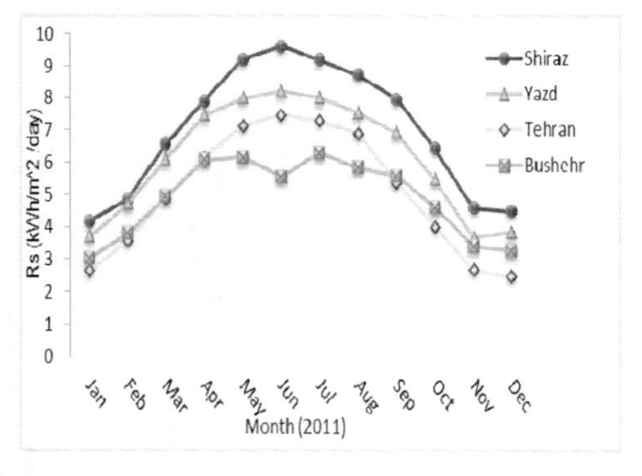

Fig. 9. Daily average solar radiation (2011)

Fig. 10. Total annual solar radiation

5 Summary and Conclusion

To harness the energy of the sun and installing solar power plants, stand alone solar thermal systems and solar collectors, sizing and economic evaluation should be performed for location where the system works at. Therefore, calculating solar radiation (R_s) must be considered as the preliminary step of employing solar energy to provide the essential information of system sizing. Recording R_s data needs meteorological devices. Purchasing, installing and maintenance of such instruments is costly. In the other hand, R_s data is not accessible in many locations due to meteorological station limitations in terms of quantity and equipment. Temperature is the most readily available data in every location. Different suggested models of estimating R_s discussed in the literature with a focus on temperature based models and their parameters. Using the famous temperature based method of Hargreaves and Samani, different amount of R_s discussed and estimated for 4 cities in Iran. The variety of climatal conditions, latitudes and numbers of cold and warm seasons thorough the year discussed as influencing parameters of the results.

The estimated R_s for these different cities shows the high potential of solar energy as a power generating source in Iran. The net consumption and total energy primary consumption in 2008 for Iran reported 166.24 billion kWh and 8.099 quadrillion Btu respectively by U.S energy information administration [29]. With 179 TWh predicted consumption for 2015 Just 4% of the generated energy is generating by solar energy now with 2600 shining hours per year [30]. Fig 10. illustrates total annual estimated R_s in this study for each city. The sum of estimated solar energy potential for the 4 investigated cities ends up to 8.43 MWh per year which is a high solar energy potential to employ in Iran. Due to high energy demands, population growth and very high fossil resources consumption (considering the exportings) of the country, this intensive solar energy should be considered as an important source of power generating.

References

1. Thorton, P.E., Running, S.W.: An improved algorithm for estimating daily solar radiation from measurements of temperature, humidity and precipitation. Agric. Forest Meteorol. 93, 211–228 (1999)
2. Rehman, S., Mohandes, M.: Artifficial neural network estimation of global solar radiation using air temperature and relative humidity. Energy Policy 63, 571–576 (2008)
3. Duzen, H., Aydin, H.: Sunshine-based estimation of global solar radiation on horizontal Surface at Lake Van region (Turkey). Energy Conversion and Management 58, 35–46 (2012)
4. Angstrom, A.: Solar and terrestrial radiation. Journal of the Royal Meteorological Society 50, 121–126 (1924)
5. Bakirci, K.: Correlations for estimation of daily global solar radiation with hours of bright sunshine in Turkey. Energy, doi:10.1016/j.energy.2009.02.005
6. Rehman, S.: Solar radiation over Saudi Arabia and comparison with empirical models. Energy 23(12), 1077–1082 (1998)
7. Aguiar, R., Collares-Pereira, M.: A time dependent autoregressive, Gaussian model for generating synthetic hourly radiation. Solar Energy 49, 167–174 (1992)
8. Jamil Ahmad, M., Tiwari, G.N.: Solar radiation models - review. International Journal of Energy and Environmrnt 1(3), 513–532 (2010)
9. Iris, M., Tiris, C., Erdalli, Y.: Water heating system by solar energy. Marmara research centre, Institute of Energy systems and environmental research, NATO-TU-COATING, Gebze, Turkey (1997)
10. Zabara, K.: Estimation of global solar radiation in Greece. SolarWind Technology 7, 267–272 (1986)

11. Almorox, J., Hontoria, C.: Global solar radiation estimation using sunshine duration in Spain. Energy Conversion and Management 45(9-10), 1529–1535 (2004)
12. Newland, F.J.: A study of solar radiation models for the coastal region of South China. Solar Energy 31, 227–235 (1988)
13. Togrul, I.T., Togrul, H., Evin, D.: Estimation of monthly global solar radiation from sunshine duration measurement in Elazig (Turkey). Renewable Energy 19, 587–595 (2000)
14. Trabea, A.A., Mosalam Shaltout, M.A.: Correlation of global solar radiation with meteorological parameters over Egypt. Renewable Energy 21, 297–308 (2000)
15. Ojosu, J.O., Komolafe, L.K.: Models for estimating solar radiation availability in South Western Nigeria. Nigerian Journal of Solar Energy 16, 69–77 (1987)
16. Mousavi-Baygi, M., Ashraf, B., Miyanabady, A.: The Investigation of different Models of estimating Solar Radiation to Recommend the Suitable Model in a Semi-arid Climate. Journal of Water and Soil 24(4), 836–844 (2010) (in Persian)
17. Allen, R.G.: Self-calibrating method for estimating solar radiation from air temperature. J. Hydrol. Eng. 2, 56–67 (1997)
18. Bristow, K.L., Campbell, G.S.: On the relationship between incoming solar radiation and daily maximum and minimum temperature. Agric. Forest Meteorol. 31, 59–166 (1984)
19. Meza, F., Varas, E.: Estimation of mean monthly solar global radiation as a function of temperature. Agric. Forest Meteorol. 100, 231–241 (2000)
20. Goodin, D.G., Hutchinson, J.M.S., Vanderlip, R.L., Knapp, M.: Estimating solar irradiance for crop modeling using daily air temperature. Data Agron J. 91, 845–851 (1999)
21. Mahmood, R., Hubbar, K.G.: Effect of time of temperature observation and estimation of daily solar radiation for the Northern Great Plains. USA. Agron J. 94, 723–733 (2002)
22. Hargreaves, G.H., Samani, Z.A.: Reference crop evapotranspiration From ambient air temperature. International Irrigation Center. Presentation at the 1985 Winter Meeting American Society of Agricultural Engineers, Logan, UT, USA (1985)
23. Hargreaves, G.H.: Simplified Coefficients for Estimating Monthly Solar Radiation in North America and Europe, Dept. Paper, Dept. Biol. and Irrig. Engineering, Utah State University, Logan, Utah
24. Allen, R.G.: Evaluation of procedures for estimating mean monthly solar radiation from air temperature. Rep, (FAO) Rome, Italy (1995)
25. Annandale, J.G., Jovanovic, N.Z., Benade, N., Allen, R.G.: Software for missing data error analysis of Penman–Monteith reference evapotranspiration. Irrig. Sci. 21, 57–67 (2002)
26. Burman, R.D., Jensen, M.E., Allen, R.G.: Thermodynamic factors in evapotranspiration. In: Proceedings of the Irrigation and Drainage Speciality Conference, pp. 28–30. ASCE, Portland (1987)
27. http://www.wunderground.com
28. http://www.eia.gov/countries/country-data.cfm?fips=IR
29. Manzoor, D.: Renewable Energy Development in Iran. In: Middle East and North Africa Renewable Energy Conference As a Preparatory Meeting for the "Renewables 2004", Germany (2004)

Chapter 20
A Decision Support Framework for Evaluation of Environmentally and Economically Optimal Retrofit of Non-domestic Buildings

Taofeeq Ibn-Mohammed[1,*], Rick Greenough[1], Simon Taylor[2], Leticia Ozawa-Meida[1], and Adolf Acquaye[3]

[1] Institute of Energy and Sustainable Development, De Montfort University, Leicester
tibn-mohammed@dmu.ac.uk
[2] School of Civil and Building Engineering, Loughborough University, Loughborough
[3] Centre for Energy, Environment and Sustainability, University of Sheffield, Sheffield

Abstract. Currently, the building sector has an oversized carbon footprint as it represent the single largest contributor to global greenhouse gas emissions (GHG), with approximately one third of global energy end use taking place within buildings. The challenge to successfully reduce the energy consumption in the building sector is to find effective strategies for retrofitting existing buildings. Significant emissions reductions are possible from applying low carbon retrofit intervention options to existing buildings. The choice of low carbon retrofit intervention options involves evaluation of applicability, energy end uses, environmental impact and cost of application versus energy savings. To develop energy efficiency strategies for building stock, there is the need for optimised methodologies and decision aid tools to evaluate whole-life economic and net environmental gain of the options. This paper describes the development of an integrated framework for a Decision Support System (DSS) based on the optimal ranking and sequencing of retrofit options for emissions reduction in non-domestic buildings. The DSS framework integrates economic (cost) and net environmental (embodied and operational emissions) cost or benefit parameters and an optimization scheme to produce an output based on ranking principles such as marginal abatement cost curve (MACC). The methodology developed can be used to identify and communicate trade-offs between various refurbishment options to aid decisions that are informed both by environmental and financial considerations.

Keywords: Decision Support Systems Economics, Emissions, Non-domestic building, Retrofit.

1 Introduction

Empirical evidence in different parts of the world regarding the depletion of world's oil reserves due to increased energy consumption and the resultant increase in

---

* Corresponding author.

A. Håkansson et al. (Eds.): *Sustainability in Energy and Buildings*, SIST 22, pp. 209–227.
DOI: 10.1007/978-3-642-36645-1_20

greenhouse gas (GHG) emissions have motivated the promulgation of a number of international treaties including the Kyoto Protocol. Increase in anthropogenic GHG has caused a rise in global average temperatures and triggered other climate changes such as rise in sea levels, coastal line erosion, and desertification (IPCC, 2007). As such, these treaties have been passed to help protect the global environment and promote environmental sustainability and it requires both industrialised and developing countries with market economies to reduce GHG emission on a global scale. In the UK, the Government has accepted relatively high burden-sharing commitments within the EU under the Kyoto Protocol and has been at the forefront of diplomatic solutions and policy development (DECC, 2009). Following The Royal Commission on Environmental Pollution (RCEP) report in 2000, the Government recommended that UK CO_2 emissions be reduced by 60%, of the then current level, by 2050, and has since increased the target figure to 80%. The transition of the UK to a low carbon economy as outlined in the White Paper: *UK Low Carbon Transition Plan* and underpinned by the 2008 Climate Change Act is expected to be driven by maintaining secure energy supplies, maximising economic opportunities and more importantly cutting emissions from every sector including the decarbonisation of energy intensive sectors of the economy.

The building sector is one of the key sectors targeted for energy efficiency improvement and emissions reduction in the UK. This sector contributes nearly 47% of the UK's total emissions (Carbon Trust, 2008). In 2006, emissions in buildings and industry were reported to be 400 $MtCO_2$, accounting for 70% of total UK CO_2 emissions (CCC, 2008). Within this, emissions from non-domestic buildings (public sector and commercial buildings) were around 78 $MtCO_2$, with emissions from residential buildings accounting for 149 $MtCO_2$, and the rest of the industry accounting for the remaining 155 $MtCO_2$ (CCC, 2008). The UK's stock of 1.8 million non-domestic buildings is highly varied in size, form and function, and proposing carbon-saving solutions to such a diverse group of buildings is non-trivial (Jenkins *et al.*, 2009). These buildings use around 300TWh of energy a year, (equivalent to the entire primary energy supply of Switzerland (IEA, 2006)); predominantly for heating, ventilation and lighting. Currently, annual emissions from existing non-domestic buildings in the UK are estimated to be over 100 $MtCO_2$ (Caleb, 2008). Despite these levels of emissions, multiple studies such as Taylor et al. (2010), McKinsey, (2008), CCC (2008) and IPCC (2007) have all shown that there is a large potential for carbon emissions reduction from non-domestic buildings, much of which is cost-effective using low-cost technologies and solutions which exist today.

As a result, a great deal of attention has been focused on buildings, and the UK government has signalled its intention to make all newly built non-domestic buildings carbon-neutral by 2019 (Energy Saving Trust, 2009). However, Hinnells *et al.*, (2008) reports that over 90% of the UK's building stock beyond 2030 will consist of buildings already existing today. It is therefore clear that the largest potential for improving energy performance of the building stock lies in the existing buildings. It is estimated that retrofitting existing buildings could save 15 times more CO_2 by 2050 than their demolition and replacement (Jowsey and Grant, 2009). Refurbishment minimises the time and cost involved in improving the energy efficiency of a building

(Carbon Trust, 2009; Energy Saving Trust, 2009). In addition, it can reduce energy use in buildings in both short- and long-term (Corus, 2010). To this end, in the past decades, there have been significant efforts towards designing, operating, refurbishing and maintaining energy efficient buildings with low environmental impact. It is therefore essential to adopt low carbon intervention measures in the most effective and optimal manner towards the minimisation of energy consumption and environmental impact of buildings.

Fortunately, recent technological advances offer promising retrofit solutions including renewable energy technologies, energy efficiency measures and inducements to change behaviour, to improve energy efficiency of buildings. Improving energy efficiency, through, for example, the implementation of voltage optimisation is one option that can reduce energy consumption in buildings. However, there are numerous technically feasible options with varying costs and different energy-saving potentials available for energy retrofit of buildings. Measures adopted to improve the energy performance of buildings using technically feasible technologies are therefore often the result of an optimization process measured across mainly two key performance indicators (KPIs) namely: environmental (energy efficiency improvement and emissions reduction potential) and economic (cost-effective measures) (De Benedetto and Klemeš, 2009). In fact, the selection of an environmentally and economically optimal set of retrofit options requires a robust decision-making methodology which will allow these options to be compared in a consistent manner by evaluating both their economic cost and their environmental benefits. This paper therefore presents the principles and foundation for the development of an integrated framework for a Decision Support System (DSS) based on a techno-economic evaluation methodology for energy retrofit of buildings with the view to obtaining the environmentally and economically optimal set of retrofit options for emissions reduction in non-domestic buildings. The DSS framework integrates economic (cost) and net environmental (embodied and operational emissions) cost or benefit parameters and an optimization scheme to produce an output based on ranking principles derived from a marginal abatement cost curve (MACC).

2 Decision Support Systems

The term Decision Support Systems (DSS) is a context-free expression (De Kock, 2003) which may mean different thing to different people. Turban (1993) as cited by De Kock (2003) asserts that there is no universally accepted model or definition of DSS, because many different theories and approaches have been proposed in this broad field. Because there are numerous working DSS theories, DSS can be defined and classified in many ways. Gorry and Scott-Morton were the first to coin the phrase 'DSS' in 1971 and they define a DSS *as an interactive computer based system that helps decision makers utilize data and models to solve unstructured problems*. Another definition put forward by Keen and Scott-Morton (1978) is "*A DSS couples the intellectual resources of individuals with the capabilities of the computer to improve the quality of decisions. It is a computer-based support system for management decision makers who deal with semi-structured problems."* Essentially, a DSS is used to collect, process and analyse data in order to make sound decisions or construct strategies from the analysis. Decision Support Systems exist to help people make decisions. DSS do not make decisions by themselves (Mallach, 1994) but attempt to automate several tasks of the decision-making process, the core of which is modelling (Turban *et al.*, 2001). DSS therefore provides the framework that allows decision-makers to view alternatives and make informed decisions (De Kock, 2003). DSS differs in their scope, the decisions they support and their targeted users (Mallach, 1994). DSS for instance have been developed for agricultural production (Jones *et al*,1998), medical diagnosis (Fitzgerald *et al.*,2008), forest management (Kangas *et al.*,2008), product supply chain (CCaLC, 2010; SCEnAT, 2011) and many other sectors.

Within the building sector, various decision aid tools have also been developed to support and advice building stock owners with respect to retrofitting decisions for energy conservation (Kumbaroglu and Maslener, 2011). Recent research includes studies by Costa et al. (2012), Chidiac et al., (2011), Yin and Menzel (2011), Diakaki et al. (2010), Loh et al. (2010), Doukas et al. (2009) etc, which focus on development of DSS based on a number of variables and techniques for energy consumption and energy efficiency improvements in buildings. Diakaki et al., (2010) for instance, developed a multi-objective decision model for the improvement of energy efficiency in buildings. The model was constructed to allow for the consideration of a potentially infinite number of alternative options according to a set of criteria. However, the model yielded no optimal solution due to the competition between the disproportionate decision criteria involved. Chidiac et al., (2011) also developed a decision-making tool in the form of a screening methodology for cost-effective energy retrofit measures in Canadian office buildings. The methodology contained there-in assesses the profitability of an energy efficiency measure using discounted payback period rule. Although, the investment appraisal technique employed is appropriate since it accounts for the time value of money, it remains inaccurate due to fixed assumptions for interest rates, energy price scenarios, inflation and the rate of changes of these variables, thus indicating the need for a more robust approach.

Similarly, Doukas *et al.*, (2009) identified the need for intervention and further evaluation of energy-saving measures in an existing building using innovative intelligent decision support model, based on the systematic incorporation of building energy management system data. Consequently, the building's energy efficiency status is identified and energy-saving measures are proposed, including various retrofit options. The proposed solutions are then evaluated using appropriate investment appraisal techniques but economic parameters, such as interest rates, energy prices etc., are fed into the model using fixed assumption, thereby ignoring the uncertainty associated with high fluctuations in historical data of energy prices. The foregoing analysis therefore shows the need for a comprehensive techno-economic evaluation methodology that will take into account the aforementioned crucial factors and uncertainty in the environmental and economic analysis of retrofit options for buildings.

Furthermore, decision making of building stakeholders when facing the specification and uptake of alternative building refurbishment options is typically economically driven. However, recent trends towards environmentally conscious design and refurbishment have resulted in a focus on the environmental merit of these options, with emphasis on a life cycle approach. Also, the most significant carbon impact of buildings has historically been attributed to operational energy consumption. But due to the advent of energy efficient equipment and appliances along with more advanced and effective insulation materials; improvements in building fabric design; reduced air permeability; benign sources of renewable energy; etc, the potential for curbing operating energy has improved. Consequently, the relative proportion of carbon emissions embodied within the building increases and its contribution to total life cycle emissions becomes more significant (Thormark, 2006).

Currently, there is an increasing focus on the reduction of embodied emissions either through optimisation of building fabric to reduce material use or through the specification of materials with a lower embodied carbon. However, methodologies employed to reduce embodied and operational emissions are considered in isolation when making decisions about energy conservation in buildings. This paper therefore presents the foundation and principles for the development of an integrated framework for a Decision Support System (DSS) based on the optimal ranking and sequencing of retrofit options for emissions reduction in buildings. The novelty of this DSS framework presented in the paper lies in the whole-life environmental and economic assessment approach taken to the integration of financial cost, embodied and operational emissions within an optimisation scheme that consist of an integrated data input, optimisation, sensitivity analysis and MACC modules. The DSS therefore allow trade-offs between various refurbishment options to be identified and communicated and ensure decisions that are informed both by environmental and financial considerations.

3 Consideration of Embodied Energy in Decision Making

Analysis of the source of emissions shows that, across the EU, at least 40% are from buildings (UNEP, 2007) with a corresponding figure of 44% (17% from non-domestic buildings and 27% from domestic buildings) in the UK (Carbon Trust, 2008). However, these data measure operational emissions only- that is emissions related to the maintenance of comfort conditions and day-to-day running of the buildings by operating processes such as heating and cooling, lighting and appliances, ventilation and air conditioning- and ignore other important life-cycle components such as maintenance, demolition and, possibly most significantly, building-related embodied energy and emissions (Acquaye, 2010; Healey, 2009). Embodied energy of a product such as a building is the total energy consumed by all the processes associated with the construction of the building. Embodied CO_2 of the building however, is the CO_2 that is emitted as a result of all the energy that is consumed during the construction of the building and it can represent a significant proportion of the building's total lifecycle energy requirements (Crawford, 2005). The Climate Change Committee (2008), also reports that a typical new 2 bed home built with traditional materials (brick, concrete foundations etc.) embodies around 80 tCO_2 with a carbon payback time (through lower operational CO_2 emissions) of several decades.

The energy used and consequent CO_2 emissions associated with building construction materials and processes are usually calculated using the concept of embodied energy and CO_2 analysis, albeit with significant variations in methodology (Acquaye,2010). Traditionally, embodied CO_2 has been deemed optional in lifecycle emissions assessment of buildings because it was estimated to be of small magnitude compared with operational CO_2, however contemporary research have changed this view. Many studies have estimated varying proportions of embodied energy to total lifecycle energy. The differences is mainly due to the type of building been assessed, the use of building, construction methods employed and buildings materials used, geographic differences, etc. Hamilton-MacLaren *et al.*, (2009) for instance estimated that less than one-fifth of the whole-life energy usage in buildings can be attributed to embodied energy. He further stated that, as energy efficiency for new-build improves towards the zero carbon targets in 2016, with a corresponding increase in building refurbishment rates, the embodied energy will assume an increasingly greater proportion, approaching 100% of the lifetime energy use and emissions.

Sartori et al. (2006) also reported that for conventional buildings the ratio of embodied energy to total lifecycle energy can be as high as 38% and for buildings matching the definition of low carbon buildings, this ratio can increase to 46% (Thormark,2006). A comprehensive assessment of building related emissions which highlights the increasing importance of embodied emissions in building energy consumption through the review of the specific relationship between embodied emissions and operational emissions over the life cycle of buildings is detailed in Ibn-Mohammed *et al.*,(2012). A brief summary, which quantitatively illustrates the overall importance of embodied energy in the total life cycle for different buildings and infrastructure, as measured by the ratio of embodied energy to lifecycle energy is shown in Figure 1.

Embodied energy analysis has been identified to be an important part of life cycle energy assessment (Crawford, 2005) and is used to determine the energy-related environmental impacts such as CO_2 emissions of a product such as a building. Despite the importance of embodied energy emissions, policies targeted at the building sector have focused historically on promoting operational energy efficiency, the deployment of renewable energy supply (RES) technologies and have failed to directly target embodied CO_2 equivalent (ECO_2e). For instance, the 2007 Energy White Paper for UK (DTI, 2007) which sets out the energy policy framework from 2007 to 2020 reported the need to reduce total energy consumption by optimising energy efficiency, reducing operational energy use but overlooked the significant energy reductions that can be achieved through considerations to embodied energy in the UK. Given the recognised importance of lifecycle assessment in evidence-based decision making (Kenny *et al.,* 2010), this is a significant omission.

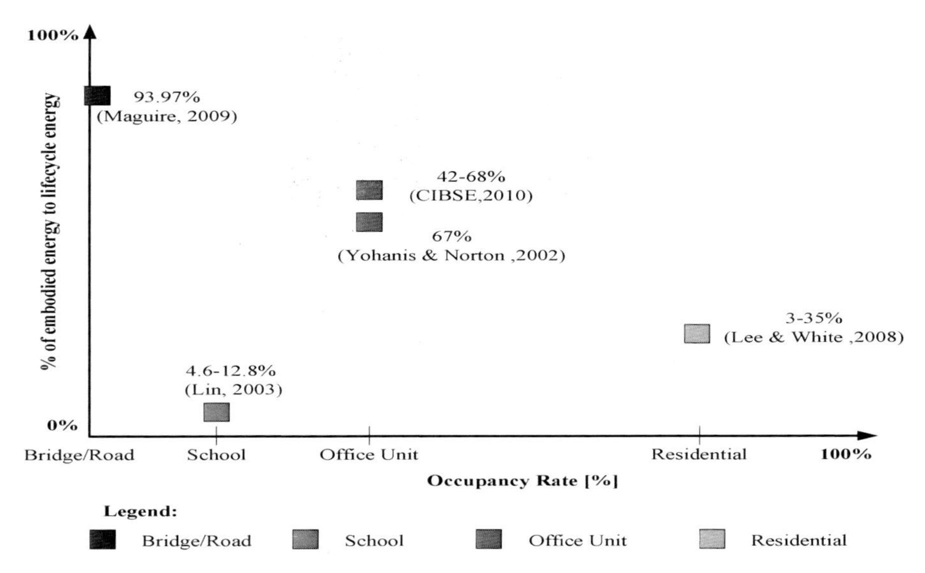

Fig. 1. Percentage ratio of embodied to life cycle energy in different infrastructure projects. See: Ibn-Mohammed et al (2012) for details. [Concept adapted from Acquaye,2010]

For many years, embodied energy has been part of the sustainability debate, but there is currently a lack of incentive integrating the calculation of embodied energy in construction decision making (Hamilton-MacLaren *et al.*, 2009). The primary reasons are partly due to methodological challenges, the focus of regulations on in-use energy and carbon, the lack of appropriate legislation and a lack of interest in the impacts of embodied energy by the public and industry stakeholders (Hamilton-MacLaren *et al.*, 2009; Rawlinson and Weight, 2007). In addition, the lengthy and demanding data collection process needed for quantification makes accounting for embodied CO_2 difficult, since tracking material sources from all their origins requires reliable data on established manufacturing processes and supply chains (Engin and Frances, 2010). The time-consuming nature of embodied energy calculation and the varying accuracy of the obtained results has to a very large extent restricted its use in the decision making process in building energy assessment and performance analysis. Furthermore, the complexity and uncertainty in embodied energy and CO_2e analysis is made worse by problems with data collection, variations in technologies and the number, diversity and interactions of processing steps (Acquaye, 2010). Pacca and Horvath (2002) also noted that uncertainties in embodied energy analysis can also arise from economic boundary and methodological constraints. Such uncertainties and variability affects decision making.

As such, there is currently no generally accepted method available to compute embodied energy accurately and consistently (Acquaye, 2010), although a general framework exist in the ISO 14000 series standards, and as a result, wide variations in measurement figures are inevitable, owing to various other factors responsible for variation and inconsistency in embodied energy results as detailed in Dixit et al., (2010) and Hamilton-MacLaren et al. (2009). However, in the pursuit of zero-carbon buildings, embodied emissions are becoming more significant, as operational emissions of buildings fall in response to regulations. Also, with the new Government's definition of useful benchmarks in the traded/non-traded price of carbon[1], which reflects the global cost of the damage a tonne of carbon causes over its lifetime, and which have been used to appraise policies and proposals set out in the UK Low Carbon Transition Plan (DECC, 2009), embodied CO_2 is likely to become one of the key metrics to address in whole-life building sustainability (Engin and Frances, 2010). Its inclusion in the decision-making process is therefore of utmost importance.

If, as forecasted, embodied energy becomes a target for emissions reduction, it will become necessary for traditional construction companies to quantify the emissions associated with their projects. This will potentially allow the emissions of the sector as a whole to be evaluated and allow a more accurate proportioning of responsibilities for the overall emissions of the country (Hamilton-MacLaren *et al.*, 2009). It will also encourage construction companies to reuse and repurpose their technology to develop renewable energy, re-use construction materials and improve social services. Consideration of embodied energy will assist in putting operational emissions savings in context and can trigger well considered improvement initiatives with a positive carbon reduction profile (Rawlinson and Weight, 2007).

At a macro-level, taking embodied and operating emissions into account will contribute to data and information required to create an energy economy that accounts for indirect and direct contributions (Dixit et al., 2010). Better awareness of the embodied energy content of building materials will encourage not only the production and development of low embodied energy materials, but also their preference among building designers to curb energy use and carbon dioxide discharge (Ding, 2004). In addition to the benefit to the environment, consideration of embodied energy at the design stage of construction or refurbishment projects will enable significant contributions to sustainable development of the nation's building stock. It is therefore pertinent to acknowledge the importance of embodied emissions when making decisions regarding carbon reduction strategies, and to calibrate the performance of buildings in terms of both embodied and operating energy in order to reduce lifecycle energy consumption. In the wake of increase global awareness on sustainable design and the strong link between global warming and CO_2 emissions, the role of new and improved DSS models to evaluate whole-life economic and net environmental gain is crucial as it can play an important role, especially in the early stage of future design processes in refurbishment projects to ensure best practise and sustainable designs.

4 Methodological Framework of the DSS

4.1 System Definition and Structure

For every refurbishment project relating to non-domestic buildings, questions such as: "*What options are applicable to reduce building emissions now and in the future?*

[1] A short term traded price of carbon of **£25 in 2020, with a range of £14 - £31.** A short term non-traded price of carbon of **£60 per tonne CO2e in 2020, with a range of +/- 50% (i.e. central value of £60, with a range of £30 - £90).**

How cost effective are these measures? What will be the return on investment? What is cheapest option now and in the long-term? What is the best combination of options and what strategy should be adopted?" will be considered in a different way by an investor and an environmentalist. The sole desire of the investor is a high financial return, whereas the environmentalist will prioritise GHG emission reduction. These are questions that require an engineering solution as much as an economic one. Answering the questions effectively depends not only on the level of expertise available but also on the capability of decision aid tools available to the analyst. The aim of the DSS presented in this paper is therefore to determine the optimal order in which a range of retrofit intervention options should be implemented to reduce greenhouse gas emissions in non-domestic buildings, taking into account both operational and embodied emissions and the cost of each option. Choosing the right options can be achieved by following the basic steps illustrated in Figure 2.

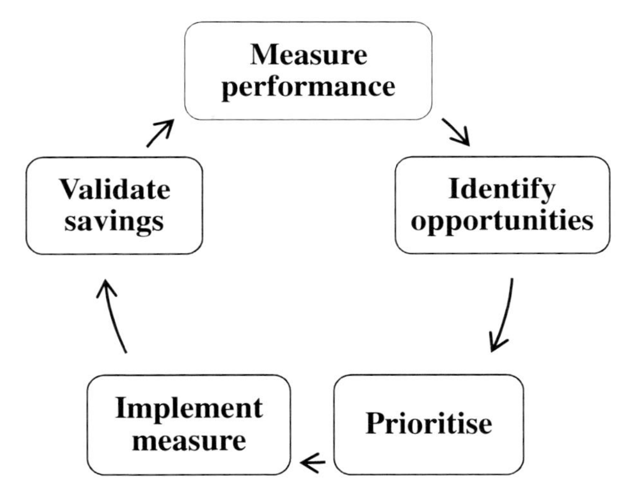

Fig. 2. Steps to choosing right options in building performance analysis

The methodological framework for the DSS is based on 5 modules which include: (i) a module for the computation of the baseline energy consumption of the building. This involves measuring energy use and energy intensity of the building at a determined level of detail for the purpose of establishing a benchmark for future comparison to itself; (ii) a module that include technically feasible low carbon intervention measures, their potential energy and CO_2 savings, investment and operating cost estimates;(iii) a module for the computation of the embodied emissions related to each low carbon intervention measure; (iv) an economic evaluation module which is based on an appropriate investment appraisal technique which incorporates sensitivity analysis under different energy price scenarios, interest rates and changes in policies; (v) an optimisation module which will integrate the measures of financial cost, operational and embodied emissions into a robust method of ranking and sequencing building energy retrofit options taking into account the interdependency of measures. The overall structure of the DSS is depicted in Figure 3.

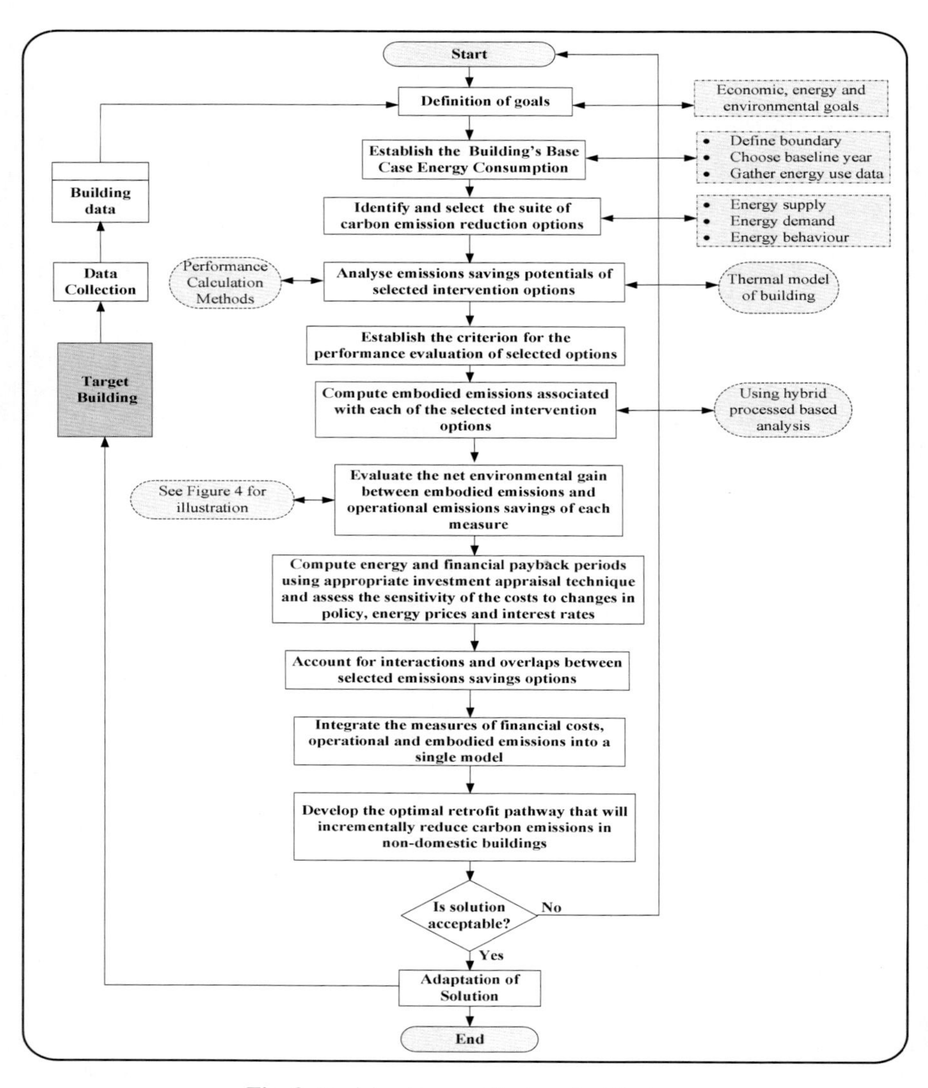

Fig. 3. Decision Support System Architecture

4.2 System Requirements

As shown in the preceding section of this paper, the information requirements for the DSS are cost, operational emissions (OE) and embodied emissions (EE). For each of the identified low carbon intervention options, including energy efficiency measures, renewable energy generation technologies and inducements to change behaviour; analysis of their potential operational emissions savings will be undertaken. This will be based on performance calculation methods using standard algorithms for low carbon energy sources, post implementation analysis and evaluations from an existing computer thermal model of the target building. Furthermore, the embodied emissions related to each low carbon intervention measure will also be evaluated. This will allow for the evaluation of the net environmental gain in terms of the embodied emissions of a low carbon intervention measure and the corresponding operational emissions savings after its implementation. This is illustrated in the data flow diagram shown in Figure 4.

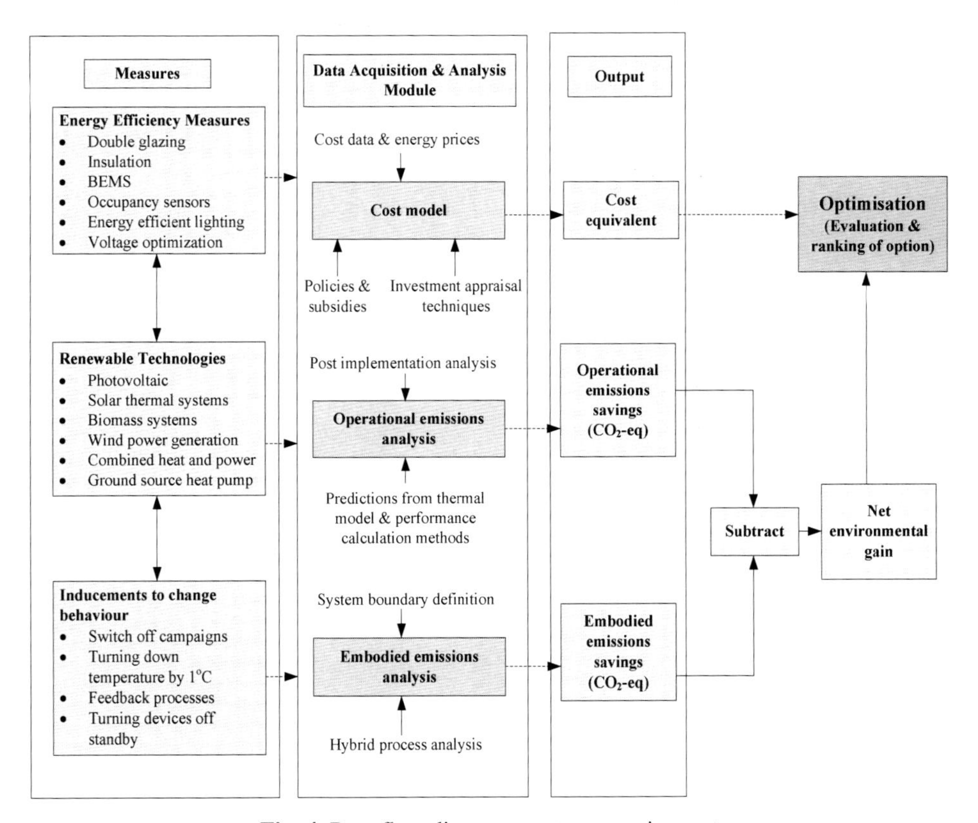

Fig. 4. Data flow diagram: systems requirement

The output will consist of a detailed set of results incorporating the expected CO_2 savings from each low carbon intervention options. A lifecycle costing method that incorporates the use of appropriate investment appraisal techniques will also be employed to assess and compare the performances of building retrofit options against the defined criteria. This will give an indication of financial benefits and carbon payback periods over the life span of the measures. The planned output will indicate the scenarios where measures that lead to net carbon reduction and also save money, and will put into perspective measures where the investment cost cannot be recovered. This will be in the form of a marginal abatement cost curve (MACC) as discussed in section 4.3.

4.3 System Output: MACC

The DSS is intended to identify the optimal order in which a range of retrofit intervention options should be implemented to reduce greenhouse gas emissions in non-domestic buildings, taking into account operational and embodied emissions as well as cost. Prioritising is important in practice because due to financial costs, project timelines and other constraints, the implementation of the building energy retrofit options is unlikely to be achieved in a single operation. One of the ranking methods employed to prioritise CO_2 emissions reductions options of an abatement project (i.e. a project to reduce net GHG emissions) based on a set criteria is the use of Marginal Abatement Cost Curves (MACC).

Marginal Abatement Cost (MAC) is the cost per tonne of GHG emissions of the abatement project while a MACC is a graphical device that combines the MACs of available abatement projects to facilitate decision making. A MACC shows the relationship between the marginal quantity of CO_2 reduced and the associated marginal costs per unit of CO_2 by introducing given energy technologies into the energy system, substituting parts of the baseline development (Morthorst, 1993). The cost curve illustrates the technological options for CO_2 emissions reduction by considering different technologies and the associated costs. The associated costs are computed using conventional investment appraisal techniques such as payback, net present value (NPV) or internal rate of return (IRR). Given a specific target for CO_2 emissions reduction, it is possible from the MACC to prioritise the economically most attractive CO_2 emissions reduction options.

A MACC normally ranks emissions reduction options according to their cost effectiveness or cost of CO_2 abatement (cost per unit of CO_2), starting by the one with the lowest reduction cost per unit of CO_2. The cost per unit of CO_2 is computed for each emissions reduction options as follows:

$$\textbf{Cost of CO}_2\textbf{ abatement} = \frac{\textbf{Full cost of abatement option}}{\textbf{CO}_2\textbf{ emissions from baseline options}} \quad \ldots\ldots (1)$$

Equation (1) can be re-written as

$$\textbf{Cost of CO}_2\textbf{ abatement} = \frac{\Delta \mathbf{C_i}}{\boldsymbol{\delta}_\mathbf{i} \times \mathbf{CO_2}} = \frac{\{\mathbf{C_{a,j}(i)} \times \mathbf{R_d(i)}\} - \mathbf{F_i}}{\boldsymbol{\delta}_\mathbf{i} \times \mathbf{CO_2}} \quad \ldots\ldots\ldots\ldots\ldots (2)$$

Where:

ΔC_i = Cost of the emissions reduction option *i*

$C_{a,j}(i)$ = Total cost of the emission reduction option during year *j* for energy related component I (costs include initial investment cost plus energy costs, operational costs, periodic or replacement costs, maintenance costs and added costs)

$R_d(i)$ = Discount rate for year *i*

F_i = Financial savings in energy cost (£) that occur from the implementation of the emissions reduction options. It takes into account fuel savings and any other changes in operation cost. It is given by the product of the price of energy (P in £/kWh) and the energy saved (E in kWh)

δ_i = The emission reduction rate of abatement option *j*, i.e. % reduction in CO_2 baseline or base case

CO_2 = Base case CO_2 of the building

$\delta_i \times CO_2$ = Effective emissions savings from an intervention option

To apply equation 2, all key parameters need to be determined beforehand.

As an example, assuming that the capital cost of implementing voltage optimisation is £20,000 and the NPV of the annual energy savings is £35,000. If the total CO_2 abatement resulting from voltage optimisation is 1,200t CO_2, by applying equation 2, we have:

$$\text{Cost of } CO_2 \text{ abatement} = \frac{£20{,}000 - £35{,}000}{1{,}200tCO_2} = \frac{-15000}{1{,}200} = -£12.5/tCO_2$$

The calculation is repeated for all options being considered. Some measures have negative costs (i.e. the NPV of the financial savings in energy cost exceeds the capital cost), so that that their implementation results in a net profit/savings over the period of interest as shown in the above example. Some other measures will also show positive costs; which means that they do not pay back their investment even if they do save CO_2.

Given a basket of intervention options, the marginal changes in CO_2 emissions (i.e. the total emissions reduction (tCO_2) achievable from an option over the period of interest) and cost-effectiveness in £/tCO_2 or equivalent are calculated. A rectangular block is then plotted for each option. The width and height of the block respectively corresponds to these values. To generate a true marginal cost curve for the investment, the blocks are lined up from the one with the lowest marginal abatement cost on the left to the largest on the right and the optimum outcome is obtained by implementing the measures in order from left to right with reference to a base case. The total width of the blocks represents the total emissions reductions achievable. As illustrated in Figure 5, option 1 is chosen as the most economically attractive, implying lower costs and a substantial CO_2 reduction compared to the baseline.

MACCs such as the one illustrated in Figure 5 are used in many carbon policy briefs. For example, they have been applied to several sectors such as waste (Hogg et al 2008), transport (Spencer and Pittini 2008), higher education (SQW Energy 2009) and many more areas. With respect to buildings, the UK and the US are two particular countries which have adopted MACCs for macro-analysis of their respective building stocks. However, it should be clear that the cost and CO_2 savings depicted by these macro curves are primarily the product of broad statistical-based approximations and not on rigorous engineering analysis (Rysanek, 2010). Also, the curve can be interpreted in a meaningful manner only if the given macroeconomic assumption such as interest rates, energy prices and policies are unchanged (Morthorst, 1993). More importantly, the costs depicted are technically not 'marginal'. The marginal cost of option 7, for instance, is not based on the implementation of just the preceding options of the curve, but is based on the implementation of all measures of the curve at once. Furthermore, a major weakness of the MACC method is that interdependencies between options are not shown. All options are considered independently concerning their costs as well as CO_2 reduction, an assumption which of course does not apply to the real world. For example, savings from more efficient boilers might be lower if the building insulation is improved first as the benefit of the former depends on sequence of application.

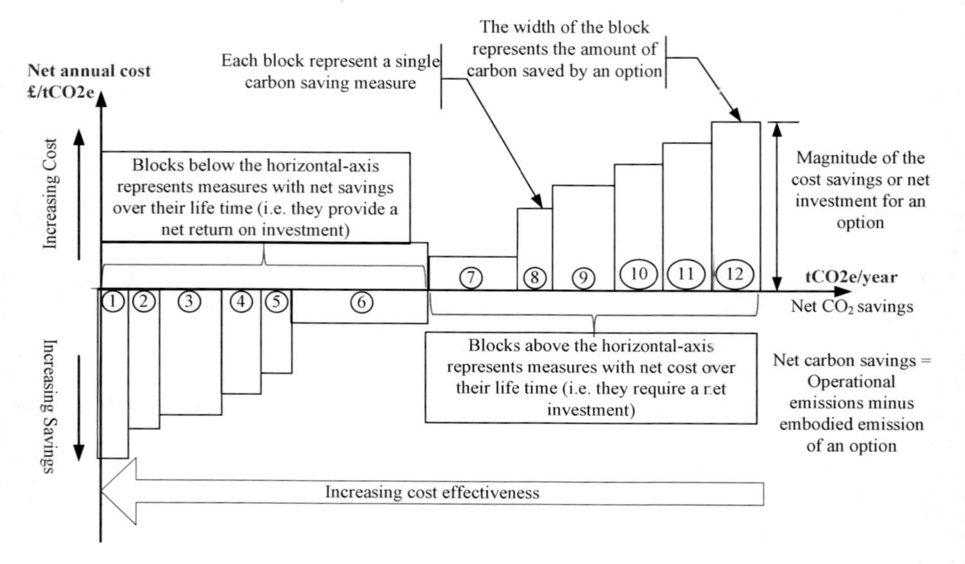

Fig. 5. MACC for ranking carbon abatement options

Other than the issues affecting MACCs enumerated above, another serious problem associated with them is the mathematical problem of considering options with negative CO_2 abatement costs (i.e. cost-effective options). The mathematical flaw can be illustrated using an example with data shown in Table 1.

Table 1. Comparison of two abatement options illustrating a flaw in mathematical formula

Abatement options	Option 1	Option 2
Net cost of CO_2 emissions saved (£)	-200	-100
CO_2 reduction (tCO_2)	20	4
Cost of abatement (£/tCO_2)	-10	-25

By physical inspection of Table 1, Option 1 should ordinarily be the preferred Option in that the economic net benefit and the CO_2 emissions savings are higher compared to Option 2. However, the CO_2 reduction criterion as stated in equation 2 leads to incorrect ranking and consequently a faulty decision, namely to the selection of Option 2. This is a serious issue because incorrect ranking implies a potential failure to achieve the optimum outcome in terms of emissions savings. The mathematical flaw shows that the standard cost-effectiveness calculation is inadequate for ranking negative-cost measures and therefore restricts the CO_2 reduction cost concept to the economically unattractive options, i.e. those that have positive net costs. For energy efficiency options with economic net benefits, the concept leads to wrong priorities. In particular, a meaningful comparison between heat-based and electricity-based options is not possible, as Taylor (2012) shows. A comprehensive analysis, including a detailed explanation and mathematical proofs showing that no figure of merit is possible for negative-cost measures is provided by Taylor (2012). There is therefore the need for an alternative approach for ranking negative cost measures. An alternative ranking method based on Pareto principles within a multi-objective optimisation framework was adopted by Taylor (2012) to rank negative-cost measures. The approach leads to a fairly clear ranking and identifies incorrectly ranked measures.

Against this backdrop, a proposed optimisation scheme which will incorporate sensitivity analysis under different price scenarios, interest rates and changes in policies is shown in Figure 6. The optimisation-assisted framework which will integrate the measures of financial cost, operational and embodied emissions into a robust method of ranking and sequencing building retrofit options taking into cognisance the issues with negative-cost measures and interdependency of measures.

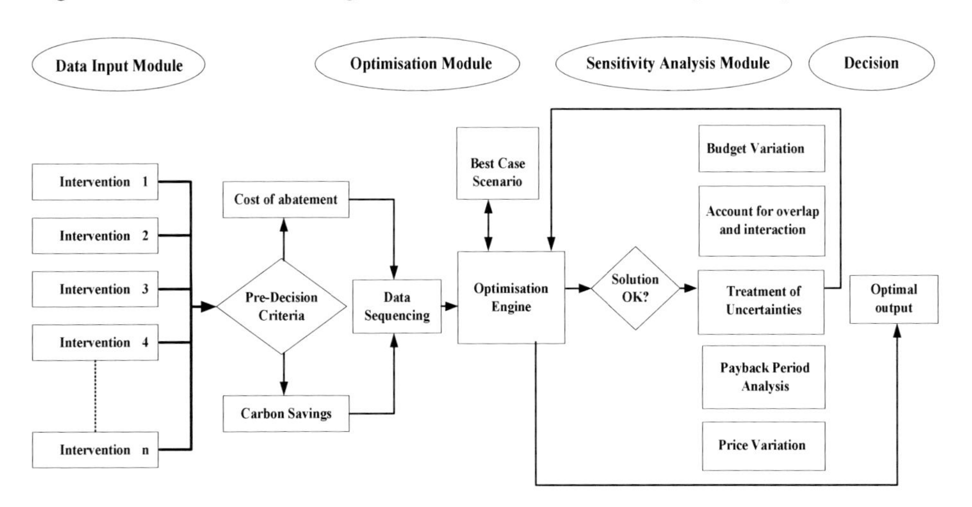

Fig. 6. Proposed optimisation scheme

5 Conclusion and Future Work

This paper describes the framework for the development of an integrated Decision Support System (DSS) based on the optimal ranking and sequencing of retrofit options for emissions reduction in non-domestic buildings. A review of existing DSS which have been developed to support and advice building stock owners with respect to retrofitting decisions for energy conservation was carried out. It was observed that most DSS have mainly focused on economics and operational emissions savings, and have neglected embodied emissions. The difficulties associated with the inclusion of embodied energy in decision making within the building sector were highlighted. Some of the reasons for this were identified as methodological challenges, the focus of regulations on in-use energy and carbon, a lack of appropriate legislation and a lack of interest in the impacts of embodied energy by the public and industry stakeholders. Despite these setbacks, it was shown that it is valuable to include embodied emissions when making decisions regarding carbon reduction strategies for buildings. The use of MACC as a useful tool to identify options which deliver the most economically efficient reductions in GHG and prioritize mitigation options within the building sector is also presented. Underlying limitations of the MACC approach and the points to be aware of, such as, the recognition of interactions between mitigation options, effects of macroeconomic assumptions and the mathematical flaw associated with the ranking of cost-effective options, before applying the results of MACC for decision making is also highlighted. This suggests that the MACC approach requires further development which the present DSS seeks to address. A robust methodological framework was presented that integrates the three variables of financial cost, embodied emissions and operational emissions. The methodology developed can be used to identify and communicate trade-offs between various refurbishment options to aid decisions that are informed both by environmental and financial considerations. The next steps include model implementation and validation.

References

1. Acquaye, A.: A Stochastic Hybrid Embodied Energy and CO2-eq Intensity Analysis of Building and Construction Processes in Ireland. Ph.D. Thesis. Dublin Institute of Technology, Republic of Ireland (2010)
2. Caleb, Non-Domestic Buildings -the missed opportunity", Caleb Management Services Ltd. (2008), http://www.chrispearson.net/epic_downloads/non_domestic_buildings.pdf (accessed March 10, 2012)
3. Carbon Trust, Low Carbon Refurbishment of Buildings- A guide to achieving carbon savings from refurbishment of non-domestic buildings (2008)
4. Carbon Trust, Building the future, today: transforming the economic and carbon performance of the buildings we work in (2009)
5. CCaLC, Carbon Footprinting the Life Cycle of Supply Chains -the CCaLC Tool (2010), http://www.ccalc.org.uk (accessed June 15, 2011)
6. CCC, Building a low-carbon economy - the UK's contribution to tackling climate change (2008), http://www.theccc.org.uk/reports/building-a-low-carbon-economy (accessed August 25, 2011)
7. Chidiac, S.E., Catania, E.J.C., Morofsky, E., Foo, S.: A screening methodology for implementing cost effective energy retrofit measures in Canadian office buildings. Energy and Buildings 43, 614–620 (2011)

8. Corus, Broken contract leads to mothball of Teesside plant", press release 04 December 2009 (2009), http://www.corusgroup.com/en/news/news/2009_tcp_mothball (accessed November 15, 2011)
9. Crawford, R.H.: Validation of the use of input-output data for embodied energy analysis of the Australian Construction industry. Journal of Construction Research 6(1), 71–90 (2005)
10. De Benedetto, L., Klemeš, J.: The Environmental Performance Strategy Map: an integrated LCA approach to support the strategic decision-making process. Journal of Cleaner Production 17(10), 900–906 (2009)
11. De Kock, E.: Decentralising the codification of rules in a decision support expert knowledge base. MSc thesis. Faculty of Engineering, Built Environment and Information Technology. University of Pretoria (2003)
12. Department of Energy and Climate Change (DECC, 2009). Carbon Valuation in UK Policy Appraisal: A Revised Approach. Climate Change Economics
13. Department of Energy and Climate Change (DECC,2009). The UK Low Carbon Transition Plan. National strategy for climate and energy, http://www.decc.gov.uk/assets/decc/white20papers/uk%20low%20carbon%20transition%20plan%20wp09/1_20090724153238_e_@@_lowcarbontransitionplan.pdf
14. Department of Energy and Climate Change [DECC], (2009) Climate Change Act 2008: Impact Assessment, http://www.decc.gov.uk (accessed September 28, 2011)
15. Department of Trade and Industry [DTI], Meeting the Energy Challenge-A White Paper on Energy (2007), http://www.dti.gov.uk/files/file39387.pdf (accessed September 20, 2011)
16. Diakaki, C., Grigoroudis, E., et al.: A multi-objective decision model for the improvement of energy efficiency in buildings. Energy 35(12), 5483–5496 (2010)
17. Ding, G.: The development of a multi-criteria approach for the measurement of sustainable performance for built projects and facilities, Ph.D. Thesis, University of technology, Sydney, Australia (2004)
18. Dixit, M.K., Fernandez-Solis, J.L., Lavy, S., Culp, C.H.: Identification of parameters for embodied energy measurement- A literature review. Journal of Energy and Buildings 42, 1238–1247 (2010)
19. Doukas, H., Nychtis, C., et al.: Assessing energy-saving measures in buildings through an intelligent decision support model. Building and Environment 44(2), 290–298 (2009)
20. Energy Saving Trust. Sustainable refurbishment: Towards an 80% reduction in CO2 emissions, water efficiency, waste reduction and climate change adaptation (2010)
21. Engin, A., Frances, Y.: Zero Carbon Isn't Really Zero: Why Embodied Carbon in Materials Can't Be Ignored (2010), http://www.di.net/articles/archive/zero_carbon/ (accessed February 27, 2011)
22. Fitzgerald, M., Farrow, N., et al.: Challenges to Real-Time Decision Support in Health Care. In: Advances in Patient Safety: New Directions and Alternative Approaches, vol. 2 (2008), http://www.ncbi.nlm.nih.gov/books/NBK43697/pdf/advances-fitzgerald_108.pdf (accessed May10, 2012)
23. Hamilton-MacLaren, F., Loveday, D., Mourshed, M.: The calculation of embodied energy in new build UK housing. Loughborough University, UK (2009)
24. Hinnells, M.: Technologies to achieve demand reduction and micro generation in buildings. Energy Policy 36, 4427–4433 (2008)
25. Hogg, D., Baddeley, A., Ballinger, A., Elliott, T.: Development of Marginal Abatement Cost Curves for the Waste Sector, Eunomia Research & Consulting (2008), http://www.theccc.org.uk/pdfs/Eunomia%20Waste%20MACCs%20Report%20Final.pdf (accessed May 15, 2012)

26. Ibn-Mohammed, T., Greenough, R., Taylor, S., Ozawa-Meida, L., Acquaye, A.: Optimal Ranking of Retrofit Options for Emissions Reduction options-A Review. In: CIBSE-ASHRAE Conference on Sustainable Systems and Services for the 21st Century, April 18-19, Imperial College, London (2012), http://www.cibse.org/content/cibsesymposium2012/Paper073.pdf
27. Intergovernmental Panel on Climate Change, IPCC (2007a) Climate Change 2007: The Physical Science Basis. Contribution of Working Group I to the Fourth Assessment Report of the Intergovernmental Panel on Climate Change. Cambridge University Press, Cambridge (2007) http://www.ipcc.ch/pdf/assessment-report/ar4/wg1/ar4-wg1-spm.pdf (accessed September 20, 2011)
28. International Energy Agency [IEA] World Energy Outlook (2006), http://www.iea.org/textbase/nppdf/free/2006/weo2006.pdf (accessed September 26, 2011)
29. Jenkins, D.P., Singh, H., Eames, P.C.: Interventions for large-scale carbon emission reductions in future UK offices. Journal of Energy and Building 41, 1374–1380 (2009)
30. Jones, J.W., Tsuji, G.Y., et al.: Decision support system for agro technology transfer: DSSAT v3. p. 157 - 177. Kluwer Academic Publishers (1998), http://www.aglearn.net/resources/isfm/DSSAT.pdf (accessed May 10, 2012)
31. Jowsey, E., Grant, J.: Greening the Existing Housing Stock, Faculty of Development and Society, Sheffield Hallam University (2009), http://www.prres.net/papers/Jowsey_Greening_The_Existing_Housing_Stock.pdf
32. Kangas, A., Kangas, J., Kurttilla, M.: Decision Support for Forest Management. Managing Forest Ecosystem 16 (2008)
33. Kemp, M., Wexler, J.: Zero Carbon Britain 2030 A New Energy Strategy - The second report of the Zero Carbon Britain project. CAT Publications, Machynlleth (2010)
34. Kenny, R., Law, C., Pearce, J.M.: Towards real energy economics: energy policy driven by life-cycle carbon emission. Energy Policy 38(4), 1969–1978 (2010)
35. Kumbaroglu, G., Madlener, R.: Evaluation of economically optimal retrofit investment options for energy savings in buildings. Energy and Building (2012)
36. Morthorst, P.E.: The cost of CO2 reduction in Denmark-methodology and results. UNEP Greenhouse Gas Abatement Costing Studies Phase Two (1993) ISBN 87-550-1954-4
37. Pacca, S., Horvath, A.: Greenhouse Gas Emissions from Building and Operating Electric Power Plants in the Upper Colorado River Basin. Environmental Science and Technology, ACS 36(14), 3194–3200 (2002)
38. Rawlinson, S., Weight, D.: Sustainability-Embodied carbon (2007), http://www.building.co.uk/data/sustainability-%E2%80%94embodiedcarbon/3097160.article (accessed June 13, 2011)
39. RCEP, Energy - The Changing Climate. London, Royal Commission on Environmental Pollution (2000), http://webarchive.nationalarchives.gov.uk, http://www.rcep.org.uk (accessed September 15, 2011)
40. Rysanek, A.: Optimized marginal abatement cost curves of individual buildings (2010), http://www.eprg.group.cam.ac.uk/wp-content/uploads/2010/05/8-Rysanek.pdf (accessed May 15, 2012)
41. Sartori, I., Hestnes, A.G.: Energy use in the life cycle of conventional and low-energy buildings: A review article. Energy and Buildings 39(3), 249–257 (2007)
42. SCEnAT Supply Chain Environmental Analysis: A new system for delivering a low carbon supply chain (2011), http://www.lowcarbonfutures.org/assets/media/2579_Low_Carbon_Report_Nov2011.pdf (accessed June 10, 2012)

43. Spencer, C., Pittini, M.: Building a marginal abatement cost curve (MACC) for the UK transport sector, Committee on Climate Change (March 2008), http://www.theccc.org.uk/other_docs/Tech%20paper%20supply%20side%20FINAL.pdf (accessed May 15, 2012)
44. SQW Energy, Research into a carbon reduction target and strategy for Higher Education in England, SQW Consulting (2009), http://www.hefce.ac.uk/media/hefce1/pubs/hefce/2010/1001/10_01a.pdf (accessed May 15, 2012)
45. Taylor, S., Peacock, A., Banfill, P., Shao, L.: Reduction of greenhouse gas emissions from UK hotels in 2030. Journal of Building and Environment 45, 1389–1400 (2010)
46. Taylor, S.: The ranking of negative-cost emissions reduction measures. Energy Policy (2012), http://dx.doi.org/10.1016/j.enpol.2012.05.071
47. Thormark, C.: The effect of material choice on the total energy need and recycling potential of a building. Building and Environment 41(8), 1019–1026 (2006)
48. UNEP (United Nations Environmental Programme), Buildings and Climate Change: Status, Challenges, and Opportunities. UNEP (2007)
49. Yin, H., Menzel, K.: Decision Support Model for Building Renovation Strategies. World Academy of Science, Engineering and Technology 76 (2011)
50. Yohanis, Y.G., Norton, B.: Life-cycle operational and embodied energy for a generic single-storey office building in the UK. Journal of Energy 27, 77–92 (2002)

Chapter 21
Modeling, from the Energy Viewpoint, a Free-Form, High Energy Performance, Transparent Envelope

Luis Alonso[1], C. Bedoya[1], Benito Lauret[1], and Fernando Alonso[2]

[1] Department of Construction and Technology in Architecture, School of Architecture. UPM. Avda. Ramiro de Maeztu, s/n, 28040, Madrid, Spain
[2] Languages and Systems and Software Engineering Department, School of Computing, UPM, Campus de Montegancedo, s/n, 28660 Boadilla del Monte, Madrid, Spain

Abstract. This article examines a new lightweight, slim, high energy efficient, light-transmitting, self-supporting envelope system, providing for seamless, free-form designs for use in architectural projects. The system exploits vacuum insulation panel technology. The research was based on envelope components already existing on the market and patents and prototypes built by independent laboratories, especially components implemented with silica gel insulation, as this is the most effective transparent thermal insulation there is today. The tests run on these materials revealed that there is not one that has all the features required of the new envelope model, although some do have properties that could be exploited to generate this envelope, namely, the vacuum chamber of vacuum insulation panels, the use of monolithic aerogel as insulation in some prototypes, and reinforced polyester barriers. These three design components have been combined and tested to design a new, variable geometry, energy-saving envelope system that also solves many of the problems that other studies ascribe to the use of vacuum insulation panels.

1 Introduction

Energy efficiency is coming to the forefront in the architecture, as, apart from the significance of a reduced environmental impact and increased comfort for users, the current energy crisis and economic recession has bumped up the importance of the financial cost of energy.

Since the Kyoto Protocol was signed in 1997, governments all over the world have been trying to reduce part of the CO2 emissions by tackling building "energy inefficiency". In Europe today, the tertiary and housing sectors account for 40.7% of the energy demand, and from 52 to 57% of this energy is spent on interior heating. The new world energy regulations, set out at the European level by the Commission of the European Communities in the First Assessment of National Energy Efficiency Action Plans as required by Directive 2006/32/EC on Energy End-Use Efficiency and Energy Services, indirectly promote an increase in the thickness of outer walls, which, for centuries, have been the only way of properly insulating a building.

A. Håkansson et al. (Eds.): *Sustainability in Energy and Buildings*, SIST 22, pp. 229–237.
DOI: 10.1007/978-3-642-36645-1_21

The use of vacuum insulation panel (VIP) systems in building aims to minimize the thickness of the building's outer skin while optimizing energy performance. The three types of vacuum chamber insulation systems (VIS) most commonly used in the construction industry today –metallized polymer multilayer film (MLF) or aluminium laminated film, double glazing and stainless steel sheet or plate –, have weaknesses, such as the fragility of the outside protective skin, condensation inside the chamber, thermal bridges at the panel joints, and high cost, all of which have a bearing on on-site construction.

Apart from overcoming these weaknesses and being a transparent system, the new F^2TE^3 (free-form, transparent, energy efficient envelope) system that we propose has two added values. The first is the possibility of generating a structural skin or self-supporting façade. The second is the possibility of designing free-form architectural skins. These are research lines that the Pritzker Architecture Prize winners Zaha Hadid, Frank Gehry, Rem Koolhaas, Herzog & de Meuron, among many other renowned architects, are now exploring and implementing.

To determine the feasibility of the new envelope system that we propose, we compiled, studied and ran laboratory tests on the materials and information provided by commercial brands. We compared this information to other independent research and scientific trials on VIPs, such as Annex39, and on improved core materials, such as hybrid aerogels and organically modified silica aerogels, conducted by independent laboratories like Zae Bayern in Germany, the Lawrence Berkley Laboratory at the University of California or the Technical University of Denmark.

After studying the results, we discovered valuable innovative ideas that we exploited to design the new high energy efficient envelope that should outperform the elements now on the market.

2 Theoretical Study of the System

From the viewpoint of energy performance, of the three types of translucent insulations that there are on the market (plastic fibers, gas and aerogel), we found that the insulation that more effectively meets the needs of the new system that we propose is aerogel.

Aerogel and nanogel (granular silica gel) have four advantages for use as thermal insulation in translucent panels:

a) Transparency: Monolithic aerogel light transparency can be as high as 87.6%.
b) Insulation: on top of transparency, it is an excellent insulator, the thermal performance of a 70 mm nanogel-filled vacuum insulated panel (VIP) is better than a 270 mm-thick hollow wall.
c) Lightness: aerogel is three times as heavy as air.
d) Versatility: monolithic aerogel can be shaped as required.

Let us note that although there are many prototypes and patents for the transparent and translucent high energy efficient aerogel-implemented façade panels under study, they are extremely difficult to analyze because there is not a lot of information available and it is not easy to get physical samples of these panels. For this reason, this research has focused on commercial products that are on the market. Most of these products use granular silica aerogel (nanogel).

In the following, we analyze these translucent and transparent panels, setting out their strengths and weaknesses and our findings as a result of this study.

2.1 Translucent Systems

In this type of systems we have analyzed systems composed of granular silica gel-filled polycarbonate, reinforced polyester and double glazed vacuum insulated panels.

1. Nanogel-filled cellular polycarbonate panels are the most widespread system on the market. They have the following strengths and weaknesses:

   Strengths: Thanks to its low density 1.2g/m³, this is a very lightweight material. It has a high light transmission index ±90% (almost the transparency of methacrylate). It is a low-cost material for immediate use. And, at the competitiveness level, it is the least expensive envelope assembly.

   Weaknesses: Durability is low. Most commercial brands guarantee their polycarbonate panels for only 10 years (as of when they start to deteriorate), whereas nanogel has a very high durability. These panels are very lightweight but very fragile to impact. Even though nanogel is an excellent acoustical insulator, the slimness of these panels means that they have acoustic shortcomings.
2. No more than two types of reinforced polyester panels are commercialized despite the potential of this material. They have the following strengths and weaknesses:
3. Strengths: Good mechanical properties: glass fiber reinforced polyester resin core composites offer excellent flexibility, compressibility and impact resistance. Good malleability: they could be shaped according to design needs but no existing system offers this option. Durability is good, as there are methods to lengthen the material's service life considerably (twice that of polycarbonate), like gelcoat coatings or protective solutions with an outer layer composed of a flexible "glass blanket".

Weaknesses: There is no self-supporting (structural) panel that is standardized and commercialized worldwide. Existing systems have design faults, as they include internal aluminum carpentry or substructures, whereas there is, thanks to the characteristics of reinforced polyester, potential for manufacturing a self-supporting panel (as in the case of single-hull pleasure boats). It is also questionable ecologically, as the polyester is reinforced with glass fiber, which has detracted from its use in building. However, this could change with the advent of new plastic and organic fibers and resins. Economically speaking, reinforced polyester manufacturing systems are very expensive, because either processes are not industrialized or, on the other hand, they rather technology intensive like, for example, pultrusion.

Double glazed vacuum insulated panels (VIP) are still at the prototype stage. Although research and prototypes abound (HILIT+ y ZAE BAYERN, for example), there are only a couple of commercial brands:

Strengths: Thanks to the combination of vacuum and aerogel insulation (both monolithic and granular), they provide the slimmest and higher insulation system in the building world (0.5W/m²K). Transparency levels for some prototypes using monolithic aerogel are as high as 85% for thicknesses of 15 to 20 mm. Additionally, the service life of the glazing and the gel is very similar.

Weaknesses: Product of a combination of vacuum core and double glazing, this component is r fragile, especially prone to impact-induced breakages. The high cost of molding glass into complex geometries rules out its use as a free form system. It is a system that depends on substructures and other components for use.

Findings: After a comparative analysis of over one hundred and forty seven (147) commercial products, and the detailed evaluation of the eight (8) which offer better performance (Figure 1), we can confirm that fiber reinforced polyester resin panels perform better than any of the polycarbonate panels studied. But these improvements are unable to offset their high production and environmental costs, generating commercially uncompetitive products.

These panels have two unexploited design lines, such as adaptation to new less harmful natural cellulose resins and fibers, the design of insulation for variable geometry translucent skins, or structural improvement for use as a self-supporting component.

As regards energy efficiency, these products improve the energy-saving performance that offers a 27cm thick traditional wall. With only 7 cm, they have an U value of 0.28W/m²K, which amounts to 7% better energy performance compared to a traditional wall and with 4% less thickness.

Looking at double glazed VIP panels; the data indicate that, although still at the prototype stage, panels like these are the most efficient commercial solution, as they offer the higher thermal and acoustical insulation performance and optimal light transmission.

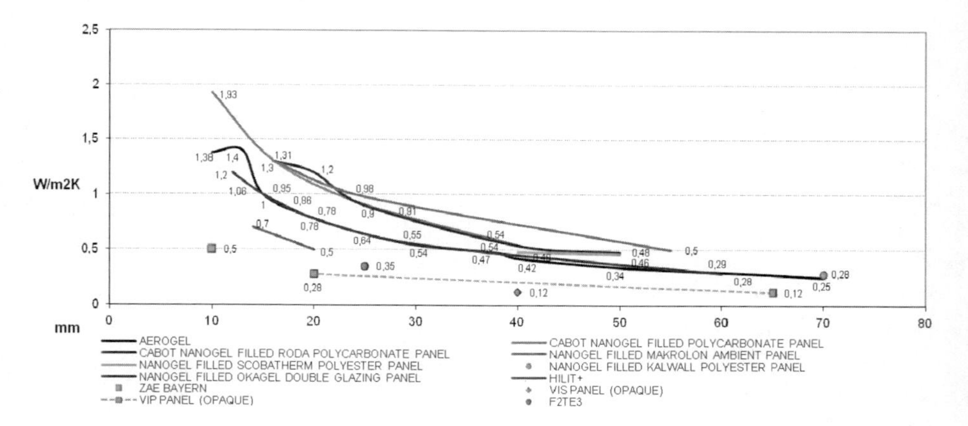

Fig. 1. Comparison of commercial systems and prototypes

At the acoustical and thermal level, the VIP panel is the most efficient of the envelopes examined, as a 15 mm panel insulates equally as well as a traditional mass wall in terms of energy expenditure (saving). We find that, with a thickness of just 60 mm, this product improves the energy efficiency performance of a 271.5 mm cavity wall.

2.2 Transparent Systems

All panels implemented with aerogel instead of nanogel are transparent. They have a high solar transmittance and low U value. At present all these systems are non-commercial prototypes, about which little is known. Noteworthy are two aerogel-insulated double-glazed vacuum insulation panels (VIP):

4-13.5-4 / 21.5 mm double-glazed vacuum panels filled with monolithic aerogel with a pressure of 100hPa in the aerogel chamber. The heat transfer coefficient Ug has a U value of 0.7 W/m²K for 14 mm and 0.5 W/m²K for 20 mm compared to the 1.2W/m²K offered by 24 mm commercial nanogel-filled double glazing VIP. This almost doubles the insulation performance of any of the commercial translucent panels. Light transmission depends on the angle of incidence, but varies from 64.7 to 87.5%. The sound attenuation index is 33dB for a panel thickness of 23 mm and noise reduction is expected to be improved to 37 dB. The energy saving compared with a dwelling that is glazed with gas-insulated triple glazing (argon and krypton) is from 10% to 20% greater.

10 mm double-glazed vacuum insulation panels with aerogel spacers inside the core (unlikely to be commercialized for another two or three years). The heat transfer coefficient Ug for 10 mm panels has a U value of 0.5 W/m²K. This is the lowest U value of all the panels studied so far, where light transmission is equal to glass.

Findings: From the analysis of the transparent panels, VIPs unquestionably come the closest to what we are looking for in this research. The only arguments against VIPs are that they are at the prototype stage. This means that they are not on the market, nor have they been tested, approved or industrialized. All this has an impact on cost. Also being the product of evacuating double glazing panels, VIPs are very fragile. Versatility is limited because, owing to the panel generation process, the maximum dimensions to date are 55 x 55 cm.

From the materials technology analysis, we find that transparent monolithic aerogel-insulated VIPs are the material that best conforms to the goals of transparency, insulation and lightness that we are looking for, provided that we accept that the panels are fragile, non-commercialized prototypes and are not very versatile in terms of size.

3 Experimental Study

Following up the results of the theoretical study outlined in Section 2, we now compare these findings with the results of an empirical experiment and computer-simulations of the real commercial panels to which we had access.

The trials are designed especially to examine the energy performance of the material. These trials were run using the methodology developed by the Department of

Building and Architectural Technology of the UPM's School of Architecture that is based on MoWiTT type tests designed by the Lawrence Berkeley laboratories at the University of California. These empirical tests have been set according to known environmental and materials conditions, in order to compare them with the simulation.

3.1 Computer Simulation

We use the DesignBuilder program for the computer simulation. This program uses EnergyPlus as a calculation engine and it has been validated as a building energy simulation program, with the BESTEST title, by the International Energy Agency (IEA) on year 2011. As an additional measure, and because Spanish law recognizes only two energy simulation programs (LIDER and CALENER), on this research we has been conducted the calibration and validation of the DesignBuilder simulations based on the MoWiTT empirical trials by the "black box" method imposed by the Ministry of Energy of Spain.

Finally, we compared the simulations data with the data from the empirical tests, and we find a discrepancy of ±6.2% between them. Taking into account the kind of the empirical tests, this discrepancy is an acceptable margin of error.

Because of the shortage of information about transparent monolithic aerogel and the impossibility of acquiring a sample, we decided to use the data of commercial translucent granular aerogel (that behaves very similar than the monolithic aerogel) and the DesignBuilder program to conduct a trial by computer simulation under the same environmental conditions as the empirical trials run on the other panels, describes the behavior of a 25 mm monolithic aerogel sheet. We find that the test space has a uniform inside temperature of between 18 and 37°C.

3.2 Empirical Trials

The empirical trials are made with boxes with an inner volume of 60 x 60 x 60 cm, insulated with 20 cm of expanded polyurethane. One of the box faces is left open by way of a window. The study elements are placed in this opening using a specially insulated frame. The trial involves exposing two such boxes to a real outside environment to study their behavior. The two boxes have two different windows: one is fitted with 6+8+6 double glazing with known properties as a contrast element and the other is fitted with the panel that we want to study. Data loggers are placed inside each box for monitoring purposes to measure and compare their inside temperature. The boxes are also fitted with a thermal sensor on the outside to capture the temperature to which they are exposed. The boxes are set in a south-facing position as this is the sunniest exposure.

We ran twenty-eight (28) temperature-measuring trials using this system, and compared the performance of different thicknesses of commercial panels with 6+8+6 double glazing. Four (4) of these panels deserve a special mention:

These four trials were evaluated and compared with the computer-simulated aerogel data (Figure 2) and data from the theoretical study. We found that, like the data output by the theoretical study, the real trials suggest that the materials behavior is suitable for designing the new envelope system. The very flat loss curves in the plot describe a very low U value. In terms of capture, there is a thermal difference of almost 30°C between the Okagel (VIP) panel and the worst of the tested panels. The difference between Okagel and the best-performing panel is almost 10 °C in terms of loss and capture. We have confirmed the experimental datum that likens the behavior of the Okagel panel to that of the computer-simulated aerogel.

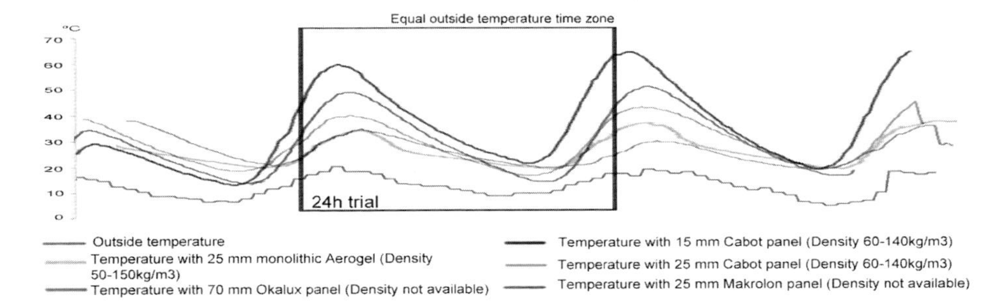

Fig. 2. Comparison of empirical data of commercial systems with computer simulation of 25mm thick sheet of silica aerogel with density 50-150kg/m³ over 72 hours (temperatures inside the test boxes)

From our computer-simulated experimental study, the data on organic aerogels supplied by the CSIC and the University of Barcelona, and the data from trials run at the University of Denmark on envelopes implemented with monolithic silica gels and the empirical trials conducted in this research, we arrive at the following conclusions:

The most efficient panel from the energy saving viewpoint is 70 mm Okalux that has thermal differences with respect to the other panels ranging from 5 to 20°C. Also striking is the disparity in the results of the 25 mm and 15 mm Cabot panels with thermal differences of 15°C.

Although the specific temperatures and factors on each test day differed from one trial to another, the 70 mm nanogel-filled Okalux VIP panel performed similarly to the 25 mm aerogel sheet.

These are key data that are useful for designing a new lightweight, slim, high energy efficient, light-transmitting envelope system, providing for seamless, free-form designs for use in architectural projects.

4 Proposal for a Free-Form Transparent Energy Efficient Envelope (F²TE³)

We propose a free-form design envelope system fabricated with cellulose fibres and polyester resin (or acrylic-based organic resin), and a vacuum core insulated with monolithic aerogel at a pressure of 100hPa. Being a self-supporting component, the system can perform structural functions, and seams between panels are concealed by an outer coating applied in situ.

F²TE³ dimensions: The minimum thickness of the modelled system panels will be 25 mm, and the sheet width, although variable, will be at most 600x600 mm. The weight per unit area will range from 15 to 7 kg/m2, and the minimum admissible flexion radius will be approximately 4000 mm. Other features are:

Light transmittance, τD65: 59%- 85% approx.

Horizontal and vertical U-value: 0.50 W/m2 K

Weighted sound reduction value: estimated at 26-45 dB

System assembly: The F²TE³ system is composed of the dry-seal connection of male and female edged panels (two female sides and two male sides on each panel that fit together seamlessly) as previously designed by the draughtsperson. Once the construction is in place, it is given an outer coating of fibres and resins and finally a gelcoat coating to protect the assembly from external agents.

Testing: A 25 mm thick prototype F²TE³ system was computer simulated to examine its energy-saving behaviour compared with a computer-simulated aerogel envelope of the same thickness (Figure 3).

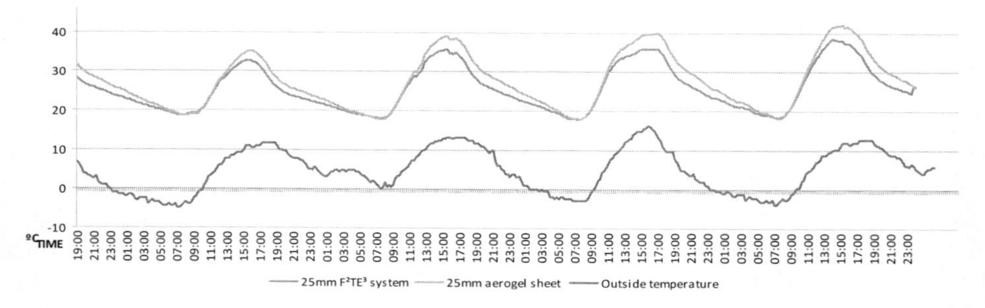

Fig. 3. Comparison of a computer simulation of a 25mm thick sheet of silica aerogel with a density of 50-150 kg/m³ with the F2TE3 system over 96 hours

As shown in Figure 10, the F²TE³ returns a better result than what would be achieved with monolithic aerogel without a barrier envelope (not feasible due to aerogel hydroscopy). The flat heat loss curve indicates that the U-value is very small and eventually equals the values for aerogel, whereas F²TE³ has a 5°C edge over aerogel for capture.

5 Conclusions

F²TE³ is a slim façade system that provides high energy efficiency. It has a seamless surface, providing for variable geometry and the option of building self-supporting structures into the same transparent system skin. The study conducted as part of this research has shown that the prototype F²TE³ system outperforms other systems existing on the market, offering added value in terms of structure, transparency and variable geometry.

Also, thanks to the validation and calibration of the simulation methodology used on this research, not only by the Spanish and international regulatory energy simulation standards, but also by the comparison of data obtained by simulation with the monitoring, we can affirm that the data obtained on this research by the simulation method has a very low error margin (6.2%)

References

Baetens, R., Jelle, B.P., Thueb, J.V., Tenpierikd, M.J., Grynninga, S., Uvsløkka, S.: Vacuum insulation panels for building applications: A review and beyond. Energy and Buildings 42(2) (2010)

Fricke, J.: Title of conference paper. In: Proceedings of 7th International Vacuum Insulation Symposium, Empa, Duebendorf/Zurich, Switzerland (September 2005) ISBN 3-528-14884-5

Heinemann, U., Weinläder, H., Ebert, H.-P.: Energy efficient building envelopes: New materials and components (ZAE BAYERN,Energy Working Group (EPA)) German Research Centre in terms of energy applications (2009)

Jensen, K., Kristiansen, F., Schultz, J.: HILIT+. Highly Insulating and Light Transmitting Aerogel Glazing for Super Insulating Windows, Department of civil engineering, BYG·DTU (2005)

Klems, J.H.: Measurements of fenestration net energy performance: considerations leading to development of the mobile window thermal test (MoWiTT) facility. Journal of Solar Energy Engineering 8(3), 165–173 (1984)

Moner-Girona, M., Martínez, E., Roig, A.: Micromechanical properties of carbon–silica Aerogel compositesAerogeles. Appl. Phys. A(74), 119–122 (2002) ISSN 0947-8396

Mukhopadhy, P., Kumaran, K., Ping, F., Normandin, N.: Use of Vacuum Insulation Panel in Building Envelope Construction: Advantages and Challenges. In: Proceedings of 13th Canadian Conference on Building Science and Technology, N R C C - 5 3 9 4, Winnipeg, Manitoba (May 2011)

Selkowitz, S., Winkelmann, F.: New models for analyzing the thermal and daylighting performance of fenestration. ASRAE/DOE Conference on Thermal Performance of the Exterior Envelopes of Buildings II, Las Vegas, December 6-9 (1982)

Simmler, H., Heinemann, U., Kumaran, K., Quénard, D., Noller, K., Stramm, C., Cauberg, H.: Vacuum Insulation Panels. Study on VIP-components and Panels for Service Life Prediction of VIP in Building Applications (Subtask A) HiPTI, IEA/ECBCS Annex 39 (2005)

Chapter 22
A Mathematical Model to Pre-evaluate Thermal Efficiencies in Elongated Building Designs

Alberto Jose Fernández de Trocóniz y Revuelta[1], Miguel Ángel Gálvez Huerta[2], and Alberto Xabier Fernández de Trocóniz y Rueda[3]

[1] Doctor Arquitecto/ Lcdo. C. Físicas/ Profesor de Instalaciones de la Escuela Téc. Sup. Arquitectura de la UPM
albertotroconiz@mac.com
[2] Doctor Arquitecto/ Prof. Titular de la Escuela Téc. Sup. Arquitectura de la UPM
miguelangel.galvez@upm.es
[3] Collaborator, Architecture Student
Common Mailing Address: Escuela Técnica Superior de Arquitectura de la UPM. Avda. Juan de Herrera 4, 2840 Madrid
alberto.fernandeztroconiz@upm.es

Abstract. This paper exposes the basic structure and results of a mathematical model that treats in a simple way the evaluation of the energy demand for certain building typologies: those annular Courtyard types, and also long Blocks. It profits from the proven fact that the Form Factor of their three-dimensional shape is identical to the Form Factor of the mean cross section or "Perimeter to Section Area Ratio".

This model constitutes a useful application of the concepts exposed by us on a previous paper at this same forum with the title "A Simple way to Assess and Compare the Thermal Efficacy in Elongated Building Designs"[4]. Now, in continuation to it, we give here one of its more interesting practical consequences; it is the formalisation of a mathematical model that takes full advantage of the potentialities of the concepts, laws, criteria & methodology already exposed that allows to treat 3D problems just in two dimensions.

The inputs for the model are 2D geometrical and physical parameters such as width, height, thermal transmittances, etc. The outputs are 2D (& 3D) thermal efficiencies expressed as the unitary and specific flow of transmission heat through the envelope, and also some comparison percentages between different possibilities.

Keywords: Area to Volume Ratio, Perimeter to Area Ratio, Form Factor, Coefficient of Susceptibility, Building Design, Building Energy, Thermal Efficiency, Sustainable Building Design, Sustainable Urban Planning.

1 Theoretical Basis

We assume as starting point the contents exposed in previous part [4] of the present work whose concepts are briefly resumed in the following facts for buildings:

A. Håkansson et al. (Eds.): *Sustainability in Energy and Buildings*, SIST 22, pp. 239–249.
DOI: 10.1007/978-3-642-36645-1_22

The heat transmitted to the outside through the surface of the envelope is an important part of the energy consumption (and corresponding contamination), that take place throughout the span of the building lifetime. The amount of this heat depends (besides the external-internal temperature differences) on the Morphological characteristics of the envelope, mainly due to two factors:

a) the insulating capabilities (reflected in thermal transmittances), and
b) the geometric configuration expressed in the ratio of the envelope area to volume, what is usually known (in Spain) as the Form Factor (FF =A/V); its inverse being a measure of the compactness of the geometry of the body [6].

These two properties have been usually treated separately; focusing either on one or the other [5], and although both have an equal importance, nowadays the professional practise and even some official standards [1], seem to forget the second geometric one. In order to regain the full picture, we have proposed a new parameter that can integrate the two into a single one; it is a *Weighted Form Factor* (FF* =A*/V); where the transmittance coefficients multiply the various areas of the envelope; in the denominator, the volume remains the same [2].

This parameter is equal to the unitary and specific flow of transmission heat through the envelope, and proves to be an adequate tool to compare the preliminary efficiency of different design solutions on homogeneous basis.

We can apply the aforementioned concepts with a distinctive advantage to a special type of buildings, to those geometrically closed such as annular courtyards, and also long Blocks and Towers (provided their end tips are relatively unimportant in relation to the total envelope). Geometrically can be described as *Quasi-Anextremic Horizontal Extrusion Building Blocks,* as shown in (Fig. 1).

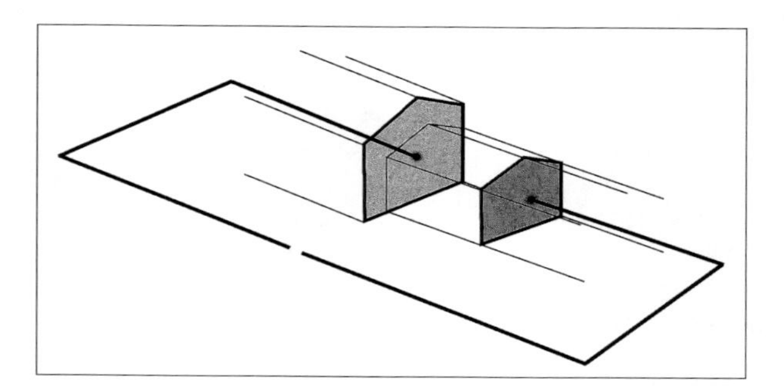

Fig. 1. Quasi-Anextremic Horizontal Extrusion Building Block

This class of body forms can have a number "n" of generating cross sections extruded along joined stretches of horizontal guidelines, closing a loop or not; in the most general case, besides the surfaces of the perimetral envelope, there can be a number of end tips that are also exposed to the exterior temperatures. It is to be remarked that the effective amount of surface tips can be low, not just only because of geometrical reasons, but also (as the model includes material properties), because of an increased insulation at the exposed extremes of the envelope.

Now we make full use of a relatively unknown Geometrical Law (formalized by us in an Identity Theorem) that establishes that for those typologies; the Form Factor of their three-dimensional shape is identical to the Form Factor of the mean cross section or the "Perimeter to Section Area Ratio"[3].

We profit from this dimensional reduction from 3D to 2D which implies that many problems of analysis and design become extremely easy in a variety of fields: urban planning, building projects, and general morphology.

The Practical Implications are easy to grasp: applied for example to the task of achieving a greater energy efficiency (by thermal transfer) of long building blocks and towers, there is no need for considering the whole exterior envelope; it is enough to ensure the optimisation of the cross section (in the tower is the floor plan), in the most adequate, compact and large 2D figure, and the most insulated perimeter for it. To achieve those goals, we have proposed new criteria and methodology, and gave hints of a two-dimensional morpho-thermal model, which is now exposed in more extension applied to elongated blocks. As an illustration of these ideas we can consider the following image (Fig. 2).

Fig. 2. Linear Office Building Block

Here is apparent in size & shape of the generation section of the extrusion for the volume. According to our theorem the complexities of the 3D envelope can be reduced to that "mean" 2D section (very explicit here) where we can initially assess the preliminary thermal adequacy for the whole building.

2 Morpho-Thermal 2D Model

Based on the foregoing basis, a modeling procedure can now be established for a variety of morpho-energetic studies, where we have to deal only in the plane. We of course focus on the problems of evaluating (in a preliminary way) the efficacies of transmission of some standard building sections, knowing that the results obtained for the two-dimensional cases, are entirely equivalent to those that can be derived from a similar analysis done (much more laboriously), in three-dimensions. It is also possible to complete the thermal picture including the effect of ventilation (but neither long or short wave radiation is considered). All this is illustrated here for blocks (but not for towers).

The model is constructed using a calculation spread sheet with the appropriate input data and formulas that throw the desired quantitative results & graphics to visually asses and compare between different possible solutions. As for the typologies treated by their sections, some of the possibilities are: Settled on the ground, Exempt (pilotis), with a Pitched Roof, a Complex Façade, more or less Insulated or Percentage of Windows, forming a Glazed Atrium, with Standard or Surplus Ventilation (Fig. 3).

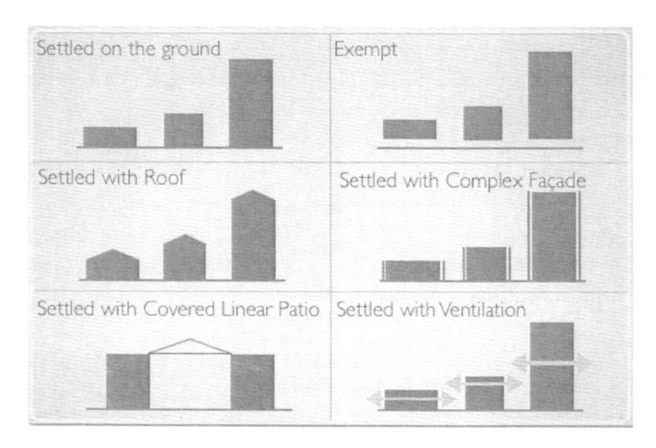

Fig. 3. Typical Sections of Building Blocks treated by the Model

As for the input factors, the common and most important ones are of course width & height and the amount of insulation & ventilation rates; we can also parameterize various others to fit each particular case (Fig. 4).

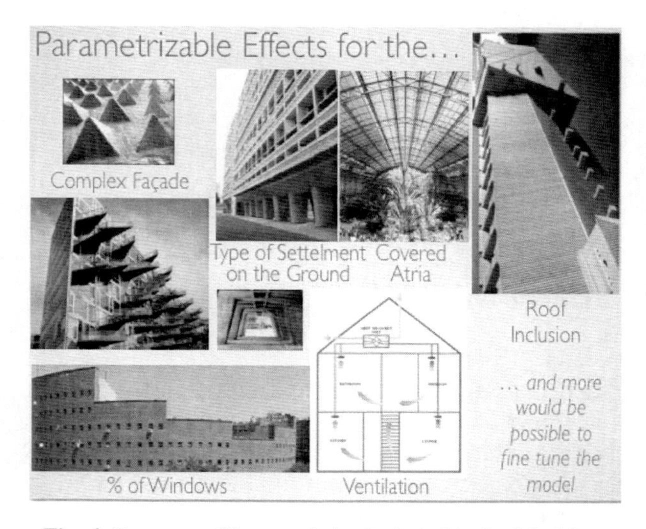

Fig. 4. Parameter Characteristics included in the Model

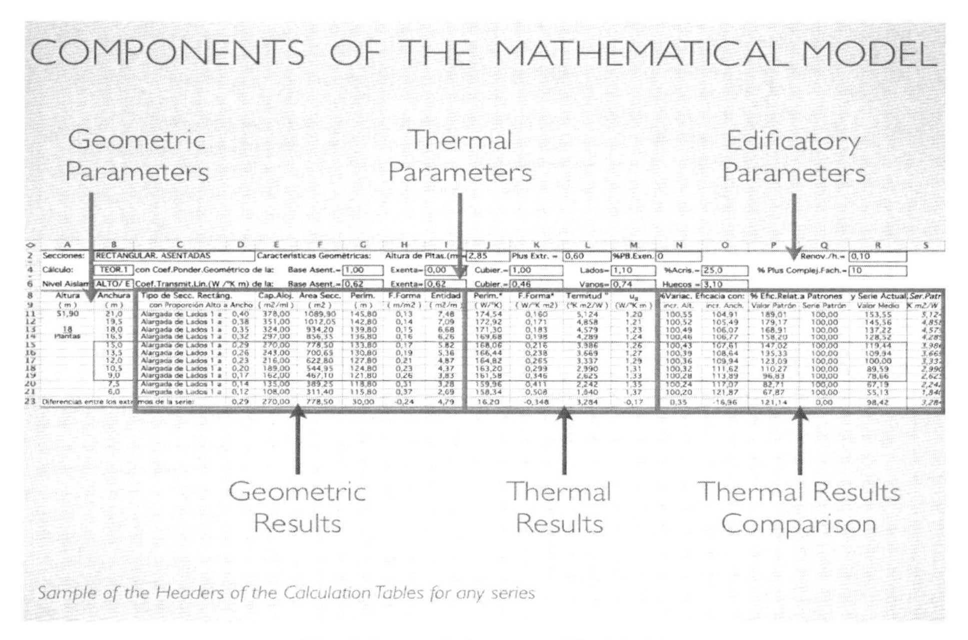

Fig. 5. Inputs & Outputs of the Model

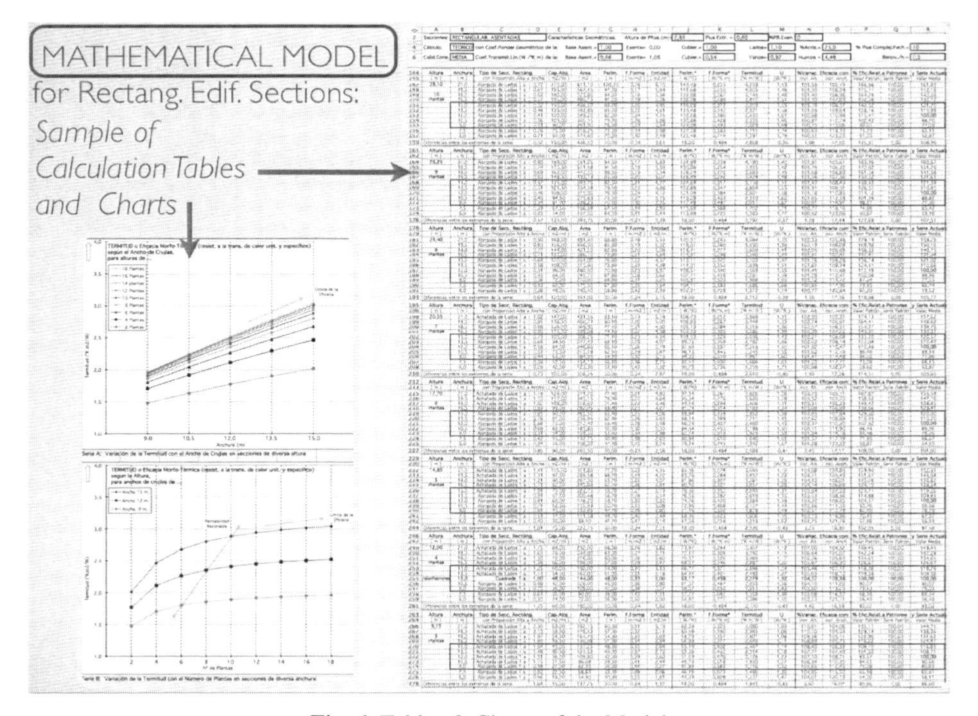

Fig. 6. Tables & Charts of the Model

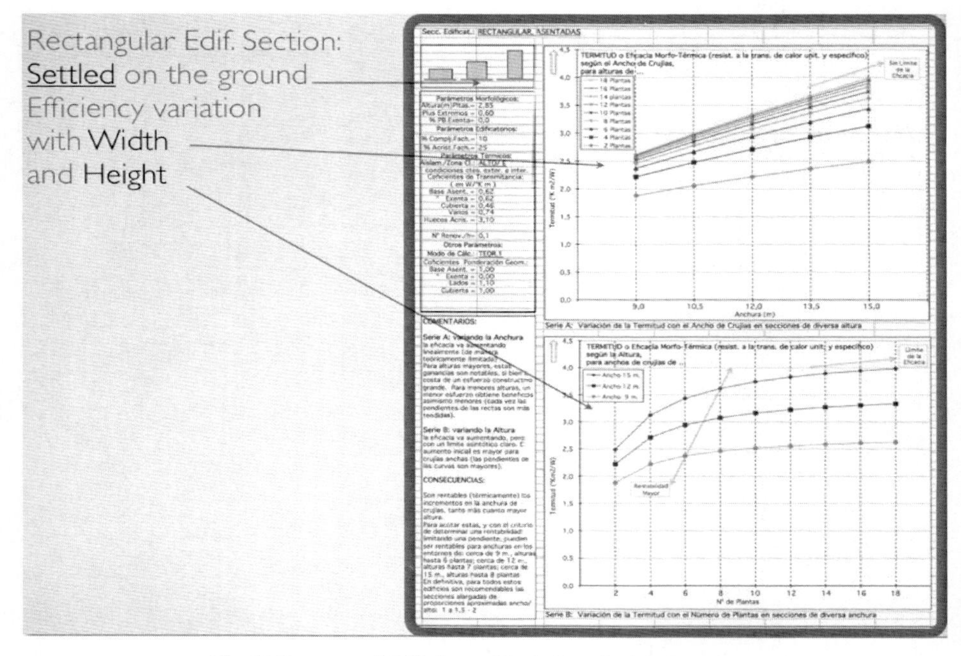

Fig. 7. Patterns of Efficiency Variations in one Typology

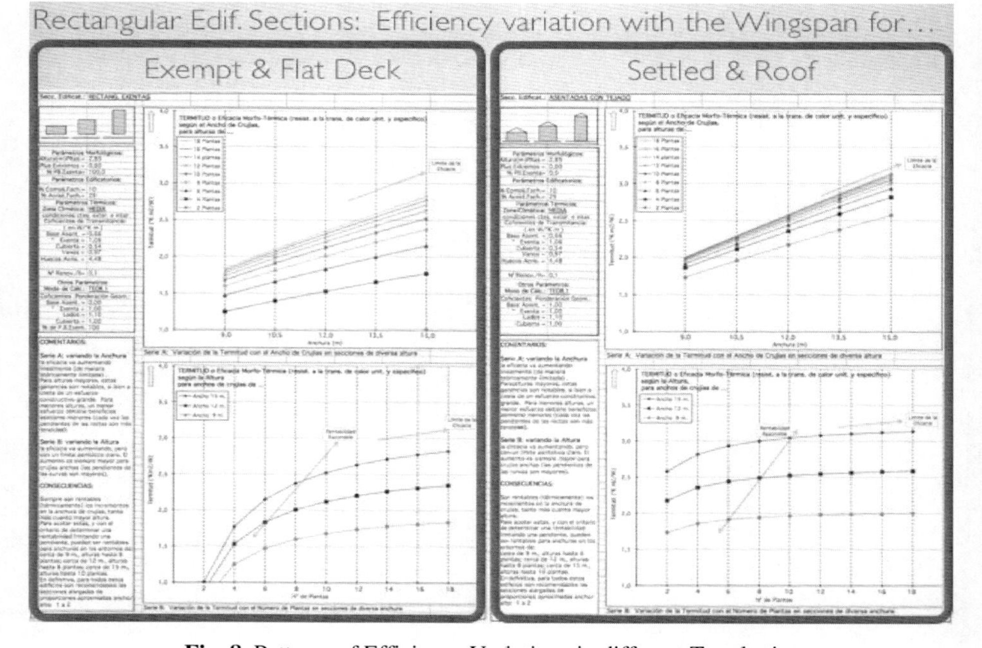

Fig. 8. Patterns of Efficiency Variations in different Typologies

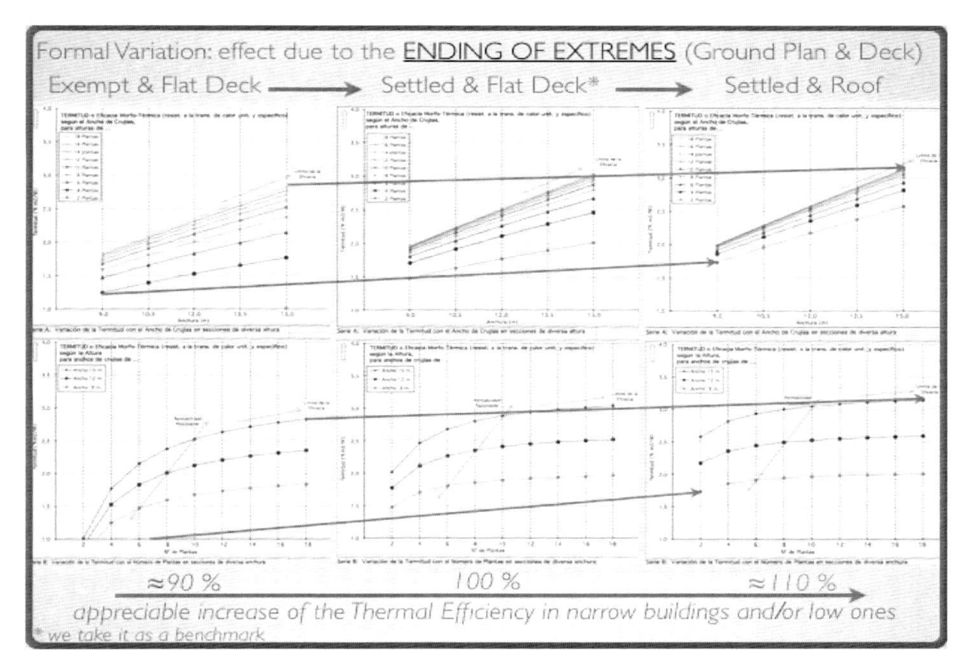

Fig. 9. Patterns of Efficiency Variations due to Building Extremes

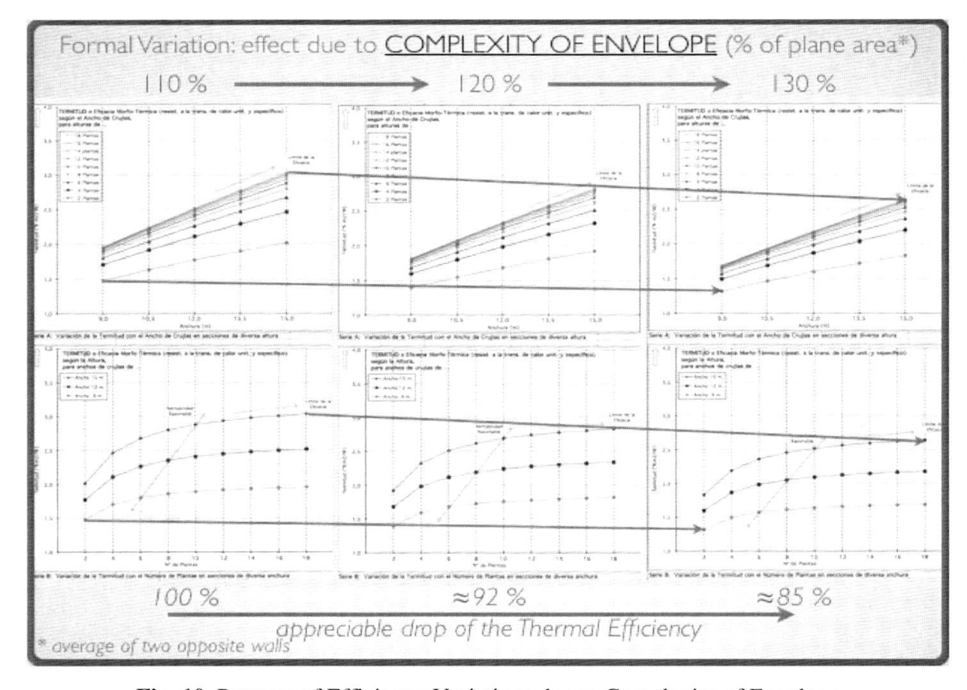

Fig. 10. Patterns of Efficiency Variations due to Complexity of Envelope

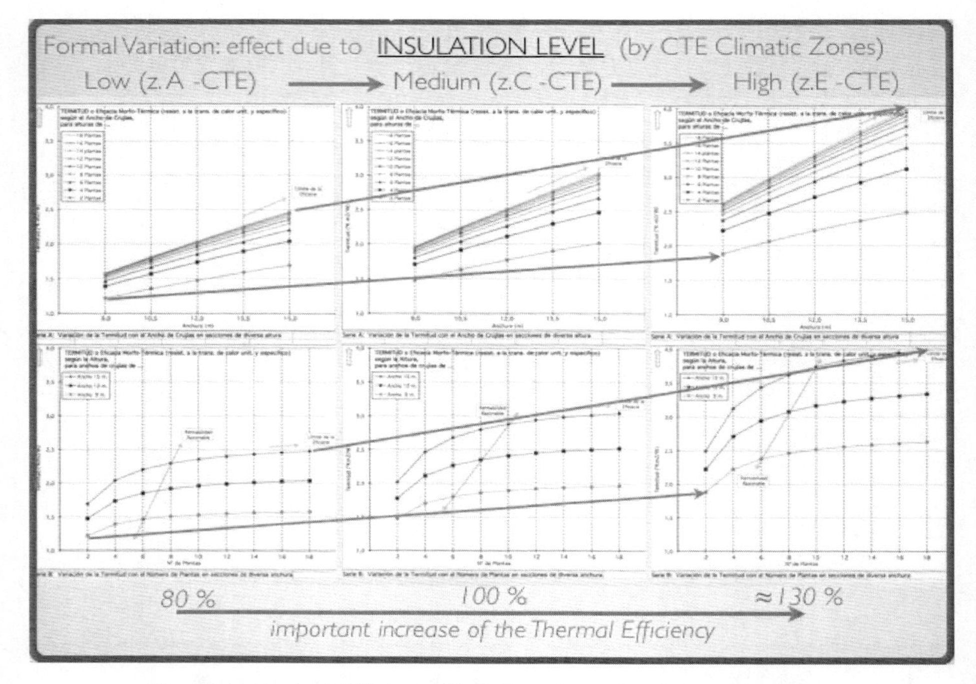

Fig. 11. Patterns of Efficiency Variations due to Insulation of Envelope

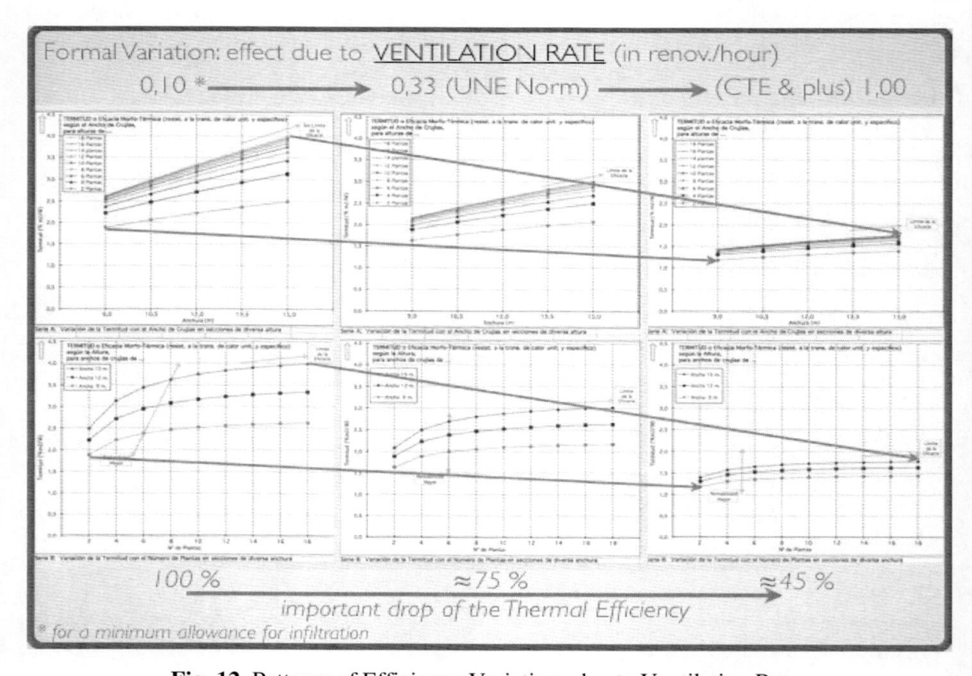

Fig. 12. Patterns of Efficiency Variations due to Ventilation Rates

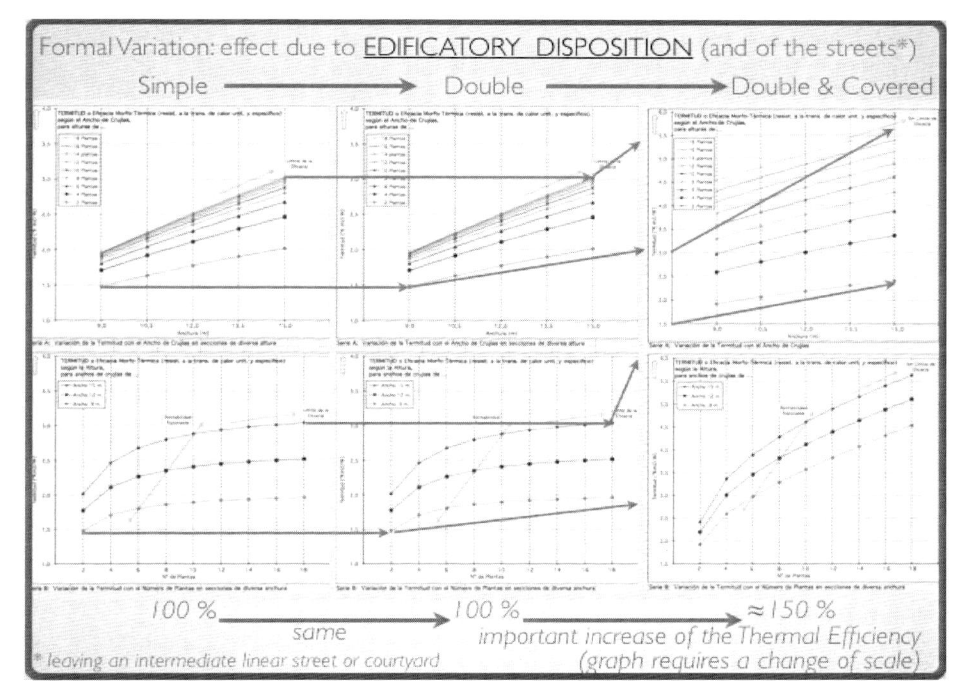

Fig. 13. Patterns of Efficiency Variations due to a Covered Courtyard

The output results, numbers and graphs for the efficiencies in relation to parameters are obtained. The graphics of thermal efficacies show important differences for width & height & configurational variations, and also the big influence of the rate of ventilation. Here we can just briefly show its appearance based on some of the graphic charts (Figs. 1.5, 1.13); for a full description and results see [3].

3 Conclusions

Based on this model, we can conclude the following facts for the variations of the THERMAL EFFICIENCY or "TERMITUDE of the building according to changes in its morphology:

A/ For each and every one of the aforementioned typologies taken separately:

1°- with an increasing width, efficiency always increases (linearly) in a theoretically unlimited way. For higher heights, these gains are significant, albeit at the expense of a greater building effort;

2°- with increasing height, efficiency increases, but with a clear asymptotic limit. The increase is always greater for those with wide bays.

B/ For the various typologies, taken with a same number of floors with increasing width efficiency increases linearly, and also in a quite similar mode in all of them, so that those gains (in efficiency) by increasing the width are similar in percentage terms, and do not have here a theoretical limit (but of course it is necessary in each case to evaluate many other consequences this might have).

C/ For the various typologies, taken with a same bay width, efficiency increases with the number of floors but at different rates of gains, and tending towards an asymptotic limit.

D/ In all the different types taken with the same bay width and the same number of floors, and in comparison to the benchmark curve (the totally settled on the ground section), we can say that:

1°- To settle the building on the ground, bears important efficiency gains, the more so in the range of low heights.
2°- The inclusion of a pitched roof has a significant gain in efficiency, especially in the range of low heights.
3°- It is very unfavourable to include elements that can increase the exchange heat on the surface of the façade, such as textures, balconies, overhangs, etc. (unless cold bridges are prevented).
4°- Duplicating the same building (with an intermediate linear street or courtyard) has no relevance in itself, but with the interposition of a covered space in between, efficiency rises dramatically.
5°- The increase of the level of insulation traduces in a linear increase in efficiency. An increased glass area has the reverse effect.
6°- The worst effect of all is the introduction of surplus ventilation.

E/ Moreover if we make a comparison of the physical model considered here with a purely geometric one (based on the conventional Form Factor NOT thermally weighted), we can conclude that to take as a preliminary (qualitative) guide a purely geometric criteria although it is simpler presents some shortcomings:

1º- With respect to the actual physical increase of efficiency by increasing width, it is not a dangerous assumption (does not induce misleading expectations).
2°- Regarding the potential increase of efficiency by increasing heights, it is a dangerous assumption (it may induce erroneous conclusions) as it does not grow forever, but in fact it has an asymptotic limit as the true physical model exhibits.

Using procedures as those here exposed the design task not restricted by volume, can focus only on the section of elongated building blocks, and their plan layouts can thus be freely adapted to what is more appropriate in relation to other important aspects: street guidelines, sunlight & wind orientation, lighting, etc. This can promote sustainable design just from the first (urban planning) stages.

References

1. AAVV, Código Técnico de la Edificación. Ministerio de la Vivienda, Madrid (2006)
2. Fernández de Trocóniz, A.J.: Factores de Forma Ponderados. Actas 1º Congreso Internacional de Arquitectura Sostenible, Valladolid (2009)

3. Fernández de Trocóniz, A.J.: Implicaciones Energéticas de la Tipología de Manzana. Tesis Doctoral, Universidad Politécnica de Madrid (2010)
4. Fernández de Trocóniz y Revuelta, A.J., Huerta, M.Á.G., López, T.G.: A Simple Way to Assess and Compare the Thermal Efficacy in Elongated Building Designs. In: M'Sirdi, N., Namaane, A., Howlett, R.J., Jain, L.C. (eds.) Sustainability in Energy and Buildings. SIST, vol. 12, pp. 285–294. Springer, Heidelberg (2012)
5. Givoni, B.: Man, Climate and Architecture. Applied Science, 319–335 (1976)
6. Knowles, R.L.: Energy and Form, pp. 66–70. The MIT Press, Cambridge (1974)

Chapter 23
Effect of Reaction Conditions on the Catalytic Performance of Ruthenium Supported Alumina Catalyst for Fischer-Tropsch Synthesis

Piyapong Hunpinyo[1,*], Phavanee Narataruksa[1], Karn Pana-Suppamassadu[1], Sabaithip Tungkamani[2], Nuwong Chollacoop[3], and Hussanai Sukkathanyawat[2]

[1] Department of Chemical Engineering, Faculty of Engineering, King Mongkut's University of Technology North Bangkok
1518 Pibulsongkram Road, Wongsawang, Bang-Sue, Bangkok, 10800, Thailand
piyapong.hp@hotmail.com

[2] Department of Industrial Chemistry, Faculty of Applied Science, King Mongkut's University of Technology North Bangkok

[3] National Metal and Materials Technology Center (MTEC) a member of National Science and Technology Development Agency (NSTDA), 111 Thailand Science Park (TSP), Paholyothin Road, Klong Nueng, Klong Luang, Pathumthani 12120, Thailand

Abstract. A Ru/γ-Al_2O_3 catalyst was prepared using by sol-gel technique in order to study its conversion and selectivity in the Fischer-Tropsch Synthesis (FTS). The effects of reaction conditions on the performance of a catalyst were carried out in a fixed bed reactor. The variation of the steady-state experiments were investigated under reaction temperature of 160-220°C, inlet H_2/CO molar feed ratio of 1/1-3/1, which both atmospheric pressure and gas space hour velocity of 1061 hr^{-1} were restricted. The influence of changing factors on CO conversion and on the selectivity of the formation of different hydrocarbon products in the reaction conditions was performed and compared to assess optimum operating conditions. In terms of FTS results, the increase of reaction temperatures led to increase of CO conversion and light hydrocarbon, while higher H_2/CO ratio has strongly influenced to increase the selectivity to higher molecular weight hydrocarbons and chain growth probability (α). Moreover, our catalyst was also markedly found to maintain selectivity to diesel faction for a wide range of H_2/CO molar feed ratios from BTL application.

Keywords: Biomass-to-Liquid, Fischer-Tropsch Synthesis, Ruthenium Supported Alumina.

1 Introduction

In recent years, Biomass-to-Liquid (BTL) process is one of the attractive options for producing green fuel, and there is a little prospect to develop a sound economic

* Corresponding Author.

A. Håkansson et al. (Eds.): *Sustainability in Energy and Buildings*, SIST 22, pp. 251–259.
DOI: 10.1007/978-3-642-36645-1_23

decision in Thailand. In principle, main process configurations, for the BTL are gasification, gas cleaning and Fischer-Tropsch Synthesis (FTS). For FTS process, the performance of catalyst is a crucial part of the heterogeneously catalyzed system for the production of liquid fuel hydrocarbon. Selection of catalyst depends on appropriate applications and conditions. Unfortunately, a major shortcoming from the biomass derived bio-syngas on this attention is considered an inappropriate synthesis gas composition, like H_2/CO ratio which can lead to low conversion in the FTS route, and then increase in an undesirable range of hydrocarbons.

The enhancement of catalyst performance is an alternative route, which can help to achieve the optimum product selectivity with matching of the BTL conditions. Several literatures have been proposed the efficiency of reaction conditions on the performance of Ruthenium (Ru) catalyst. Ruthenium catalyst was presented as the high active catalyst for running the reaction under low operating conditions to obtain the desired hydrocarbon products [1]. This may be advantage in the case of bio-syngas which H_2/CO ratios are much lower than the recommended ratio of the conventional FT (required at least two times more hydrogen than carbon monoxide). In particular, they are quite good for the formation of higher molecular weight products with the desired olefin and paraffin selectivity [2].

Efficiency of FTS depends not only on the selected catalyst but also on thermodynamic conditions. Optimum thermodynamic conditions can promote achievement of high conversion and product selectivity. Some researches effort targeted in this direction. M.E. Dry, 2002 [3] proposed that the viability of the FT process depends on three key factors such as temperature, feed gas composition, and chemical and structural catalysts. A similar series of literature was stressed by C.N. Hamelinck et al, 2004 [4] that selectivity and conversion in FTS are a function of temperature, feed stream composition, pressure, catalyst and reactor type and size. They tried to form linear relations which selectivity dependency on temperature, molar H_2/CO ratio, and pressure, respectively.

This paper investigated the influence of reaction temperatures and H_2/CO ratios coupling with a ruthenium based FT catalyst performance under BTL applications. The results in terms of the variation of changing these two parameters on the catalytic performance of promoted Ru catalyst were reported via reaction conversion and product distribution (using ASF distribution model) under steady-state reaction and reasonably explained.

2 Experimental Sections

2.1 Catalyst Preparation

In our group study, the amount of $10\%Ru/Al_2O_3$ catalyst was prepared by the sol-gel technique using ruthenium trichloride hydrate ($RuCl_3.xH_2O$) (Acros organics Company) and aluminium isopropoxide ($Al(OC_3H_7)_3$) (Acros organics Company) as precursors. Nitric acid solution (Carlo Erba Reagents Company) was added and mixed in the precursor solution. The mixed solution was then refluxed, stirred at the temperature of 90-95°C for 12 hr. The catalyst was dried and calcined in air at 400°C by controlling heat rate (10°C/min). Afterwards, the dried gel was crushed and sieved into 355-600 μm in size.

2.2 Tubular Fixed Bed Reactor System

A Tubular Fixed Bed Reactor (TFBR) with the length of 300 mm, internal diameter (ID) of approximately 8 mm made from stainless steel (SS316) tube was adopted for this work. The diagram of FTS system is shown in Fig. 1. The TFBR was heated by an electric tube furnace of 220V, 1500W, and equipped with constant thermo-couple (K-Type) indicators located inside and outside of the catalytic bed. The reactive gases were supplied from three cylinders consisting of hydrogen (H_2), carbon monoxide (CO) and nitrogen (N_2) and the mass flow rates of each component were controlled by three separate mass electronic mass flow controllers, enabling the desired H_2/CO ratio to be obtained. System pressure was regulated by a spring-load type back pressure regulator located at the bottom of TFBR. Two pressure transducers were also installed at the top and bottom of TFBR to monitor the pressure difference across the bed.

After the product gas had left the TFBR, it was passed through a condensing section in order to separate rather heavy components before entering the gas detector. The liquid hydrocarbon products were collected in which the temperature was maintained at around -5°C by silicon oil media, and then the non-condensable hydrocarbons were analyzed for the compositions by Gas Chromatography (GC). The gas flow rate of the effluent stream was measured by a bubble gas meter.

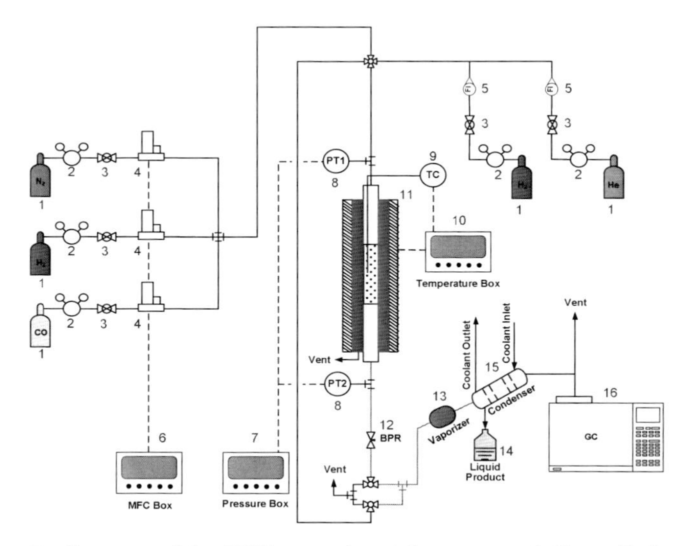

Fig. 1. Schematic diagrams of the FTS's experimental apparatus. 1-Gas cylinders, 2-Pressure regulators (Swagelok®), 3-Ball valves (Swagelok®), 4-Mass flow controllers (Aalborg®), 5-Gas Rota-meter (Aalborg®), 6-MFC box, 7-Pressure box, 8-Pressure Transducer (GENSPEC©), 9-Thermo-couple (Type-K), 10-Temperature box, 11-Electric furnace (Carbolite®), 12-Back pressure regulator (Spring type, Swagelok®), 13-Vaporizer, 14-Liquid Product, 15-Condenser (Ice bath), 16-Gas chromatography (Agilent® Model GC-6890N).

2.3 Experimental Designs

The experimental program was planned for a total of 12 runs. Among them, the experimental sets for run number 1 through run number 12 were arranged according to the percent of conversion with respect to CO and hydrocarbon distributions, while the reaction pressure and space velocity were kept constant. For this studies, a stabilization period of CO conversion (%), as a function of reaction time (36 hours) was maintained to ensure that the stable catalytic reactions were established. Descriptive samples were cumulatively collected during a typical period of 20-36 hr to ensure the steady state behavior of the reaction. The data from this series of investigation were provided in Table 1.

Table 1. Effect of reaction conditions on catalytic performance under atmospheric pressure and GHSV=1061 hr^{-1}

Runs.	Temperature (°C)	H_2/CO ratio	CO conversion (%)	S_{C1-C4} (%)	S_{C5+} (%)
1	160	1/1	N/A	N/A	N/A
2		2/1	25.0	0.04	99.96
3		3/1	31.1	0.07	99.93
4	180	1/1	15.9	0.40	99.60
5		2/1	31.5	0.51	99.49
6		3/1	41.2	0.71	99.29
7	200	1/1	16.4	1.10	98.90
8		2/1	33.4	2.85	97.15
9		3/1	42.4	8.30	91.70
10	220	1/1	17.9	3.92	96.08
11		2/1	34.8	8.24	91.76
12		3/1	43.0	14.55	85.45

The description in terms of CO conversion, selectivity, chain growth probability and yield of products are given below. The CO conversion (%) is evaluated according to the normalization method by using in Eq (1).

CO conversion (%) = [Moles of CO converted to hydrocarbon product / Moles of CO fed into reactor] x100 (1)

The selectivity (%) towards the individual components on carbon basis is calculated according to the same principle by using in Eq (2).

Selectivity of j product (%) = [Moles of j product / Moles of all product] x100 (2)

To describe the experimental deviation of hydrocarbon form, the "Anderson-Schultz-Flory" distribution model was applied in this study. The distributions followed the sort of exponential function as stated in Eq (3).

$$W_n = n \cdot (1-\alpha)^2 \cdot \alpha^{n-1} \tag{3}$$

After taking logarithm on both sides of in Eq (4), it obtains

$$\log (W_n/n) = n \cdot \log (\alpha) + \log [(1-\alpha)^2/\alpha] \tag{4}$$

where is the weight fraction of the products with carbon number (n), obtained from experimental results involving hydrocarbons from C1 to C24. In addition, the chain growth probability (α) depends upon the reaction conditions and the type of catalyst [5]. In the experiments, a majority of the FTS products were taken into account, whereas the trace of oxygenated contents was not included.

3 Results and Discussion

3.1 Characterization of Catalysts

The results of BET and porosity tests with three different adding promoters over ruthenium particle catalyst are summarized in Table 2. The specific surface area of alumina was found to be 252 $m^2 \cdot g^{-1}$ while its pore volume was 3.18 $cm^3 \cdot g^{-1}$ with pore diameter of 0.201 nm.

Table 2. BET and porosity data of 10%Ru/γ-Al_2O_3 catalysts prepared by the sol-gel method

Properties	10%Ru/γ-Al_2O_3	Unit
BET surface area	252	$m^2 \cdot g^{-1}$
Pore volume	3.18	$cm^3 \cdot g^{-1}$
Pore diameter	0.201	nm

3.2 Effect of Inlet H_2/CO Molar Ratios

Studies on the effect of temperature and H_2/CO ratio were conducted for 10%Ru/Al_2O_3 catalysts with two different gas and liquid phases FT synthesis. At the beginning, the temperature was set constant in each point at 160, 180, 200, and 220°C, and experiments with three different H_2/CO ratios of 1/1, 2/1 and 3/1 were varied to complete the number of 12 experimental runs. The gas hour space velocity for all these experiments were maintained at a value of 1061 hr^{-1}.

The performance of the catalyst via CO conversion was exhibited in Fig. 2. It can be observed that CO conversion against time-on-stream was continuously enhanced by increasing the value of H_2/CO molar ratio. Fig. 2a illustrated that the reaction temperature is imposed, namely 220°C, CO conversion remained smoothly the trend (approximately 43%) for H_2/CO molar ratio as high as 3/1 and dropped to approximately 35% and 18% when hydrogen partial pressure was continuously diminished. Similar patterns of curve in Fig. 2b and c indicate that a further increase in the H_2/CO ratio resulted in a significant increase in the CO conversion, since conversion decreased dramatically when the H_2/CO ratio fell to unity. Meanwhile, temperature of 160°C (Fig. 2d) was the least conversion for all of experiments. Both 31% and 25% of CO conversion were obtained from H_2/CO ratio of 3/1 and 2/1 except for H_2/CO ratio of 1/1 could not be measured, which could be correlated with the relative partial pressures of hydrogen present in reaction condition.

In the part of the results of hydrocarbon distributions under three different H_2/CO ratios was illustrated in Fig. 3. It can be found that the gas product rates were raised due to the increase of the hydrogen partial pressure. The selectivities for gaseous hydrocarbons increased from 0.04% to 14.55% (not shown here) while those of liquid hydrocarbons decreased from 99.96% to 85.45%. At the same time, the higher H_2/CO ratio in feed leads to obtain higher molecular weight hydrocarbons as compared with that of the deficient ratio in all four temperature experiments. Especially, at H_2/CO ratio equals to 3/1 in Fig. 3a and b, a bar level performed a significant shift towards more hydrocarbons which could be observed a median carbon number of C7 to C24. Meanwhile, Fig. 3c and d contain the data on the liquid hydrocarbon distribution as a function of temperature of 180 °C and 160 °C. No distinct pattern was observed in two figures due to the increase in H_2/CO ratio. These results reflect that the reaction conditions i.e., H_2/CO ratio and temperature play a very important role in FTS reaction, and directly influence the CO conversion and product selectivity [6, 7].

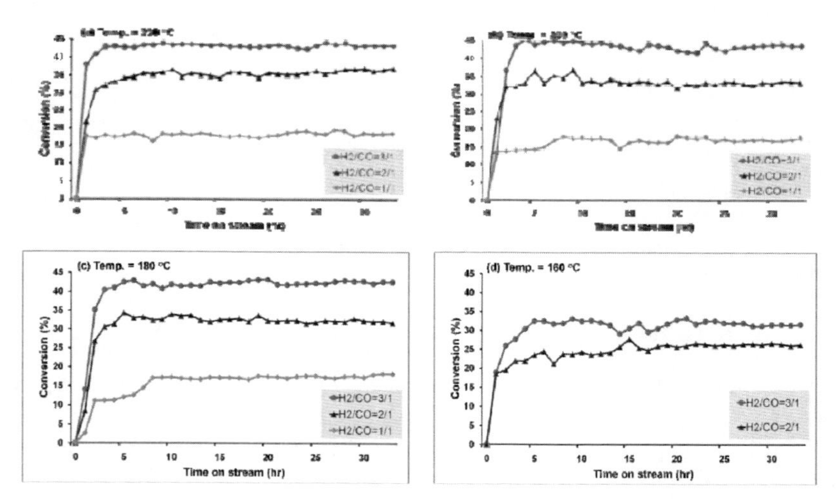

Fig. 2. Carbon monoxide conversion (%) against time on stream restricted each temperature; (a) 160°C, (b) 180 °C, (c) 200 °C and (d) 220 °C for the FTS by considering the effect of different H_2/CO ratios at H_2/CO ratio = 1/1(Green line), H_2/CO ratio = 2/1 (Blue line) and H_2/CO ratio = 3/1 (Red line) over 10%Ru/Al_2O_3 catalyst under pressure 1 atm and GSHV 1061 hr^{-1}.

As resulted above, it seemed to be that both H_2/CO ratio and temperature strongly influence the performance of the catalyst and the selectivity of products. A few studies have been conducted for elucidating effect depends both of them that higher hydrogen partial pressure led to an enhancement of hydrogen species on the catalyst surface, which abundant of these species in order to less heavy hydrocarbon [8].

Because the coverage of surface hydrogen was increased with increasing H_2/CO ratio, it may provide more chance for those species for further hydrogenation to CH_4 or other gaseous hydrocarbons. Meanwhile, the effect of reaction temperature might be due to enhanced desorption of the heavier hydrocarbons on the surface catalyst at higher temperatures [9]. As mentioned above, from our results, there was a clear correlation between H_2/CO ratio and temperature corresponding to liquid (C5+) hydrocarbon selectivity. It was found a clear depletion of C5+ when both of them were increased in the same direction. Consequently, the formation of gaseous hydrocarbons selectivity a similar trend occurred as following previously reported with increasing H_2/CO ratio and temperature.

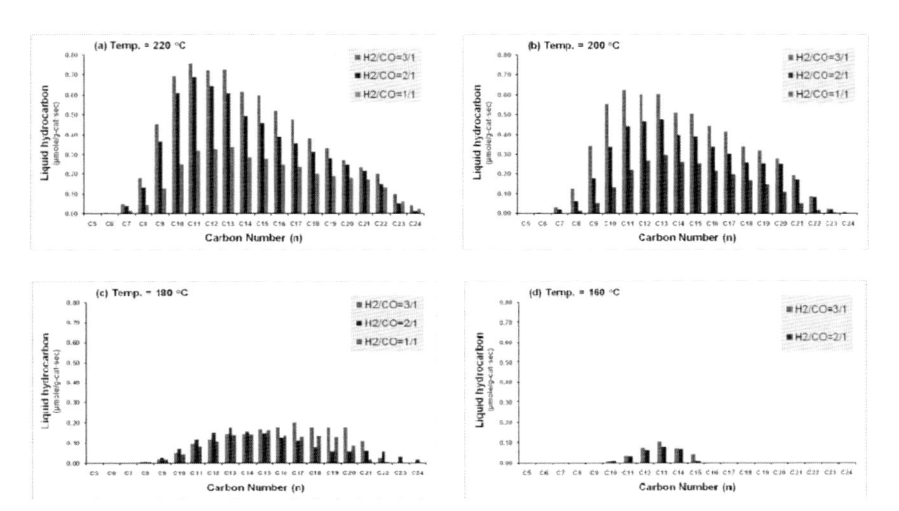

Fig. 3. Liquid hydrocarbon distribution (C5+) against time on stream restricted each temperature; (a) 160 °C, (b) 180 °C, (c) 200 °C and (d) 220 °C for the FTS by considering the effect of different H2/CO ratios at H_2/CO ratio = 1/1(Green bar), H_2/CO ratio = 2/1 (Blue bar) and H_2/CO ratio = 3/1 (Red bar) over 10%Ru/Al_2O_3 catalyst under pressure 1 atm and GSHV 1061 hr^{-1}.

When the description of the FTS product distributions was established, it was often interpreted by the Anderson-Schulz-Flory (ASF) equation. The ASF scheme was used to attribute a phenomenon of the chain propagation and the termination probabilities which are independent of carbon number. The ASF plot of the product distributions was drawn for comparison in each reaction condition, while a distinct line was separately fit to determine the chain growth probabilities (α).

For experimental data in Fig. 4, H_2/CO ratio of 3/1 was chosen to identify a suitable the ASF plot. As noticed from this figure, the combination of the gaseous and the liquid hydrocarbon were used to calculate the values of W_n/n. There was a spread deviation of ASF distribution curves between the light gas (C1-C4) and the liquid product (C5+) because of separately analyzed by GC. To compare in a quantitative basis the effect on the alpha (α) value, the slope of the best straight line for the range of C10+ was estimated with the R^2 value up to 95%. The product distributions have a slight declining tendency with the increase of the carbon number, while the lower H_2/CO ratio led to higher heavy hydrocarbons (C10+) and lower light ones (C1-C9). Moreover, in case of chain growth probability a decrease in the value of α from 0.90 to 0.88, occurred when H_2/CO ratio was increased from 1/1 to 3/1.

From the results, clearly, higher H_2/CO ratios trend to favor for the formation of light hydrocarbons while lower H_2/CO ratios is favorable for the production of heavy hydrocarbons [7]. Therefore, the selectivity to C5+ slightly decreased with increasing H_2/CO ratio. This trend may be attributed to the increased H_2 enrichment (high hydrogen partial pressure) corresponding to the increasing H_2/CO ratio in feed, in order to increase the possibility of the CO reactant react more on the catalyst surface by the dissociation of C-atoms and participated in the chain propagation with H_2. Consequently, the monomer hydrocarbon reacted with H_2 incoming and a new monomer free radical species, leading to the release of chain polymerization of hydrocarbons in the chain termination step on the active surface areas of ruthenium catalyst, which induce to more obtain the hydrocarbon product when compare with the deficient H_2/CO ratio.

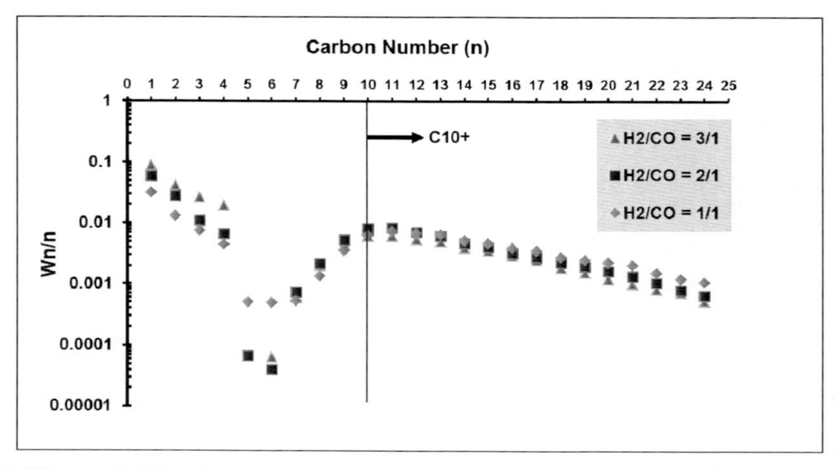

Fig. 4. Effect of H_2/CO ratio on the distribution of hydrocarbon against carbon number (Temperature = 220°C, pressure = 1 atm, GSHV = 1061 hr^{-1})

4 Conclusions

In an effort to increase both the conversion and selectivity under the BTL conditions via FTS toward green liquid fuel hydrocarbons, operating parameters concerning the reaction fed conditions for BTL processes were investigated for the developed ruthenium-alumina supported catalyst. Reaction temperature and H_2/CO molar feed ratio were two priority parameters for considering the effect of reaction conditions in the FT catalytic reactors.

The ruthenium catalyst markedly presented a higher activity towards the formation of hydrocarbons, especially the formation of higher molecular weight hydrocarbon with the desired products. The elevation of reaction temperatures was found to be influenced not only by the increase of CO conversion but also with increasing hydrocarbon products distribution on the performance of the catalyst, while higher H_2/CO ratio has strongly influenced the chain growth probability and the same time it favors the enhancement of amount of C5+ products. Hence, selecting a suitable both reaction temperature and inlet H_2/CO ratio can result in optimal catalytic performance of the ruthenium catalyst, achieving an optimal diesel fraction (C5-C24).

Acknowledgements. The authors would like to thank the Thailand Graduate Institute of Science and Technology (TGIST), the Graduate College of KMUTNB, and annual government statement of expenditure and budgets for financial support on research studies.

References

[1] Van Der Laan, G.P., Beenackers, A.A.C.M.: Kinetics and Selectivity of the Fischer-Tropsch Synthesis: A Literature Review. Catalysis Reviews - Science and Engineering 41(3-4), 255–318 (1999)

[2] Schulz, H.: Short history and present trends of Fischer–Tropsch synthesis Applied Catalysis A: General 186, 3–12 (1999)

[3] Dry, M.E.: The Fischer–Tropsch process: 1950–2000. Catalysis Today 71, 227–241 (2002)

[4] Tijmensen, M.J.A., Faaij, A.P.C., Hamelinck, C.N., van Hardeveld, M.R.M.: Biomass Bioenergy 23, 129–136 (2002)

[5] Dry, M.E.: Catalytic Aspects of Industrial Fischer-Tropsch Synthesis. Journal of Molecular Catalysis 71, 133–144 (1982)

[6] Bukur, D.B., Lang, X., Akgerman, A., Feng, Z.: Effect of Process Conditions on Olefin Selectivity during Conventional and Supercritical Fischer-Tropsch Synthesis. Industrial & Engineering Chemistry Research 36, 2580–2587 (1997)

[7] Liu, Y., Teng, B.-T., Guo, X.-H., Li, Y., Chang, J., Tian, L., Hao, X., Wang, Y., Xiang, H.-W., Xu, Y.-Y., Li, Y.-W.: Effect of reaction conditions on the catalytic performance of Fe-Mn catalyst for Fischer-Tropsch synthesis. Journal of Molecular Catalysis A: Chemical 272, 182–190 (2007)

[8] Tian, L., Huo, C.-F., Cao, D.-B., Yang, Y., Xu, J., Wu, B.-S., Xiang, H.-W., Xu, Y.-Y., Li, Y.-W.: Effects of reaction conditions on iron-catalyzed Fischer–Tropsch synthesis: A kinetic Monte Carlo study. Journal of Molecular Structure: THEOCHEM 941, 30–35 (2010)

[9] Dasgupta, D., Wiltowski, T.: Enhancing gas phase Fischer–Tropsch synthesis catalyst design. Fuel 90, 174–181 (2011)

Chapter 24
Integration of Wind Power and Hydrogen Hybrid Electric Vehicles into Electric Grids

Stephen J.W. Carr[1], Kary K.T. Thanapalan[1], Fan Zhang[1], Alan J. Guwy[1], J. Maddy[1], Lars-O. Gusig[2,*], and Giuliano C. Premier[1]

[1] Sustainable Environment Research Centre (SERC), University of Glamorgan, UK
[2] University of Applied Sciences and Arts Hannover, Dept of Mechanical Engineering, Ricklinger Stadtweg 120, D-30173 Hannover, Germany
lars.gusig@fh-hannover.de

Abstract. Integrating wind energy and electrical car fleets to electrical grids can result in large and erratic fluctuations from additional power sources and loads. To identify the potential use of hydrogen storage in hydrogen-hybrid-electric vehicles, a grid-to-vehicle model (G2V) with three different scenarios has been modeled. The target is to maximize hydrogen supply to vehicles, and to facilitate more renewable energy onto the grid. Daily analysis for an existing network shows that under passive demand conditions extra wind is allowed onto the network, but some wind must be curtailed, while not all the hydrogen demand can be satisfied. With active demand, all of the wind is utilized and all hydrogen demand can be met. In addition a significant amount of hydrogen remains in store at the end of a day.

Keywords: Hydrogen storage, wind energy, electric-vehicles, power management.

1 Introduction

The effort to increase the amount of renewable energy we use is important to a great variety of sectors; especially in the area of i) transportation and mobility, ii) in housing and building, and iii) process and production industries. Focus lies both on the increase of efficiency and on the change from fossil and nuclear fuels to more sustainable solutions. All three sectors consume a significant amount of primary energy each leading to a specific end use of thermal, mechanical or electrical power. As all three are connected by the electrical power grid, apart from optimizing every subsystem, an alternative approach is to look at the interactions between the different areas as a whole. One way to integrate renewable energy into the grid is through wind farms. The amount of wind power currently installed will soon reach 200 GW worldwide. EU-regulations motivate estimated growth rates between 14% and 33% for the next 8 years and beyond (Tab.1).

* Corresponding author.

A. Håkansson et al. (Eds.): *Sustainability in Energy and Buildings*, SIST 22, pp. 261–270.
DOI: 10.1007/978-3-642-36645-1_24

Table 1. Selected regional overview of installed wind capacities and annual growth rates

Installed Capacity	Wales	UK	Lower Saxony	Germany	EU	US	World
Currently installed	0.5 GW	6 GW	7 GW	29 GW	86 GW	44 GW	197 GW
Target capacity [1]	2.5 GW	22 GW	15 GW	63 GW	-	-	-
Annual growth rate [2]	33%	22%	14%	15%	-	-	-

Increasing the amount of wind power has significant effects on the electrical grid that has to carry the additional load. Large erratic fluctuations of wind power are a common characteristic and have to be compensated by spinning reserve or increased storage facilities. Besides conventional storage systems (e.g. pump storage stations, battery/flywheel storage) in recent years a particular research emphasis has been placed on hydrogen storage systems and associated consumption in electric drive vehicle fleets.

The generation of hydrogen from renewable energy intermitancy's peaks is one possible means of storing power for future needs. After production by electrolysis hydrogen can be stored as pressurized gas, liquid, or in metal hydride storage amongst others [1]. It can later be used in fuel cells to produce electrical energy at times of great demand or it can be utilized in FC vehicles or to fuel internal combustion engine vehicles (ICEV). When used to produce electrical energy again, the round trip storage efficiency could be around 40%, taking into account the efficiency of the electrolyser, the storage device itself and the fuel cell. The connection of hydrogen storage systems to electrical grids and corresponding network management concepts have been investigated for example in [2].

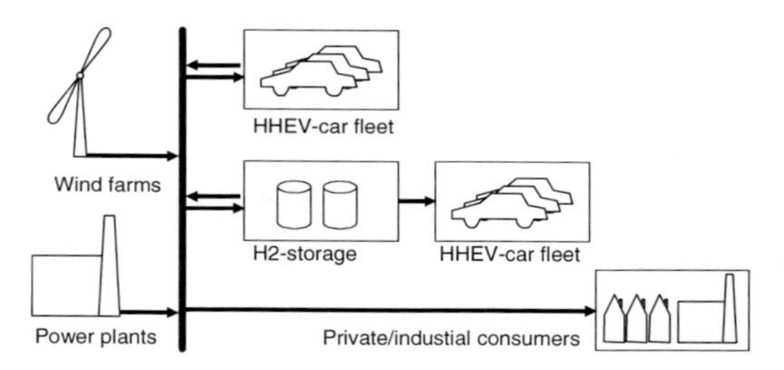

Fig. 1. Different power sources and loads combined for G2V network simulation

Using car fleets as a means of storage has been termed vehicle-to-grid (V2G, [3]) Using great number of plug-in electric vehicles (PEV) could increase the storage capacity of electrical networks significantly but on the other hand increases the load on the grid requiring new control mechanisms. Combining hydrogen storage with hydrogen hybrid electrical vehicles (HHEV) could result in a workable G2V approach. In

---

[1] Estimates based on different national/regional data sources: UK Wind Energy Database (www.bwea.com/ukwed/index.asp), Windpower–Wales (www.windpower-wales.com), Bundesverband Windenergie – BWE (www.wind-energie.de/statistiken/international), 2020 vision - How the UK can meet its target of 15% renewable energy (Renewables Advisory Board, 2008).

[2] Estimates calculated based on target years 2020 and 2025 as stated in sources as of [1)]

this case the cars are used as an additional load on the grid, such that additional wind power may be accommodated.

It is the target of this paper to determine to what extent the integration of wind energy and the multiple hydrogen-storage facilities in HHEV can increase the amount of renewables accepted by the grid.

2 Grid Model and Demand Scenarios

A network with wind farms, consumer load, and hydrogen vehicle demand supplied by electrolysis is studied in the three configurations outlined in Table 2.

Table 2. Overview of configuration sets for simulation

Set	Power sources	Loads
Base	Grid supply point	Consumers
I	Grid supply point	Consumers + H2 vehicle demand
II	Grid supply point + wind farms	Consumers + H2 vehicle demand, passive
III	Grid supply point + wind farms	Consumers + H2 vehicle demand, active

According to transport statistics presented by the Department for Transport, in 2009, the total vehicle traffic volume for cars was 250 billion vehicle miles[3]. There were 28.2 million cars licensed for use on the roads in Great Britain[4], among them, 1.4 million cars were used in Wales[5]. Therefore, the average mileage is 24 miles per day for a passenger car. Probability of passenger cars not in use during a weekday can be gathered from the United Kingdom Time Use Survey 2000[6]. It shows significant characteristics similar to Fig.2: most of the cars are used during the rush hour between 8:00 and 17.00, yet there are still 88% of cars parked and available to be used as energy storage.

In order to create a hydrogen demand profile, the fuel economy needs to be taken into consideration. For hydrogen fuel cell passenger car, the consumption of hydrogen for e.g. a Mercedes-Benz B-Class fuel cell car is 0.97 kg of Hydrogen per 100 km[7]. Taking into consideration the average mileage, and a factor of 39.4kWh/kg H_2, the profile of the power resulting from the refueling demand for hydrogen in Wales can be estimated as shown in Fig. 2, which also shows the wind power profile for a typical day. The wind and load time series are obtained from the Centre for Sustainable Electricity Distributed Generation[8].

---

[3] Transport Statistics Great Britain: 2010. Department for Transport. `http://webarchive.nationalarchives.gov.uk/20110218142807/dft.gov.uk/pgr/statistics/datatablespublications/tsgb/`

[4] Licensed vehicles by body type, Great Britain, annually: 1994 to 2010. Department for Transport. `http://www.dft.gov.uk/statistics/tables/veh0102/`

[5] Licensed cars, by region, Great Britain, annually: 2000 to 2010. Department for Transport. `http://www.dft.gov.uk/statistics/tables/veh0204/`

[6] United Kingdom Time Use Survey, 2000. Economic and Social Data Service. `http://www.esds.ac.uk/findingData/snDescription.asp?sn=4504`

[7] Mercedes-Benz B-Class F-Cell Datasheet. Mercedes-Benz.

[8] `http://www.sedg.ac.uk/`, "UKGDS," 2007.

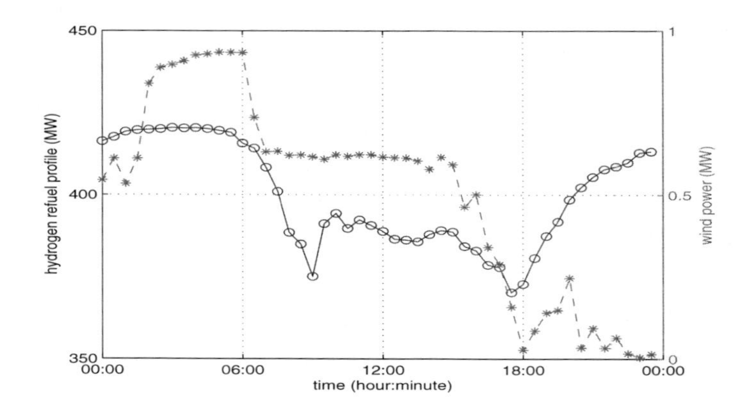

Fig. 2. The hydrogen refueling profile (circles) and wind power (stars) for Wales

2.1 Description of Network Studied

The example network studied, Fig.3, is obtained from the Western Power Distribution Long Term Development Statement[9] and is part of the South Wales electricity distribution network. It consists of a 66 kV network with a 132 kV network as its grid supply point. The network supplies 51.9 MVA of load.

Four wind farm sites are chosen to be at locations remote from the main grid connection point in order to represent plausible locations for wind farms. The wind farms are sized such that when considered individually their maximum output can be accepted at maximum load on the network. This gives wind farm capacities which are suitable at each individual node, but cause curtailment when considered together. The capacity allocation is carried out by running an optimal power flow (OPF) routine with generators operating at unity power factor.

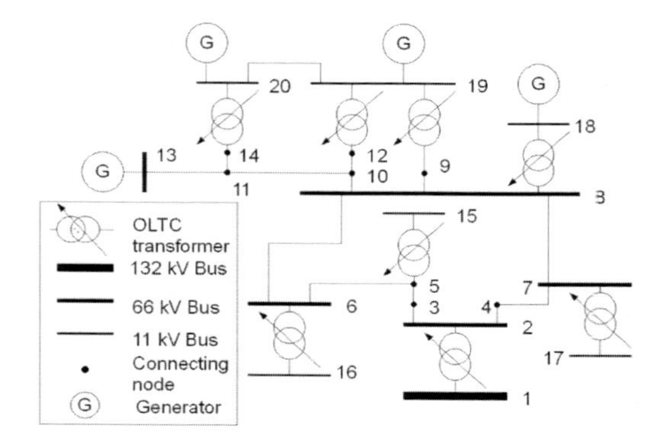

Fig. 3. Diagram of Network: Based on a section of South Wales distribution network

[9] Long Term Development Statement (South Wales) plc´s Electricity Distribution System 2005.

Table 3 presents the wind farm capacity allocations [2]. Hydrogen demand for vehicles is added to the system. This hydrogen demand is supplied by electrolysers, operating at an assumed efficiency of 81%.

Table 3. Wind farm capacity allocation scenarios

Wind farm bus	13	18	19	20	Total
Capacity (MW)	22.44	31.94	25.88	17.51	102.2

The demand for hydrogen on the network is associated with the load at each bus, and is proportional to that load. The overall demand is set by comparing the ratio of the total energy demand for electricity in Wales with that required to supply a given proportion of the passenger vehicle fleet with hydrogen. There are 1.4 million passenger vehicles in Wales. By equating the demand for electricity in Wales with the demand for electricity on the network studied it is estimated that 16,000 cars are associated with this Network. This represents ~1% of the car vehicle fleet in Wales. Five hydrogen demand levels are studied, where hydrogen demand level 1 represents 1/5th of the demand of the entire vehicle fleet considered to be converted to HFCV (3,200 cars), 2 represents 2/5^{ths}, 3 represents 3/5^{ths}, 4 represents 4/5^{ths} and 5 represents 5/5^{ths}, or the whole fleet, converted to HFCV's.

2.2 Optimisation Scenarios

Two different power management scenarios have been simulated:

a) Passive hydrogen demand: The OPF uses an objective function in which the hydrogen demand acts as a dispatchable load on the network. This load does not have to be met if network constraints do not allow it. In this case extra wind power is allowed onto the network passively, through the additional load created by the hydrogen demand. No hydrogen storage in addition to that in the vehicles is needed in this case.
b) Active hydrogen demand: The OPF uses an objective function which maximises the amount of wind energy accepted onto the network by taking energy in the form of hydrogen into storage, whilst minimising the amount of hydrogen demand not met at each time step.

The objective function used for these optimisation scenarios is defined as:

$$OF = -C_W \sum_{WF} P_{WF}(t) - C_s \sum_{el} (1-e) P_{el}(t) - C_h \sum_{h} (H_h^{dem}(t) - H_h^{\sup}(t))$$

Where C_W, C_s, and C_h are the nominal costs associated with wind power, storage and importing hydrogen. P_{WF} is the power available from each wind farm, P_{el} is the power consumed by each electrolyser, H_h^{dem} is the demand for hydrogen at each bus, and $H_h^{\sup}$ is the hydrogen supplied at each bus. For the active hydrogen demand scenario, C_W and C_s are of equal value, and the objective function works to maximise wind power onto the network whilst minimising the hydrogen demand not met. For the passive demand scenario C_s is set greater than C_W but less than C_h, so that there is a large penalty for the electrolyser to operate. In this case it will not operate to increase wind power, but only to supply hydrogen demand. The parameter e determines the priority given to minimizing wind curtailment. In this case a value of 0.7 is found to give optimal minimizing of wind curtailment. The constraints take into account the real and reactive power flows at each bus as well as the thermal, voltage, transformer and generator limits [2].The OPF is run at each half hour time-step over the course of one year in order to determine the extra energy which the hydrogen demand allows to be utilised from the wind power.

3 Hydrogen Hybrid Electric Vehicle Model (HHEV)

Recently various types of hybrid electric vehicles have been introduced to the market. The HHEV is one of the most common. Several researchers considered the development of simulation models for the analysis and performance improvement of these vehicles. Research effort[10] at the University of Glamorgan (UoG) led to the production of three HHEV's and associated simulation tools which can be use to further investigate and alleviate the problems such as energy management, fuel consumption and storage. Utilizing the HHEV system model developed by the UoG, power management of grid-connected renewable HHEV system will be investigated in this paper. The grid model described above will be linked to the HHEV system model to configure the G2V power flow mechanism. The HHEV system model included the following subsystems; electrochemical power source model, DC/DC converter subsystem, vehicle dynamic and a driver model. These are then lumped together in a systematic way and programmed in MATLAB/Simulink for a customized study. Details of these customized vehicle models can be found in [4]. This work considers an investigation of the power management of a grid connected-HHEV system, so the generic electrochemical power source models for HHEV system is described here. The HHEV system model may consist of several different power source subsystems. In this study, to simplify the system complexity and analyses, the numbers of power sources are limited to three which are; fuel cell stack, ultracapacitor and battery pack. In the fuel cell system model fuel cells are connected in series to form the stack, the total stack voltage can be calculated by multiplying the cell voltage, by the number of cells, n of the stack. Thus the stack voltage v_{st} is given by the equation $v_{st} = n \times (E - v_{act} - v_{ohm} - v_{con})$. The fuel cell voltage is calculated by subtracting the fuel cell losses or overvoltages form the fuel cell open circuit voltage. Where v_{act} represents the activation overpotential at the electrodes; v_{ohm} represents the ohmic overpotential caused by electrical and ionic conduction loss; v_{con} represents the concentration overpotential caused by mass transport limitations of the reactants to the electrodes. For the case of ultracapacitor (UC) and battery pack models, equivalent

[10] Thanapalan K, Liu GP, Williams JG, Wang B, Rees D (2009) Review and analysis of fuel cell system modeling and control. Int. J. of Computer Aided Engineering and Technology 1: 145–157.

circuit modeling approach was adopted and the governing equation is $V_{uc} = R_{uc} I_{uc} + \frac{1}{C_{id}} \int_0^t I_{uc} dt + V_{id}$. The UC model consists of an equivalent series resistance R_{uc} and an ideal capacitance C_{id}. The battery state of charge S is the only state variable of the battery system model and is given by $S = \left(Q_{max} - \int_0^t I_b.dt \right).Q_{max}^{-1}$ where I_b is the battery current, and Q battery capacity. The electrochemical power source model can be parameterized to represent the power source subsystem of the specific HHEV system for further analyses.

4 System Simulation and Results of G2V System Analyses

The flow of power in and out of any kind of electric vehicle (EV) and hybrid electric vehicle (HEV) can be valuable to the grid provided that the feed is happening as and when it is needed [5]. To cope with the increasing energy demand and renewable energy input, more energy may need to be stored on the grid. The G2V power flow mechanism describes a system in which HHEV's utilize power from the grid to generate hydrogen and allows the demand for hydrogen to be satisfied. The G2V power flow mechanism can act as additional grid energy storage to store electricity on a large scale notionally on the grid. Wind is a valuable renewable energy source, as it can produce significant amounts of electrical energy [6]. However, by its nature, wind energy is unpredictable; the amounts of electrical energy produced will vary over time and depends heavily on the weather and other factors [7]. The G2V method can facilitate the use of more wind energy in a useful way to sustain better energy management.

4.1 Case I

The network is first analysed with no wind on the network. This allows the proportion of hydrogen demand which can be met by the network to be analysed. This is done for the five hydrogen demand levels outlined in section 2.

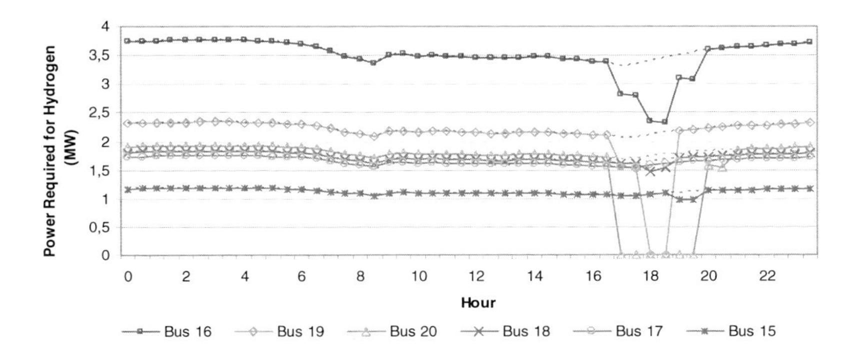

Fig. 4. Power required to satisfy hydrogen demand (dashed line) and supply (solid line)

Fig.4 shows the hydrogen demand, and the extent to which it can be met for hydrogen demand level 5. It can be seen that not all of the hydrogen demand can be met. This deficit occurs between the hours of 16:30 and 20:00, when demand for electricity is already at its highest. The extra demand for electricity due to hydrogen production can not all be met, so some of the hydrogen demand, acting as a dispatchable load, is not met. Table 4 shows the total hydrogen demand, hydrogen supplied, and hydrogen not supplied for the different hydrogen levels. In the worst case (level 5), 3.6% of the hydrogen demand cannot be supplied. The hydrogen cannot be produced due to voltage constraints reaching their limits.

Table 4. Hydrogen demand, supplied, and not supplied for different demand levels on network

H_2 Demand Level	Total Hydrogen Demand (MWh)	Hydrogen Supplied (MWh)		Hydrogen not supplied (MWh)	
	Both Cases	Case I	Case II	Case I	Case II
1	47.07	47.07	47.07	0	0
2	94.14	93.90	94.14	0.245	0
3	141.2	139.9	141.2	1.30	0
4	188.3	184.3	188.1	3.96	0.205
5	235.4	226.8	234.4	8.52	0.93

4.2 Network with Wind Farms, Case II and Case III

With wind on the network, the hydrogen demand can be controlled in two ways; passively or actively. When controlled passively, the hydrogen demand acts similarly to the system without wind, acting as an additional dispatchable load, with any extra wind being allowed onto the network by constraints being relieved by the additional loads. When controlled actively, the hydrogen is produced to allow as much extra wind onto the network as possible, with any surplus hydrogen being stored.

Case II, passive**:** It can be seen from Table 4 that more hydrogen demand is met with the wind on the network. This is due to the extra generation increasing voltage levels, allowing more demand on to the system before constraints are reached. Conversely,

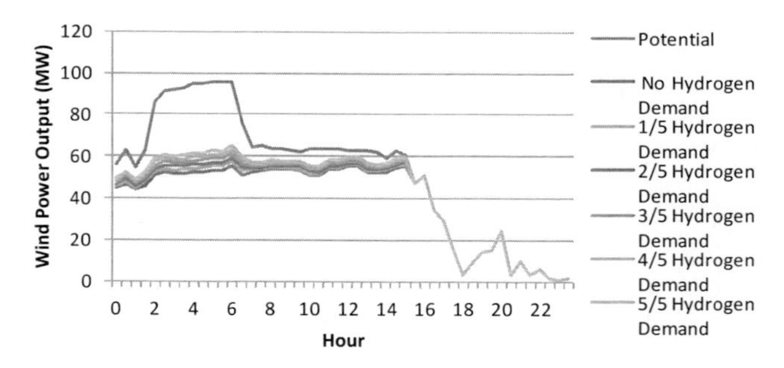

Fig. 5. Potential and actual wind power outputs for different hydrogen demand levels

the additional load from the hydrogen demand allows more wind onto the network, which can be seen from Fig.5. Despite the additional load, a large amount of the wind power is still curtailed.

Case III, active hydrogen demand: With active hydrogen demand, all the available wind power is utilized, and all of the hydrogen demand can be met for all hydrogen demand levels. However, at the end of the day's analysis a large amount of energy is left in the store. This may be partly due to the day chosen for analysis being one with a large wind power output, with a capacity factor of ~50%, compared to an expected annual capacity factor of ~30%. For hydrogen demand level 1, 269.3 MWh is left in the storage representing 5.7 days supply, whilst for demand level 5, 176.1 MWh is left in the storage representing 0.7 days supply. It is apparent that for most scenarios, the storage can be expected to fill rapidly. One solution to this could be to utilise fuel cells to reconvert the hydrogen stored back into grid electricity.

A total of 1243.3 MWh of wind energy can be produced over the course of the day. Without any additional hydrogen demand, 940.6 MWh can be accepted on to the system, whilst 302 MWh must be curtailed. With passive hydrogen demand, up to 1032.6 MWh can be accepted, with 211 MWh being curtailed. With active hydrogen demand, all of the available wind energy (1243.3 MWh) can be utilised.

5 Conclusions

This paper describes the development of a G2V optimal power flow mechanism with two purposes: To maximize hydrogen demand supplied to vehicles, and to accommodate more renewable energy onto grid energy storage system. With active demand, all of the wind is utilized, and all hydrogen demand can be met. Significant amounts of hydrogen remain in the store at the end of the day. In order to assess the overall performance of the system, a year round analysis should be carried out. This would allow component sizing to be analysed, and component costs calculated in order to determine the overall cost of the hydrogen produced. By including fuel cells in the analysis, the effect of allowing hydrogen to be converted to grid electricity can be studied. This could be either to support the grid, or to take advantage of price fluctuations in a market system.

References

[1] Sherif, S., Barbir, F., Veziroglu, T.: Wind energy and the hydrogen economy - review of the technology. Solar Energy 78, 647–660 (2005)

[2] Carr, S.: The integration of hydrogen storage with embedded renewable energy storage. Thesis, University of Glamorgan (2010)

[3] Kempton, W., Tomic, J.: Vehicle-to-grid power fundamentals: Calculating capacity and net revenue. Journal of Power Sources 144(1), 268–279 (2004)

[4] Thanapalan, K., et al.: Design and Implementation of Renewable Hydrogen Fuel Cell Vehicles. Renewable Energy & Power Quality Journal (RE&PQJ) 9, 310–315 (2011)

[5] Khayyam, H., Ranjbarzadeh, H., Marano, V.: Intelligent control of vehicle to grid power. Journal of Power Sources 201, 1–9 (2012)
[6] Linnemann, J., Steinberger-Wilckens, R.: Realistic costs of wind-hydrogen vehicle fuel production. International Journal of Hydrogen Energy 32, 1492–1499 (2007)
[7] Carr, S., et al.: Energy Storage for Active Network Management on Electricity Distribution Networks with Wind Power. In: Int. Conf. on Renewable Energies and Power Quality, Santiago de Compostela, Spain, March 28-30 (2012)

Chapter 25
Analysis of Thermal Comfort and Space Heating Strategy
Case Study of an Irish Public Building

Oliver Kinnane, M. Dyer, and C. Treacy

TrinityHaus, Trinity College Dublin, Dublin 2, Ireland

Abstract. Targets have been set to reduce the energy consumption in public buildings in Ireland by 33% by 2020. Space heating accounts for a significant portion of the energy load of public buildings. Diverse space heating strategies are often required to meet the requirements of spaces of various usage within public buildings, including within multi-purpose or event spaces. To analyse the thermal comfort of occupants and efficiency of the space heating strategy a post-occupancy evaluation was carried out on an event space at Dublin City Council local authority offices. The evaluation, based on the results of monitoring (temperature, energy), modeling and the assessment of comfort as perceived by occupants, has shown that thermal comfort is not adequately achieved. This is the case even though significant energy is being expended to achieve comfort levels via a current inefficient space heating strategy.

1 Introduction

Ireland has been assigned a target of 20% reduction in greenhouse gas emissions by 2020 (EPA, 2009). So as to present exemplar action, Ireland has set ambitious targets to decrease energy consumption in public buildings by 33% by 2020 (NEEAP, 2009).

Local authority buildings are often culpable of high energy consumption, given the wide range of functions operated within. The display of energy certificates (DEC) within these buildings (>1000m^2) has increased awareness of this high consumption. However, in-depth knowledge of operational consumption requires more intricate assessment.

For a group of buildings such as the Dublin City Council offices (used as a case study in this paper), which were built and added to over a number of decades, a greater level of characterization is required for the diverse building types. This complex of office buildings contains examples of fully mechanised, mixed mode and naturally ventilated buildings, built in the 1970s, 1990s and 2000s. A multi-purpose event space is the most recent addition to the complex of buildings yet this is a highly intensive energy consuming space.

Multi-purpose event spaces are common to public buildings, and are used for a wide range of functions including meetings, lectures, public consultations, exhibits, performances etc. These events are hosted regularly yet sporadically and without a routine schedule and common occupancy, making an efficient building operational strategy difficult to achieve.

A. Håkansson et al. (Eds.): *Sustainability in Energy and Buildings*, SIST 22, pp. 271–280.
DOI: 10.1007/978-3-642-36645-1_25

For these reasons a post occupancy evaluation (POE) was carried out. Functionally a POE performs as a diagnostic tool of building performance (Preiser 1995). It might also be viewed as a process that involves a rigorous approach to the assessment of both the technological and anthropological elements of a building in use. It is described as a systematic process guided by research covering human needs, building performance and facility management (Hadjri & Crozier 2009). In this study a POE was undertaken to gain insight into this specific event space, and to gain understanding of occupant response to this recent development. It is also an aim to identify specific issues which could be used to inform other spaces within the Dublin City Council offices and other event spaces in other public buildings.

There has been a proliferation of POE methodologies over the last decades, which exist in research and practice (Hadjri & Crozier 2009). Many are general and hence, unspecific to building type. Other methodologies have been developed in response to specific requirements of specific building types. So as to develop an appropriate methodology of POE a review of relevant literature, documenting methodologies most closely related to the type required for this study, was undertaken.

POE literature of multi-purpose event spaces is limited. Most closely related perhaps are POE studies of performance theatres (Kavgic et al. 2008) and lecture theatres (Cheong & Lau 2003). These studies collect thermal comfort data and also analyse thermal conditions using computational fluid dynamic (CFD) models. Kavgic et al. (2008) present a comprehensive assessment methodology based on a 4 step approach similar to that of Cheong and Lau (2003), for a study undertaken on two nights. Ventilation, space heating and the impact on indoor air quality are investigated by Noh et al. (2007) in a lecture theatre (Noh et al. 2007).

The methodology presented in this paper builds on the methodologies described in these studies and adapts them for the specifications of a multi-purpose event space.

2 Methodology

The event space at Dublin City Council offices (the (Wood Quay) Venue) is characterised by high energy consumption (350 kWh/m^2/yr) for a commercial building (Gething & Bordass 2006). A high proportion of this energy load is accounted for by the space heating load. However despite this, reported occupant thermal discomfort is common within the Venue. Diverse heating strategies, with heated air supplied at temperatures from 20°C to 48°C, have been used at the Venue however, complaints of discomfort continue. Energy consumption varies dramatically when heated air is supplied at high temperatures in comparison to low temperatures with almost a 50% increase in energy load from one to the other.

2.1 The Case Study Building

The Wood Quay venue is situated in the basement of a 1970s tower building. It was originally planned as a public museum space but was left undeveloped for decades when finally developed as a venue to host a wide range of public events. The Venue

(Fig. 1) is of an irregular form and contains unique features, which add to its complexity. The overall floor area is 428 m^2, with an internal volume of 1662 m^3, which includes a two-story space as shown in Fig 1.

An old limestone and mortar wall – part of the old city wall of Dublin circa 1200 AD – runs from outside the building envelope into the space, and forms a backdrop to the speaker's podium. This wall has a significant impact on the thermal conditions within the venue space and is the focus of much of the assessment of the venue.

The space is mechanically ventilated, via a system of 24 ceiling diffusers, which supply air from the two-storey space. A large air handling unit (AHU) is the source of ventilation and heated air. The heated air is introduced via circular diffusers, mounted behind a perforated ceiling. An extract plenum exists at 2 meter height above ground level. The heated air is supplied to the Venue in an on-off relay cycle with a period of 2 hours.

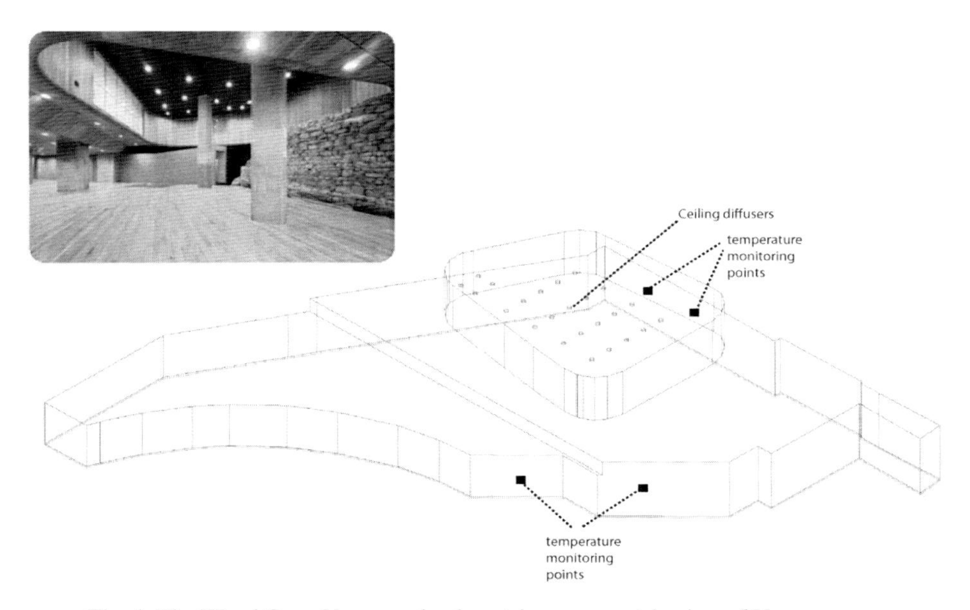

Fig. 1. The Wood Quay Venue and volumetric, axonometric view of Venue space

2.2 Research methodology

The case study described in this paper employed an assessment methodology consisting of collection and analysis of metered data, an occupant survey and building simulation based on the boundary conditions derived from field-study readings.

Preliminary stage –

- Develop an understanding of the building, its design, construction and systems.

Data collection and monitoring -

- A quantitative study of metered temperature was used to assess the indoor thermal conditions within the Venue.
- A quantitative study of metered gas and electricity data was used to assess the energy performance of the Venue.
- A qualitative study of occupant perception of the indoor Venue environment was carried out via an occupancy survey.

Analysis, modeling, and assessment -

- Simulation of operational phenomena based on recorded parameters
- Assessment and correlation of parameters of thermal conditions, energy consumption and occupant perception of thermal comfort.

3 Data Monitoring

Temperature and humidity are monitored at 4 locations within the Venue. The locations of these monitoring points are shown in Fig 1.

Surveying of occupants of the Venue was carried out on 9 occasions (19.10.2011 – 13.12.2011) within a 2 month period in the final months of 2011. This period was studied as this constitutes a significant portion of the heating season – the season during which most complaints of thermal discomfort were reported at the Venue. The 9 dates on which the survey was carried out were the dates on which events were hosted at the Venue. Each event lasted between between 2 and 3 hours and the commencement times are listed in Table 1.

A trial study undertaken in June 2011 (20.06.2011) to assess the developed questionnaire is also included in the subsequent result tables for comparative purposes.

Different respondents were surveyed on each occasion. Due to the range of activities and events hosted at the venue, data sets of different sizes were surveyed on different days. Assessment of thermal comfort and indoor environmental conditions was undertaken using a 5-point rating scale. Temperature, humidity, air movement, air quality, noise level, light level, odour level and overall impression were assessed.

This study is focused on those parameters related to thermal comfort, hence of particular interest are the occupant response to temperature, humidity and air movement.

A total of 147 people returned completed questionnaire forms.

Table 1. Date, event commencement time and number of occupants surveyed

Date (2011)	20.06	19.10	27.10	28.10	01.11	07.11	10.11	17.11	07.12	13.12
Event start time	10.00	12.00	14.30	08.30	08.30	08.30	10.00	09.00	09.00	12.30
No. of occupants surveyed	12	9	16	17	21	17	14	15	9	17

The Venue is characterised by a high energy load which includes a base load of approximately 60 kWh. Table 2 documents the energy load on the days of occupant survey.

Table 2. Date, electricity, gas and total energy in kWh

Date (2011)	20.06	19.10	27.10	28.10	01.11	07.11	10.11	17.11	07.12	13.12
Electricity (kWh)	200	319	195	216	227	279	209	244	330	336
Gas (kWh)	83	181	170	175	190	202	142	151	287	333
Total Energy (kWh)	283	500	365	391	417	481	351	395	617	669

4 Data Analysis

The relevant temperatures, including the supply, extract, room indoor and outdoor temperature are shown in Fig 2 over the 2 month period during which the POE was undertaken. The days of occupant survey are marked. Energy loads on days when air of different temperatures were supplied are also shown on the graph. The energy load is increased by almost 50% on days when air is supplied at 36 °C relative to days when the air is supplied at ~26 °C. The indoor room temperature remains relatively constant (~21 °C +/- 1 °C) during the surveyed days.

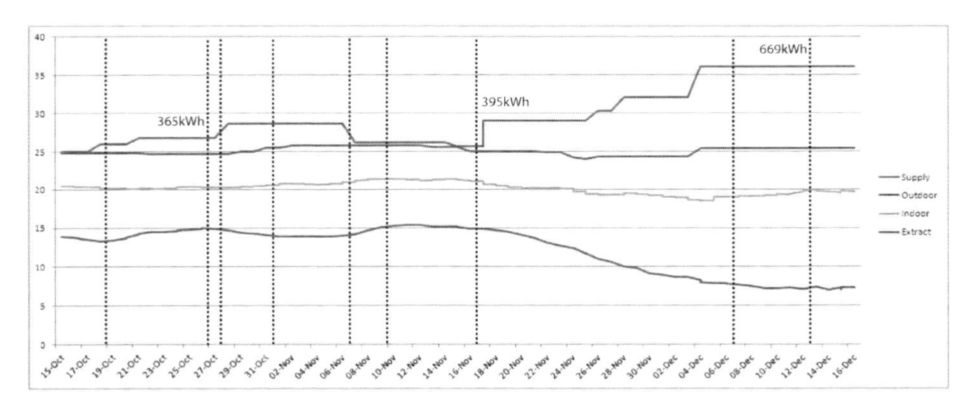

Fig. 2. Indoor, outdoor, supply and extract temperatures for the 2 months of the monitoring period. Dates of occupant surveys are marked and energy loads on certain dates.

The temperature of the air being supplied via the AHU to the space is significantly higher than the indoor air within the Venue. This implies that the supplied heated air is not mixing effectively with the internal air and not conditioning it. This phenomenon is investigated using the Space Diffusion Effectiveness Factor (*SDEF*) calculated as:

$$SDEF = \frac{T_{exhaust} - T_{room}}{T_{room} - T_{supply}}$$

An *SDEF* value of < 1 implies that some amount of heated supply air is not mixing with the room air and is leaving the conditioned space as exhaust.

The *SDEF* values on the survey days are listed below:

Table 3. Extract, supply temperature, indoor venue temp and SDEF values

Date (2011)	20.06	19.10	27.10	28.10	01.11	07.11	10.11	17.11	07.12	13.12
Extract temp.	24.1	24.7	24.7	24.5	25.5	25.8	25.8	25	25.4	25.4
Supply temp.	25.45	26	26.7	28.6	28.6	26.2	26.2	25.7	36.1	36.1
Indoor room temp.	23.8	19.8	20.3	20.6	21.5	21.8	21.5	21	20.2	20
SDEF	0.2	0.8	0.7	0.5	0.9	1.5	0.9	0.85	**0.33**	0.34
Outdoor temp.	13.1	9.5	12.2	7.3	9.4	**4.9**	12.3	11	**4.6**	**3.5**

The *SDEF* is less than 1 on all but one of the dates the survey was undertaken and for the majority of the dates within the 2 months of the heating season analysed.

Heated air of high temperature (>36°C) is supplied on cold days (<5°C, (07.12.2011, 13.12.2011)) in an effort to enhance the indoor thermal conditions. However, on days (07.11.2011) when the temperature drops below 5°C and air is supplied at lower temperatures (26.2°C) the indoor Venue temperature remains at a comfortable level (21.5 °C). These results show that the indoor Venue environment is almost unresponsive to the temperature at which the heated air from the AHU is supplied at. These results also propose that the heated air is being stratified closer to the ceiling in the two-story space and is not penetrating the occupied space below.

The results of the questionnaire surveys are presented in Fig. 3 and 4 below. Occupants were given a 5 point scale on which to rate indoor environmental quality

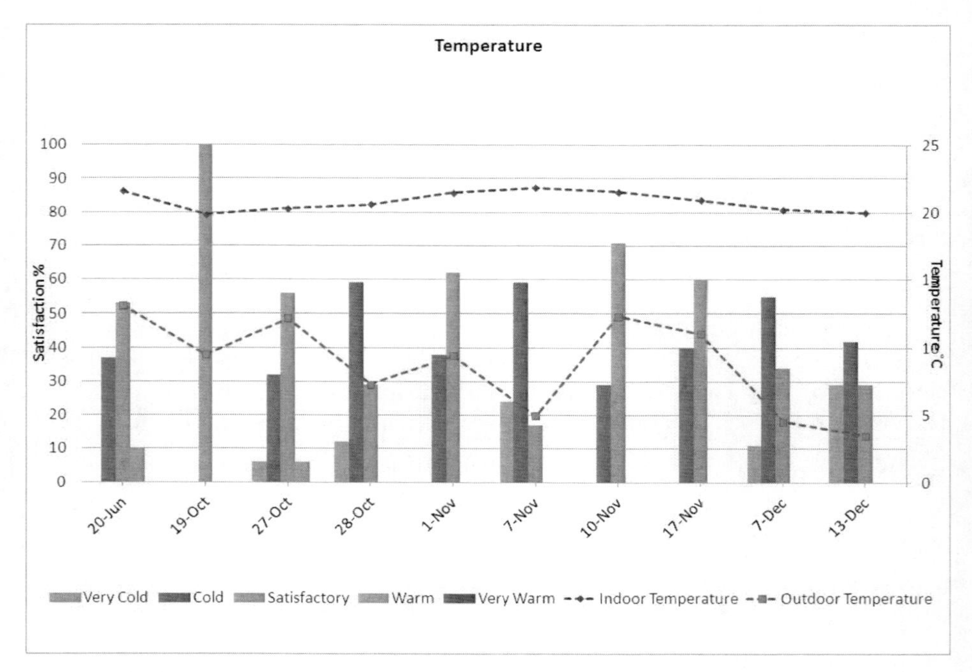

Fig. 3. Occupant response to survey of temperature, outdoor and indoor temperature

parameters. For the dates surveyed within the heating season (19.10.2011 – 13.12.2011) 51% responded as being satisfied, 48% that they were cold or very cold, while < 1% felt warm. When air movement was surveyed, 60% responded as satisfied, 35% reported draughty or very draughty conditions and 5% perceived still air.

During the 2 cold days (<5°C outdoor temperature) in December (07.12.2011 and 13.12.2011) 69% of those surveyed reported being cold or very cold and 62% reported draughty or very draughty conditions.

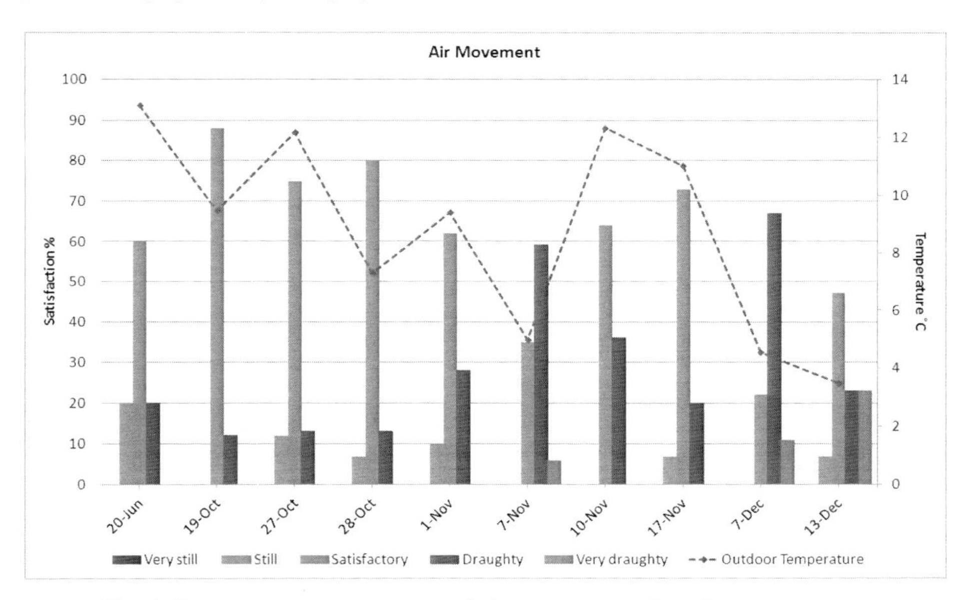

Fig. 4. Occupant response to survey of air movement, and outdoor temperature

Table 4. Percentage of responses of satisfaction to the thermal conditions of the Venue

Date (2011)	20.06	19.10	27.10	28.10	01.11	07.11	10.11	17.11	07.12	13.12
Event start time	10.00	12.00	14.30	08.30	08.30	08.30	10.00	09.00	09.00	12.30
No. occupants surveyed	12	9	16	17	21	17	14	15	9	17
% satisfied with temperature	53%	100%	56%	29%	62%	**17%**	71%	60%	34%	29%
% satisfied with air movement	60%	88%	75%	80%	62%	35%	64%	73%	**22%**	47%
Outdoor temp.	13.1	9.5	12.2	7.3	9.4	**4.9**	12.3	11	**4.6**	**3.5**

Occupant reaction of thermal discomfort are high particularly on days of low outdoor temperature (07.11.2011, 07.12.2011, 13.12.2011). This strong correlation between outdoor thermal conditions and occupant expression of thermal discomfort propose that the building envelope of the Venue does not enable proper delineation between the indoor and outdoor environmental conditions. Occupants are also not receiving the benefits of the supplied heated air most likely due to the fact that the air is stratifying in the upper 2-story portion of the Venue.

For the purpose of visualization and presentation of the monitored phenomena a CFD modeling study was undertaken. The surface temperatures of the Venue were measured with an infra-red sensor. An average of these readings were used as the boundary conditions for input to the CFD model. Surface temperatures of the walls, ceiling and floors were recorded as 22, 25 and 20 respectively and these are used as the boundary conditions in the model shown.

The venue was modeled using EnergyPlus software. A geometrical model was first developed in Design Builder with accurate representation of the Venue space.

To enable the CFD model of the space a 3D grid was defined with 150x150x75 points in the 3 –dimensions.

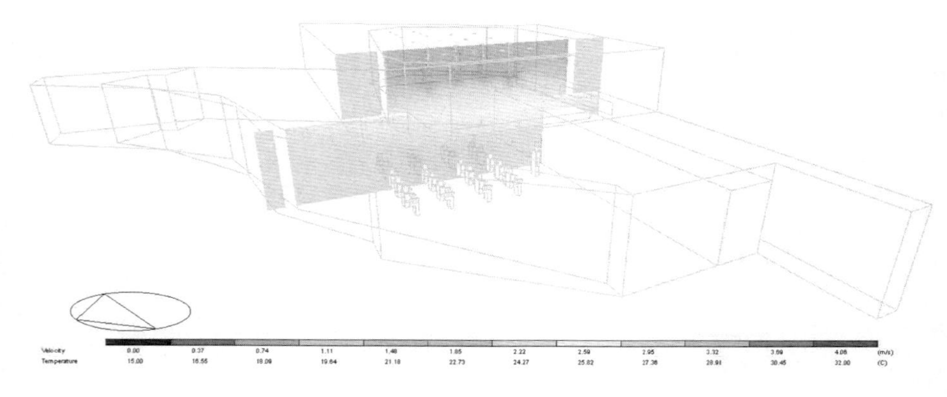

Fig. 5. CFD modelling study of thermal stratification

5 Discussion and Results

The Venue is a multi-purpose event space within the Dublin City Council office complex that is characterised by high energy consumption. A significant proportion of this energy consumption is accounted for by space heating. Yet complaints of thermal discomfort are common amongst occupants of the Venue.

From the review of indoor, outdoor and supply temperatures it is shown that on days of low outdoor temperature, the indoor temperature is not significantly affected by the supply temperature. The indoor temperature remained in the range of 19.8 °C to 21.8°C, irrespective of the supply air temperature which was varied between 26 °C and 36°C. This proposes that the monitored and occupied portion of the indoor Venue environment is unresponsive to the temperature of air supplied from the AHU via the ceiling diffusers. This result is supported by study of the diffusion effectiveness of the heated air system which proposes that the supply air is not affecting the temperature of the indoor air effectively. The heated air is not mixing with the room air and is instead being extracted before doing so. It is likely that the heated air is stratifying in the upper two-story space and is not penetrating the lower space of the Venue. The CFD modelling assessment based on experimentally recorded boundary conditions supports the thermal stratification proposal.

The results of the occupant survey compliment the results of the temperature monitoring. Complaints of thermal discomfort were common on survey days during the heating season, with 48% of those surveyed reporting to be cold or very cold. 35% of those surveyed reported draughty or very draughty conditions. The days of greatest response of thermal discomfort were on those days when the outdoor temperature dropped below 5°C even though the indoor temperature remained at relatively comfortable levels.

The following are the significant results and conclusions of this study:

- The current strategies of space heating are inefficient both from an energy consumption and thermal condition enhancement point of view.
- Complaints of thermal discomfort were common although indoor thermal conditions were generally in the comfort region.
- Space diffusion ineffectiveness and thermal stratification characterise the space heating of the Venue.
- Significant energy savings are available with changes to the current operational heating strategy.
- Alternative methods of space heating should be investigated however, fluid mixing requires much less energy than surplus heating and is, therefore, a more economical method in raising the temperature of the bottom region (Tanny & Teitel 1998).

In collaboration with Dublin City Council Building Management changes are currently being made to the space heating strategy based on this investigation. Monitoring and assessment of the new strategy will be undertaken in the next heating season. Further building assessment including investigation of the building envelope has enabled the identification of sources of heat loss and infiltration.

References

Cheong, K.W.D., Lau, H.Y.T.: Development and application of an indoor air quality audit to an air-conditioned tertiary institutional building in the tropics. Building and Environment 38(4), 605–616 (2003)

Department of Communications Energy and Natural resources, The National Energy Efficiency Action Plan 2009-2020 (2009)

Environmental Protection Agencey, Ireland's Greenhouse Gas Emissions in 2009 (2010)

Gething, B., Bordass, B.: Rapid assessment checklist for sustainable buildings. Building Research & Information 34(4), 416–426 (2006)

Hadjri, K., Crozier, C.: Post-occupancy evaluation: purpose, benefits and barriers. Facilities 27(1/2), 21–33 (2009)

Kavgic, M., et al.: Analysis of thermal comfort and indoor air quality in a mechanically ventilated theatre. Energy and Buildings 40(7), 1334–1343 (2008)

Noh, K.-C., Jang, J.-S., Oh, M.-D.: Thermal comfort and indoor air quality in the lecture room with 4-way cassette air-conditioner and mixing ventilation system. Building and Environment 42(2), 689–698 (2007)

Preiser, W.F.E.: Post-occupancy evaluation: how to make buildings work better. Facilities 13(11), 19–28 (1995)

Tanny, J., Teitel, M.: Efficient utilization of energy in buildings and ponds. Applied Thermal Engineering 18(11), 1111–1119 (1998)

Chapter 26
Protection of Ring Distribution Networks with Distributed Generation Based on Petri Nets

Haidar Samet and Mohsen Khorasany

School of Electrical and Computer Engineering, Shiraz University, Shiraz, Iran
samet@shirazu.ac.ir, Mohsen.Khorasany@gmail.com

Abstract. Nowadays limitation of energy sources has increased the demands for using Distributed Generation (DG) in electricity generation. Entrance of DG to network is accompanied by some problems in network protection. DGs connection to network makes some problems in coordination between protection relays. Wrong operation of relays interns irreparable shocks to network. So for increasing the protection of DG in distribution network we should solve protection problems. In this article at first the problems that caused by entrance of DGs in protection system, are presented. After that an error detection system in function of relays based on Petri net is introduced. In the past, Petri net was used for radial system protection in presence of DG. The main goal of this article is explanation of Petri net structure corresponding to ring distribution network and use of backup relay for insurance of DG separation from network.

Keywords: Petri nets, Relay, DG, Protection.

1 Introduction

The presence of a significant dispersed generation (DG) capacity in existing distribution systems would cause in most cases some conflicts with correct network operation. This is due to the fact that the distribution system is typically designed as a passive and radially operated network, which is conceived with neither generators operating in parallel nor power flow control. In general, the impact of DG depends on its penetration level and connection point in distribution networks, as well as on DG technology (e.g. synchronous generators, asynchronous generators and static converter interfaced generation systems). In order to ensure correct distribution system operation and adequate service quality to customers, various technical issues have to be tackled, such as voltage control, power quality, lines thermal condition, and system protection [1].

A typical distribution network (without dispersed generators) and its protection system are represented in Fig. 1. MV distribution networks are supplied by primary substations, PS (HV/MV), typically equipped with two transformers, each supplying section of the MV bars. The two sections can be put in parallel by closing a tie switch.

A. Håkansson et al. (Eds.): *Sustainability in Energy and Buildings*, SIST 22, pp. 281–288.
DOI: 10.1007/978-3-642-36645-1_26

The radial circuits are formed by main feeders and laterals. Especially in medium/high customers density areas, the main feeders can be back-fed by a neighboring PS by closing normally open switches [2].

In this network there are not any DGs, but as already said entrance of DGs result in many problems for protection system. Such as [3]:

1. Sensitivity problems: protection function of feeders may become disturbed by entrance of DGs. Undetected faults or delayed performance of relays which undamaged the system are effects of this problem.
2. Selectivity problems: DG may result in unnecessary disconnections of the feeder it is connected to.
3. Reclosing problems: presence of DGs may disturb fast operation of auto reclosing.

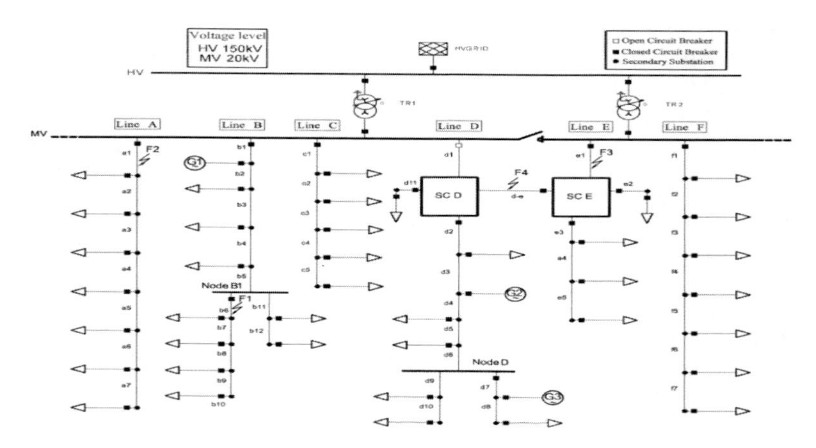

Fig. 1. Typical scheme for distributed network and its protection [2]

In a radial system we can do fault correction with opening only one switch. But in presence of DGs, since there are several sources, opening only one key does not guarantee fault correction. Consequently if a fault happens, DG should be disconnected to turn the system to a common radial system.

The disconnection from network may happen fast or slowly which lead to some problems in network voltage. As conclusion we cannot have reclosing in presence of DGs, since DG does not have enough time for disconnection from network [4].

Existence of these problems and the increasing demand for entrancing DGs to distribution networks made scholars to examine different ways for strengthening the protection structure of distribution network where DGs exist. Considering the problems that occur by DGs in protection network, the disconnection of DGs from network should be guaranteed. Since in presence of DGs the network turn to radial shape the relays should be mounted in a way that disconnect the network during fault. And by using a backup relay the disconnection from network is guaranteed. Moreover we Use a Petri net for detecting fault in the operation of relays. In the following the Petri net network is defined and protection model and equations of fault detecting are presented.

2 Petri Nets Description

As a graphical and mathematical tool, Petri nets prepare an identical environment for modeling, analyzing and designing systems with separate events. One of the most important benefits of using Petri net is that it prepares an identical model for analyzing behavioral characteristic and performance assessment of systematic structure of separate events.

A Petri net can be known as a specific kind of bipartite directed graph populated by three types of objects. These objects are named as places, transitions and directing arc. In a Petri net, places are demonstrated by circles and transition as bars of boxes. A directed arc which connects a place to a transition shows that place is an input place to a transition and a directed arc which connects a transition to a place shows that the place is an input place to a transition. In its most uncomplicated form, a Petri net can be depictured by a transition with its input and output places. This basic net can be used to show different characteristics of the modeled systems. For example input (output) places can illustrate preconditions (post conditions), the transition an event. The availability of resources can be demonstrated by input places, their utilization can be shown by transition, and the release of resources can be indicated by output places.

To study dynamic behavior of the modeled system, in terms of its states and their changes, each place may potentially hold either none or a positive number of tokens, showed by small solid dots. Attendance or absenteeism of a token in a place could exert an error in the protection system. The current state of the modeled system is quantified by distribution of tokens on places at any given time which is named Petri net marking. A marking of a Petri net with m places is shown by an (n×1) vector N, elements of which denoted as N(p), are nonnegative integers representing the number of tokens in the corresponding places [5].

In fact we can define Petri net as below [6]:

- $PN = (P, T, I, O, N_0)$
- $P = (P_1, P_2, P_3, \dots)$ is a finite set of places.
- $T = (T_1, T_2, T_3, \dots)$ is a finite set of transition.
- $I : (P * T) \rightarrow M$ direct arc from places to transitions.
- $O : (P * T) \rightarrow M$ direct arc from transitions to places.
- $N_0 : P \rightarrow M$ initial state of system.

If I(P,T) =k [O(P,T) = k] then there exist k directed (parallel) arcs which connects place p to the transition t (transition t to the place p). Mostly, in the graphical representation, parallel arc that connects a place (transition) to a transition (place) are illustrated by a single directed labeled with its multiplicity, or weight k.

By altering dispensation of tokens on places, which may reflect the occurrence of events or execution of operations, for example, one can study dynamic behavior of the modeled system.

3 Protection Modeling Using Petri Net

Existence of DG in distribution network result in radial structure in network (Fig. 2) and as it is said it makes some problems in protection network, consequently in this structure as fault happens, DG should be disconnected from network. In suggested structure when fault occurs DG is separated from network by its related relay and if its relay does not work, backup relay will be used for opening the line that DG is connected to.

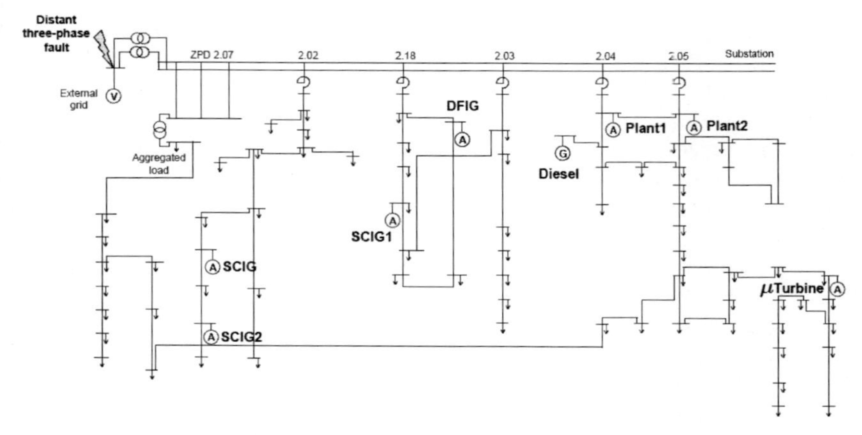

Fig. 2. A typical distribution network with DG [9]

In Fig.3 protection structure is presented in simple manner. In this figure R_{L1} and R_{L2} are line relays which are connected to DG. And R_{dg} is relay that is related to DG.When fault happens in line 2, R_{L2} relay will operate and R_{dg} will disconnect DG from network by its operation. If relays of DG do not work, then R_1 and R_2 will operate as backup relay and will disconnect DG from network. Consequently integral function in this situation is really important. In [7] a method for detecting error in function of relay based on Petri net is presented, but in this method radial network and backup relay are not considered.

In the following a Petri net method line protection with DG that is equipped by relay and backup relay is presented.

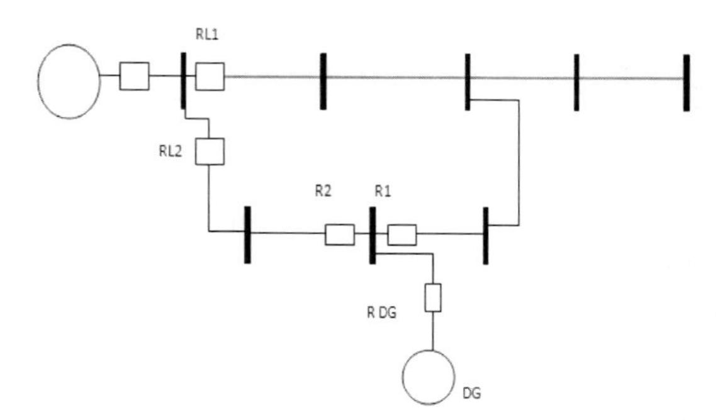

Fig. 3. Proposed structure for protection of ring distribution network

Petri Net Modeling

In this pattern a protected line includes DG in which for disconnecting DG a CB (circuit breaker) that is equipped with a relay has been considered, and CBs of lines are as backup system for disconnecting DG.

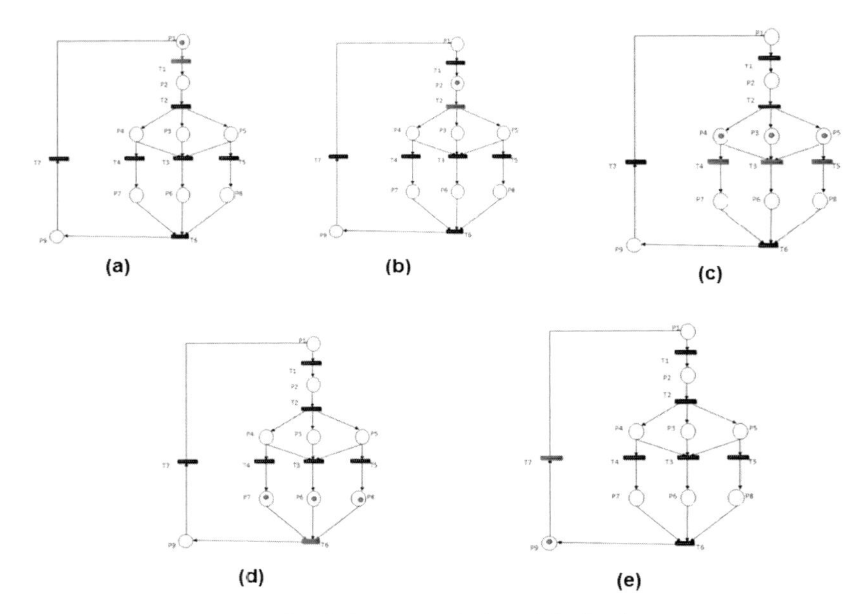

Fig. 4. Correct evolution of the Petri net

In Fig.4 p1 shows that there is no fault, p2 shows a situation in which relay can detect fault, p3 is for detecting fault by DG relays. P4 and p5 identify the possibility of detecting fault for CB relays that are connected to DG (R_1, R_2). P6 shows the separation of DG by its relay and p7 and p8 represent separation of DG by line relays (R_{L1},R_{L2}). P9 shows that system is ready for accepting initial conditions. Also t1 shows occurrence of fault, t2 demonstrates sending of trip signal, function of DG relay is shown by t3, t4 and t5 are indicant for trip of R_{L1} andR_{L2} and t6 is demonstrator for error correction and t7 shows recovery of system.The initial marked Petri net indicates that the line relay is read to sense a fault (token in $p1$) .If a fault occurs, the correct evolution of the Petri net is shown in Fig. 6(a-e). The movement of tokens models the execution of the operations. It can be noted if the line relay does not trip, the protection of DG operates. For detecting fault we use the procedure that is explained in [8] and it is described briefly here:

We suppose that R is Petri net matrix with P place and T transition. If P^- determines weight of arcs that exist from place p to transition t and P^+ determines arcs from transition t to place p then evaluation of matrix R is as follow:

$$S[x+1] = S[x] + (P^+ - P^-).\partial = S[x] + C.\partial[x] \quad (1)$$

In which s[x] is R marking in second t and O is indicant for fire of transitions. By Considering mentioned terms two kinds of faults are possible in Petri net: transition fault, place fault, transition fault consists of precondition and post condition and is modeled by doubling e_T^+ and e_T^- vectors which their dimension is 1*m and the number of theirelements shows the total number of precondition and post condition .

A transition fault takes place if transition t_j and not all tokens are deposited to output places even though tokens from the inputplaces have been used (post condition failure).If the tokens which are supposed to be removed from the input places of the faulty transition are not removed, a precondition failure is happened.

$$S_f[x] = S[x] - P^+.e_T^+ + P^-.e_T^- \tag{2}$$

Place fault happens when the numbers of tokens in the place are incorrect and error vector is clarified bye_P:

$$S_f[x] = S[x] - P^+.e_T^+ + P^-.e_T^- + e_p \tag{3}$$

In particular, in order to identify faults in a Petri net a redundant Petri net RH can be constructed. In particular, d places are added to the original Petri net Q such as:

$$S_H[x] = [\begin{smallmatrix} I_n \\ G^* \end{smallmatrix}]S[x] \tag{4}$$

In which I_n is identity matrix, G* is d*n matrix and should be designed. Based on the method that has beenpresented in [4] for detecting error, these two matrixes should be obtained:

$$\begin{cases} \underline{s_T}[x] = \underline{D}.\underline{e_T} \\ \underline{S_p}[x] = \left[-\underline{G^*}\underline{I_d}\right].\underline{e_P} = \left[-\underline{G^*}\underline{I_d}\right].\underline{S_f} \end{cases} \tag{5}$$

In which $\underline{S_T}$[X] is transition fault matrix, $\underline{S_P}$[X] is represents for possible fault matrix and sf is transition indication vector of network after error and d is a matrix which its dimension is d*n and should be designed. In practice designing of G*d is based on GF (r) and is as follow:

$$[-G^*I_d] = \varphi^{-1}.H_d \tag{6}$$

In which H is checking parity matrix and is defined as follow:

$$H = \begin{bmatrix} 1 & \alpha^1 & \alpha^2 \cdots & \alpha^{q-2} \\ 1 & \alpha^2 & \alpha^4 & \alpha^{2(q-2)} \\ \vdots & & \ddots & \vdots \\ 1 & \alpha^d & \cdots & \alpha^{d(q-2)} \end{bmatrix} \tag{7}$$

In which 'α' is initial element of GF (r) and Ω with d last columns form H_d. After development of these matrixes by using above terms possible fault and transition fault are calculated and a central control by evaluation of this matrix in any stage can correct relay function error. The Petri net structure models the protection systems and provides a simulation environment of the network for several inputs. A collection of static structures, transition and places explain sequence of events and controls status of network operation by monitoring the data from switches and sensors distributed on the power network.

4 Analysis of the Proposed Method

As it was said before detecting of error performance using modeling of protection system by Petri net is possible. For this purpose e_P and e_T matrix should be calculated.e_P Matrix is a matrix with p*1 dimensions in which p is the number of places. e_P Matrix determines the errors occurred in places.

e_T Matrix is a matrix with t*1 dimensions in which t is the numbers of petri net transient.e_T Matrix determines the errors occurred in transients. These matrixes are calculated by equations 1 to 6. Each -1 element of this matrix indicates error in the performance of protection system. In the following some probable outputs for e_P and e_T matrix and their related description is presented. As determined by calculating e_P and e_T matrixes, the performance of protection system determined and probable errors could be detected.

Table 1. Analysis of the proposed method

Output Matrix	Description
$e_P = [0\ \ 0\ \ -1\ \ 0\ \ 0\ \ 0\ \ 0\ \ 0\ \ 0]^T$	DG relay cannot detect error
$e_P = [0\ \ 0\ 0\ 0\ 0\ 0\ \ -1\ \ 0\ \ 0]^T$	Error is detected by relay but relay has not worked
$e_T = [0\ \ -1\ 0\ 0\ 0\ 0\ 0]^T$	Error is detected by DG relay but trip signal is not sent
$e_T = [0\ 0\ 0\ 0\ \ -1\ \ 0\ \ 0]^T$	Error signal for RL1 relay is not sent

5 Conclusion

In this article at the beginning it was shown that entrance of DG leads to change in distribution network structure and this structure should not be considered as radial network. This change in structure makes some problems in network especially in distribution network protection, and then an appropriate structure for distribution network protection that includes DG with a backup breaker for assurance of DG disconnection from network when an error occurs was presented. Moreover petri net algorithm- a graphical algorithm for detecting error in system function- for check of error existence in function of suggested protection system was used and appropriate structure for Petri net algorithm network was presented.

References

1. Jenkins, N., Allan, R., Crossley, P., Kirschen, D., Strabac, G.: Embedded Generation, The institution of electrical engineers, London (2000)
2. Conti, S.: Analysis of distribution network protection issues in presence of dispersed generation. Electric Power Systems Researc 79 (2009)
3. Maki, K., Repo, S., Jarventausta, P., Karenlampi, M.: Definition of DG protection planning methods for network information systems. In: IEEE Proceeding on Smart grids for Distribution Conference, Frankfurt, June 23-24 (2008)
4. Kumpulainen, L., Kanhaniemi, K.: Analysis of distributed generation on automatic reclosing. In: Proceeding of IEEE Power System Conference and Exposition, New York, October 10-13 (2004)
5. Zurawski, R., Zhou, M.: Petri Nets and Industrial Applications; A Tutorial. IEEE Transaction on Industrial Electronics 41(6) (December 1994)
6. Ren, H., Mi, Z.: Power system fault diagnosis modeling techniques based on encoded petri nets. IEEE in Power Engineering Society General Meeting (2006)
7. Calderaro, V., Galdi, V., Piccolo, A., Siano, P.: A petri net based protection monitoring system for distribution networks with distributed generation. IEEE in Electric Power Systems Research (2009)
8. Hdjicostis, C.N., Verghese, G.C.: Power system monitoring using petri net embedding. In: IEEE Proceedings on Generation, Transmission and Distribution, vol. 147 (September 2000)
9. Kincaid, D.: The Role of Distributed Generation in Competitive Energy Market, Gas Research Institute, Distributed Generation Forum (1999)

Chapter 27
Real-Time Optimization of Shared Resource Renewable Energy Networks

Stephen Treado and Kevin Carbonnier

The Pennsylvania State University,
Department of Architectural Engineering,
University Park, PA, USA
streado@engr.psu.edu

Abstract. Shared resource renewable energy networks allow for the burden of high capital cost to be managed by sharing the cost and benefits of renewable energy use. In order to maximize the benefit gained from shared renewable energy, we propose a methodology to optimize the use of renewables via scheduling of energy use. By offering reduced energy rates, residents will be encouraged to run heavy energy consumers such as clothes dryers at times which improve the load generation and energy demand matching as deemed by a designed and optimized decision engine.

Keywords: Decision engine, Load scheduling, Optimization, Energy utilization efficiency, Renewable energy network, Residential shared resources.

1 Introduction

Other researchers have explored the optimization of renewable energy systems through energy storage [1]. Demand manipulation work has been completed with respect to electrical cost schedules by [2]. This paper will focus on the demand side of the issue primarily in terms of renewable energy utilization.

The driving principle behind load manipulation is the nature of non-mandatory loads which allow for some shaping of the demand curve if energy consumers are willing to alter the times in which they consume their energy. Non-mandatory loads are loads that are flexible in time. Some examples are dish washing, clothes washing and drying, and bathing. If consumers can be convinced to shift the time when they wash their dishes by a few hours for example, then peak energy costs can be avoided and utilization efficiency of community renewable energy can be maximized. Optimizing the system to directly match demand and generation is preferred as it will result in the most efficient energy use by eliminating distribution and storage costs. The two main objectives of load shifting are either utilization efficiency maximization, or energy cost minimization. Other potential objectives include minimizing emissions and avoiding the number of demand changes.

A. Håkansson et al. (Eds.): *Sustainability in Energy and Buildings*, SIST 22, pp. 289–298.
DOI: 10.1007/978-3-642-36645-1_27

The optimization will be based on a decision engine which users will query with a request for energy usage. The user interface will then respond with either a go ahead for the task requested, or offer an alternative time with the associated benefits in terms of cost and fuel type used. In this manner, the energy consuming tasks will be scheduled on a first come first serve basis. This will encourage users to put in requests in the beginning of the day, so that a daily schedule can be mapped for the community. Knowing the potential loads for the day beforehand is beneficial for the system since in some cases loads need to be met earlier rather than delayed.

Ultimately, demand manipulation for optimal use of energy can yield great societal benefits in terms of savings with a huge potential area of application given the simplicity of the concept. Although best suited for powerful machines highly networked into home appliances, the core notion of demand shifting can still be implemented via less sophisticated means such as predetermined schedules for users to simply follow, requiring only initial analysis on load and generation trends.

Creating change is societal views towards energy usage is a critical step towards achieving environmental sustainability. Introducing this change with a community effort and financial rewards is a promising way to bring the green movement down to a personal scale and motivate consumers to be more mindful of the impacts of energy usage habits. Feedback in terms of contribution to the community effort to reduce fossil fuel usage as well as emissions should also be implemented to positively reinforce individual efforts.

2 Background

2.1 Load Structure

The sample rate structure shown below for various load types is an example of the end user incentive method used to motivate building occupants to shift their energy consumption times.

Symbol	Load Type	Description	% of Grid Rate
M	Mandatory	Met immediately	100
R	Rush	Met ASAP	90
D	Discretionary	Met within given time window	75
F	Flexible	Met whenever possible	50

The rates can be changed dynamically based on external factors such as energy availability and predicted changes in economic conditions. As part of the system feedback, users may be given the option to alter the urgency of their energy demand for an adjustment in cost. In some cases, the system may suggest a reduction in the urgency of demands to satisfy unexpected increases in loads.

A more sophisticated system will allow users to create a profile of load types based on the specific job (dish washing, clothes drying, etc.). This would reduce the amount of input required to the system from the individual users.

2.2 System Layout

In order to automate the system energy usage, each of the major consumers must be connected to a master network. Via the network, the system will be able to remotely start the devices for which the users input queries to the decision engine. A system wide diagram of the renewable energy network is shown below in Fig. 1.

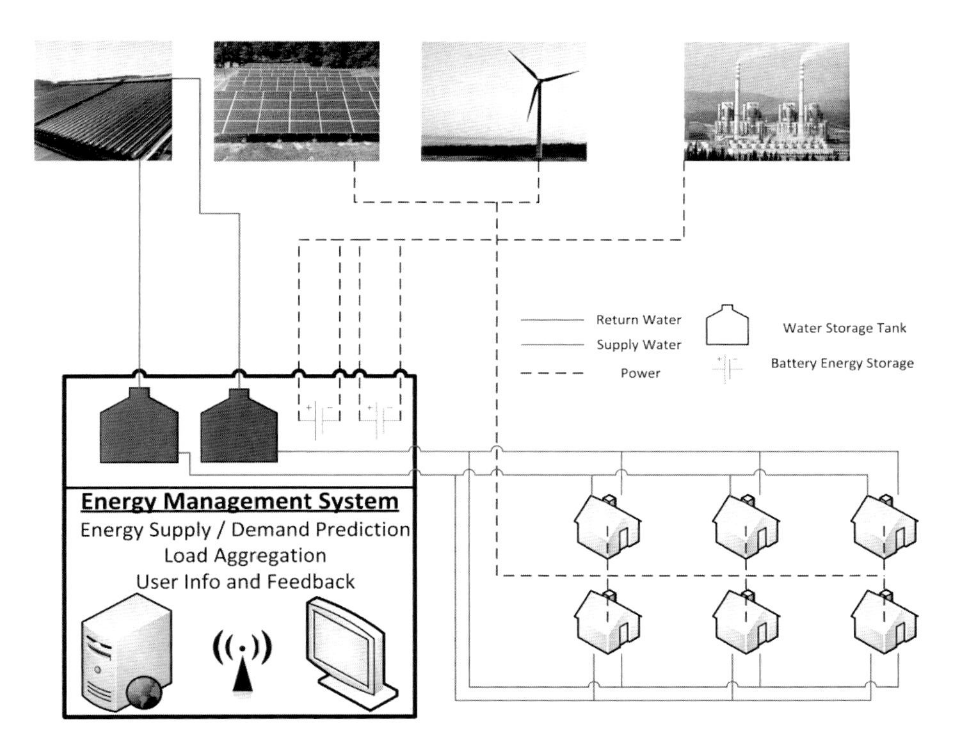

Fig. 1. Shared Resource Renewable Energy System Layout

3 Methodology

3.1 System Sizing

A major factor in the optimal solution is the sizing of the community energy generation. Given the average aggregate thermal and electrical energy demand for the community, the renewable energy production will typically be sized for the base, average, or peak load. In order for the optimization problem to be non-trivial, the system will be assumed to be sized to the average aggregate load for both thermal and electrical demand.

3.2 Performance Evaluation

Renewable energy utilization efficiency is a function of waste energy in the form of losses. In this example, losses occur due to energy storage and energy fed back to the grid. There exists other losses such as transmission losses, however these losses are unavoidable. The energy utilization efficiency is based on the maximum potential utilization efficiency after unavoidable losses. This is to strictly evaluate the decision engine on a scale from 0 to 100 percent for clarity, although the maximum thermodynamic efficiency will be less than 100 percent.

3.3 Model Inputs

In order to assist the decision engine, demand and generation forecasting will be implemented in order to make better decisions which will lead to either increased renewable energy utilization efficiency or reduced overall cost. The demand forecasting will be based on historical data of energy consumption for the community which will be updated over time to improve the accuracy of the forecasting by keeping up with trends in energy usage in the community. In terms of generation, forecasting in this example will be done with historical data as well, although a more sophisticated system would take into account local weather forecasts to predict the potential generation more accurately.

The first step to generating an effective decision engine for community energy management is to collect and interpret the generation and demand load profiles for thermal and electric energy. For this example, sample data is used to demonstrate the methodology of load manipulation. The aggregate thermal and electrical demand profiles used for this sample case study are shown in Fig. 2 and Fig. 3 below. The source of data for all the demand profiles is [2].

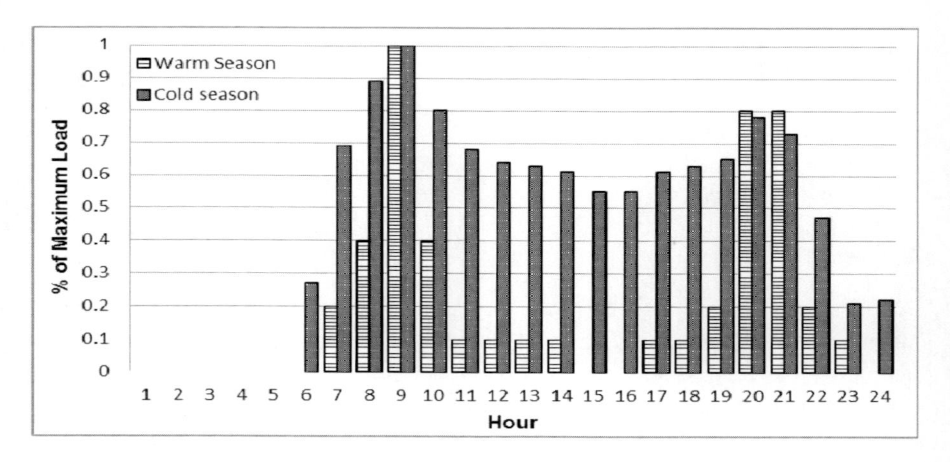

Fig. 2. Sample Thermal Energy Demand Profile

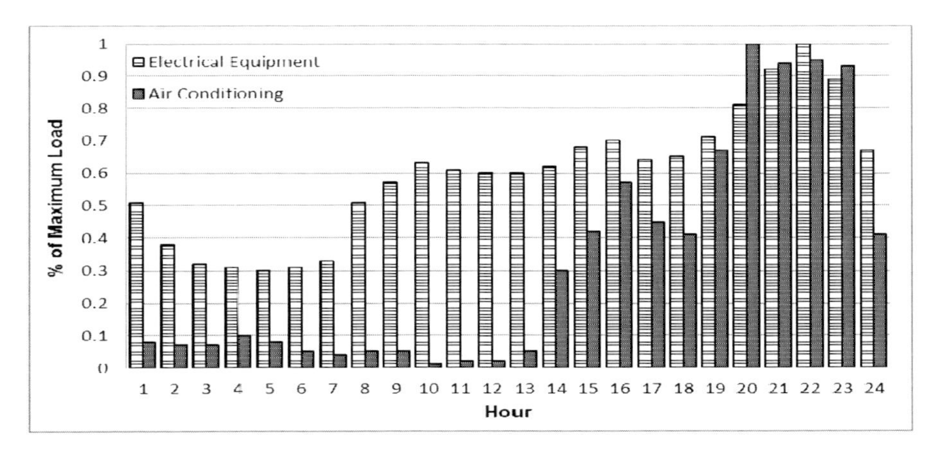

Fig. 3. Sample Electrical Energy Demand Profile

Now that the instantaneous demand as well as the predicted demand for the day is known, the generation side of the system can be analyzed in the same way as shown in Fig. 4 and Fig. 5. The source of data for the generation profiles is [3].

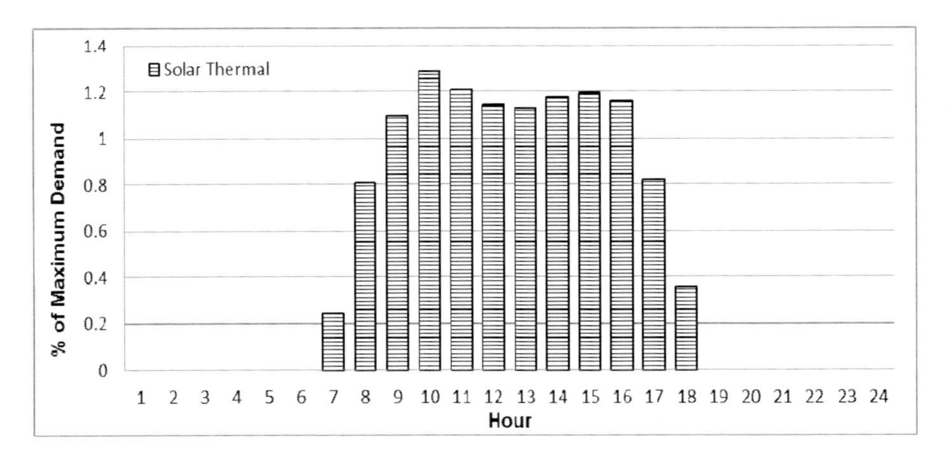

Fig. 4. Typical Thermal Energy Generation Profile

The thermal and electrical generation plots have been adjusted to be sized to the average demand for a cold season. This will facilitate the optimization problem, giving the system the possibility of achieving 100% energy utilization efficiency.

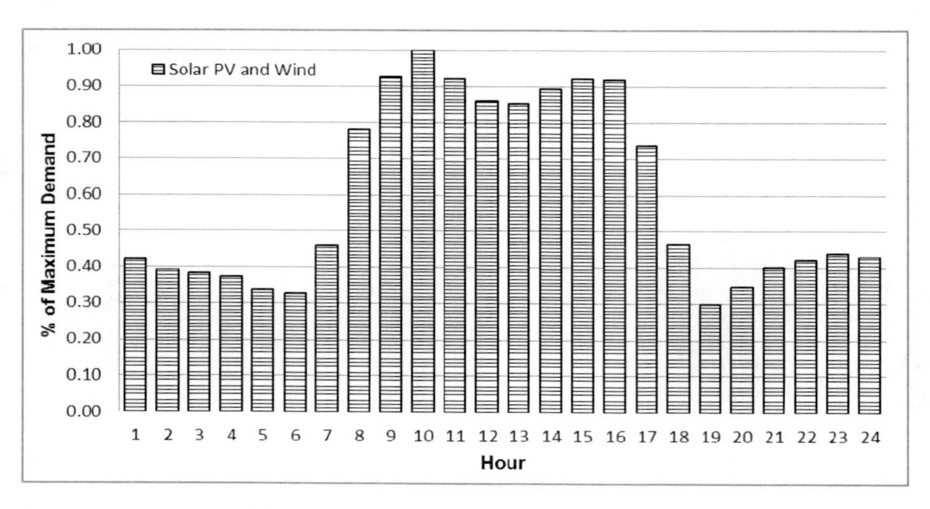

Fig. 5. Typical Electrical Energy Generation

3.4 Load Matching

Separating the mandatory from the non-mandatory loads is the first step to optimizing the system. Then, by superimposing the loads and the generation, the areas which need to be adjusted to reduce cost or increase utilization efficiency become clear. An example of this superposition is presented in Fig. 6 and Fig. 7 for the electrical and thermal loads, respectively. The optimization in this example is for a cold season, meaning that the thermal demand is highest and the air conditioning electrical demand is omitted.

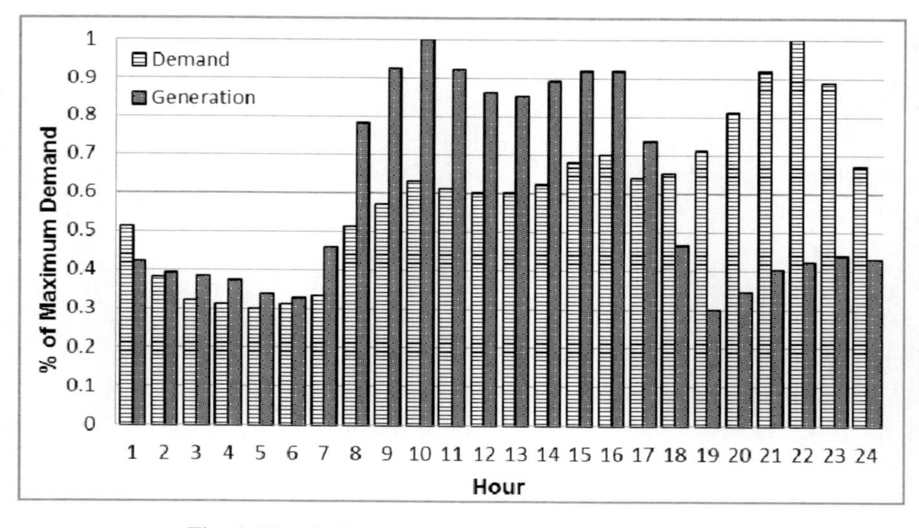

Fig. 6. Electrical Energy Matching Prior to Optimization

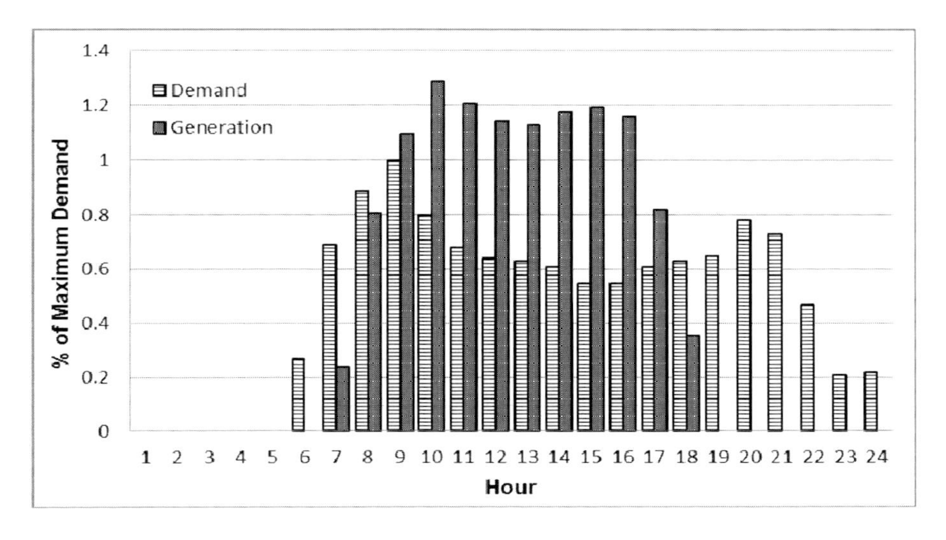

Fig. 7. Thermal Energy Matching Prior to Optimization

The above figures highlight which source of energy is used to supply energy to the demands, either renewable or grid energy. The area in which generation is greater than demand, all renewable energy is used to meet the load. However, when demand is greater than generation, the difference is met by the grid and auxiliary thermal energy source.

Using this graph, the percentage of non-mandatory loads, and the storage capacity of the renewable system, designers are able to schedule non-mandatory loads to optimize the renewable energy management and usage. System optimization can be done is many ways. The primary factors controlling the optimal solution are aggregate load by type (rush, flexible, etc.), energy availability (generating and stored), predicted upcoming loads, and objective (cost, emissions, efficiency, etc.). Correlating all of these factors mathematically to an objective function allows for designers to optimize the energy usage with time for any conditions.

The flowchart below in Fig. 8 shows the hierarchy of load types. The interesting part of the figure is the decision as to whether or not storage is needed. This decision would be based on the predicted upcoming loads from both the schedule and weather data.

Suppose we are at hour 21 (Refer to Fig. 6) and have just depleted the electrical storage. Depending upon the percentage of each load type on the system, there may be enough generation for the mandatory, rush, and discretionary loads. In this case, given that the system is predicting an increase in loads at hour 22, it would be beneficial for the system to not supply the discretionary load and store electrical energy for the upcoming increase in mandatory loads. With this strategy, the use of grid energy can be avoided by stocking up on stored energy beforehand.

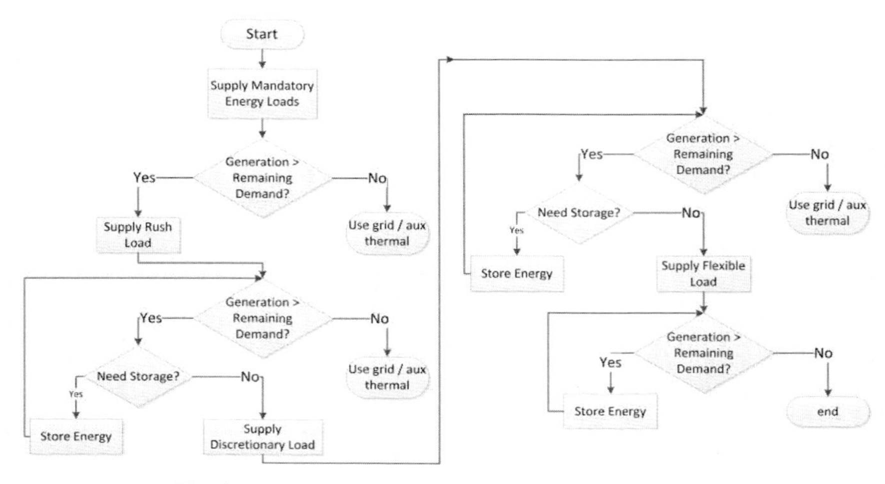

Fig. 8. Proposed energy management decision flowchart

4 Results

We have investigated the potential for optimization of the sample data provided in the previous figures for a specific set of varying conditions. The variable conditions are: load type distribution, willingness of residents to revise load types, and loss coefficients. We evaluated the solutions only in terms of *thermal* energy utilization efficiency to condense the results. The methodology for other metrics such as cost is very similar. The assumptions made for the example optimization are listed below. First, the parameters of the system are defined in words for clarity.

Unutilized Energy = storage loss + conversion loss + grid use or aux thermal
Energy utilization efficiency = (Generation – Unutilized Energy) / Generation

Each of the equations above in fact represent two equations, one for thermal and one for electrical. The more rigorous mathematical equations are listed below:

$$\dot{q}_{loss,t} = \int_0^T C_1 * \dot{q}_{ts}\, dT + \left(C_2 * \dot{q}_{ts,in}\right) + \left(C_3 * \dot{q}_{ts,out}\right) + \dot{q}_{auxiliary\ heat}\ [W] \quad (1)$$

$$\dot{q}_{loss,e} = \int_0^T D_1 * \dot{q}_{es}\, dT + \left(D_2 * \dot{q}_{es,in}\right) + \left(D_3 * \dot{q}_{es,out}\right) + \dot{q}_{grid\ feed}\ [W] \quad (2)$$

$$\eta_t = \left[\frac{\int_0^T (\dot{q}_{st} - \dot{q}_{loss,t})\, dT}{\int_0^T \dot{q}_{st}\, dT}\right] * 100 \quad (3)$$

$$\eta_e = \left[\frac{\int_0^T (\dot{q}_{pv} + \dot{q}_w - \dot{q}_{loss,e})\, dT}{\int_0^T (\dot{q}_{pv} + \dot{q}_w)\, dT}\right] * 100 \quad (4)$$

Constants

C_1 and X_1= storage loss coefficient over time
C_2 and X_2= conversion loss coefficient into storage
C_3 and X_3= conversion loss coefficient out of storage

Subscripts

t = thermal	ts = thermal storage	st = solar thermal	w = wind
e = electrical	es = electrical storage	pv = photovoltaic	T = time

Assumptions for sample case

$C_1 = 0.04$	$C_2 = 0.05$	$C_3 = 0.03$	(Loss Coefficients)
M = 60%	R = 10%D = 25%	F = 5%	(Load Types)

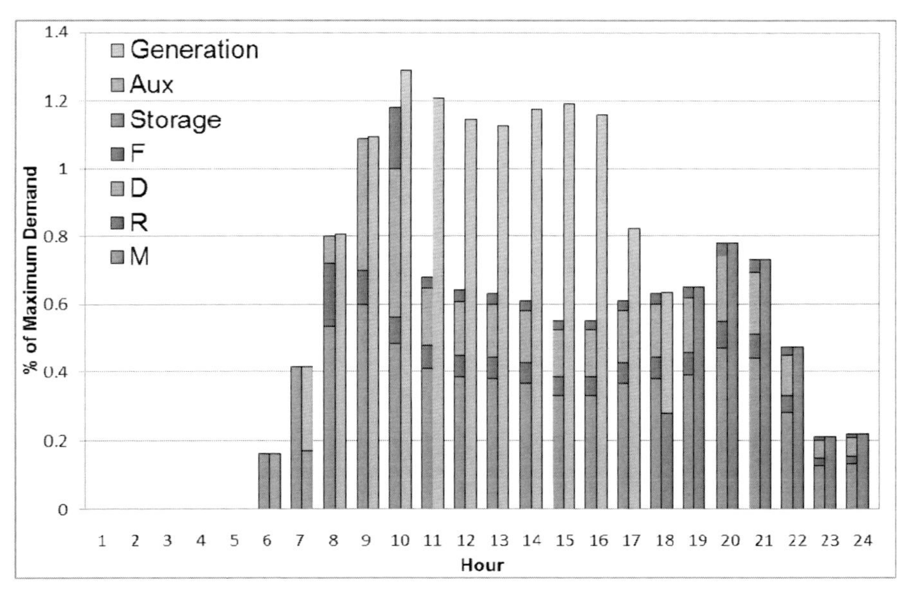

Fig. 9. Simple Thermal Optimization of Load Matching via Load Delay

Fig. 9 is the ideal optimal solution for this system, given the assumptions. In addition to the assumptions listed above, no initial storage capacity and no ability to move loads earlier in time was assumed for simplicity. As a result of this analysis, the energy utilization efficiency for the non-optimized and simply optimized solution in this case is 78.7% and 85.5%, respectively. It is important to note that having the requests for energy usage early in the day will allow for a reduction in storage use by applying loads during hours of high excess generation. Applying this with these sample data and assumptions yields an efficiency of 92.9%.

The solution has several uncontrollable variables separating the theoretical optimal solution from the actual solution. First, the prediction of the loads will not be accurate given the many variation of hourly load based on resident behavior which cannot well anticipated. Second, the prediction of the energy generation has its own variables which will further increase the difficulty of reaching the optimal solution.

The answer to these challenges is energy storage. Although it comes at a cost, both financially and thermodynamically, energy storage allows for the system to handle unanticipated variations on both the demand and generation side of the system. Losses beyond inefficiencies will only come when the storage is either full (energy must be input into the grid) or when the storage is depleted (energy from the grid must be used). From this, the optimal level at which to maintain storage levels becomes an interesting problem. This optimization is dependent on the predicted loads of the system and is therefore dynamic.

5 Conclusions

With the use of a decision engine which can optimize the energy usage of a community using a shared energy renewable resource network, the overall efficiency can increase while the fuel costs of the system can decrease significantly. Properly matched load, given the cooperation of residents results in fewer losses due to storage, as well as less use of grid electricity and auxiliary thermal energy generation via natural gas.

References

[1] Dagdougui, Minciardi, Ouammi, Robba, Sacile: A Dynamic Decision Model for the Real-Time Control of Hybrid Renewable Energy Production Systems. IEEE Systems Journal 323 (2010)

[2] Pedrasa, Spooner, MacGill: Robust Scheduling of Residential Distributed energy Resources Using a Nocel Energy Service Decision-Support Tool. Innovative Smart Grid Technologies (ISGT), IEEE PES (2011)

[3] Barbieri, Melino, Morini: Influence of the thermal energy storage on the profitability of micro-CHP systems for residential building applications. Applied Energy, 3–4 (2011)

[4] Vick, Clark, Mehos: Improved electrical load match in California by combining solar thermal power plants with wind farms. In: The SOLAR, Conference. American Solar Energy Society, San Diego (2008)

Chapter 28
Evaluation of the LCA Approaches for the Assessment of Embodied Energy of Building Products

Ayşen Ciravoğlu and Gökçe Tuna Taygun

Yıldız Technical University Faculty of Architecture Barbaros Bulvarı 34349 Beşiktaş Istanbul Turkey
{aysenc,gokcetunataygun}@gmail.com

Abstract. In this study among current approaches that involve usage of LCA that are employed in different countries and established by different institutions, Athena, BEAM, BEES, BREEAM, DGNB, EcoHomes, GaBi, Green Star and LEED are studied. Findings related to General Information, Usage Features, LCA Processes, Environmental Impact Assessment Criteria, Features of Assessment and Energy in Environmental Impact Assessment Criteria of Evaluated Tools are given and tools are compared related to these subjects. Comparison and evaluation of embodied energy criteria in selected tools that evaluate environmental impacts that have occurred throughout the lifespan of building products present the importance of product decision processes in building design. Appropriate product decision requires an adequate level of product information and this necessitates an information database. Besides, development of weightings, grading, and/or calculation methods that are appropriate to countries and regions where the evaluation tool is used is important in terms of effectiveness and productivity of the evaluation.

Keywords: embodied energy, building products, LCA, environmental assessment tools.

1 Introduction

We are living in a century where natural resources on earth are lessening and production and consumption processes inevitably result in pollution of air, water and soil. In this framework, it is a well known fact that buildings enormously affect the environment. According to the Worldwatch Institute, the building sector globally uses 40% of stone, aggregate and sand, and 25% of wood. Also, annually, buildings are responsible for 40% of energy and 16% of water consumption in the world. When we evaluate the fact that most of the consumption and environmental effects are caused by the building sector, sustainable design of buildings stands out to be a significant subject. Even though sustainability has environmental, social and economic dimensions, undoubtedly the primary subject of building design today is energy.

It is obvious that the projects issued for the reduction of energy consumption in buildings are limited with operational energy. However research shows that the

A. Håkansson et al. (Eds.): *Sustainability in Energy and Buildings*, SIST 22, pp. 299–312.
DOI: 10.1007/978-3-642-36645-1_28

embodied energy of a building can be equal to its 30 years of operational energy. As known embodied energy is the energy consumed by all of the processes associated with the production of a building, from the mining and processing of natural resources to the manufacturing, transport and product delivery. Embodied energy is an important element in Life Cycle Assessment (LCA). Life Cycle Assessment (LCA) is the evaluation of the possible environmental effects that have occurred throughout all life cycle processes of products. This article explores the place of embodied energy in LCA approaches.

1.1 Aim of the Study

During the 1990s, the concepts of sustainable design and high performance buildings, as well as the increasing adoption of these concepts in the marketplace, have been furthered by the development of assessment tools. There are international studies, checklists and computer programs evaluating building products on energy and environmental effect levels. LEED (Leadership in Energy and Environmental Design), BREEAM, CASBEE and Green Star in assessment of energy and environmental design issues in buildings, GaBi software in evaluating life cycle of building products are significant evaluation models. However considering the geographical data and its application in different localities in the assessment tools that are stated above, inadequacies can be seen. Local data to support such assessment tools are also lacking.

It is obvious that there are limited data and research on production of building materials and their environmental behavior. What the paper puts forth is an overview of the current situation on LCA approaches. The outcomes that will be generated from evaluation of the methods and software which form the aim of this paper, is expected to propose an easily applicable, understandable model which takes into account geographical and other contextual data to calculate embodied energy of building products. Eventually the findings will encourage producers in the building sector to evaluate their products and services on an energy consumption basis.

1.2 Method of the Study

In order to develop a model which will help the decision making process of material selection for designers, first of all, a selection of current available models which assess LCA is studied.

As known, there are many tools that evaluate life cycle processes of buildings. However it is not easy for both the designers and other actors of the sector to select the most suitable assessment tool from a wide range of possibilities due to the context of the design/building. The aim, content and weighting of the tools designed for LCA differ from each other related to different needs, geographies, and localities. For instance, the building stock in Europe is old, and therefore the maintenance and refurbishment of the existing buildings are critical issues for sustainable building construction. As the urban areas grow rapidly, the situation is different in North America (Kohler and Moffatt, 2003).

If we want to classify environmental assessment tools, the following classification derived from the ATHENA classification (Haapio and Vitaniemi, 2008) and the IEA Annex 31 Classification (Haapio and Vitaniemi, 2008) can be state as follows:

a. Energy Modeling Software (APACHE, TSol, etc.)
b. Environmental LCA Tools for Buildings and Building Stocks
    a. Product Comparison Tools and Information Sources (BEES, TEAM, etc.)
    b. Whole Building Design and Decision Support Tools (ATHENA, BEAT 2002, BeCost, Eco-Quantum, Envest 2, EQUER, LEGEP, PAPOOSE, etc.)
    c. EcoEffect, ESCALE, etc.
c. Environmental Assessment Frameworks and Rating Systems (BREEAM, Eco-Profile, Environmental Status Model, DGBN, LEED, etc.)
d. Environmental Guidelines or Checklists for Design and Management of Buildings (ECOPROP, LEGOE, etc.)
e. Environmental Product Declarations, Catalogues, Reference Information, Certifications and Labels (EcoSpecifier, Swan Eco-label, etc.).

Environmental LCA Tools for Buildings and Building Stocks consist of interactive software, Environmental Assessment Frameworks and Rating Systems which rely more on guidelines and questionnaires than on databases and are passive tools. Interactive software tools "provide calculation and evaluation methods which enable the user or decision maker to take a pro-active approach (to explore a range of options in an interactive way)". Passive tools support decision making, they do not allow interaction with the user (Haapio and Vitaniemi, 2008).

In this paper, tools from categories Environmental LCA Tools for Buildings and Building Stocks and Environmental Assessment Frameworks and Rating Systems are selected for comparison and evaluation. As can be followed from the list below, the selected tools are 3 LCA tools (ATHENA, BEES, GaBi) and 6 environmental assessment frameworks and rating systems (BEAM, BREEAM, DGBN, ECOHOMES, GREEN STAR, LEED). The reason for making such a comparison is to see the differences of approaches among varied environmental assessment systems. The paper aims to also review differences among interactive software and passive tools and globally used models and local approaches.

The selected tools evaluated in this study are:

- ATHENA (Athena Sustainable Materials Institute)
- BEAM (The Building Environmental Assessment Method Society)
- BEES (Building for Environmental and Economic Sustainability),
- BREEAM (Building Research Establishment Environmental Assessment Method)
- DGBN (Deutsche Gesellschaft für Nachhaltiges Bauen)
- ECOHOMES (BRE Building Research Establishment)
- GaBi (PE International)
- GREEN STAR (Green Building Council of Australia)
- LEED (Leadership in Energy and Environmental Design)

The study is based on literature review; none of the tools have been tested in the study. The information is gathered from web sites of the available tools.

2 Embodied Energy and Life Cycle Assessment

2.1 Embodied Energy

Embodied energy is the energy consumed by all of the processes associated with the production of a building, from the acquisition of natural resources to product delivery, including mining, manufacturing of materials and equipment, transport and administrative functions. The embodied energy of a building is a significant multiple of the annual operating energy consumed, ranging from around 10 for typical dwellings to over 30 for office buildings. Making buildings such as dwellings more energy efficient usually requires more embodied energy thus increasing the ratio even further. Choice of material and design principles has a significant, but previously unrecognized, impact on energy required to construct a building. Embodied energy is one measure of the environmental impact of construction and the effectiveness of any recycling, particularly CO_2 emissions. CO_2 emissions are highly correlated with the energy consumed in manufacturing building materials. On average, 0.098 tonnes of CO_2 are produced per gigajoule of embodied energy (Tucker, 2000).

The scenario of balancing operating energy is quite familiar and requires no further clarification. The concept of life cycle energy balance might not, however, be quite so obvious. Here, energy inputs comprise a building's net life cycle embodied energy plus its total operational energy inputs. The gross life cycle embodied energy is the embodied energy of all the materials used in initially constructing the building, plus energy used in its maintenance, repair and refurbishment, including all the transport energy that that entails. It also includes the energy used to dismantle the building and transport the materials to the landfill or recycling facility. To get the net figure, the embodied energy recovered from the recycled materials must be taken away from the gross embodied energy figure. The operational energy consists of the total energy used to maintain the habitability of the building throughout its life. A figure for a building's total energy debt is found by adding together the figure for its net embodied energy and its total operating energy. Our aim then is to design buildings which will produce enough energy during their lifetime to recover this energy debt (Storey and Baird, 1999).

The Gross Energy Requirement (GER) is a measure of the true embodied energy of a material. In practice this is usually impractical to measure. The Process Energy Requirement (PER) is a measure of the energy directly related to the manufacture of the material. This is simpler to quantify. Consequently, most figures quoted for embodied energy are based on the PER. This would include the energy used in transporting the raw materials to the factory but not energy used to transport the final product to the building site. In general, PER accounts for 50 to 80% of GER. Even within this narrower definition, arriving at a single figure for a material is impractical as it depends on (Reardon et al., 2010):

- Efficiency of the individual manufacturing process.
- The fuels used in the manufacture of the materials.

- The distances materials are transported.
- The amount of recycled product used, etc.

Each of these factors varies according to product, process, manufacturer and application. They also vary depending on how the embodied energy has been assessed. Factors that have considerable bearing on the final energy coefficient are firstly whether heat, and how much, is needed in the manufacturing process, followed by the amount of physical force needed in the manufacturing process. Transport, especially if it is by an efficient means such as sea transport, tends to have only a small influence on the final result. Fuel type, internal wastage and efficiency of the manufacturing plant all have a noticeable, but usually rather less significant, bearing on the final result.

However assessment of embodied energy of a building material alone cannot determine the building's environmental properties. For instance the reuse of building materials commonly saves about 95% of embodied energy which would otherwise be wasted. Some materials such as bricks and tiles suffer damage losses up to 30% in reuse. The savings by recycling of materials for reprocessing varies considerably, with savings up to 95% for aluminum but only 20% for glass. Some reprocessing may use more energy, particularly if long transport distances are involved. On the other hand, materials such as concrete and timber have the lowest embodied energy intensities but are consumed in very large quantities; whereas the materials with high energy content, such as stainless steel, are used in much fewer amounts. Steel can be re-used and/or recycled in the building industry. Nevertheless the greatest amount of embodied energy in a building is often in concrete and steel. However, using these values alone to determine preferred materials is inappropriate because of the differing lifetimes of materials, differing quantities required to perform the same task, different design requirements (Tucker, 2000) and different reuse, recycle properties.

Further more embodied energy must always be considered in the context of the total energy requirement over the lifetime of a building. Choice of materials can influence operating energy requirements as well as embodied energy. For example, a high mass material such as concrete, although having a larger embodied energy than timber, has the potential for reducing HVAC energy requirements due to its good heat storing properties. In the case of glass fiber insulation, the energy savings over the building's life can be many times that of its initial energy cost (Baird et al., 1998). In choosing between alternative building materials or products on the basis of embodied energy, not only the initial materials should be considered but also the materials consumed over the life of the building during maintenance, repair and replacement (Tucker, 2000). Therefore a life cycle assessment tool which includes embodied energy as well as other relevant data should be taken into consideration.

2.2 Life Cycle Assessment

Life Cycle Assessment (LCA) is the evaluation of the possible environmental effects that have occurred throughout all life cycle processes of products. Life Cycle processes of a product are:

- Acquisition of raw material: Acquisition of raw material and energy sources from soil and transportation of raw material to the processing unit.
- Production of the product:

  - Production of the material: Processing of the raw material in order to be used in manufacturing of a product.
  - Production of the product: Production of a finished product (e.g. component, element, unit, fragment).
  - Packaging and distribution of the product,
- Installation of the product to the building.
- Usage of the product, repair, maintenance and re-usage.
- Completing of the lifespan of the product, recycle or disposal of the product.

Transportation between the processes is also a part of LCA. The concepts of open loop and closed loop can be explained as follows:

- Closed loop: usage of a product that has completed its lifespan in the production of the same material.
- Open loop: usage of a product that has completed its lifespan in the production of a different material (Ciambrone, 1997; Curran, 1996; Horne et al., 2009; Keoleian et al., 1994; Tuna-Taygun, 2005; Vigon et al., 1994).

LCA consists of four steps which are related to each other (Ciambrone, 1997; Curran, 1996; Horne et al., 2009; Tuna-Taygun, 2005; Vigon et al., 1994):

- Completion of the study,
- Inventory analysis,
- Impact analysis,
- Interpretation of the evaluation.

There are many methods to calculate LCA and differing methods offering different options. One of the most comprehensive studies about broadening and deepening LCA methods and tools is the CALCAS project funded by the EU 6th framework program. The study includes extensive research on current methods of Environmental Impact Assessment: Material flow analysis, substance flow analysis, cost benefit analysis, life cycle costing, total cost assessment (total cost accounting), total cost of ownership, environmental input-output analysis/environmentally extended input-output analysis, hybrid analysis, integrated hybrid analysis, environmental risk analysis, environmental impact assessment, computable general equilibrium model, input-output analysis (including social accounting matrices), eco-efficiency analysis, material intensity per service unit, external costs (externe), ecodesign product oriented environmental management systems, energy/exergy analysis, multicriteria analysis, life cycle activity analysis, partial equilibrium modeling, carbon footprint, social life cycle assessment or societal LCA, strategic environmental assessment, sustainability assessment, life cycle optimization, sustainable process design and green accounting are all methods evaluated within the scope of the project. SWOT analysis performed due to the content of the project on these methods and models mainly serves for the identification of opportunities for combinations and/or integrations, towards a deepened and broadened life cycle analysis. According to the study each of these models has its own path of development and its research needs, which have been briefly described in the project. However the challenge is to understand how all these advances can make available the necessary knowledge for making the framework operational, i.e. how they can be fit into the framework in a way so that they complement each other and provide coherent, relevant, reliable and complete output (Zamagni et al., 2009).

According to Baird et al. (1997), while not covering all considerations, energy analysis is one method which can be used, albeit crudely, to estimate the environmental impact of different activities. Statistical analysis utilizes published statistics to determine energy use by particular industries; input-output analysis captures every dollar transaction, and hence every energy transaction, across the entire national economy and process analysis involves the systematic examination of the direct and indirect energy inputs to a process. The most effective one seems to be hybrid energy analysis which attempts to incorporate the most useful features of the three analysis methods outlined above, especially input-output analysis and process analysis.

Input–output analyses generally suffer from a lack of detail in calculating direct contributions simply because the data on which they are based are spread over a limited number of industrial sectors. Hence, the particular process of interest may only be contained in an aggregated classification, with a corresponding loss of accuracy. Process analysis of the direct requirements is therefore often more accurate. However, in determining indirect requirements, the results of process analysis are dependent on the choice of the system boundary, that is, on how many of the first and higher order processes are included. Process analysis therefore suffers from truncation error, as it is not possible to include all of the higher order contributions. In contrast, input–output analysis does account for all higher order terms, subject to the limits of aggregation of the sectors. Nevertheless, it suffers from errors of a different nature (Lenzen and Dey, 2000).

A number of researchers have suggested and used a hybrid LCA approach combining process with input–output analysis. Input–output analysis is a top–down economic technique, which uses sectoral monetary transaction matrices describing complex interdependencies of industries in order to trace resource requirements and pollutant releases throughout a whole economy. In a tiered hybrid LCA, the direct and downstream requirements (for construction, use, and end-of-life), and some important lower-order upstream requirements of the functional unit are examined in a detailed process analysis, while remaining higher-order requirements (for materials extraction and manufacturing) are covered by input–output analysis. In this way, advantages of both methods, completeness and specificity, are combined. Moreover, the selection of a boundary for the production system becomes obsolete. An input–output technique called structural path analysis can be employed to extract a preliminary ranking of the most important input paths into the functional unit. This ranking can be used to prioritize the inventory list and to systematically delineate the process and input–output part of the hybrid LCA according to the required level of specificity and accuracy. Furthermore, it can provide data to fill gaps in the existing incomplete life cycle inventories (Lenzen and Treloar, 2002).

Another hybrid approach has been suggested by Treloar (1996), which involves extracting embodied energy paths and selecting the most important energy requirements (which are not necessarily the direct or first order requirements). Process analysis is carried out for these important paths, and the remainder is covered by input–output analysis (Lenzen and Dey, 2000). Treloar et al. (2000) argue that existing techniques such as LCA do not account adequately for upstream processes. Thus, there is a need for a more comprehensive LCA method, so that the direct and indirect environmental impacts of design and engineering decisions can be assessed. They acknowledge previous input-output LCA methods, but suggest that a hybrid LCA method is more appropriate. Currently the authors are developing the proposed hybrid LCA method by: (a) investigating the best available input-output LCA models; and (b) application of the proposed hybrid LCA method to different building types and other non-building products. The proposed hybrid LCA method is claimed to enable informed decision making with regard to the collection of case specific LCA data. The proposed hybrid LCA method is assumed to enable potentially a large increase in framework completeness, and hence its overall reliability, at the cost of only a small increase in research time compared with a traditional LCA. The environmental impact of the initial and recurring construction phases for buildings is pursued to be able to be calculated more comprehensively.

3 Evaluation of the Life Cycle Assessment Approaches

In the following section of the paper, among different tools that make environmental assessments, which are proposed by different institutions and scientists in different geographies, the selected tools are analyzed and compared according to the priorities related to assessment of embodied energy.

3.1 General Information on Evaluated Tools

First of all, general information related to the evaluated tools is given in the table below. Here, the establishment that proposed the tool, the country that the tool is proposed in, and the usage dates of the tools are given. The establishments are generally

Table 1. General Information on Evaluated Tools

NAME OF THE TOOL	THE ESTABLISHMENT THAT PROPOSED THE TOOL	THE COUNTRY THAT THE TOOL IS PROPOSED IN	USAGE DATE OF THE TOOL
ATHENA	Athena Sustainable Materials Institute	Canada	1997
BEAM	Building Environmental Assessment Method Society	China	1996
BEES	National Institute of Standards and Technology	USA	1994
BREEAM	Building Research Establishment	England	1990
DGNB	German Green Building Council	Germany	2007
ECOHOMES	Building Research Establishment	England	2000
GaBi	PE International	Germany	no data
GREEN STAR	Green Building Council of Australia	Australia	2002
LEED	The U.S Green Building Council (USGBC)	USA	1998

Green Building Councils, research centers or institutes that work on a volunteering basis. The only exception is GaBi which is offered by a commercial establishment. The listing of the countries where the tools are proposed is important information as evaluation criteria and weightings depend on local specialties. Here another important piece of information is whether the tool is used locally or globally. ATHENA, BEES, BREEAM, LEED and GaBi are used globally.

3.2 Comparison Related to Usage Features

Table 2 brings forth a general evaluation related to usage of the tools. The questions that are answered in the table are whether the usage is obligatory, which tools are used, what are the levels of evaluation and finally who are the users. It should be stated that all of the tools evaluated in this study work on a volunteer basis. The only exception is EcoHomes, whose usage has become obligatory in social housing projects starting from 2003. As known tools make the evaluation through either checklists or software. This is important data as it defines whether the model is interactive or passive. LCA tools, ATHENA, BEES and GaBi, use software. The table also gives information on evaluation levels of the tools. Even though the subject of this paper is LCA of building products, here, in order to compare and generate solutions, the tools that evaluate the building or all products are also evaluated.

Table 2. Usage Features of Evaluated Tools

USAGE FEATURES OF TOOLS	NAME OF TOOL	ATHENA	BEAM	BEES	BREEAM	DGNB	ECOHOMES	GaBi	GREEN STAR	LEED
IS USAGE OF THE TOOL OBLIGATORY?	**Yes**						X			
	No									
EVALUATION TOOL	**Checklist**									
	Software									
EVALUATION LEVEL OF THE TOOL	**Building**									
	Building Product									
USERS OF THE TOOL	**Designer**									
	Product Producers									
	Building Users									
	Not Defined									

X: the model is obligatory for social housing projects starting in 2003.

Users of the tools also give information about the best practices of the tool. Here the users are designers, product producers, and building users. Besides, in ATHENA engineers and researchers, in BEAM engineers, contractors, building managers, in BREEAM building managers, in DGNB auditors, planners, in ECOHOMES building managers can use the program and lastly in BEES usage is open to everyone.

3.3 Comparison Related to LCA Processes

Table 3 brings forth a comparison of the selected models related to LCA processes included in the tools. As can be followed from the table, except for ATHENA, BEAM and BEES, tools do not define LCA processes. It is obvious that in order to receive an effective and sufficient evaluation, definitions of the processes are of great importance.

Table 3. LCA Processes of Evaluated Tools

NAME OF TOOL / LCA ASSESSMENT PROCESSES		ATHENA	BEAM	BEES	BREEAM	DGNB	ECOHOMES	GaBi	GREEN STAR	LEED
ACQUISITION OF THE RAW MATERIAL				■						
PRODUCTION	Production of the Material	■	■	■						
	Production of the Product	■	■	■						
	Packaging and Distribution of the Product	■	■	■						
INSTALLATION OF THE PRODUCT		■	■	■						
USAGE OF THE PRODUCT		■	■	■						
RECYCLING OF THE PRODUCT				■						
DISPOSAL OF THE PRODUCT			■	■						
NOT DEFINED					■	■	■	■	■	■

3.4 Comparison Related to Environmental Impact Assessment Criteria

In Table 4 a comparison of the selected tools according to the environmental impact assessment criteria can be followed. In the table below various criteria that appear in the tools such as energy, building product, pollution (air, water, soil), human health, waste management are brought together in common titles. Along with the criteria explained in the table below in ATHENA; quality of indoor environment, renewal, design process; in BEAM quality of indoor environment, noise pollution, lighting pollution; in BEES quality of indoor environment, economical performance; in BREEAM management, roads, transportation; in DGNB qualities related to socio-cultural, functional, economic processes; in ECOHOMES transportation; in GaBi evaluation of the total cost; in GREEN STAR quality of indoor environment, transportation, usage of the land, ecology; and in LEED quality of indoor environment, renewal, design process are also taken into account. When the results of Table 4 and Table 1 are compared, which means the information of the geographies where the tools are used and the criteria related to EIA are compared, diversities due to local differences appear as important information where special attention is needed.

Table 4. Environmental Impact Assessment Criteria in Evaluated Tools

ENVIRONMENTAL IMPACT ASSESSMENT CRITERIA \ NAME OF TOOL		ATHENA	BEAM	BEES	BREEAM	DGNB	ECOHOMES	GaBi	GREEN STAR	LEED
ENERGY		■	■		■	■	■	■	■	■
BUILDING PRODUCT		■	■	■	■	■	■	■	■	■
POLLUTION	Air	■		■	■	■	■	■		
	Water	■	■	■	■	■	■		■	
	Soil		■			■			■	
HUMAN HEALTH				■	■		■			
WASTE MANAGEMENT						■				

3.5 Comparison Related to Features of Assessment

Table 5 shows three important features of assessment of selected tools. These are whether the assessment includes regional/local usage or not, if original databases are used in the assessment, and if weightings are paid attention to. These data are important because regional/local information paves the way for the model to be used in different geographies. Original database usage is another important feature because it provides coordination in assessment. Lastly weightings are of great importance because differences in usage of environmental impact criteria can be evaluated by weightings. Weighting is inherent to these systems although it might not be addressed explicitly; those systems without an explicit weighting method give all criteria equal weighting or implicitly weight the criteria by points allocated (Todd et al., 2001). In the absence of scientifically based weights, some organizations use 'consensus-based' weighting. In this approach, groups of experts or users rank various elements, such as environmental issues, in terms of their relative importance or assign points to these elements. This ranking or scoring is then used to establish weights (Dickie and Howard, 2000).

Table 5. Features of Assessment of Evaluated Tools

FEATURES of ASSESSMENT \ NAME OF TOOL	ATHENA	BEAM	BEES	BREEAM	DGNB	ECOHOMES	GaBi	GREEN STAR	LEED
REGIONAL/LOCAL USAGE		■		■	■	■		■	■
ORIGINAL DATABASES USED	■		■				■		
WEIGHTINGS		■		■	■	■		■	

3.6 Embodied Energy in LCA of Building Products

When we look at the tools that have been evaluated in the context of this paper in terms of embodied energy, we can define that except for BEES, in ATHENA, BEAM, BREEAM, DGNB, EcoHomes, GaBi, GREEN STAR and LEED energy issues appear in evaluation criteria. However evaluations of the energy issues differ from each other. As stated previously, models either use software or checklists as evaluation tools. In the tools that use software (Table 2), it is not possible to find information related to energy calculations. When checklists are used as an evaluation method, points assigned to different energy criteria can be followed. Therefore information regarding the tools that are used in LCA Tools is given in Table 6. The percentage of energy issues with respect to all evaluation criteria is given in the tools that use checklists. Still this does not give definite information related to the effect of energy in LCA. Points that are used in tools differ among different regional conditions, in other words the results differ if weightings are applied or not.

Here it can be stated that as in LCA tools processes are not defined openly it is not possible to evaluate the approach to embodied energy. In checklists and/or frameworks processes are more openly shared. However in those approaches currently embodied energy doesn't seem to be a primary concern.

Table 6. Energy in Environmental Impact Assessment Criteria of Evaluated Tools

NAME OF TOOL	ATHENA	BEAM	BEES	BREEAM	DGNB	ECOHOMES	GaBi	GREEN STAR	LEED
PERCENTAGE OF ENERGY IN ENVIRONMENTAL IMPACT ASSESSMENT CRITERIA	-	35	-	19	No data	22	-	No data	32

4 Conclusion

In this study among current tools that involve LCA that are used in different countries and established by different institutions, ATHENA, BEAM; BEES, BREEAM, DGNB, EcoHomes, GaBi, GREEN STAR and LEED are studied. In this paper findings related to:

- General Information on Evaluated Tools (Table 1)
- Usage Features of Evaluated Tools (Table 2)
- LCA Processes of Evaluated Tools (Table 3)
- Environmental Impact Assessment Criteria of Evaluated Tools (Table 4)
- Features of Assessment of Evaluated Tools (Table 5)
- Energy in Environmental Impact Assessment Criteria of Evaluated Tools (Table 6)

are given and tools are compared related to these subjects.

Inappropriate usage of production techniques in buildings and building products, irresponsible usage of natural resources, waste generation related to production and building results in negative effects to the natural environment. Production, recycling, transportation and usage of energy, the main inputs in the production of buildings and building products may result in environmental pollution.

Comparison and evaluation of embodied energy criteria in the LCA tools that evaluate environmental impacts occurring throughout the lifespan of the building products present the importance of product decision processes in building design. Appropriate product decision requires an adequate level of product information and this necessitates an information database. This system can be realized with product databases that are generated from local information. Development of weightings, grading, and/or calculation methods that are appropriate to countries and regions that the evaluation tool is used in is important in terms of effectiveness and productivity of the evaluation.

It is thought that by taking into consideration energy consumption in the design process not only in usage of buildings and/or products but also in all life cycle processes will result in prevention of environmental pollution, protection of natural resources, economy and therefore implementation of obligatory energy policies.

Acknowledgments. This paper originates from an ongoing research project funded by Yıldız Technical University.

References

Baird, G., Alcorn, A., Haslam, P.: The energy embodied in building materials - updated New Zealand coefficients and their significance. The Institution of Professional Engineers New Zealand Transactions 24(1/CE), 46–54 (1997)

Baird, G., Alcorn, A., Wood, P., Storey, J.B., Jacques, R.: Progress Towards the Specification of Embodied Energy Performance Criteria for New Zealand Buildings. In: Proceedings of the Green Building Challenge 1998 Conference, Vancouver, October 26-28, pp. 154–161 (1998)

Ciambrone, D.F.: Environmental Life Cycle Analysis. Lewis Publishers, USA (1997)
Curran, M.A.: Environmental Life-Cycle Assessment. McGraw-Hill, USA (1996)
Dickie, I., Howard, N.: Assessing Environmental Impacts of Construction – Industry Consensus, BREEAM and UK Ecopoints, BRE Digest #446 (2000); Todd, J. A., Crawley, D., Geissler, S., Lindsey, G.: Comparative Assessment of Environmental Performance Tools and the Role of the Green Building Challenge. Building Research & Information 29(5), 324–335 (2001)
Haapio, A., Vitaniemi, P.: A Critical Review of Building Environmental Assessment Tools. Environmental Impact Assessment Review 28, 469–482 (2008)
Horne, R., Grant, T., Verghese, K.: Life Cycle Assessment: Principles, Practice and Prospects. Csiro Publishing, Australia (2009)
Kohler, N., Moffatt, S.: Life-Cycle Analysis of the built Environment. UNEP Industr Environ 2-3, 17–21 (2003); Haapio, A., Vitaniemi, P.: A Critical Review of Building Environmental Assessment Tools, Environmental Impact Assessment Review 28, 469–482 (2008)
Lenzen, M., Treloar, G.: Rejoinder to Börjesson and Gustavsson Embodied energy in buildings: wood versus concrete reply to Börjesson and Gustavsson. Energy Policy 30, 249–255 (2002)
Lenzen, M., Dey, C.: Truncation error in embodied energy analyses of basic iron and steel products. Energy 25, 577–585 (2000)
Reardon, C., Milne, G., McGee, C., Downtown, P.: Your Home Technical Manual, Australian Government Department of Climate Change and Energy Efficiency (2010)
Storey, J.B., Baird, G.: Towards the self-sufficient city building. The Institution of Professional Engineers New Zealand Transactions 26(1/GEN), 1–8 (1999)
Todd, J.A., Crawley, D., Geissler, S., Lindsey, G.: Comparative assessment of environmental performance tools and the role of the Green Building Challenge. Building Research & Information 29(5), 324–335 (2001)
Treloar, G.J.: Extracting embodied energy paths from input–output tables: towards an input–output-based hybrid energy analysis method. Economic Systems Research 9(4), 375–391 (1996); Lenzen, M., Dey, C.: Truncation error in embodied energy analyses of basic iron and steel products, Energy 25, 577–585 (2000)
Treloar, G.J., Love, P. E.D., Faniran, O.O., Iyer-Raniga, U.: A hybrid life cycle assessment method for Construction. Construction Management and Economics 18, 5–9 (2000)
Tucker, S.: Embodied Energy (2000), http://www.dbce.csiro.au/indserv/brochures/embodied/embodied.htm (date of visit June 19, 2000)
Tuna-Taygun, G.: Yapı Ürünlerinin Yaşam Döngüsü Değerlendirmesine Yönelik Bir Model Önerisi , unpublished PhD thesis, Yıldız Technical University, Institute of Science, İstanbul (2005)
Vigon, B.W., Tolle, D.A., Cornaby, B.W., Latham, H.C., Harrison, C.L., Boguski, T.L., Hunt, R.G., Sellers, J.D.: USEPA Risk Reduction Engineering Laboratory, Life-Cycle Assessment Inventory Guidelines and Principles. Lewis Publishers, USA (1994)
Zamagni, A., Buttol, P., Buonamici, R., Masoni, P., Guinée, J. B., Huppes, G., Heijungs, R., van der Voet, E., Ekvall, T., Rydberg, T.: D20 Blue Paper on Life Cycle Sustainability Analysis, Deliverable 20 of Work Package 7 of the CALCAS project (2009), http://www.calcasproject.net/default.asp?site=calcas&page_id=E2669B0F-9DB7-4D1E-95B0-407BC7949030 (date of visit: June 24, 2012)

Chapter 29
Exergetic Life Cycle Assessment: An Improved Option to Analyze Resource Use Efficiency of the Construction Sector

Mohammad Rashedul Hoque[1], Xavier Gabarrell Durany[1], Gara Villalba Méndez[1], and Cristina Sendra Sala[2]

[1] Sostenipra research group, Institute of Environmental Science and Technology (ICTA), Department of Chemical Engineering, Reference Network in Biotechnology (XRB), Universitat Autónoma de Barcelona, Bellaterra, 08193 Barcelona, Spain
[2] eco intelligent growth, Avinguda Diagonal, 523, 5è 1ª, 08029 Barcelona, Spain
MohammadRashedul.Hoque@uab.cat, Xavier.Gabarrell@uab.cat, Gara.Villalba@uab.cat, c.sendra@ecointelligentgrowth.net

Abstract. This article presents an effort to pinpoint how efficiently resources are used in the construction sector applying exergetic life cycle assessment methodology in a cradle-to-grave life cycle approach. Polypropylene (PP) and polyvinyl chloride (PVC), two widely used thermoplastics in construction applications, are chosen as case study materials in this analysis involving raw material extraction, resin manufacturing, and post-consumer waste management life-cycle stages. Overall life cycle exergy efficiency of PP and PVC is quantified 27.1% and 9.3%, respectively, characterized by a low efficiency of manufacturing and recycling processes for both materials. Improving the efficiency of manufacturing and recycling processes will thus reduce exergy losses from the system. From resource conservation point of view, mechanical recycling can be the viable option for end-of-life plastic waste management, since it loops materials back directly into new life cycle, and thus reduces primary resource inputs in the production chain and associated environmental impacts.

Keywords: Cradle-to-grave, exergy efficiency, exergy loss, thermoplastic, waste management.

1 Introduction

The sustainability of the construction sector is largely dependent on efficient use of natural resources, since it is the largest user of non-renewable energy and material resources. More recently, increasing awareness about the impacts associated with resource use, as well as their scarcity, has led to efforts to reduce energy dependency and to shift industrial activities towards improved technologies. One of the major efforts in achieving this goal is to increase the efficiency of present industrial processes. Adequate evaluation of resource consumptions and environmental impacts

A. Håkansson et al. (Eds.): *Sustainability in Energy and Buildings*, SIST 22, pp. 313–321.
DOI: 10.1007/978-3-642-36645-1_29

throughout the overall life cycle is critical for the proper evaluation of technologies. Life Cycle Assessment (LCA) is one of the most popular tools to compare different scenarios of products' end-of-life. An LCA consists of three main steps: determination of mass and energy in- and out flows through all stages of the life cycle; evaluation of the environmental impacts associated with resource consumption; and ways to decrease the environmental, economic and social burdens [9]. Apart from LCA, a more recent approach in order to assess the sustainability of technological options is the thermodynamic or exergetic life-cycle assessment (ELCA). The ELCA examines exergy flows and seeks to reduce exergy destructions and to improve the efficiency of processes [5]. In conventional energy efficiency (refers to first law analysis) only identifies losses of work; however, the exergy analysis (known as second law analysis) takes the entropy production into consideration by including irreversibilities [9]. Therefore, exergy analysis offers useful insights for the correct assessment of the process itself. The thermodynamic analysis of life cycle shows a cumulative loss of exergy due to entropy generation. Resource extraction from the ecosphere, conversion of the resources into products and wastes, and irreversibility can be analyzed in exergy terms, showing the role of process efficiency in sustainability. The ELCA uses the same framework as the LCA, but the additional criterion is the life-cycle irreversibility, the loss of exergy during the complete life cycle [5]. Life cycle irreversibility can be used as the measure of inefficient use of natural resources. The ELCA has been applied to account the depletion of natural resources [9], exergy input to production system [21], resource consumption in the built environment [6]; however, it can also be a useful tool to assess the inefficiency of a system since it shows in which component the losses of natural resources take place [8]. With this information better proposals for reducing the loss of natural resources can be obtained. Consequently, exergy-based analysis can be used as a powerful tool for process optimization.

Construction sector is the second largest consumer of plastics after packaging sector. In 2010, the sector consumed 9.5 million tons of plastics in Europe, 21% of total European plastics consumption [16]. The construction industry uses them for a wide and growing range of applications including insulation, piping, window frames, and interior design. This growth is mainly due to plastics' unique features, which include durability and resistance to corrosion, better insulation property, cost efficiency, minimum operation and maintenance, hygienic, and sustainability [16]. Despite recent advances on closed loop industrial ecology concept, 49% of all the plastic wastes generated in Europe are still disposed of to landfill [11], a management alternative that generates serious environmental problems due to their low density, resistance to biological degradation and combustible nature. Consequently, extraction of finite natural resources is increasing to meet the excess demand of primary material manufacturing. From energy conservation point of view, disposal of plastic waste to landfill also mean substantial exergy losses as these materials contain significant embodied exergy. Therefore, recycling of post-consumer plastic wastes provides opportunities to reduce oil usage, carbon dioxide emissions and the quantities of waste requiring disposal [13] Reuse and recycling of plastic waste will therefore have a major implication on both efficient resource use and reduced environmental impact. In recent years, plastic recycling has become more popular in the European Union

(EU) due to regulatory limitations. In 2010, the overall recovery of building and construction plastic waste increased 4.3% compared to that in 2009, including 44.7% PVC and 5.1% PP recovery. Fig. 1 shows the building and plastic wastes recovery in European countries in 2010, from which we observe that the recycling of plastic waste, generated from building and construction sector, is still very low compared to energy recovery and landfill. It is also noticed that feedstock recycling within the European countries is negligible.

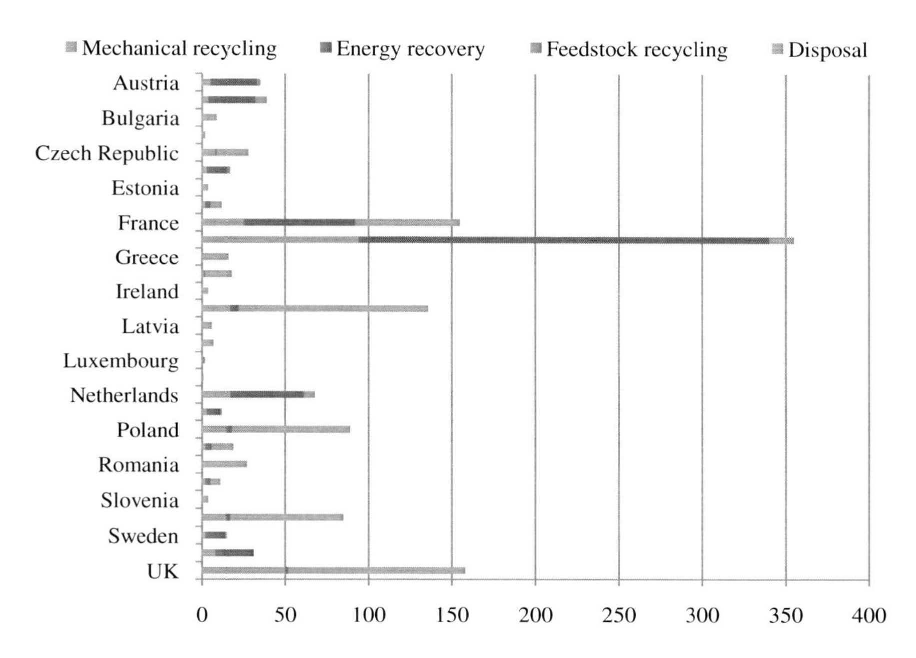

Fig. 1. Recovery and disposal of construction plastic waste in in Europe in 2010, all numbers are in kilotons (kt) (Source: [9])

The Waste Framework Directive, 2008/98/EC, aims to protect human health and the environment against harmful effects caused by the collection, transport, treatment, storage and landfilling of waste. The Directive sets new recycling targets to be achieved by the EU Member States by 2020, including 70% recycling rates for construction and demolition waste. The Directive will therefore have an influence on the disposal of plastics used in the construction sector.

In this study, we have applied ELCA to polypropylene (PP) and polyvinyl chloride (PVC), two widely used plastic materials in construction applications, as a case study in a cradle-to-grave life cycle approach. The ELCA is applied not as a substitute for LCA but rather an improvement options for potential exergy efficiency improvements of systems. The purpose of this analysis is to give the reader an understanding how efficiently resources are used during the whole life cycle of PP and PVC. The results from this analysis will provide deeper insights of resource use inefficiencies for the selected materials and will thus offer opportunities to reduce the exergy loss by improving process efficiency.

2 Materials and Methods

The exergy efficiency, e_2, of a system is the relation between the process products' exergy and the input exergy. It can be calculated as the useful product output divided by the total exergy input of a process, as expressed in equation 1 [20].

$$e_2 = \frac{Ex_{\text{product/s}}}{Ex_{\text{materials}} + Ex_{\text{utilities}}} \quad (1)$$

where, $Ex_{\text{materials}}$ and $Ex_{\text{utilities}}$ are the exergy of input materials and utilities, respectively; utilities include water, steam, electricity, and fuels required for the process.

Exergy efficiency calculation requires detailed and disaggregated analysis of material and energy input to each process step of a production chain. Depending on simple chain or branched chain production process we can calculate the exergy of the whole process as explained by Ayres et al [4]. Exergetic efficiency of resource (crude oil and natural gas) extraction, resin manufacturing process, and mechanical recycling of PP and PVC waste have been calculated using equation 1. This study does not include indirect exergy consumptions associated with the materials and utilities necessary for manufacturing of fixed capitals, such as machineries and production facilities. From life cycle perspective, it would generally be favorable to increase the amount of recycled plastics entering new life cycles. Among the three different recycling or recovery processes (mechanical, feedstock, and energy recovery), mechanical recycling (which loops the material back directly into new life cycles) is the European plastic industry's preferred recycling techniques [10].To certain extent, this technique substitutes the processes of resource extraction, intermediate production, and polymerization during the production of virgin material. Thus, this study focuses on mechanical recycling to account the exergy efficiency of recycling process. Exergetic analysis requires quality data on chemical exergy of materials and utilities involved in the process, and a proper mass balance of the process. We used published literature data to account material and energy balance for unit processes of the case study materials, data sources are illustrated in table 1.

Table 1. Data sources related to the ELCA of PP and PVC

Parameter	Data source
Extraction and processing of crude petroleum and natural gas	[12]
Transportation of petroleum and natural gas through pipeline	[9,12]
Chemical exergy of materials and utilities involved in the production of PP and PVC	[3,20]
PP and PVC resin manufacturing process	[3,14,17]
Recycling of PP and PVC	[1,7,13,18]

3 Case Studies

Consumption of plastic materials is becoming popular mainly because of their stability and long life, 10-50 years of life time depending on the types of use [16]. In addition, these materials require minimum care and maintenance during the use phase. Plastic manufacturing processes consist of various industrial segments and are related in a complex manner. For example, chlorine, a major raw material for PVC, is manufactured by the soda industry and then converted to PVC by the plastics industry via vinyl chloride monomer (VCM) production. Energy resources, raw materials, and manufacturing processes used in plastic manufacture vary from country by country depending on the availability of feedstock materials and energy sources. Plastic resin production from virgin resources is analyzed in this study in order to assess the overall life cycle exergy efficiency.

3.1 Polypropylene

Polypropylene has become one of the most versatile bulk polymers due to its good mechanical and chemical properties. Global consumption of PP in 2010 is estimated 48.4 million tons with global capacity utilization 82%. In recent years, PP consumption in the construction sector is increasing especially for piping and fittings. Other uses include window profile, wall covering, and filler fibers in concrete production. It is produced in a low-pressure process, polymerized from propylene. Ethylene and propylene are co-produced by cracking naphtha or gas oil in a steam-cracker [14]. For PP production, many different polymerization processes exist, such as solution polymerization, bulk polymerization in liquid propylene, and several gas-phase processes. Exergy inputs and outputs in different steps of the production chain to produce 1 kg virgin PP is illustrated in table 2. Assuming a branched chain production process [3], we estimated the exergy efficiency of PP production process, is 53.4%.

Table 2. Exergy inputs and outputs for the production of 1 kg PP in present industrial practices

Process step	Input exergy (GJ)	Output exergy (GJ)	Exergy efficiency (%)
Naphtha production	81.16	61.40	75.7
Ethylene production	13.41	8.74	65.1
Propylene production	56.68	45.19	79.7
Propylene polymerization	50.60	46.31	91.5

One of the major problems of recycling plastic waste is greater inhomogeneity of the polymers present in the waste. However, it is noticed that recycling systematically generates higher output compared to primary production process. It is assumed that 1 kg of PP recycling require 13.44 MJ of utility (including transport and processing) [7], representing exergy efficiency of 70.4% (Fig. 2).

3.2 Polyvinyl Chloride

Pure PVC is hard, brittle material which degrades at around 100°C and is sensitive to deterioration under the influence of light and heat. Pure PVC is therefore supplemented with additives which improve its service life properties and allow it to be processed. Use of pure PVC in building materials depends on the product characteristics and varies 35-98% [16]. Nearly 57% of the total PVC consumption goes to the construction industry. Global demand of PVC is expected to an average growth of around 4.7% per year from 2010 to 2015, and 4.2% from 2015 to 2020 [16]. PVC production is more complex rather than that for PP, because the processes involve multiple reactions such as naphtha to ethylene, ethylene to ethylene dichloride (EDC) followed by vinyl chloride monomer (VCM) production, and PVC from VCM polymerization. EDC is one of the precursors for VCM in direct chlorination process. It is assumed that HC1 generated from thermal decomposition of EDC is used as the raw material for the oxy-chlorination process. Exergy inputs and outputs in different process steps for the production of 1 kg virgin PVC is illustrated in table 3, from which the overall exergy efficiency is accounted for 23.7%.

Table 3. Exergy inputs and outputs for the production of 1 kg PVC in present industrial practices

Process step	Input exergy (GJ)	Output exergy (GJ)	Exergy efficiency (%)
Naphtha production	83.76	63.42	75.7
Ethylene production	77.21	50.28	65.1
Ammonia (NH_3) production	0.24	0.16	67.0
Chlorine (Cl_2) production	5.33	2.30	43.1
VCM production	36.74	21.99	59.8
Vinyl chloride polymerization	24.38	19.66	80.6

Due to the high chlorine content, some of the recycling techniques are not favorable for PVC. In particular, landfilling and composting are not suitable because of known and unknown hazards associated with the oxidative degradation of PVC in the environment. Incineration and pyrolysis may also be disfavored because of the production of large amounts of hydrogen chloride and other toxic chemicals. Thus, mechanical and chemical recycling are the two main processes of PVC recycling, the former is preferred when the provenance of PVC waste is known. In this analysis, we calculated the exergy efficiency of mechanical PVC recycling, of 54.2%, assuming an average yield of 90% PVC [19]

4 Results and Discussions

The study shows that within the PP and PVC chains, the resin production plays a major role in overall resource use efficiency. Exergy consumptions associated with manufacturing processes are relatively high and major inefficiency takes place in the production stages for both PP and PVC. We used 1MJ of electrical energy equals 1MJ of exergy in this analysis [20]. However, if we consider the efficiency of electricity production (assuming 31.8% efficient [8]), the manufacturing or recycling efficiency would be even lower. For example, exergy efficiency of PVC and PP production becomes 20.0% and 50.1%, respectively, considering the electricity production efficiency. Exergetic efficiency of different life-cycle stages of PP and PVC is illustrated in Fig. 2, representing the overall life cycle efficiency of PVC and PP, 9.3% and 27.1%, respectively. Thus, PP has better exergetic efficiency for both production and recycling processes. On the other hand, the gross CO_2 emission from the manufacture of plastics is 1.4 and 1.7 kg-CO_2/kg-PP and PVC, respectively [12,19]. Improving process efficiency will thus reduce the cumulative exergy consumption throughout the material life cycle and associated carbon emissions. This analysis is not intended to make specific proposals for improving a products' life cycle efficiency but to assess how inefficiently today's systems perform, and to evaluate the losses and exergy consumptions throughout the life-cycle stages of the case study materials. Improvement opportunities can be assessed comparing the efficiency outcomes of the processes analyzed with the theoretical exergy efficiency. As an example, theoretical exergy efficiency of VCM production from ethylene, which is an exothermic reaction (equation 2), is 91.9%; whereas, in present industrial practice the said efficiency is 59.8% (table 3).

$$CH_2{=}CH_2 + Cl_2 \rightarrow CH_2{=}CHCl + HCl + 16.4\text{kJ} \quad (2)$$

Therefore, it is logical to conclude that significant exergy improvement opportunities still exist but requires process optimizations.

Mechanical recycling is being used for 35.5% of the total plastics waste recovered in the EU27+2 in 2010 [10]. This technique directly recovers clean plastics for reuse in the manufacturing of new plastic products. The difficulties of plastic recycling are

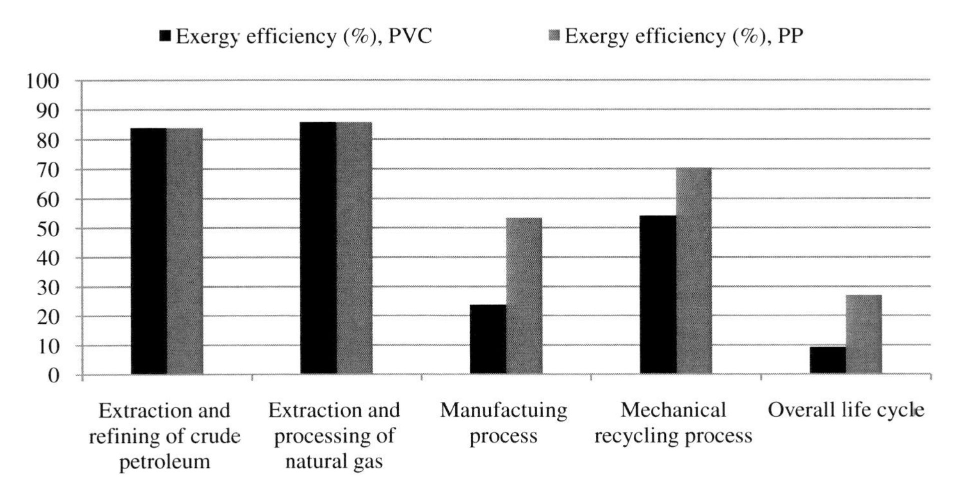

Fig. 2. Exergy efficiency of overall life cycle and different life-cycle stages of PP and PVC

mainly related to the degradation of recyclable material and heterogeneity of plastic wastes [15]. Mechanical recycling can only be performed on single-polymer plastic. The more contaminated the waste, the more difficult it is to recycle it mechanically. Separation, washing, and preparation of plastic waste are all essential to produce high quality, clean, and homogenous end-products [1]. Both mono and composite fractions can be collected separately from the waste stream by sorting. For mixed and composite plastic wastes, a suitable feedstock recycling process can be developed. By definition, energy recovery implies burning waste to produce energy in the form of heat, steam, and electricity. This is only considered a sensible way of waste treatment, when material recovery processes fail due to economical constrains.

5 Conclusions

Increased efficiency preserves exergy by reducing the exergy necessary for a process, and therefore reduces environmental damages. The results obtained from this analysis show that the exergy efficiency of manufacturing and recycling processes are low compared to raw material extraction and processing for both PP and PVC. Comparing the results of this analysis with the theoretical minimum exergy required in different life-cycle stages will offer the improvement potentials of the processes. The material choice can be discussed depending on the ELCA results as it is a straight-forward parameter that can easily be communicated even though within ELCA, material choice is simply one parameter among many inter-dependent parameters. In addition, the results of this analysis can also be compared with other quantitative tools covering the technological aspects such as social, financial, and environmental parameters to strengthen sustainable resource use.

References

1. Al-Salem, S.M., Lettieri, P., Baeyens, J.: Recycling and recovery routes of plastic solid waste (PSW): A review. Waste Manage 29, 2625–2643 (2009)
2. Amini, S.H., Remmerswaal, J.A.M., Castro, M.B., Reuter, M.A.: Quantifying the quality loss and resource efficiency of recycling by means of exergy analysis. Journal of Cleaner Production 15, 907–913 (2007)
3. Ayres, R.U., Ayres, L.W.: Accounting for resources 2: The life cycle of materials. Edward Elgar, Cheltenham UK and Northampton MA (1999)
4. Ayres, R.U., Subramanian, K., Werner, A.: Energy efficiency in the chemical industry. In: Proceedings of third international conference on energy use management, Berlin, vol. 4, pp. 181–191 (1981)
5. Cornelissen, R.L.: Thermodynamics and sustainable development: The use of exergy analysis and the reduction of irreversibility. Dissertation, Universiteit Twente (1997)
6. De Meester, B., Dewulf, J., Verbeke, S., Janssens, A., Van Langenhove, H.: Exergetic life-cycle assessment (ELCA) for resource consumption evaluation in the built environment. Build. Environ. 44, 11–17 (2009)
7. Dewulf, J., Van Langenhove, H.: Thermodynamic optimization of the life cycle of plastics by exergy analysis. Int. J. Energy Res. 28, 969–976 (2004)

8. Dincer, I.: The role of exergy in energy policy making. Energy Policy 30, 137–149 (2002)
9. Dincer, I., Rosen, M.A.: Exergetic life cycle assessment. In: Exergy: Energy, Environment and Sustainable Development, pp. 397–416. Elsevier, New York (2007)
10. ECVM (European Council of Vinyl Manufacturers): Plastic waste from building and construction, http://www.plasticseurope.org/documents/document/20120319104114-final_full_report_eol_plastics_b&c_2010_augmented_16032012.pdf (accessed May 25, 2012)
11. EuPR (European Plastics Recyclers): Plastics - the Facts 2010: An analysis of European plastics production, demand and recovery (2009), http://www.plasticsrecyclers.eu/uploads/media/20101028135906-final_plasticsthefacts_26102010_lr.pdf (accessed August 28, 2011)
12. Franklin Associates: Cradle-to-gate life cycle inventory of nine plastic resins and four polyurethane precursors. The plastics division of the American Chemistry Council, Washington, DC (2010)
13. Hopewell, J., Dvorak, R., Kosior, E.: Plastics recycling: challenges and opportunities. Phil. Trans. R. Soc. B (364), 2115–2126 (2009)
14. Narita, N., Sagisaka, M., Inaba, A.: Life cycle inventory analysis of CO_2 emissions manufacturing commodity plastics in Japan. Int. J. LCA. 7, 277–282 (2002)
15. Perugini, F., Mastellone, M.L., Arena, U.: A life cycle assessment of mechanical and feedstock recycling options for management of plastic packaging wastes. Environ. Prog. 24, 137–154 (2005)
16. PlasticsEurope: The plastic portal, use of plastics: Building and construction, http://www.plasticseurope.org/use-of-plastics/building-construction.aspx (accessed September 24, 2011)
17. Ren, T., Patel, M., Blok, K.: Energy efficiency and innovative emerging technologies for olefin production. In: European Conference on Energy Efficiency in IPPC Installations, Vienna (2004)
18. Sadat-Shojai, M., Bakhshandeh, G.-R.: Recycling of PVC wastes. Polym. Degrad. Stab. 96, 404–415 (2011)
19. Shonfield, P.: LCA of management options for mixed waste plastic. Waste resource action programme WRAP. London (2008)
20. Szargut, J., Morris, D.R., Steward, F.R.: Exergy analysis of thermal, chemical, and metallurgical processes. Hemisphere, New York (1988)
21. Talens Peiró, L., Lombardi, L., Villalba Méndez, G., Gabarrell Durany, X.: Life cycle assessment (LCA) and exergetic life cycle assessment (ELCA) of the production of biodiesel from used cooking oil (UCO). Energy 35, 889–893 (2010)

Chapter 30
Methodology for the Preparation of the Standard Model for Schools Investigator for the Sustainability of Energy Systems and Building Services

Hisham Elshimy

Architectural Engineering Department
Pharos University
Alexandria, Egypt
hisham_elshimy@pua.edu.eg

Abstract. Sustainable architecture is a major goal for all the bodies and institutions interested in architecture in various directions (planning - design - construction); especially with the lack of environmental resources required to ensure building of environmentally compatible with the built environment and the natural environment. Diverse concepts of sustainability and green architecture in the property sector: Sustainable design, Green Architecture, Sustainable Construction and Green Building are all new methods of design and construction, evoking environmental and economic challenges. This has cast a shadow over the various sectors in this era. The premises are new, design, implementation and operation of sophisticated methods and techniques contributing to reducing environmental impact, and at the same time minimizing costs, specifically operating and maintenance costs (Running Costs), as these premises contribute to a safe and comfortable physical environment. Thus, motivating the adoption of the concept of Sustainability in the property sector is not different from the motivation that led to the emergence and adoption of the concepts of Sustainable Development, Dimensions of Environmental, and Economic and Social entity.

Keywords: Building Economic, School, Hot and Humid Coastline, Green Architect, Sustainability.

1 Introduction

Educational buildings are considered one of the most important applications that can adapt to environmental architecture. These buildings meet the functional needs of the users; whether they are students, teachers or administrators, through the achievement of design requirements such as lighting, ventilation, movement and use of spaces.

A. Håkansson et al. (Eds.): *Sustainability in Energy and Buildings*, SIST 22, pp. 323–335.
DOI: 10.1007/978-3-642-36645-1_30

2 The Main Goals of Research

The goals depends on several points which form the methodology in order to obtain those goals, those points are:

1- The possibility of study design and analysis of determinants of schools in Egypt documented from the buildings to get to the educational system is a standard school design in order to achieve sustainability of school buildings
2- Identify elements in non-listed buildings to the list of education and that affect the sustainability of the design
3- Develop a new system compatible and harmonized and combine the list of buildings of educational and LEED system
4- Determine the design standard forms for schools comply with the environmental conditions and structural and geographical distribution of schools in the region hot and humid coastal.

3 The General Methodology of the Research

It depends on the merging of the design methodology certified from the Educational Constructions Association (the design guide for schools – prime education) and LEED V3 methodology into one organization; through applying the general essential conditions of designing schools, with adding two main items:

-Energy and atmosphere
-Water efficiency

This is in the organization of design with working on raising the competency of the other items and on specifying a structure for the understanding of the organization.

Through solidifying the LEED V3 methodology, we find that in order to accomplish the permanence, it is demanded to prove the compatibility amongst the requirements of the permanence through the Environmental, Social and Economical axis.

Through a general review of the methodology for Educational Buildings in Egypt, writing it according to LEED methodology, we find the following:

1- Sustainable Site: Standardization of the criteria of selecting School sites in order to achieve sustainability of the site. [1]
2- Water Efficiency: Does not exist since there is no wastewater treatment methodology.
3- Energy and Atmosphere: Does not exist since there is no methodology for raising energy[2].
4- Material Resources: Does not exist since the materials used cannot be recycled.
5- Indoor Environmental Quality: There are some criterions for internal processing.
6- Innovation and Design Process: There are some criterions for design.
7- Regional Priority Credit: There are criterions.

Through identifying the criterions for choosing schools, which will be used in the research study analysis, we find that the median percentage for School projects according to LEEDS is 53%, Silver Grade. Therefore, one of the objectives of the research is to raise the efficiency of the medium of the business to reach the Gold grade (60%.79%).[3]

This is done through a standard methodology for designing a standard form for the schools located in the hot and humid costal region. This raises the efficiency of the designed schools to the Gold Grade.

4 Principles of Sustainable Design [4]

Economy of Resources is economizing resources where the architect reduces the use of non renewable resources in the construction and operation of buildings. There is a continuous flow of resources, natural and manufactured, in and out of a building. This flow begins with the production of building materials and continues throughout the building's life span to create an environment for sustaining human well .being and activities. After a building's useful life, it should turn into components for other buildings.

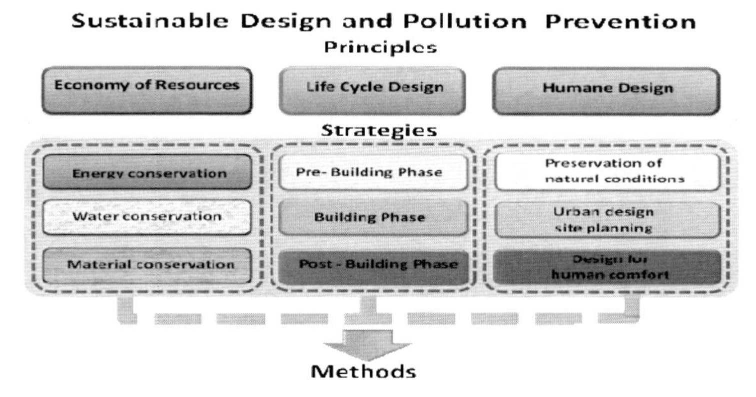

Fig. 1. Conceptual Framework for Sustainable Design and Pollution Prevention in Architecture

4.1 Economy of Resources

Conserving energy, water, and materials can yield specific design methods that will improve the sustainability of architecture. These methods can be classified as two types

A. Input-Reduction Methods: Reduce the flow of non renewable resources input to buildings. A building's resource demands are directly related its efficiency in utilizing resources. [5]

B. Output-Management Methods: Reduce environmental pollution by requiring a low level of waste and proper waste management. [6]

Energy Conservation:

Energy Conservation is an input-reduction method and depends on some points:

• Water Conservation • Materials Conservation • Life Cycle Design [7]

5 The Process of Sustainable Design of Educational Buildings in Egypt[8]

Based methodology of Sustainable Building design education in Egypt on a group of the main axes including:

1. How to choose the school site to achieve the requirements of environmental sustainability in terms of lighting, ventilation and proximity of services and facilities [9]
2. How to distribute the components of the educational building (chapters . administration . library . hobbies . gymnasiums) in harmony with the site conditions in order to achieve calm manner, natural ventilation and natural lighting [10]
3. How to set up the building in terms of choice of construction materials used, working to facilitate the use of environmental recycling and methods of environmental manipulation (temperature . humidity . noise) so that the building fits the environmental site conditions and functional requirements. This is the methodology in terms of consecutive methodology configuration and overlapping in terms of structure through the availability of common elements of a link between the three components of the methodology [11]

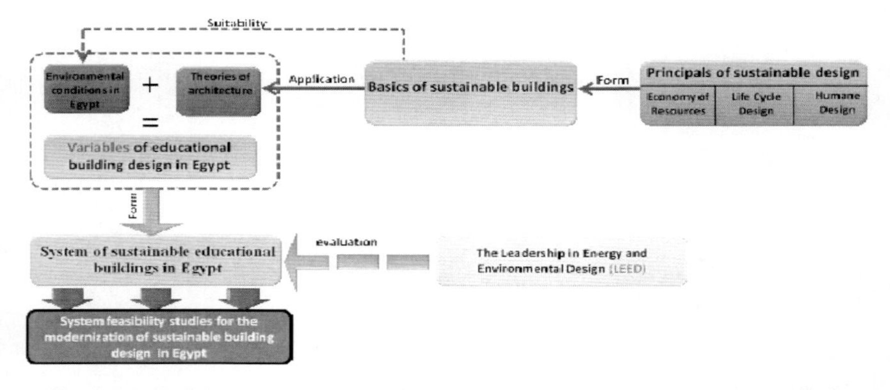

Fig. 2. Methodology of Sustainable Design of Educational Buildings in Egypt[12]

6 Evidence Used for the Study and Analysis[13]

Based research study on the list of educational buildings in Egypt as an essential reference for the mechanism design schools through a range of determinants of design and developed to reach the best conditions are achieved comfort in the arrival and be compatible with the function of schools, but it's formula and designed by did not take into account the foundations of sustainability and the determinants of energy-saving components of the system for LEED

A- The features in the Selection of Educational Building Site [14]
B- The Characteristics of the Natural and Cultural Environment at the Educational Building Site Selection [15]

7 Measure the Efficiency of Sustainable Buildings [16]

There are different methodologies to measure the Sustainable Building but the Leadership in Energy and Environmental Design (**LEED**) methodology is a distinctive one [17]

Different LEED versions have varied scoring methodologies based on a set of required "prerequisites" and a variety of "credits" in the six major categories listed above. **USGBC LEED 2009 (v3)** [8]:**Certified** . 40.49 points **Silver** . 50.59 points **Gold** . 60.79 points **Platinum** . 80 points and above [18]:

8 Determinants of Projects' Selection for the Study Applying the Methodology Proposed for Educational Projects Located in the Province of Applied Hot and Humid Coastal[19]

Case study examples were selected for a public and private schools on the nomination of the distinguished body of educational buildings in Alexandria. They are new, created in accordance with the requirements of the Authority and located within the territory of one climate, Alexandria. The distribution of these schools varies within the city of Alexandria as follows:

Table 1. Comparison between LEED V3 2009 and Requirements for the Design of Buildings, Educational Projects Located in the Province of Applied Hot and Humid Coastal

LEED V3 2009	Points	Requirements for the design of buildings, educational projects located in the province of applied hot and humid coastal
First : Sustainable Sites	24	The foundations of the site selection First: Topography Second : Physical Characteristics Third : Soil Fourth : Protection from pollution
Second : Water Efficiency . special item.	11	No item equivalent
Third :Energy and Atmosphere . special items.	33	No item equivalent
Fourth : Materials and Resources	13	Founded by the outer membrane of building and finishing material
Fifth : Indoor Environmental Quality	19	Founded by the outer membrane of building and finishing material
Sixth : Innovation and Design Process	6	Absorptive capacity of the site
Seventh : Regional Priority Credits	4	All items: -The foundations of the site selection. -Founded by the outer membrane of building and finishing material. -Absorptive capacity of the site.

Zahran School is located in Smouha district, nearby the airport area. El Awayed School is located in El Awayed area and Corona School located in Backus area. The area will vary from large area such as Zahran School and a small one such as Corona School.

It also varies between smoothed urban, rural urban and country level. There is also an economic and social diversity.

8.1 Evaluation of El Awayed School by LEED V3 2009 [20]

Awayed School is considered one of the distinctive schools' models. It was nominated for the School Buildings' award in Alexandria. This is because it is built in accordance with the requirements of the Educational Buildings' Authority, which takes into account the Environmental Design requirements of the hot and humid coastal region. This means that it passed the first stage of the methodology of Sustainable Design, meaning it has been approved but has not been evaluated according to the requirements of LEED.

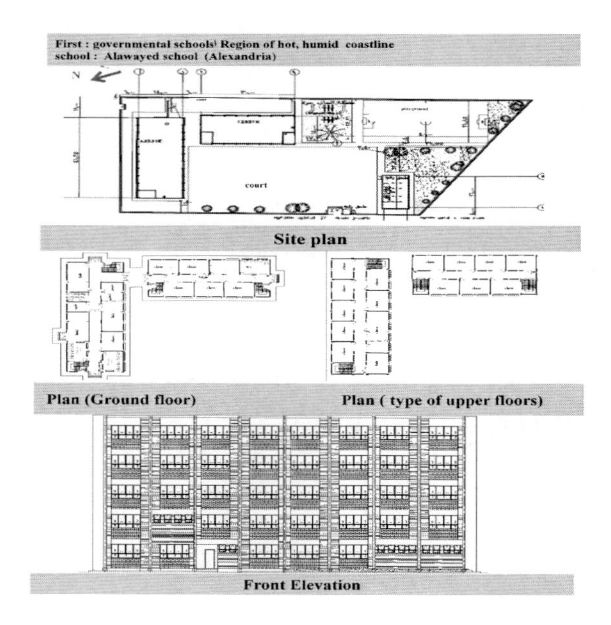

Fig. 3. Drawings of Al Awayed School according to LEED V3 2009

Table 2. Evaluation of El Awayed School according to LEED V3 2009

LEED 2009 for Schools New Construction and Major Renovation
Project Checklist

Project Name
28/3/2010

Y	N	?			Possible Points
		11	**Sustainable Sites**		**Possible Points: 24**
Y			Prereq 1	Construction Activity Pollution Prevention	
			Prereq 2	Environmental Site Assessment	
			Credit 1	Site Selection	1
		1	Credit 2	Development Density and Community Connectivity	4
		1	Credit 3	Brownfield Redevelopment	1
		2	Credit 4.1	Alternative Transportation—Public Transportation Access	4
			Credit 4.2	Alternative Transportation—Bicycle Storage and Changing Rooms	1
		1	Credit 4.3	Alternative Transportation—Low-Emitting and Fuel-Efficient Vehicles	2
		1	Credit 4.4	Alternative Transportation—Parking Capacity	2
		1	Credit 5.1	Site Development—Protect or Restore Habitat	1
		1	Credit 5.2	Site Development—Maximize Open Space	1
			Credit 6.1	Stormwater Design—Quantity Control	1
			Credit 6.2	Stormwater Design—Quality Control	1
			Credit 7.1	Heat Island Effect—Non-roof	1
		1	Credit 7.2	Heat Island Effect—Roof	1
		1	Credit 8	Light Pollution Reduction	1
			Credit 9	Site Master Plan	1
		1	Credit 10	Joint Use of Facilities	1
		5	**Water Efficiency**		**Possible Points: 11**
Y			Prereq 1	Water Use Reduction—20% Reduction	
		2	Credit 1	Water Efficient Landscaping	2 to 4
			Credit 2	Innovative Wastewater Technologies	2
		2	Credit 3	Water Use Reduction	2 to 4
		1	Credit 4	Process Water Use Reduction	1
		11	**Energy and Atmosphere**		**Possible Points: 33**
Y			Prereq 1	Fundamental Commissioning of Building Energy Systems	
Y			Prereq 2	Minimum Energy Performance	
Y			Prereq 3	Fundamental Refrigerant Management	
		10	Credit 1	Optimize Energy Performance	1 to 19
			Credit 2	On-Site Renewable Energy	1 to 7
		1	Credit 3	Enhanced Commissioning	2
			Credit 4	Enhanced Refrigerant Management	1
			Credit 5	Measurement and Verification	2
			Credit 6	Green Power	2
		5	**Materials and Resources**		**Possible Points: 13**
Y			Prereq 1	Storage and Collection of Recyclables	
		1	Credit 1.1	Building Reuse—Maintain Existing Walls, Floors, and Roof	1 to 2
		1	Credit 1.2	Building Reuse—Maintain 50% of Interior Non-Structural Elements	1
			Credit 2	Construction Waste Management	1 to 2
			Materials and Resources, Continued		
			Credit 3	Materials Reuse	1 to 2
			Credit 4	Recycled Content	1 to 2
		1	Credit 5	Regional Materials	1 to 2
		1	Credit 6	Rapidly Renewable Materials	1
		1	Credit 7	Certified Wood	1
		7	**Indoor Environmental Quality**		**Possible Points: 19**
Y			Prereq 1	Minimum Indoor Air Quality Performance	
Y			Prereq 2	Environmental Tobacco Smoke (ETS) Control	
Y			Prereq 3	Minimum Acoustical Performance	
			Credit 1	Outdoor Air Delivery Monitoring	1
			Credit 2	Increased Ventilation	1
			Credit 3.1	Construction IAQ Management Plan—During Construction	1
			Credit 3.2	Construction IAQ Management Plan—Before Occupancy	1
		2	Credit 4	Low-Emitting Materials	1 to 4
			Credit 5	Indoor Chemical and Pollutant Source Control	1
			Credit 6.1	Controllability of Systems—Lighting	1
			Credit 6.2	Controllability of Systems—Thermal Comfort	1
			Credit 7.1	Thermal Comfort—Design	1
			Credit 7.2	Thermal Comfort—Verification	1
		2	Credit 8.1	Daylight and Views—Daylight	1 to 3
		1	Credit 8.2	Daylight and Views—Views	1
		1	Credit 9	Enhanced Acoustical Performance	1
		1	Credit 10	Mold Prevention	1
		2	**Innovation and Design Process**		**Possible Points: 6**
		1	Credit 1.1	Innovation in Design: Specific Title	1
			Credit 1.2	Innovation in Design: Specific Title	1
			Credit 1.3	Innovation in Design: Specific Title	1
			Credit 1.4	Innovation in Design: Specific Title	1
			Credit 2	LEED Accredited Professional	1
		1	Credit 3	The School as a Teaching Tool	1
		1	**Regional Priority Credits**		**Possible Points: 4**
		1	Credit 1.1	Regional Priority: Specific Credit	1
			Credit 1.2	Regional Priority: Specific Credit	1
			Credit 1.3	Regional Priority: Specific Credit	1
			Credit 1.4	Regional Priority: Specific Credit	1
		42	**Total**		**Possible Points: 110**

Certified 40 to 49 points Silver 50 to 59 points Gold 60 to 79 points Platinum 80 to 110

8.2 Evaluation of Corona School by LEED V3 2009[20]

Corona School is considered one of the distinctive schools' models. It was nominated for the School Buildings' award in Alexandria. This is because it is built in accordance with the requirements of the Educational Buildings' Authority, which takes into account the Environmental Design requirements of the hot and humid coastal region. This means that it passed the first stage of the methodology of Sustainable Design, meaning it has been approved but has not been evaluated according to the requirements of LEED(second stage)

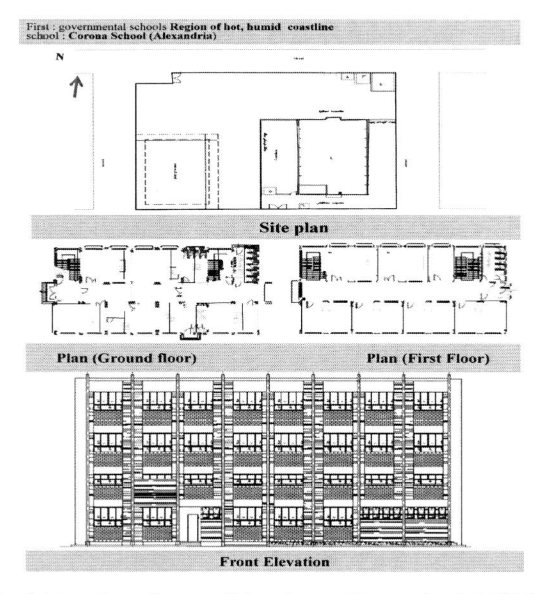

Fig. 4. Drawings Corona School according to LEED V3 2009

Table 3. Evaluation of Corona School according to LEED V3 2009

LEED 2009 for Schools New Construction and Major Renovation
Project Checklist

Project Name
25/3/2010

Y	N	?	Credit	Item	Points
		13		**Sustainable Sites**	Possible Points: 24
Y			Prereq 1	Construction Activity Pollution Prevention	
			Prereq 2	Environmental Site Assessment	
		1	Credit 1	Site Selection	1
		2	Credit 2	Development Density and Community Connectivity	4
		1	Credit 3	Brownfield Redevelopment	1
		2	Credit 4.1	Alternative Transportation—Public Transportation Access	4
			Credit 4.2	Alternative Transportation—Bicycle Storage and Changing Rooms	1
		1	Credit 4.3	Alternative Transportation—Low-Emitting and Fuel-Efficient Vehicles	2
		1	Credit 4.4	Alternative Transportation—Parking Capacity	2
		1	Credit 5.1	Site Development—Protect or Restore Habitat	1
		1	Credit 5.2	Site Development—Maximize Open Space	1
			Credit 6.1	Stormwater Design—Quantity Control	1
			Credit 6.2	Stormwater Design—Quality Control	1
			Credit 7.1	Heat Island Effect—Non-roof	1
		1	Credit 7.2	Heat Island Effect—Roof	1
		1	Credit 8	Light Pollution Reduction	1
			Credit 9	Site Master Plan	1
		1	Credit 10	Joint Use of Facilities	1
		6		**Water Efficiency**	Possible Points: 11
Y			Prereq 1	Water Use Reduction—20% Reduction	
		3	Credit 1	Water Efficient Landscaping	2 to 4
			Credit 2	Innovative Wastewater Technologies	2
		2	Credit 3	Water Use Reduction	2 to 4
		1	Credit 3	Process Water Use Reduction	1
		12		**Energy and Atmosphere**	Possible Points: 33
Y			Prereq 1	Fundamental Commissioning of Building Energy Systems	
Y			Prereq 2	Minimum Energy Performance	
Y			Prereq 3	Fundamental Refrigerant Management	
		11	Credit 1	Optimize Energy Performance	1 to 19
			Credit 2	On-Site Renewable Energy	1 to 7
		1	Credit 3	Enhanced Commissioning	2
			Credit 4	Enhanced Refrigerant Management	1
			Credit 5	Measurement and Verification	2
			Credit 6	Green Power	2
		5		**Materials and Resources**	Possible Points: 13
Y			Prereq 1	Storage and Collection of Recyclables	
		1	Credit 1.1	Building Reuse—Maintain Existing Walls, Floors, and Roof	1 to 2
		1	Credit 1.2	Building Reuse—Maintain 50% of Interior Non-Structural Elements	1
			Credit 2	Construction Waste Management	1 to 2
				Materials and Resources, Continued	
			Credit 3	Materials Reuse	1 to 2
			Credit 4	Recycled Content	1 to 2
		1	Credit 5	Regional Materials	1 to 2
		1	Credit 6	Rapidly Renewable Materials	1
		1	Credit 7	Certified Wood	1
		6		**Indoor Environmental Quality**	Possible Points: 19
Y			Prereq 1	Minimum Indoor Air Quality Performance	
Y			Prereq 2	Environmental Tobacco Smoke (ETS) Control	
Y			Prereq 3	Minimum Acoustical Performance	
			Credit 1	Outdoor Air Delivery Monitoring	1
			Credit 2	Increased Ventilation	1
			Credit 3.1	Construction IAQ Management Plan—During Construction	1
			Credit 3.2	Construction IAQ Management Plan—Before Occupancy	1
		2	Credit 4	Low-Emitting Materials	1 to 4
			Credit 5	Indoor Chemical and Pollutant Source Control	1
			Credit 6.1	Controllability of Systems—Lighting	1
			Credit 6.2	Controllability of Systems—Thermal Comfort	1
			Credit 7.1	Thermal Comfort—Design	1
			Credit 7.2	Thermal Comfort—Verification	1
		1	Credit 8.1	Daylight and Views—Daylight	1 to 3
		1	Credit 8.2	Daylight and Views—Views	1
		1	Credit 9	Enhanced Acoustical Performance	1
		1	Credit 10	Mold Prevention	1
		3		**Innovation and Design Process**	Possible Points: 6
		1	Credit 1.1	Innovation in Design: Specific Title	1
		1	Credit 1.2	Innovation in Design: Specific Title	1
			Credit 1.3	Innovation in Design: Specific Title	1
			Credit 1.4	Innovation in Design: Specific Title	1
			Credit 2	LEED Accredited Professional	1
		1	Credit 3	The School as a Teaching Tool	1
		1		**Regional Priority Credits**	Possible Points: 4
		1	Credit 1.1	Regional Priority: Specific Credit	1
			Credit 1.2	Regional Priority: Specific Credit	1
			Credit 1.3	Regional Priority: Specific Credit	1
			Credit 1.4	Regional Priority: Specific Credit	1
		46		**Total**	Possible Points: 110

8.3 Evaluation of Zahran School by LEED V3 2009 [20]

Zahran School is considered one of the distinctive schools' models. It was nominated for the School Buildings' award in Alexandria. This is because it is built in accordance with the requirements of the Educational Buildings Authority, which takes into account the Environmental Design requirements of the hot and humid coastal region. This means that it passed the first stage of the methodology of Sustainable Design, meaning it has been approved but has not been evaluated according to the requirements of LEED (second stage)

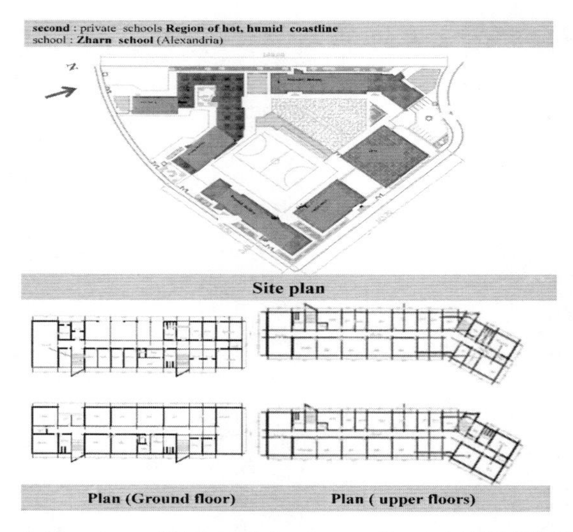

Fig. 5. Drawings of Zahran School according to LEED V3 2009

Table 4. Evaluating Zahran School according to LEED V3 2009

LEED 2009 for Schools New Construction and Major Renovation
Project Checklist

Project Name
25/3/2010

Y	?	N		Item	Possible Points
		19		**Sustainable Sites**	**Possible Points: 24**
Y			Prereq 1	Construction Activity Pollution Prevention	
			Prereq 2	Environmental Site Assessment	
		1	Credit 1	Site Selection	1
		3	Credit 2	Development Density and Community Connectivity	4
			Credit 3	Brownfield Redevelopment	1
		3	Credit 4.1	Alternative Transportation—Public Transportation Access	4
		1	Credit 4.2	Alternative Transportation—Bicycle Storage and Changing Rooms	1
		2	Credit 4.3	Alternative Transportation—Low-Emitting and Fuel-Efficient Vehicles	2
		2	Credit 4.4	Alternative Transportation—Parking Capacity	2
		1	Credit 5.1	Site Development—Protect or Restore Habitat	1
		1	Credit 5.2	Site Development—Maximize Open Space	1
			Credit 6.1	Stormwater Design—Quantity Control	1
			Credit 6.2	Stormwater Design—Quality Control	1
		1	Credit 7.1	Heat Island Effect—Non-roof	1
		1	Credit 7.2	Heat Island Effect—Roof	1
		1	Credit 8	Light Pollution Reduction	1
		1	Credit 9	Site Master Plan	1
		1	Credit 10	Joint Use of Facilities	1
		1		**Water Efficiency**	**Possible Points: 11**
Y			Prereq 1	Water Use Reduction—20% Reduction	
		1	Credit 1	Water Efficient Landscaping	2 to 4
			Credit 2	Innovative Wastewater Technologies	2
			Credit 3	Water Use Reduction	2 to 4
			Credit 3	Process Water Use Reduction	1
		18		**Energy and Atmosphere**	**Possible Points: 33**
Y			Prereq 1	Fundamental Commissioning of Building Energy Systems	
Y			Prereq 2	Minimum Energy Performance	
Y			Prereq 3	Fundamental Refrigerant Management	
		15	Credit 1	Optimize Energy Performance	1 to 19
		2	Credit 2	On-Site Renewable Energy	1 to 7
		1	Credit 3	Enhanced Commissioning	2
			Credit 4	Enhanced Refrigerant Management	1
			Credit 5	Measurement and Verification	2
			Credit 6	Green Power	2
		7		**Materials and Resources**	**Possible Points: 13**
Y			Prereq 1	Storage and Collection of Recyclables	
		2	Credit 1.1	Building Reuse—Maintain Existing Walls, Floors, and Roof	1 to 2
		1	Credit 1.2	Building Reuse—Maintain 50% of Interior Non-Structural Elements	1
		1	Credit 2	Construction Waste Management	1 to 2
				Materials and Resources, Continued	
		1	Credit 3	Materials Reuse	1 to 2
			Credit 4	Recycled Content	1 to 2
		1	Credit 5	Regional Materials	1 to 2
			Credit 6	Rapidly Renewable Materials	1
		1	Credit 7	Certified Wood	1
		14		**Indoor Environmental Quality**	**Possible Points: 19**
Y			Prereq 1	Minimum Indoor Air Quality Performance	
Y			Prereq 2	Environmental Tobacco Smoke (ETS) Control	
Y			Prereq 3	Minimum Acoustical Performance	
		1	Credit 1	Outdoor Air Delivery Monitoring	1
		1	Credit 2	Increased Ventilation	1
		1	Credit 3.1	Construction IAQ Management Plan—During Construction	1
		1	Credit 3.2	Construction IAQ Management Plan—Before Occupancy	1
		1	Credit 4	Low-Emitting Materials	1 to 4
		1	Credit 5	Indoor Chemical and Pollutant Source Control	1
		1	Credit 6.1	Controllability of Systems—Lighting	1
		1	Credit 6.2	Controllability of Systems—Thermal Comfort	1
		1	Credit 7.1	Thermal Comfort—Design	1
		1	Credit 7.2	Thermal Comfort—Verification	1
		1	Credit 8.1	Daylight and Views—Daylight	1 to 3
		1	Credit 8.2	Daylight and Views—Views	1
		1	Credit 9	Enhanced Acoustical Performance	1
		1	Credit 10	Mold Prevention	1
		5		**Innovation and Design Process**	**Possible Points: 6**
		1	Credit 1.1	Innovation in Design: Specific Title	1
		1	Credit 1.2	Innovation in Design: Specific Title	1
		1	Credit 1.3	Innovation in Design: Specific Title	1
			Credit 1.4	Innovation in Design: Specific Title	1
		1	Credit 2	LEED Accredited Professional	1
		1	Credit 3	The School as a Teaching Tool	1
		3		**Regional Priority Credits**	**Possible Points: 4**
		1	Credit 1.1	Regional Priority: Specific Credit	1
		1	Credit 1.2	Regional Priority: Specific Credit	1
		1	Credit 1.3	Regional Priority: Specific Credit	1
			Credit 1.4	Regional Priority: Specific Credit	1
		67		**Total**	**Possible Points: 110**

Certified 40 to 49 points Silver 50 to 59 points Gold 60 to 79 points Platinum 80 to 110

8.4 Evaluation of House of English School by LEED V3 2009 [20]

House of English School is considered one distinctive school's models. It was nominated to the School Buildings' award in Alexandria. This is because of being built in accordance with the requirements of the Educational Buildings' Authority, which takes into account the Environmental Design requirements of the hot and humid coastal region. This means that it passed the first stage of the methodology of Sustainable Design, meaning it has been approved but has not been evaluated according to the requirements of LEED (second stage)

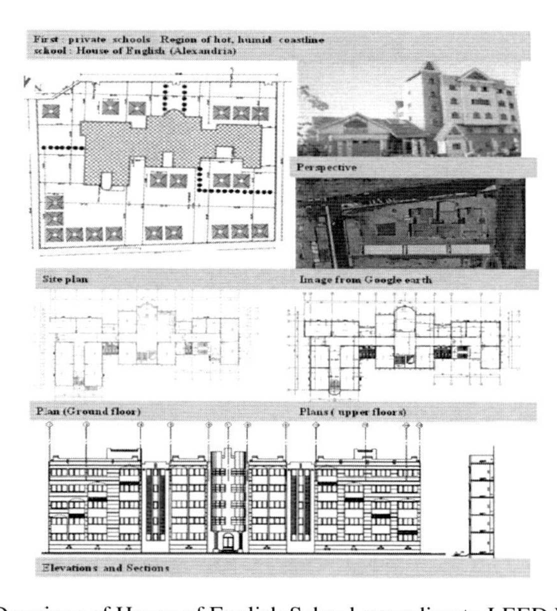

Fig. 6. Drawings of House of English School according to LEED V3 2009

Table 5. Evaluation of House of English School according to LEED V3 2009

LEED 2009 for Schools New Construction and Major Renovation
Project Checklist

Project Name
25/3/2010

Y	?	N	Code	Item	Possible Points
		10		**Sustainable Sites**	Possible Points: 24
Y			Prereq 1	Construction Activity Pollution Prevention	
			Prereq 2	Environmental Site Assessment	
		1	Credit 1	Site Selection	1
		2	Credit 2	Development Density and Community Connectivity	4
			Credit 3	Brownfield Redevelopment	1
			Credit 4.1	Alternative Transportation—Public Transportation Access	4
		1	Credit 4.2	Alternative Transportation—Bicycle Storage and Changing Rooms	1
		1	Credit 4.3	Alternative Transportation—Low-Emitting and Fuel-Efficient Vehicles	2
			Credit 4.4	Alternative Transportation—Parking Capacity	2
		1	Credit 5.1	Site Development—Protect or Restore Habitat	1
			Credit 5.2	Site Development—Maximize Open Space	1
		1	Credit 6.1	Stormwater Design—Quantity Control	1
			Credit 6.2	Stormwater Design—Quality Control	1
			Credit 7.1	Heat Island Effect—Non-roof	1
		1	Credit 7.2	Heat Island Effect—Roof	1
		1	Credit 8	Light Pollution Reduction	1
		1	Credit 9	Site Master Plan	1
			Credit 10	Joint Use of Facilities	1
		2		**Water Efficiency**	Possible Points: 11
Y			Prereq 1	Water Use Reduction—20% Reduction	
		1	Credit 1	Water Efficient Landscaping	2 to 4
			Credit 2	Innovative Wastewater Technologies	2
		1	Credit 3	Water Use Reduction	2 to 4
			Credit 4	Process Water Use Reduction	1
		18		**Energy and Atmosphere**	Possible Points: 33
Y			Prereq 1	Fundamental Commissioning of Building Energy Systems	
Y			Prereq 2	Minimum Energy Performance	
Y			Prereq 3	Fundamental Refrigerant Management	
		14	Credit 1	Optimize Energy Performance	1 to 19
		3	Credit 2	On-Site Renewable Energy	1 to 7
		1	Credit 3	Enhanced Commissioning	2
			Credit 4	Enhanced Refrigerant Management	1
			Credit 5	Measurement and Verification	2
			Credit 6	Green Power	2
		7		**Materials and Resources**	Possible Points: 13
Y			Prereq 1	Storage and Collection of Recyclables	
		3	Credit 1.1	Building Reuse—Maintain Existing Walls, Floors, and Roof	1 to 2
		1	Credit 1.2	Building Reuse—Maintain 50% of Interior Non-Structural Elements	1
		1	Credit 2	Construction Waste Management	1 to 2
				Materials and Resources, Continued	
		1	Credit 3	Materials Reuse	1 to 2
			Credit 4	Recycled Content	1 to 2
		1	Credit 5	Regional Materials	1 to 2
			Credit 6	Rapidly Renewable Materials	1
		1	Credit 7	Certified Wood	1
		14		**Indoor Environmental Quality**	Possible Points: 19
Y			Prereq 1	Minimum Indoor Air Quality Performance	
Y			Prereq 2	Environmental Tobacco Smoke (ETS) Control	
Y			Prereq 3	Minimum Acoustical Performance	
		1	Credit 1	Outdoor Air Delivery Monitoring	1
		1	Credit 2	Increased Ventilation	1
		1	Credit 3.1	Construction IAQ Management Plan—During Construction	1
		1	Credit 3.2	Construction IAQ Management Plan—Before Occupancy	1
		1	Credit 4	Low-Emitting Materials	1 to 4
		1	Credit 5	Indoor Chemical and Pollutant Source Control	1
		1	Credit 6.1	Controllability of Systems—Lighting	1
		1	Credit 6.2	Controllability of Systems—Thermal Comfort	1
		1	Credit 7.1	Thermal Comfort—Design	1
		1	Credit 7.2	Thermal Comfort—Verification	1
		1	Credit 8.1	Daylight and Views—Daylight	1 to 3
		1	Credit 8.2	Daylight and Views—Views	1
		1	Credit 9	Enhanced Acoustical Performance	1
		1	Credit 10	Mold Prevention	1
		4		**Innovation and Design Process**	Possible Points: 6
		1	Credit 1.1	Innovation in Design: Specific Title	1
		1	Credit 1.2	Innovation in Design: Specific Title	1
			Credit 1.3	Innovation in Design: Specific Title	1
			Credit 1.4	Innovation in Design: Specific Title	1
		1	Credit 2	LEED Accredited Professional	1
		1	Credit 3	The School as a Teaching Tool	1
		3		**Regional Priority Credits**	Possible Points: 4
		1	Credit 1.1	Regional Priority: Specific Credit	1
		1	Credit 1.2	Regional Priority: Specific Credit	1
		1	Credit 1.3	Regional Priority: Specific Credit	1
			Credit 1.4	Regional Priority: Specific Credit	1
		58		**Total**	Possible Points: 110

Certified 40 to 49 points Silver 50 to 59 points Gold 60 to 79 points Platinum 80 to 110

8.5 Collected Analytical Study of Case Studies

The Collected Analytical Study of Case Studies depends on measurements that compare items and determine the positive and passive sides in each case.

Table 6. Assess and Ranks the Efficiency of Selected Examples Applied (Governmental sector. Private sector)

Compare Items / Type of School	Points	Governmental Sector		Private Sector	
		El Awayed School	Corona School	Zahran School	House of English School
First: Sustainable Sites	24	11	13	19	10
Second: Water Efficiency	11	5	6	1	2
Third: Energy and Atmosphere	33	11	12	18	18
Fourth: Materials and Resources	13	5	5	7	7
Fifth: Indoor Environmental Quality	19	7	6	14	14
Sixth: Innovation and Design Process	6	2	3	5	4
Seventh: Regional Priority Credits	4	1	1	3	3
Total	110	42	46	67	58
Grade		Certified	Certified	Gold	Silver

First: Sustainable Sites

The Governmental sector has low points since the site selection is bad as it is far from transportation and located in Low-.Emitting and Fuel-Efficient Vehicles. So, it is not compatible from the Social and Environmental side. On the other hand, the Private sector has high points in since the site selection is good one as it is near from transportation and located in high .Emitting and Fuel-Efficient Vehicles. So, it is compatible from the Social and Environmental side.

Table 7. Selected Schools' Level According to LEED Assessment Configuration

El Awayed School	Corona School	Zahran School	House of English School
Certified	Certified	Gold	Silver

9 Conclusions

1. Sustainable Design is one of the most important determinants that should be addressed in the designing of schools so that the design will be compatible with the functional requirements and to ensure the sustainability of the established educational building.
2. Sustainable Architecture depends on two grounds; the first is related to the stages of implementation and initial operation in order to recycle the building materials used in the implementation phase and the second is managing the waste resulting from the building, such as solid, liquid and gas waste and utilize them in the operational phase.
3. Sustainable Design is based on a set of bases, like the economy in materials, which sets out a series of strategies that in turn shapes the ways of Sustainable Design.
4. Criteria for Sustainable Building material products consist of three phases; pre-building phase, building phase and post building, which are the main elements in life cycle of the design.
5. Human Design depends on a strategy that consists of three items; Preservation of Natural Conditions, Urban Design Site Planning and Design for Human Comfort, all of which lead to some of the applications methods.
6. Design process of green building depends on the study of the internal and external environment in order to match between the green building and environment [external that is naturally built and internal] according to the comparison between the requirements of the educational building authority in Egypt and the theories of Architecture.

   We can conclude that all of them did not put into consideration the conditions of Sustainable Design Building like Economy of Resources and Life Cycle Design, although they comply with Human Design.
7. By applying the basic of Sustainable Buildings on the Educational Buildings Components (classes. administration. Services-.facilities. play grounds. parking), we can conclude that all the components need this check list of basic Sustainable Building.
8. LEED 2009 (V3) Methodology is a good methodology for measurement and evaluation, which is a suitable tool for Design Process, Pre-Construction Process and Constriction Process in Egypt.
9. After using LEED 2009 (V3) Methodology in evaluating the design of Schools (El Awayed – Corona.Zahran.House of English), we can conclude the following:

- Natural materials, like wood and stone, are environmental assessment.
- Select the suitable site for the Education Building must be far from sound and visual pollution.
- Using a smart technique for saving energy, Solar and Wind Power, is an important aspect in the Equipment Stage.
- The Smart Design is a compact which means that the spaces between the buildings are limited.
- The main facilities and infrastructure are important for choosing the location of the Education Building in order to save energy and use alternative energy tools such as electric energy and water energy.
- Linking green area and buildings in Educational Buildings provide suitable ventilation and a good view.
- Recycling the waste water is a good way where the treated water can be used in agriculture.

10 Recommendations

1. The criteria should be more explicit; the language should be clarified and made more proscriptive.
2. Existing criteria should be expanded to specifically addressing the environmental consequences of Architectural Design decisions. In many cases, adding "environmental impact" to the elements listed in a given criteria can accomplish this goal.
3. A higher level of technical and environmental knowledge is required. Students must be capable of integrating environmental knowledge into the design process.
4. Since Ecological Design is required and integrated partly in the entire design process, not merely an area of specialization; therefore, accreditation should require Environmentally Sustainable Design principles.
5. It is our goal to have this compendium used widely among Architecture Educational Buildings in the United States. To this end, a range of viewpoints and feedbacks from architectural educators and practitioners have been incorporated into the development of the compendium. At present, Compendium modules cover Sustainable Design, Sustainable Building Materials, Recycling and Reuse of Architectural Resources and Case Studies.
6. Training and vocational rehabilitation and human development in the area of Sustainable Green Architecture is a feasible investment, as the architect and engineer who has experience in this area will inevitably be transferred to professional surroundings and thereby can be created to our national experience in practical applications and practices of Sustainable Green Architecture.
7. Developing standard criteria for the application of best practices, Integrated Design, Identification of systems and Applications of Sustainable Green Architecture.

References

[1] Wikipedia articles needing reorganization, Low-energy building, Architecture, Sustainable building, Environmental design, Sustainable development

[2] http://www.greenhomebuilding.com

[3] From Our Common Future, pp. 23–27. Oxford University Press, London (1987)
[4] Kim, J.J., Rigdon, B., Graves, J.: Introductory Module, Project Interns;College of Architecture and Urban Planning, pp. 4-10. University of Michigan (2004)
[5] Gipe, P.: Wind Power: Renewable Energy for Farm and Business, pp. 50–52 Chelsea Green Publishing (2004)
[6] U.S. Department of Energy, Energy Efficiency and Renewable Energy, Solar Water Heaters, 35-40 (2009)
[7] Saleh, H.A.: Saudi Engineering Conference VI . King Fahd University of Petroleum and Minerals, Activate the health and environmental dimension in the design of construction projects, Dhahran, vol. 1, pp. 23–26 (October 2002)
[8] `http://www.usgbc.org`
[9] CIB TG 16,Sustainable Construction, Proceedings of the First International Conference, CIB TG 16, Tampa, Florida, USA, pp. 33–44 (1994)
[10] Mendler, S.F., Odell, W.: The hok Guide Book to Sustainable Design, pp. 11–16. John Wiley & Sons, Inc., New York (2000)
[11] `http://www.buildings.../about/leed.htm`
[12] Chiara, J., Callender, J.H.: Time.Saver Standards for Building Types, pp. 23–33. McGraw.Hill Companies (2001)
[13] Mills, E.D.: Planning:The Architect's Handbook. The British Library press (1995)
[14] Educational Buildings Authority,Arab Republic of Egypt National Presidency of the Council of Ministers to ensure the quality of education And dependence (1):document quality standards for buildings Basic education educational buildings, Arab Republic of Egypt, 22–33 (2008)
[15] Regional Office for West Asia and North Africa, Population Council International, The educational building environment in Egypt, 35–37(2001)
[16] National Center for Research on Housing and Construction, Egyptian Code 2007; 23 of the foundations of design and implementation requirements for the protection of installations from fire, 16–18 (2007)
[17] The Public Authority for Housing and Building Research and Urban Planning and The Ministry of Education, Design Criteria for the Educational Buildings of Basic Education in Greater Cairo, 12–22 (1990)
[18] The General Authority for Educational Buildings, Assessment of the functionality of the architectural models for educational buildings, 3–25 (1992)
[19] Beynon, J.: UNESCO and The Ministry of Education, in collaboration with the Planning and Monitoring Unit, Physical Facilities for Education; What Planners Need to.Know, 40–50 (1997)
[20] The General Authority for Educational Buildings, Alexandria (2011)

Chapter 31
Latin-American Buildings Energy Efficiency Policy: The Case of Chile

Massimo Palme[1], Leônidas Albano[2,3], Helena Coch[2], Antoni Isalgué[2], and José Guerra[1]

[1] Universidad Católica del Norte, Antofagasta, Chile
mpalme@ucn.cl
[2] Universitat Politècnica de Catalunya, Barcelona, Spain
[3] Universidade de Brasília, Brasília, Brazil
leonidas.albano@gmail.com

Abstract. In the last years many Countries implemented energy efficiency strategies in their policy. An important sector affected by the new regulations is the building sector. Buildings construction and use represent the 30% of the total CO_2 emissions of the world. It is one of the most important sectors to regulate. EPBD directive of the European Union is one of the actions already token in this way. Low emissions buildings construction only will be a goal if an appropriate policy will be structured world-width. Latin-American area is an important region of the world respect to growth and development. For this reason, it is very important to consider the implementation and discussion of the local energy efficiency norms. This paper reviews actual norm in Chile and discuss critically the effectiveness of the past policy. Future ways to improve the contribution of the construction sector reducing emissions are suggested, by comparison among policies of other Countries, especially the policies of Spain and Brazil. Special attention is focused on the dynamical simulation by software, which is the base of the energy label certification in many Countries. Examples of the Lider-Calener Spanish certification are presented and used to justify the possible future emissions reduction scenario for Chile.

1 State of the Question

The technical norm in Chile is still recent. At the moment only affect the minimum values of the main parameters of the energy building evaluation, such as the walls and windows transmission coefficients. It has to be improved and this process is just started. Analysis is structured on five topics: construction material environmental costs, thermal demand of heating and cooling, natural lighting and ventilation, solar radiation use (solar thermal and photovoltaic), systems efficiencies. For each one, actual norms are described and possible improvements are discussed, taking in to account the other Countries experiences. Desired indices of sustainability and efficiency are searched by comparison of the consumption among Countries during the past 20 years, especially after the Kyoto Protocol signature.

A. Håkansson et al. (Eds.): *Sustainability in Energy and Buildings*, SIST 22, pp. 337–346.
DOI: 10.1007/978-3-642-36645-1_31

1.1 Construction Costs

At this moment in Chile environmental cost are not considered by norms. No strategies are proposed to reduce the production costs of materials. Production costs of materials in Chile have not been evaluated yet. The only reference to the embodied energy available at the moment is the page 21 of the "Guide of design for energy efficiency in the social dwelling" [1], published by the Dwelling and Urbanism Department.

1.2 Thermal Demand

Energy efficiency considerations in Chile are always related to the thermal demand. In the year 2000, the thermal norm was considered for the first time in the general norm of construction and urbanism. In the first phase of the thermal norm development, thermal transmittance of the entire building (average) was considered as the representative efficiency coefficient. Maximum values for this coefficient were established for the different climatic regions of Chile. The climatic zones were defined using the average degree day of heating and cooling concepts. In a second stage of the norm development, maximum thermal transmittances for each wall, roof or ventilated floor were established, in order to equilibrate the thermal dissipation of the buildings. Once terminated the second stage, a technical manual was published [2]. A third phase of the norm development is still started. This phase will consider the dynamic simulation to ass the total energy demand of buildings. Probably, specific software will be developed to this scope. At the moment, only exists a software to minimum requests evaluation. The program names CCTE and can be downloaded from the web site of the Dwelling and Urbanism Department. Figure 1 shows the bases of the thermal norm in Chile.

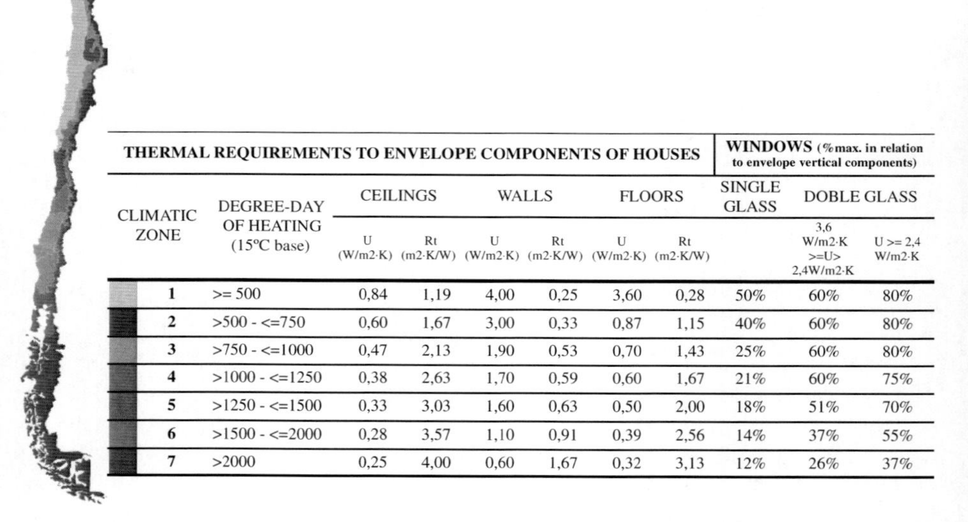

THERMAL REQUIREMENTS TO ENVELOPE COMPONENTS OF HOUSES								WINDOWS (%max. in relation to envelope vertical components)		
CLIMATIC ZONE	DEGREE-DAY OF HEATING (15°C base)	CEILINGS		WALLS		FLOORS		SINGLE GLASS	DOBLE GLASS	
		U (W/m2·K)	Rt (m2·K/W)	U (W/m2·K)	Rt (m2·K/W)	U (W/m2·K)	Rt (m2·K/W)		3,6 W/m2·K >=U> 2,4W/m2·K	U >= 2,4 W/m2·K
1	>= 500	0,84	1,19	4,00	0,25	3,60	0,28	50%	60%	80%
2	>500 - <=750	0,60	1,67	3,00	0,33	0,87	1,15	40%	60%	80%
3	>750 - <=1000	0,47	2,13	1,90	0,53	0,70	1,43	25%	60%	80%
4	>1000 - <=1250	0,38	2,63	1,70	0,59	0,60	1,67	21%	60%	75%
5	>1250 - <=1500	0,33	3,03	1,60	0,63	0,50	2,00	18%	51%	70%
6	>1500 - <=2000	0,28	3,57	1,10	0,91	0,39	2,56	14%	37%	55%
7	>2000	0,25	4,00	0,60	1,67	0,32	3,13	12%	26%	37%

Fig. 1. Climatic zones, degree-day of heating, maximum transmittance values for Chile

1.3 Natural Lighting and Ventilation

No ventilation standards are defined in the actual thermal norm. Moreover, infiltration and wind pressure are not considered in the energy demand calculation. Roof ventilation strategy to reduce the sun-air temperature in summer was not considered by the law. In the illumination sector, things are the same. Natural lighting exigency is not considered to reduce CO_2 emissions at the moment. In general, there are no lighting standards in the technical norm. It can be said that this is one of the analysis points that more has to be improved.

1.4 Solar Radiation (thermal and photovoltaic)

Solar energy use is discussed and partially regulated in specific laws like the technical norm for the solar thermal panels [3], but it is not related directly with the construction sector. Obligation to install solar panels in the new buildings is not present yet. Minimum solar energy contribution is considered only in order to assign the government economic contribution to installers.

1.5 Systems Efficiencies

Systems efficiency is not considered in the norm. However, simulation program CCTE considers an efficiency of 0.75 for the heating systems and a COP of 2.6 for the cooling systems. These values are fixed and representative. CCTE uses the efficiencies to calculate the energy consumption by the thermal demand estimation. Results are clearly only indicatives.

2 Other Experiences

Energy savings and CO_2 emissions are problems that have to be resolved at the global level. For this reason governments of all around the world cites first in Kyoto (1997) and then in Copenhagen (2009) with the objective to accord the action that each country has to take. Problems of emergent economies to respect the desired emissions values were one of the most important items on the work table of Copenhagen cite. Chile, for many reasons, took an intermediate position in this discussion, but it appears very clear that at one moment country will have to take side. In this way the experiences of countries that are still in process of implementation of new norms is especially interesting for Chile. In this paper we select Spain and Brazil cases, which are close to Chile implementing energy norms in their laws.

2.1 Spain

European directive EPBD (2003) and Kyoto protocol (1997) obligated the signer countries to implement an energy certification process of buildings in their laws. In Spain, implementation of the directive was done by tree stages: first, the development of a new Thermal Norm [4]; second, the publication of a new Installation Manual [5];

and third, the obligation to have an energy label for all new buildings. Government established (RD 47/2007) two certification options: a simplified certification (that does not permit to obtain the best certification results) or a simulation process based on the Lider and Calener tools (DOE-II based simulators adapted to Spanish norms).

2.1.1 Construction Costs

Spanish law does not obligate to reduce the construction costs, and is difficult to obtain information about real production costs of materials. However, a lot of researchers are looking for a reasonable solution, asking to the Government the inclusion of this kind of evaluation in the norms. It is probable that embodied energy represents more than 40% of the total energy consumed by a building [6]. Moreover, embodied energy is more difficult to reduce than consumption energy during use.

2.1.2 Thermal Demand

Like in the case of Chile, thermal demand represents the central body of the norm. The new thermal norm, published in 2006, define climatic zones and maximum transmittance values. Thermal demand evaluation is also described in the document. Lider software is suggested as the better option to obtain the norm verification. Additionally, Spanish government decided to obligates constructors and architects to obtain energy labels of buildings. This can be done by software Calener or by alternative processes. The climatic zones are defined by an index of climatic severity, which combines degree-day concept with solar radiation. Result is a winter-summer classification of 12 climate typologies. For each zone, maximum transmittances of each wall, window, floor and roof are established. Moreover, average maximum values of transmittances are defined. Lider software evaluates the energy demand and verifies the norm. Thermal bridges and condensation are other limitations considered by the norm. The energy label assignment process is analyzed in the follow because it considers the systems efficiency in dynamical simulation. Results are expressed by comparison among the analyzed building and a reference building that has the same form, orientation and dimension of the simulation object, and the minimum values of the energy saving norm for transmittances. The consequence is that the architectural evaluation is only partially satisfied. Spanish norm evaluates materials and structures solutions, but not the design of the building, its orientation and climatic study of the project.

2.1.3 Natural Lighting and Ventilation

Indoor air quality is defined by the installation manual (RITE). Minimum values of ventilation are also established. Natural lighting is not an obligation, but it is recommended. At the other hand, illumination system has to respect the efficiency systems standards. Infiltration is considered as propriety of the window and expressed by the experimental values of m^3air/m^2window with a pressure of 100 Pa during the experiment. The norm fixes the maximum value 27 or 50 m^3/m^2 depending on the climate zone.

2.1.4 Solar Energy Use

Minimum of solar energy contribution to hot water and electricity production are considered in the document. New buildings have to have a minimum of the 30% of solar contribution in hot water production. Depending on the climatic zone, the minimum value can rise up to the 70% of solar contribution requirement. Non residential buildings have to have a minimum of photovoltaic production. CTE define the calculation of the minimum power of the photovoltaic system, which depends on climate and use of the building. In any case it will be more than 6.25 kWp.

2.1.5 System Efficiency

Thermal systems minimum efficiencies are established by the installation manual (RITE). Illumination system efficiency is defined by the thermal norm (CTE). Illumination efficiency is assed as the VEEI coefficient, which relates the illuminated area with the electrical power consumed to produce 100 lux.

2.1.6 Energy Efficiency Certification Process

Once defined thermal demand of the building and decided the systems typologies and efficiencies, Spanish law requires a dynamical simulation by software or an alternative certification of efficiency. The Calener program was proposed as the official instrument to do the simulation. It has to be notice that Lider-Calener simulation considers a lot of parameters, but that not all these parameters have the same influence on the final result. Recent studies demonstrates that some parameters are excessively incident on the final certification [7], [8].

TYPOLOGY	PARAMETER	UNIT	EFECTIVENESS
Climatic	Climatic zone	/	Very low
	Altitude	m	Low
	Air renewal coefficient	1/h	High
User-dependent	Glass heat loss corrected by use of blinds in winter	%	Low
	Solar factor by use of blinds in summer	%	High
	Boiler typology	/	High
	Boiler efficiency	%	High
	Boiler power	kW	Very high
System-dependent	Combustible typology	/	Very high
	Final unit typology	/	Medium
	Final unit power	kW	Medium
	Solar contribution to hot water	%	High
	Building use	/	Very low
	Internal humidity production	/	Low
	Orientation	/	Very low
	Linear thermal conductivity of materials	W/mK	High
	Specific heat of materials	J/kgK	Medium
Building characteristics	Water diffusion resistance	/	Low
	Glass solar factor	%	Medium
	Glass transmittance	W/m^2K	High
	Fixed protection transmission	%	Medium
	Fixed protection absorption	%	Medium
	Density of materials	kg/m^3	Low

2.2 Brazil

In 2003, the Electric Energy Conservation National Program (Procel) which was created in 1985, launched the Energy Efficiency in Buildings National Program called Procel Edifica. In 2009, together with the National Institute of Metrology, Standardization and Industrial Quality (INMETRO), Procel began the first regulation for certification of buildings as part of the Brazilian Program for Certification (PBE). In 2010, although not required, the Technical Regulation of Quality – RTQ for the Energy Efficiency Level of commercial, public and services buildings [9] was approved and is the focus of this article. In November of the same year, the RTQ for residential buildings [10] was approved. Thus, the certification process is performed differently for the two types of buildings.

2.2.1 Regulation

The RTQ objective is to create conditions for labeling the level of energy efficiency of buildings and is applicable to buildings with minimum total useful area of 500m2 and/or when the energy demand is greater than or equal to 2,3kV, including air-conditioned buildings, partially air conditioned buildings, and naturally ventilated buildings. There are two approaches to achieve compliance: prescription method and simulation method. The prescription method evaluates three systems, the building envelope system (ENV), the lighting system (LPD) and the air conditioning system (AC). The simulation method compares its results with similar building models that comply with the prescription requirements. The certificate can be granted partially, as long as the building envelope system is also evaluated. After the evaluation of the three systems, together with the natural ventilation simulation, the general classification is obtained, with accordance to three different allowed combinations of the two methods. Each system evaluation receives a different weight: ENV=30%, LPD=30%, AC=40%. And the general classification is calculated according to this weight distribution using a general equation. All individual systems have efficiencies ranking from A (more efficient) to E (least efficient) and their evaluation uses numeric equivalents that correspond to a given efficiency ranking from 5 to 1. In addition to the prescription and simulation methods, the building must meet minimum general prerequisites for electrical circuits, water heating and lifts, and minimum specific prerequisites for the lighting and air conditioning systems, in order for it to be eligible for labeling. In addition, any initiatives that increase the efficiency of the building could receive bonus points, when justified and the energy savings proven. These can include, systems and equipment that provide water use reduction, renewable energy sources, combined heat and power systems and technology or systems innovations like natural lighting which are proven to increase the energy efficiency of a building.

2.2.2 Building Envelope System (ENV)

For the Building Envelope System, in addition to the calculation and efficiency determination procedures, there are specific prerequisites established by the RTQ. Those include, thermal transmittance, absorptance of external surface, and skylights, in accordance with the bio-climactic zone where the building is located and the desired labeling level (A to E). The Brazilian Bio-climactic Zones are established under the Brazilian Standard NBR 15220-3 (ABNT, 2005) [11]. The method for classifying the energy efficiency of the envelop system is based on a consumption indicator that is derived from equations. For each bio-climactic zone, there is an equation for buildings with a projection roof area of less than 500m2 and another for a projection area greater than 500m2. The first equation represents the maximum limit allowed Shape Factor and the second represents the minimum limit. These limits represent the range in which the building can be modeled. This scale is divided into four intervals that represent different levels of classification ranging from A to E [12].

2.2.3 Lighting System (LPD)

For the lighting system classification, in addition to the installed power limits, there are specific control systems that must be complied with. Use of available natural light is achieved through the installation of an independent control system from that of the rows of lights closest to exterior openings. The system evaluation will be done using the building area and the buildings activities methods. The building area method evaluates by combining all of the environments of the building and assigning them a single limit value. In the building activities evaluation, the environments are assessed individually. The necessary luminance level (lux) is defined for each environment by using the Brazilian Standard NBR 5413 (ABNT, 1992) [13].

2.2.4 Air Conditioning System (AC)

The specific prerequisites of the air conditioning system refer the protection of the condensation units, insulation of the air ducts and that of the air conditioning systems from artificial heating. In regards to the procedures for determining the efficiency, the systems must have an energy efficiency which is recognized by the PBE / INMETRO, adopting the classification of the National Label of Energy Conservation (ENCE). If not regulated by INMETRO, the minimum requirements set out in RTQ based on ASHRAE 90.1 (2001) [14] must be met. The calculation of heat load must be carried out in accordance with generally accepted manuals and rules such as the latest version of the ASHRAE Handbook of Fundamentals and the Brazilian Standard NBR 16401 (ABNT, 2008) [15]. Moreover, they shall meet the RTQ requirements for temperature control and automatic switch off systems, isolation between thermal zones, control and dimensions of hydraulic and ventilation systems, as well as heat recovery equipment.

2.2.5 Simulation

The RTQ establishes the minimum requirements for the weather file and that of the thermal energy simulation software to be used, as an example, validated by the ASHRAE Standard 140. However, the final technical report from the Institute for Technological Research of São Paulo (IPT, 2004) [16], recommends the software BLAST and primarily Energy Plus be used in the evaluation process for thermal performance of non air conditioned buildings and mentions the use of DOE-2 for completely air conditioned buildings. The RTQ also defines the procedures that must be complied with in the simulation, as well as the parameters regarding naturally ventilated or non air conditioned environments.

2.2.6 Construction Costs

Up to now, construction costs have not been mentioned by the RTQ. In Brazil, there is some difficulty in calculating these costs due the nonexistence of a database for the consumption of materials and energy involved in the processes related to building construction. However, there are initiatives, such as the creation of the Association of Life Cycle in 2002 and the Community LCA which carry out projects that aim to create inventories, methodologies and libraries, supported by the ISO 14000 international standard, and specifically by the Brazilian Standard NBR ISO 14040 (ABNT, 2009) [17], started in 2001, which describes the structure of a life cycle assessment (LCA).

2.2.7 Certification Process

There are two phases in the certification process. The first, Design and Documentation, gives a certificate with a label stating the efficiency level met. The second is completed by an accredited professional who audits the building once it is occupied and with all systems installed. The certificate may be displayed in the building.

3 Comparison and Future Perspective

The first impression is that technical norm in Chile is at the moment really insufficient. Comparison among the equivalents norms of Spain (cultural reference for Chile) and Brazil (local reference country) shows that many aspects of energy evaluations have to be already investigated. Only the embodied energy evaluation is little structured by both Spanish and Brazilian Norms like Chile. Systems efficiency seems to be one of the most important aspects to improve, because of the effectiveness over the final results. Spain and Brazil are working hard in this direction. Chile at the moment needs the codification of standards for the ventilation, lighting and heating-cooling systems. Thinking specifically on the architecture, the tree countries have a comparable level of developing in the technical norms. Maybe Spain norms use more attention to details, like infiltration cross the windows or thermal bridges of walls and floors. Both Spain and Brazil separate residential buildings from other use buildings. It will be a goal the development in Chile of specific norms for no-residential buildings, which have an important impact over the final consumption and related CO_2 emissions of the Country. As conclusion, it can be said that a certification process appears as an urgent need in Chile, a pretending reference country in the local area. Application of the new norms in the construction sector is a fact that is difficult to evaluate at the moment. In Brazil like in Chile the norm introduction is still recent, but development level of these Countries suggests that results will be available soon. Spain was involved in the global crisis and this signified a stop in the massive construction of the country respect to the last years. A little analysis for Spain results can be tried looking to few examples of building projected after the norm introduction in 2007. Table 1 shows the CO_2 emissions and the obtained energy labels of 9 new buildings (3 familiar houses, 3 little blocks of dwellings and 3 medium blocks of dwellings).

Table 1. Emissions and energy labels for example buildings

	Studied building emissions ($kgCO_2/m^2$)	Studied building energy label	Reference building emissions ($kgCO_2/m^2$)	Reference building energy label
Campolier	53.5	E	31.2	C
Mas Torrent	15.2	B	32.9	D
Llinars	14.2	C	19.7	D
Alta cortada	12.7	B	23.2	D
Miriana	17.3	C	27.4	D
Les franqueses	5.3	B	16.9	D
Cambrils	4.2	B	16.7	D
Blanes	4.9	A	14.1	D
Figueres	9.5	B	17.0	D

Energy saving and CO_2 emissions reduction can be clearly an important effect of massive introduction of norms. If any new building reduces emissions by the 40%, the total emissions reduction can be about the 8% for a country on development like Chile. A comparable emissions reduction in the old buildings by retrofit can signify a global reduction of the 20%. For this reason, appears very important the introduction of a certification system in Chile. If this system has to be similar to the Spanish, it has to be discussed. Lider-Calener certification process is not perfect. Especially, does not consider the architectural effects like form, dimension, and orientation. Brazil, at the other hand, suggests the use of more complete and complicate software to obtain the energy labels, like Blast and Energy Plus. Moreover, dynamical simulation always has the problem of transparency and repeatability of results. Recent studies show that sensitive analysis and robustness concepts have to be applied to the certification processes [18].

References

1. Bustamante, W., et al.: Guía de diseño para la eficiencia energetica en la vivienda social, Ministerio de Vivienda y Urbanismo, Santiago de Chile (2009)
2. AA. W., Manual de aplicación reglamentación térmica, Ministerio de Vivienda y Urbanismo, Santiago de Chile (2006)
3. AA. VV., Norma Técnica para la contribución solar minima, Ministerio de Energía, Santiago de Chile (2010)
4. AA. VV., Código Técnico de la Edificación, Ministerio de Fomento, Madrid (2006)
5. AA. VV., Reglamiento de Instalaciones Térmicas de Edificios, Ministerio de Fomento, Madrid (2007)
6. Pagés, A., Palme, M., Isalgué, A., Coch, H.: Energy consumption and CO2 emissions in construction and use of buildings according to floor area. In: Proceedings of the World Renewable Energy Congress X, Glasgow (2008)
7. García, X.: Efecto del dimensionado de los equipos: certificación energetica de los edificios. Era solar: energías renovables 149, 40–50 (2009)

8. Palme, M., Isalgué, A., Coch, H., Serra, R.: Relevant factors in Spanish energy certification process. In: Proceedings of the Passive and Low Energy Cooling Conference III, Rode Island (2010)
9. AA. VV., Brazil: Ministry of Development, Industry and Foregin Trade. National Institue of Metrology, Standardization and Industrial Quality, Portaria 372, INMETRO (2010)
10. AA. VV., Brazil: Ministry of Development, Industry and Foregin Trade. National Institue of Metrology, Standardization and Industrial Quality, Portaria 449, INMETRO (2010)
11. AA. VV., NBR 15220-3 Thermal Performance in Buildings Part 3: Brazilian bioclimatic zones and building guidelines for low-cost houses, Brazilian Association for Technical Standards, ABNT (2005)
12. Lamberts, R., Goulart, S., Carlo, J., Westphal, F.: Regulation for energy efficiency labeling of commercial buildings in Brazil. In: Proceedings of the Passive and Low Energy Cooling II, Crete Island (2007)
13. AA. VV., NBR 5413 Interior lighting Specification, Brazilian Association for Technical Standards, ABNT (1992)
14. AA. VV., ASHRAE Standard 90.1 Energy Standard for Buildings Except Low-rise Residential Building, American Society of Heating, Refrigerating and Air-conditioning Engineers, Atlanta (2004)
15. AA. VV., NBR 16401 Central and unitary air conditioning systems Part I, II and III, Brazilian Association for Technical Standards, ABNT (2008)
16. AA. VV., Final Technical Report 72 919-205 Validação de softwares aplicativos para simulação do comportamento térmico de habitações, Instituto de Pesquisas Tecnológicas do Estado de São Paulo, IPT (2004)
17. AA. VV., NBR ISO 14040 Environmental Management – Life cycle assessment – Principles and framework, Brazilian Association for Technical Standards, ABNT (2009)
18. Palme, M.: Energy Sensitivity of Buildings, PhD Thesis, Barcelona (2010)

Chapter 32
Thermal Performance of Brazilian Modern Houses: A Vision through the Time

Leônidas Albano[1,2], Marta Romero[2], and Alberto Hernandez Neto[3]

[1] Universitat Politècnica de Catalunya, Barcelona, Spain
leonidas.albano@gmail.com
[2] Universidade de Brasília, Brazil
[3] Universidade de São Paulo, Brazil

Abstract. The aim of this paper is to assess the influence of envelope components on the thermal performance of modern naturally-ventilated houses built in the 1950's and 1960's in Goiânia, located in middle-west of Brazil. The study is based on a vision through the time. This study allows the analysis of how passive strategies have been treated before the establishment of the sustainable concept, helping us to reflect on the future from a historical perspective. Seven rooms in two houses were selected for in situ measurements and simulations. Indoor and outdoor air temperatures and relative humidity were measured with data loggers HT-500, during 91 days in three different months in 2011: June (low air temperatures and low relative humidity); September (high air temperatures and low relative humidity); and December (medium air temperatures and medium relative humidity). The measured dates are analyzed allowing the comparison among different building comfort zones. Two different building energy simulation programs, EnergyPlus and AnalisisBio, are used to solve the energy balances and to evaluate the thermal performance and the comfort of the cases. The main conclusion is that it is possible to identify, at the same time, technical limits of applied solutions that result in a thermal performance below its potential, and, a considerable reduction on the thermal demand of the modern naturally-ventilated houses.

1 Introduction

The sustainability in the architecture is a contemporary concept created after discussions realized since the last energy revolution, started with the energy crisis in 1973. Besides it, the paper intents treat this theme through a historical view, identifying considerable information about the influence of envelope components on the thermal performance of houses built before the establishment of the sustainability concept. Before the establishment of this concept, the passive strategies were not treated as an important point in the sustainable roll, but it was just an architectural aspect inside on the architectural roll. This general vision is important to integrate and to understand technical themes in the architecture. The passive design means a reduction of the thermal demand and the energy consumption. The relation between the historical focus and the

A. Håkansson et al. (Eds.): *Sustainability in Energy and Buildings*, SIST 22, pp. 347–355.
DOI: 10.1007/978-3-642-36645-1_32

contemporary concept is direct because the same effort is necessary to create comfort spaces reducing thermal demands. In this way, the paper studies the thermal performance of modern houses thought from a purely architectural vision.

The modern architecture is known as a product of a period characterized by a rupture with the tradition, history and the nature but it cannot be generalized in the time and in the space. The major factor in the Brazilian modern architecture is the climate, as Bruand (1997) states [1]. The climate and the tradition architecture elements were always present in the Brazilian modern architecture. However, the way as the new architecture is diffused for the country produces changes how historic and nature elements is treated in the architecture produced in the interior of the Brazil, as in the city of Goiânia. In the paper, this changes is considered though its influence in the building thermal performance and in the efficiency of the passive strategies applied in the modern houses.

The thermal performance of the modern houses located in Goiânia is a theme that transcends the technical study as it is related to a very important historical moment for Brazil in the 20th century. It is necessary to consider a historical, socio political and economic contextualization of a moment and place, so the theme can be addressed in its real complexity.

Without losing this contextualized view, this paper intends to focus the study to analyze the influence of different strategies used in the different components of the building in its thermal performance, as well as their representation according to each functional zone of two modern single-family naturally-ventilated houses, built in the late 1950's and early 1960's in the city of Goiânia, capital of Goiás state, located in the Midwest of Brazil. (Lat. 16.41S, Long. 49.25W, Alt. 741m/29,173.23in). Due to be naturally-ventilated buildings, the study focuses on calculating variations of internal hygrothermal conditions and estimation the periods of lack of comfort, comparing the results with the external city conditions and also the immediate surroundings and seeking in determining the thermal loads references to individual contribution of individual architectural elements of the two houses. Were considered the actual Brazilian normalization and the singularities indicated on the ResHB method about the internal thermal improvements, use of spaces and thermal zones differentiation [2].

2 Analysis Method

Below are presented the basis for the study object definition, the methods and research procedures, the tools and instruments, also the general considerations regarding the analyses of the results.

The ABNT NBR 15575-1 (2008) [3] defines two levels of approach in assessing the thermal performance of houses: simplified normative and global informative, recommended on more detailed evaluations. In this paper we adopted the global informative evaluation that defines two procedures of requirements and criteria verification: by measurements on constructed buildings and through computer simulations.

Two houses were selected (1 and 2), starting from studies and surveys of the modern architectural heritage from the city of Goiânia [4]. To select the rooms to be measured we considered the requirements of the mentioned regulation that: relates the thermal conditions inside the houses with external conditions at shade; and focuses the study on prolonged stay rooms, living rooms and bedrooms, preferably those with the biggest surface exposed to direct sunlight and less favorable solar orientation. To define the measurement period we considered the typical conditions of the reference year in accordance to the meteorological data from 1961-1990 [5] as the summer and winter recommendations of the used regulations. This way, for each house were selected: 01 external environment shaded, 01 living room and 01 bedroom with easy access during the experiment. At the house 1, the central covered patio was also selected, for its peculiarity and strategic location. All in all, 07 environments were chosen in both houses for the measurements to be taken. Indoor and outdoor air temperatures and relative humidity were measured with thermal hygrometric Data Loggers HT-500, during 91 days in three different months in 2011: June (low air temperatures and low relative humidity); September (high air temperatures and low relative humidity); and December (medium air temperatures and medium relative humidity). The measured dates are used to calibrate the virtual model of the dynamic simulations. The environmental dates about the city of Goiânia were obtained by the measured dates of INMET (2011) [6].

All rooms of the residential units were simulated, considering the thermal exchanges between them, and the social and intimate functional sectors were also evaluated, with emphasis on the measured rooms. EnergyPlus building energy simulation program 4.0.0.024 was used to solve the energy balances and to evaluate the different contributions of each significant envelope component in the thermal performance of the houses, further allowing the comparison with the data obtained through measurements. These simulations were performed for the same period of the in situ measurements. The reference model for thermal performance simulations was based on the calibrated model, maintaining its volume and solar orientation. It was defined within the same thermal zones. The thermal properties of the materials and components used in the evaluation were determined as prescribed in the ABNT NBR 15220-3 (2005) [7], as shown in Table 1. AnalisisBio Programme was used to determine the periods of comfort and to compare the different variables allowing the analysis based on Givoni (1998) [8].

Data from the measurements and simulations have been treated to allow comparison such as: year period, territorial scales (urban/ immediate surroundings), type of spaces (internal or external), different houses (1 and 2), functional sectors (social/intimate) and internal rooms (living room/ bedroom). In this paper the data was analyzed from the thermo-hygrometric amplitude in accordance with the evaluation parameters of thermal performance existing on ABNT NBR 15575 (2008) [3], in a way to enable the analysis for surroundings correction and the thermal performance of the buildings in this scale. At the analysis of the results the particularities of residential typology in relation to other ones listed on ResHB Method were considered. The ResHB Method is an implementation of RHB Method, developed by the research project ASHRAE RP-1199, in order to adapt the HB

Method to the particularities of residential typology, characterized by: less internal heat gain, mainly from the surrounding components, although it is generally more exposed to the other typologies [2]. The internal spaces were understood from the diversity and flexibility of related uses in these rooms, therefore higher tolerance of users to internal thermal fluctuations.

Table 1. Description of the walls and roofs of the references models and its thermal properties: thermal transmittance (U) and thermal delay (φ)

Envelope components	U	φ
External Walls (thickness: 270mm) of solid brick (100x60x220mm)	2,25 W/m2K (0,39 Btu/ft2 h °F)	6,8h
Roof of fibro-cement roofing tile (thickness: 7mm) with slab of concrete of 200mm	1,99 W/m2K (0,35 Btu/ft2 h °F)	7,9h

3 Case Studies

From a brief historic contextualization of the studied cases, below are presented the climatic characteristics from the city of Goiânia, followed by the case studies presentation.

Goiânia was founded in 1937 to be the new State Capital of Goiás, located in the Brazilian Midwest. In a context of national order characterized by political and economical changes, Goiânia eventually stood out as an expression of progress in a quest for modernization of the countryside. The early 50s were characterized by the countryside development, with the construction of a new capital for the country, Brasília, and the expansion of national infrastructure. It is in that decade that the first examples of modern architecture begin to appear in Goiânia [9].

The arrival of this new architecture transformed the architectural procedures that guaranteed the continuity of local tradition, changing the way these buildings adapted to climate, the factor which most interfere in the Brazilian architecture. Among the different typologies built, the residential ones were highlighted for their relevant role in the expression of the new architectural language. Among the strategies for climate adaptation used the highlights are: expansion of the openings in the building surroundings; constant use of cross ventilation in internal rooms; replacement of external porches and corridors for stilts, terraces and balconies; use of "*brise soleil*" as main solar protection element, extensively studied by architects and engineers of the time. The architectural changes mentioned affected the transparency, porosity and solar protection of the residential surroundings and directly influenced their thermal performance, mainly for being naturally ventilated.

The region where Goiânia is located is characterized by the continental and regular cyclic process of air masses displacements, implying a clear rainfall, causing the city climate to be formed by the composition of two main seasons: wet and dry (Aw according to Köppen). An important factor in relation to the dichotomy between the dry and wet seasons arise from the combined effects of nebulosity and insolation. Because of the nebulosity of 80% in December and 43% in June, although the difference between the number of daily hours of the summer and winter solstice is approximately 2h, the insolation in December is lower than in June, 161h/month and 275h/month respectively. Even thought the solar irradiation in December is higher than in June, 3361 W/m2 per day (1,066.13 Btu/ft2 h) and 2708 W/m2 per day (859 Btu/ft2 h) respectively, causes the south facade, struck by summer insolation, can be more transparent and free of solar protections. In opposite to the North facade, exposed to winter insolation, which should be well protected and more opaque [10].

According to ABNT NBR 15220-3 (2005) [6], the buildings located in the city of Goiânia (Bioclimatic Zone 6) must have: shaded openings with ventilation surface between 15% and 25% of the pave surface; heavy external walls, light and isolated roof. The passive thermal conditioning strategies recommended are: in the summer, evaporating cooling, thermal mass for selective cooling and ventilation; in the winter, internal thermal inertia. This way the characteristics of the surroundings have a greater influence in the environmental conditions during summer time, while the internal conditions of the buildings have bigger influence during wintertime.

Among the 78 modern houses currently identified in Goiânia, 2 houses in good state of conservation were selected, as shown in Fig. 1: House Abdala Abrão, projected by David Libeskind and built in 1961 (House 1), and House Eurípedes Ferreira, projected by Eurico Godoy and built in the late 50s (House 2), both located at Setor Sul, central area of the city, far from each other 280m (306 yards) approximately. Both houses have 2 floors and a functional shed on the back, focusing

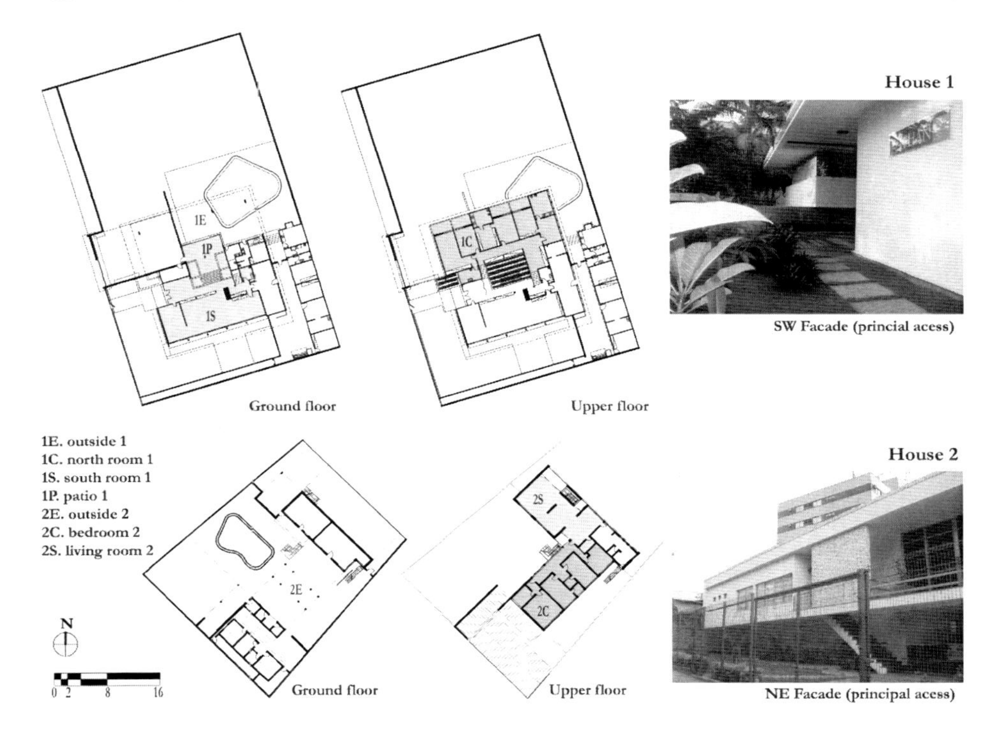

Fig. 1. Description of houses with plans and principal facade photo

the residential areas (social, intimate and service) at the upper floor and leisure activities at the ground floor. Due to the topography, House 1 has the upper floor partially resting on the ground, allowing the main building access through this pave. Its intimate sector is located at the North side, oriented Northwest (Azimuth 343º), while the social areas are located at the South side. On House 2 is the opposite, the social areas are located at Northeast (Azimuth 44º) and the intimate at Southeast. As to shape, the design of House 1 is a square, with central covered patio while House 2 has an "L" shape. Externally both have a swimming pool, partially paved areas and permeable gardens with trees and grass.

4 Results and Discussion

The main results of the study are presented below. After the results regarding the surrounding corrections the results related to buildings will be presented with emphasis on presenting the hygrothermal conditions of both houses. At last are presented the main results from the simulations evidencing the thermal performance and comfort aspects.

Even thought the houses are located in a dense urban area at the center of the city Goiânia, the surround corrections actions, such as shading the external areas using vegetation, good soil permeability covered by grasses and the presence of water with the installation of swimming pools showed a greater efficiency in the dry season. Because its major impact is in the air humidity, the surround corrections are also considerable in June.

The simulations done with AnalisisBio Programme pointed that passive solutions result in a zero energy demand in June. December presents more uncomfortable hours because the surround corrections are thought for dry seasons. Because this, the problem in December is the air humidity and it is not the air temperature. Fig. 2 shows the Givoni graphic for the three sites and the three studied months.

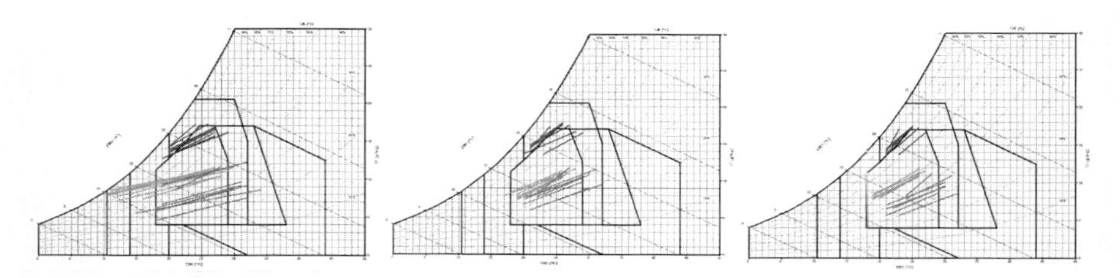

Fig. 2. Comfort graphics about the surround corrections in the different houses: on the left) Goiânia site measure dates; on the middle) House 1 outside; on the right) House 2 outside. Blue is June, red is September and lilac is December.

The simulations done with EnergyPlus Programme, as shown in Fig. 3, pointed the elements of thermal inertia as the factor of higher intervention rate on the thermal performance of the studied cases. The principal architectural components about its influence in the thermal performance of the building were: the ceiling (on red in Fig. 3b) and the walls (external and internal – on blue and lilac in Fig. 3b). The internal ceiling was pointed as the main responsible for heat accumulation during sun

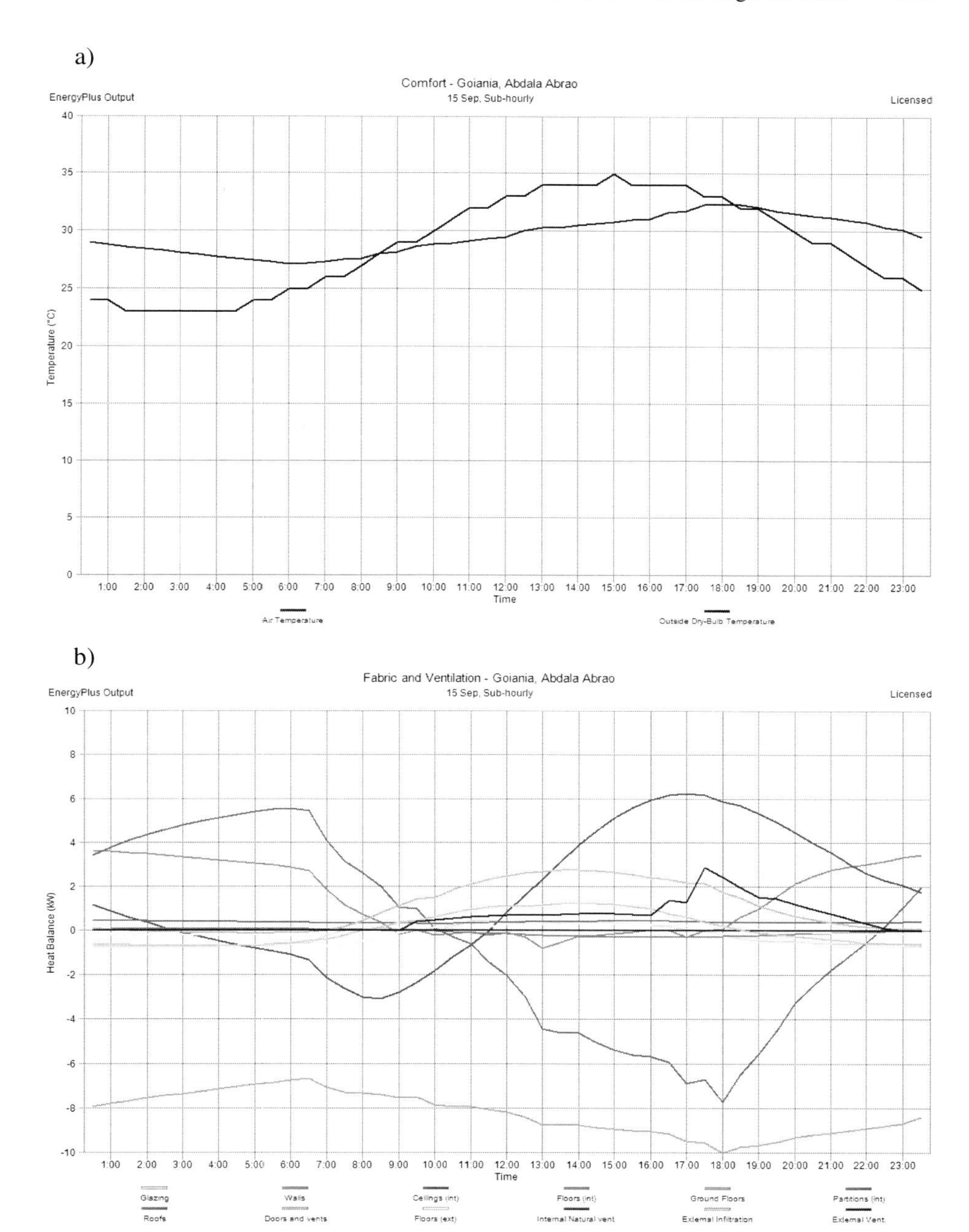

Fig. 3. Variations interior air temperature, outside dry-bulb temperature, air relative humidity and heat balance by each important building component of House 1: a) oscillation of interior air temperature and outside dry-bulb temperature in September; b) oscillation of heat balance by each important building component in September; c) oscillation of interior air temperature, outside dry-bulb temperature and air relative humidity in June

c)

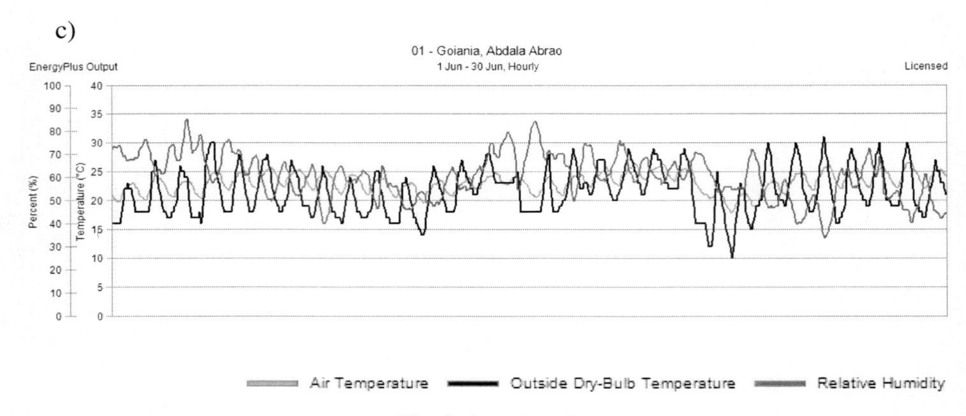

Fig. 3. (*continued*)

exposure and for the thermal delay on the buildings, resulting it to be the main re-emitter of heat during the last hours of insolation, increasing the temperatures in this period of the day. The walls were the responsible for heat emission during the first daily hours, reducing the peaks of minimum air temperature during this time of the day. This combination between horizontal and vertical elements in the building confirms the importance of the integration of shaded and solar exposure architectural components, as the internal walls and the internal ceilings.

5 Conclusions

The use of two methods of analysis, merging various computational tools and measuring instruments proven particularly adequate for the type of study, allowing: analysis of the cases from different scales, from global to specific, besides allowing the isolation, differentiation and comparison of various elements and intervening variables to the thermal performance of the studied cases. The results obtained with the different methods are complementary, indicating in both houses, that while some building components help to improve the thermal performance of the modern naturally-ventilated houses, others worsen this, resulting in a thermal performance below his potential.

As the demonstrated measure and simulated dates, June is a month with the best index with a good surround correction and good inside conditions. Fig. 3c shows a good control of the internal air temperatures in relation to outside air temperatures with oscillations between 20ºC (68ºF) and 25ºC (77ºF) during the days.

The most critic period of the year is September with higher air temperatures and low air relative humidity. The comparison between measure and simulated dates evidences the use of air conditioning in the bedroom of house 1 (1C in the Fig. 1) in some days of this month. Although, the air temperature peak does not represent an uncomfortable situation in the others studied rooms for people that live in this climate, as Givoni (1998) [8] indicates. Still thus, the air temperature peak occurs with 3 hours of delay, sufficient time: to reduce the inconvenient effects with a mix of external and internal air, and to supply the critical moment applying cooling systems with low energy demand.

The reduction of the thermal demand through the application of the passive strategies found in modern houses evidences the good relation between the historical architecture focus and the contemporary sustainable concept. The study open the theme for a large period of our architecture history understanding the contemporary concepts as a re-interpretation of past concepts as Cook (1988) [10] states when says that the passive solar design is a recent creation of a old concept.

References

1. Bruand, Y.: Arquitetura Contemporânea no Brasil, 3rd edn. Perspectiva, São Paulo (1997)
2. ASHRAE. ASHRAE Handbook-Fundamentals. Atlanta: American Society of Heating Refrigeration and Air Conditioning Engineers, Inc. (2005)
3. ABNT. ABNT NBR 15575-1:2008. Residential Buildings up to 5 storied – Performance. Part 1: General requirements. Associação Brasileira de Normas Técnicas, Rio de Janeiro (2008)
4. Moura, A. A. P.: Arquitetura residencial moderna em Goiânia: delineando um cenário. Programa de Especialização do Patrimônio – PEP. Goiânia: Instituto do Patrimônio Histórico e Artístico Nacional. Estado de Goiás – IPHAN-GO (2009)
5. INMET. Normais climatológicas do Brasil 1961-1990. Instituto Nacional de Meteorologia, Brasília (2009)
6. INMET. Rede de Estações. Superfície Automática (2011), `http://www.inmet.gov.br/sonabra/maps/pg_automaticas.php` (accessed January-December 2011)
7. ABNT. ABNT NBR 15220-3:2005. Thermal performance in buildings. Part 3: Brazilian bioclimatic zones and building guidelines for low-cost houses. Associação Brasileira de Normas Técnicas, Rio de Janeiro (2005)
8. Givoni, B.: Climate considerations in buildings and urban design. John Wiley & Sons Inc., New York (1998)
9. Vaz, M.D.A.C., Zárate, M.H.V.: A experiência moderna no cerrado goiano. Vitruvius. [e-journal] Arquitextos 067, año (December 06, 2005) (accessed December 2010)
10. Fernandes, A.M.C.P.: Clima, homem e arquitetura. Trilhas Urbanas, Goiânia (2006)
11. Romero, M.A.B.: Arquitetura bioclimática do espaço público. Editora Universidade de Brasília, Brasília (2001)

Chapter 33
Energetic and Exergetic Performance Evaluation of an AC and a Solar Powered DC Compressor

Orhan Ekren[1,*] and Serdar Çelik[2]

[1] Ege University, Izmir, Turkey
orhanekren@gmail.com
[2] Southern Illinois University, Edwardsville, IL, USA
scelik@siue.edu

Abstract. This study represents experimental performance analyses of an alternative current (AC) and a direct current (DC) refrigeration compressors implemented in a 79 liter refrigerator. Experiments were carried out at continuously running (ON) and periodically running (ON/OFF) operation modes. Data was analyzed and a comparison in terms of cooling capacity, power input, coefficient of performance (COP), Carnot COP, and exergy efficiency was conducted. The comparison showed that DC compressors can be much more efficient than AC compressors in refrigeration units.

Keywords: Direct current, compressor, solar energy, variable speed, energy saving, exergy.

1 Introduction

Reduction of energy usage and energy conservation in household appliances has been pulling more attention by manufacturers and research institutions, recently. Green technologies have been focusing on either decreasing energy usage, or creating environmentally-friendly systems, worldwide. HVAC and refrigeration systems occupy the largest portion of overall energy consumption in domestic use. Domestic refrigeration systems have an advantage of being powered by green energy technologies such as solar and/or wind energy because of their low power demand. Using renewable energy resources along with an AC refrigerator requires an inverter, while DC systems can directly be implemented into the system. There are other uses of small-sized refrigeration systems besides for domestic use. Many vehicles such as trucks, caravans, boats, cars, etc. are often equipped with portable cooling appliances. Although small refrigeration systems' energy consumption is comparatively lower, energy usage reduction on these systems has big potential for saving energy as they are widely used. Although not many, several studies in literature have focused on solar energy assisted refrigeration systems employing DC or AC compressors. Modi et al. [1] present converting procedure of a 165 liter domestic refrigerator with an AC

[*] Corresponding author.

A. Håkansson et al. (Eds.): *Sustainability in Energy and Buildings*, SIST 22, pp. 357–365.
DOI: 10.1007/978-3-642-36645-1_33

compressor to solar powered one. They used a conventional domestic refrigerator for this purpose and it was re-designed by adding a battery bank, an inverter and a transformer, and powered by photovoltaic panels. Various performance tests were carried out to study the performance of the system. It was reported that COP decreased with time from morning to afternoon with a maximum COP of 2.1 at 7:00 AM. Economic feasibility of the system for the climate conditions of Jaipur city in India was also investigated. According to the analysis the system was found to be economic only with carbon trading option. Axaopoulos and Theodoridis [2] presented a solar-powered ice-maker using four DC compressors. This system was operated without batteries. The DC compressors were controlled using a maximum power tracking (MPT) controller. The MPT controller forced the compressor power to follow available solar energy to achieve much higher energy utilization. Overall efficiency of the system was reported to be 9.2%. Kattakayam and Srinivasan [3] investigated a domestic refrigerator employing R-12 refrigerant with an AC compressor powered by PV panels, a battery bank and an inverter. They presented the cool-down, warm-up and steady-state performance of the refrigerator which has an internal volume of 165 liters and 100 W cooling capacities. According to experimental results of their work, this kind of refrigerators could be used to improve the energy efficiency of domestic refrigerators. They mention that there was no degradation in the thermal performance when it is operated with non-sinusoidal input at inverter. Also, the performance could be improved by changing to vacuum insulated panels as the heat leak reduction contributes the highest to the cooling capacity. Ewert et al. [4] investigated a solar-powered refrigeration system using batteries. Three different cooling technologies (thermoelectric, Stirling, and vapor compression) were tested in the same vacuum insulated cabinet (365 liters), experimentally. The cabinet was also coupled with phase-change thermal storage materials. Experimental setup had a PV powered DC compressor in the vapor compression unit, a PV driven free piston Stirling cooler in the Stirling unit, and a PV driven thermoelectric unit. COP was compared under steady-state conditions. The thermoelectric solar refrigerator had a COP as high as 0.04 while the Stirling cooler reached a COP value of 0.14. Vapor compression cycle had the highest COP of approximately 1.0.

Although several aspects of solely an AC or solely a DC compressor have been studied in literature, there is a gap in comparison of refrigerators employing AC or DC compressors. In this study, performances of an AC and a DC compressor implemented on a 79 liter refrigerator are experimentally investigated in terms of cooling capacity, power input, COP, COP_{Carnot} and exergy efficiency.

2 Experimental Studies

Experiments were carried out on a mini-refrigerator. AC and DC compressors were mounted on it, respectively. The refrigerator thermostat was set at 6th (maximum) yielding the operation mode as ON throughout the experimental work. Each test was conducted three times for the sake of repeatability. Average values from the three experiments were used in the analyses. Calculations were performed during the steady-state period. Refrigerant (R134a) temperatures and pressures were measured by T-type thermocouples, and ratio-metric type transducers, respectively. Power input for the compressors was measured by a wattmeter.

3 Energetic Analysis

For the thermodynamic analysis of the cooling cycle, cooling capacity is calculated by

$$\dot{Q}_{cooling} = UA(T_{amb} - T_{cabinet}) \tag{1}$$

where UA is the total heat transfer coefficient of the refrigerator cabinet which was obtained experimentally as 0.6021 W/°C. $T_{ambient}$ and $T_{cabinet}$ are ambient and cabinet temperatures, repectively.

Coefficient of performance of the compressors is

$$COP = \frac{\dot{Q}_{cooling}}{\dot{W}_{comp}} \tag{2}$$

where $\dot{W}_{comp}$ is compressor power input. This value was measured by a wattmeter. COP_{Carnot} and percentage of COP_{Carnot} are calculated by Eq.'s 3 and 4, respectively.

$$COP_{Carnot} = \frac{T_{cold}}{T_{hot} - T_{cold}} \tag{3}$$

where T_{cold} and T_{hot} are evaporation and condensation temperature of the cooling cycle, respectively. Percentage of COP_{Carnot} shows the potential that already has been used,

$$\%COP_{Carnot} = \frac{COP}{COP_{Carnot}} * 100 \tag{4}$$

4 Exergetic Analysis

Some assumptions were made during the exergy analysis for the sake of simplicity. These assumptions can be listed as;

- All processes are steady state and steady flow with negligible potential and kinetic energy also no chemical or nuclear reactions in the system.
- Heat transfer to the system and work transfer from the system are positive. Adiabatic compressor and expansion device.
- Saturated vapor at the inlet of the compressor.
- Pressure drops are neglected at evaporator and condenser since the values are too low.

Hence, exergy balances for a steady flow can be expressed as [5]:

$$\sum_{in} \dot{E}x - \sum_{out} \dot{E}x + \sum \dot{E}x^{Q} - \dot{E}x^{W} - \dot{E}x_{des} = 0 \tag{5}$$

In the refrigeration cycle, exergy balance of the compressor can be calculated as,

$$\left(\dot{E}x_{in} - \dot{E}x_{out}\right)_{comp} - \dot{W}_{comp} - \dot{E}x_{des\_comp} = 0 \tag{6}$$

where $\dot{W}_{comp}$ is the supplied actual electrical power to the compressor. For the condenser, exergy balance is formulated as,

$$\left(\dot{E}x_{in} - \dot{E}x_{out}\right)_{con} + \dot{Q}_{con}\left(1 - \frac{T_{amb}}{T_{cond}}\right) - \dot{E}x_{des\_con} = 0 \tag{7}$$

Similarly, exergy balance of the capillary tube is defined as,

$$\left(\dot{E}x_{in} - \dot{E}x_{out}\right)_{cap} - \dot{E}x_{des\_cap} = 0 \tag{8}$$

Exergy balance of the evaporator can be written as,

$$\left(\dot{E}x_{in} - \dot{E}x_{out}\right)_{ev} + \dot{Q}_{ev}\left(1 - \frac{T_{amb}}{T_{evap}}\right) - \dot{E}x_{des\_ev} = 0 \tag{9}$$

where T_{amb}, T_{cond}, and T_{evap} are the ambient (dead state), condensation, and evaporation temperatures, respectively. The current refrigerator's DC compressor can be powered by an 80 W photovoltaic panel utilizing multicrystalline cells. The inlet exergy of the photovoltaic module includes solar radiation intensity exergy. According to Petela's theorem [6],

$$\dot{E}x_{in\_pv} = S\left(1 - \frac{4}{3}\frac{T_{amb}}{T_{sun}} + \frac{1}{3}\left(\frac{T_{amb}}{T_{sun}}\right)^4\right) \tag{10}$$

where T_{sun} is the sun's temperature (~5778 K) and S is the solar radiation on the photovoltaic panel in W/m^2. Exergy destruction of the photovoltaic panel includes external and internal components. External exergy destruction occurs due to heat losses ($\dot{Q}_{loss}$) from the photovoltaic panel surface. Internal exergy destruction is caused by electrical losses, optical losses, PV-sun temperature difference, and PV-environment temperature difference. In this study, only electrical component was considered for the internal exergy destruction. Hence, total exergy destruction of the photovoltaic panel is

$$\dot{E}x_{tot\_dest,pv} = \dot{E}x_{dest\_ext} + \dot{E}x_{dest\_\mathrm{int}} = h_{pv-srf}A_{pv-srf}(T_{cell} - T_{amb}) \left(1 - \frac{T_{amb}}{T_{cell}}\right) + \left(I_{sc}V_{oc} - I_m V_m\right) \tag{11}$$

where h_{pv-srf} , A_{pv-srf} and T_{cell} are the convective heat transfer coefficient from the photovoltaic surface to ambient, area of the photovoltaic surface and cell temperature, respectively. The convective heat transfer coefficient from the photovoltaic surface to ambient can be defined by using $h_{pv-srf} = 2.8 + 3V_w$ correlation [7] for natural convection, considering wind velocity (V_w).

In Eq.11, $I_{sc}V_{oc}$ represents maximum electrical energy which can be produced by photons and I_mV_m, denotes maximum power point of the photovoltaic panel at the reference conditions as declared by the manufacturer [8]. Exergy efficiency of the overall system is formulated on the net rational efficiency basis:

$$\psi_{SYS} = \frac{\dot{E}x_{ev}}{\dot{W}_{comp\_elect}} = \frac{\dot{E}x_{in\_ev} - \dot{E}x_{out\_ev}}{\dot{W}_{comp\_elect}} \tag{12}$$

Also, exergy efficiency of the each components (ψ) equals to the ratio between $\dot{E}x_{out}$ and $\dot{E}x_{in}$.

5 Results

Measured and calculated values of the compressors were analyzed to define energetic and exergetic performances of the AC and the DC compressors. Total duration for all experiments were four hours, with a sampling frequency of 30 seconds. Comparison of the AC and DC compressors' cabinet temperatures is illustrated in Fig. 1. The refrigerator thermostat sets the cabinet temperature to -10 °C at the highest level.

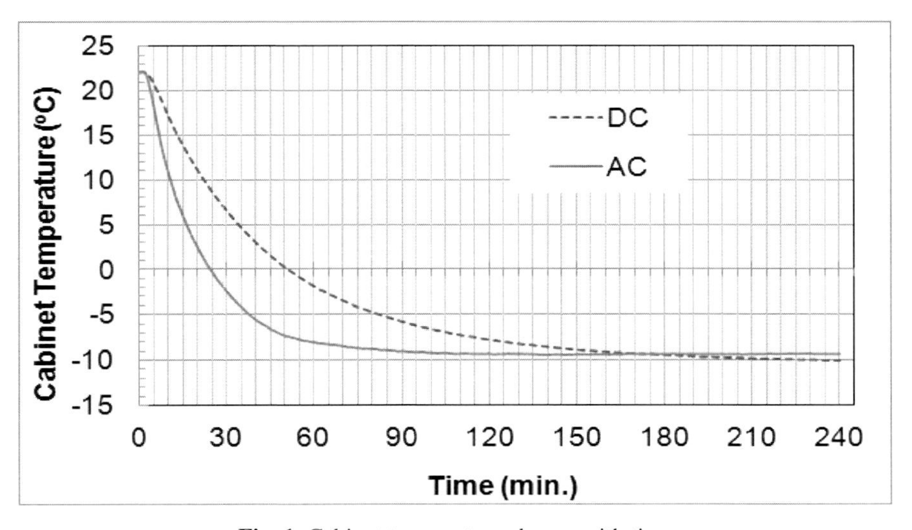

Fig. 1. Cabinet temperature change with time

Evaporation and condensation temperatures are shown in Fig. 2a and 2b, respectively. These temperatures reveal information about the Carnot efficiency, as well as the inside and outside conditions for the refrigerator.

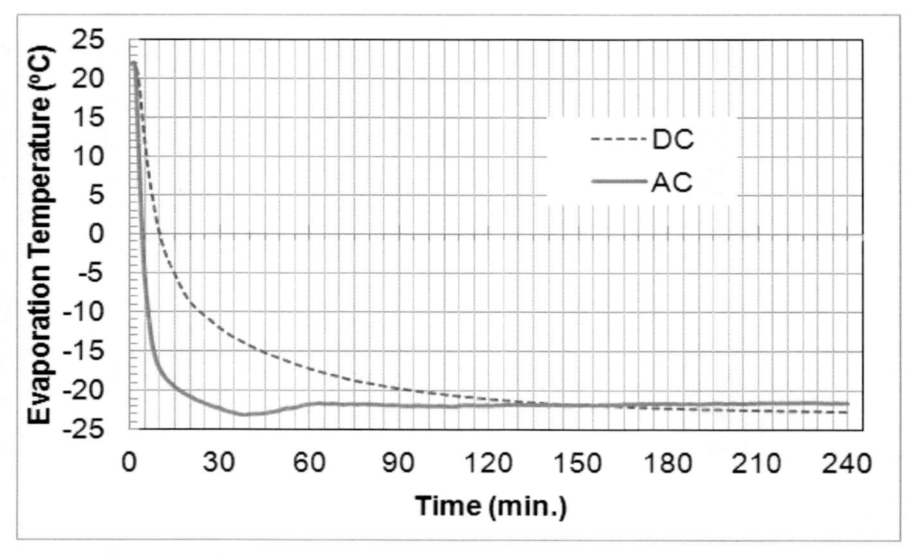

Fig. 2a. Evaporation temperatures change with time

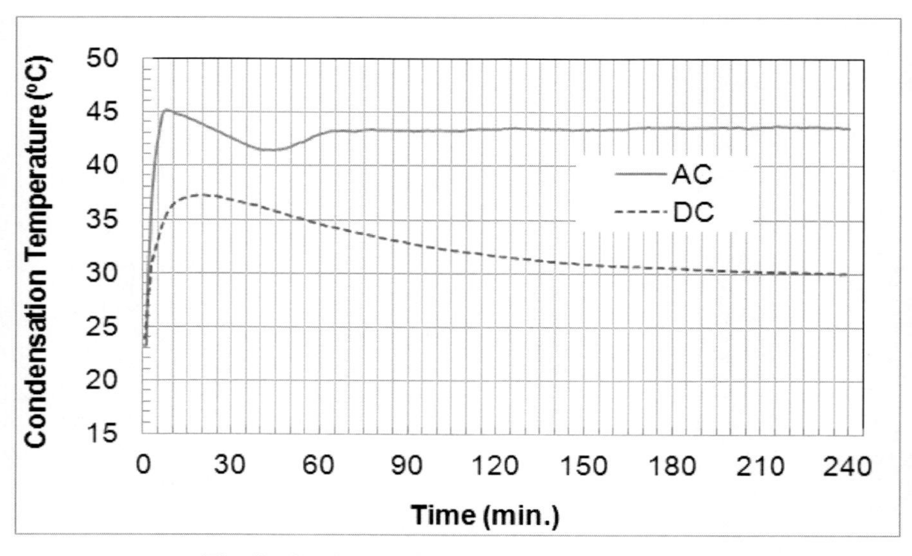

Fig. 2b. Condensation temperature change with time

To calculate the *COP* of the refrigeration system, cooling capacity is calculated via measured temperatures and power input. Power input values for both systems are illustrated in Fig. 3.

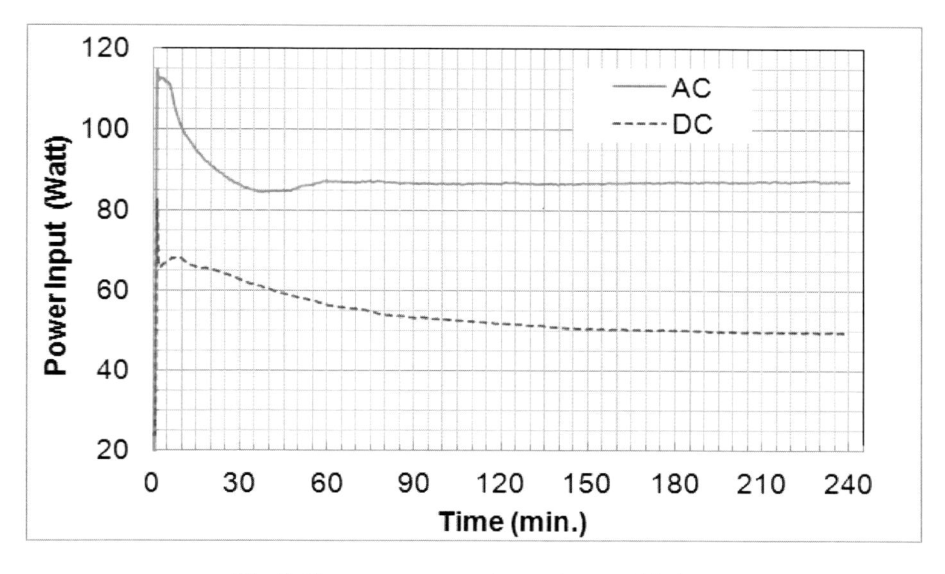

Fig. 3. Compressor power input change with time

It is seen from Fig.3 that power input to the DC compressor is lower than the AC compressor. Under these conditions, the DC and the AC compressors' cooling capacities are 19.8 W and 19.9 W, respectively. However, the DC compressor provides higher COP than the AC because of lower power input. Table 1 represents energetic and exergetic performance results for both systems.

Table 1. Energetic and exergetic performance results

DC System	$\dot{E}x_{des}$ (kW)	ψ_{SYS}	COP	COP_{Carnot}	$\dot{Q}_{cooling}$ (kW)
Compressor	0.043				
Condenser	0.001				
Capillary	0.001	7.7	0.4	4.7	0.02
Evaporator	0.007				
PV	0.036				
AC System	$\dot{E}x_{des}$ (kW)	ψ_{SYS}	COP	COP_{Carnot}	$\dot{Q}_{cooling}$ (kW)
Compressor	0.076				
Condenser	0.003				
Capillary	0.002	4.0	0.2	3.9	0.02
Evaporator	0.007				
PV	0.035				

6 Conclusion

Using solar energy without a power inverter yields an added benefit to refrigeration systems employing DC compressors. Performance of the DC and the AC type refrigeration systems were experimentally investigated. The comparisons were realized in terms of cooling capacity, *COP*, COP_{Carnot}, and exergy efficiency.

Energy analysis showed that both compressors yielded almost same cooling capacities, however, DC system yielded twice as high *COP* as the AC system because of lower power input. Also, DC compressor resulted in 23% higher COP_{Carnot} because of smaller temperature span between the condensation and evaporation temperatures. DC compressor's COP_{Carnot} percentage value was also observed to be higher, meaning that the DC compressor uses higher portion of its potential.

In addition, DC compressor had lower compressor surface and discharge temperatures than that of the AC compressor. Higher temperatures at the compressor side affect the life of lubrication oil and refrigerant in the compressor, unfavorably. Furthermore, DC compressor vibration levels were lower with respect to the AC compressor. This may be an important selection criterion for some applications where noise and system reliability are highly considered. Higher performance of DC compressors employed in refrigeration systems makes them a strong candidate for improved HVAC&R units.

Exergy analysis provided some useful information about the system such as total irreversibility distribution of the components and this analysis helps determine which component plays a key role to raise the system efficiency. According to exergy analysis, the largest exergy destruction occurs in the compressor and the photovoltaic panel, in this study. These two components should be further investigated to improve overall efficiency. The DC compressor showed 92.5% higher exergy efficiency than the AC compressor. Speed increasing and ON/OFF operation decreased energy and exergy efficiency of the systems. ON operation of the compressors resulted in about 50% higher exergy efficiency than that of the ON/OFF operation.

Acknowledgements. The authors would like to acknowledge the support of the Scientific and Technological Research Council of Turkey (TUBITAK) Reference number: B.02.1.TBT.0.06.01-219-413-3067.

References

1. Modi, A., Chaudhuri, A., Vijay, B., Mathur, J.: Performance analysis of a solar photovoltaic operated domestic refrigerator. Applied Energy 86, 2583–2591 (2009)
2. Axaopoulos, P.J., Theodoridis, M.P.: Design and experimental performance of a PV Icemaker without battery. Solar Energy 83, 1360–1369 (2009)
3. Kattakayam, T.A., Srinivasan, K.: Thermal performance characterization of a photovoltaic driven domestic refrigerator. International Journal of Refrigeration 23, 190–196 (2000)
4. Ewert, M.K., Agrella, M., De Monbrun, D., Frahm, J., Bergeron, D.J., Berchowitz, D.: Experimental evaluation of a solar PV refrigerator with thermoelectric, stirling, and vapour compression heat pumps. In: Proceedings of ASES Solar 1998 Conference, Albuquerque, USA (1998)

5. Dincer, I., Rosen, A.M.: EXERGY: Energy, Environment and Sustainable Development, 1st edn. Elsevier (2007)
6. Sarhadi, F., Farhat, S., Ajam, H., Behzadmehr, A.: Exergetic optimization of a solar photovoltaic array. Journal of Thermodynamics (2009), doi:10.1155/2009/313561
7. Watmuff, J.H., Charters, W.S., Proctor, D.: Solar and wind induced external coefficients for solar collectors. COMPLES 2, 56 (1977)
8. BP Solar,: BP380 Photovoltaic Module Data Sheet, `http://www.ecofreak.co.uk/pdf15901/BP380J%20S%20Data%20Sheet.pdf` (accessed April 12, 2012)

Chapter 34
Effectiveness of Sustainable Assessment Methods in Achieving High Indoor Air Quality in the UK

Gráinne McGill*, Menghao Qin, and Lukumon Oyedele

Queen's University Belfast, UK
{gmcgill03,m.qin,l.oyedele}@qub.ac.uk

Abstract. The use of sustainable assessment methods in the UK is on the rise, emulating the future regulatory trajectory towards *zero carbon* by 2016. The indisputable influence of sustainable rating tools on UK building regulations conveys the importance of evaluating their effectiveness in achieving *true* sustainable design, without adversely effecting human health and wellbeing. This paper reviews the potential trade-offs between human and ecological health in sustainable building design, particularly between building energy conservation and indoor air quality. The barriers to effective adoption of indoor air quality strategies in sustainable assessment tools are investigated, including recommendations, suggestions and future research needs. The consideration of occupants' health and wellbeing should be paramount in any sustainability assessment method, particularly indoor air quality, thus should not be overshadowed or obscured by the drive towards energy efficiency. A balance is essential.

Keywords: Assessment Methods, indoor air quality, sustainability.

1 Introduction

Since the introduction of BREEAM in 1990, considerable attention has been given to the development of environmental rating tools for use within the construction industry (Lee 2012). These tools provide the opportunity to assess projects environmental performance through criterion regarding the balance between the environment, energy, ecology and social and technological issues (Clements-Croome 2004). With the utilisation of these assessment methods on the rise, it is important to evaluate their effectiveness in addressing building performance, while recognising the trade-offs between human and ecological health (Levin 2005).

These trade-offs are now particularly important in the current building industry as research suggests building design strategies aimed at tackling the effects of climate change may have a negative impact on indoor air quality (IAQ) (Yu & Crump 2010; Crump et al. 2009; ECA-IAQ 1996). For example, the drive towards increased levels of airtightness in homes can be potentially dangerous if toxic finishes and materials

* Corresponding author.

A. Håkansson et al. (Eds.): *Sustainability in Energy and Buildings*, SIST 22, pp. 367–372.
DOI: 10.1007/978-3-642-36645-1_34

are not avoided indoors. This is supported by Boyd (2010), who suggests that the air quality in an energy efficient, airtight home may be worse compared to a leaky one, due to the potential for build-up of indoor air pollutants.

Thus, the drive towards energy efficiency may be unintentionally and inadvertently creating unhealthy living environments; through the generation of moisture problems, increase of toxic materials, reduction in ventilation rates, tightening of building envelopes and an over-reliance of elaborate technologies (Caroon, 2010; Crump et al, 2009; Wasley, 2000). As suggested by Clausen et al (2011) in *Reflections on the state of research*, important research questions include: 'how can we ensure that IEQ goals are met as energy consumption to operate buildings is reduced?' Furthermore, Clausen et al (2011) suggest the need for closer co-operation with green building councils to increase the awareness of indoor environmental quality and the effectiveness of meeting these needs in certification methods.

2 Building Energy Conservation Versus Indoor Air Quality

As suggested by Curwell et al (1990), "combining the highest quality air-conditioning systems, the flexibility for occupiers of opening windows, and the highest-possible level of energy efficiency cannot be achieved." Ideal solutions for indoor air quality and building energy conservation appear to be mutually incompatible. At present, the majority of environmental rating tools aimed at the built environment fail to adequately address the overall environmental performance of buildings and the necessary trade-offs involved (Levin, 2005). Instead, these tools remain highly subjective, encouraging building energy conservation and efficiency while disregarding subsequent health and wellbeing issues related to IAQ.

For instance, as suggested by Yu & Kim (2011), 'Buildings satisfying the requirements of BREEAM for airtightness would potentially enhance the indoor concentrations of pollutants as well as the risk of proliferation of moulds in these buildings.' According to Bluyssen et al (2010), BREEAM provides an assessment of building design with regards to a wide range of sustainability concerns, however does not guarantee the provision of good indoor air quality. This is supported by Yu & Crump (2010), who explain that the concept of 'health and wellbeing' is insufficiently addressed in BREEAM, which is mainly predominated by building energy conservation.

The latest version of BREEAM (2010) for multi-residential buildings disregards important IAQ considerations, such as radon, moisture control, contaminant control, pre-occupancy flush and advice on IAQ after occupancy. Furthermore, the scheme allocates only one point for low emissions of volatile organic compounds (VOC) (of which only fixtures and fittings are considered), which may heighten the problem of indoor air pollution (IAP), particularly in airtight buildings.

Similarly, the government's Code for Sustainable Homes (CSH's) section on 'health and wellbeing' does not directly address any issue relating to indoor air quality. In fact, the only aspect of IAQ that is considered by this scheme is 'adequate ventilation' in the home office, absence of HFC and EPS for blowing agents, nitrogen dioxide emissions (into the atmosphere), the use of low Volatile Organic Compound (VOC) products (only for home improvements), adequate drying space and dust pollution. As suggested by Crump et al (2009), in practice category 8 concerning guidance to the homeowner on efficient understanding and operation of heating and ventilation systems should benefit IAQ, however these benefits are difficult to quantify.

This lack of consideration on IAQ is particularly alarming considering the link between the CSH's and the UK building regulations. For instance, by 2013 the regulations for energy standards are intended to be raised in line with level 4 of the CSH's and level 6 (zero carbon) by 2016 (McManus et al. 2010). Furthermore, since May 2008, the UK government made rating against the CSH's mandatory for all new homes, with homes not assessed receiving an automatic 'nil' rating (DCLG 2009; Crump et al. 2009). Without adequate consideration of IAQ, the drive to meet zero-carbon energy standards may be detrimental to the health and wellbeing of building occupants.

International schemes such as America's LEED (Leadership in Energy and Environmental Design) consider indoor air quality considerably more than British and Irish counterparts. However, with LEED it is still possible to achieve the highest level of certification (platinum) without adhering to any criteria concerned with IAQ (Environment and Human health Inc. 2010; Levin 2012). Out of a total of 136 points, the interior environmental quality section receives only 21 points compared to 38 available points for energy and atmosphere (LEED for Homes 2010). Furthermore, according to Environment and Human Health Inc. (2010), LEED ignores widely published recommendations by the Environmental Protection Agency (EPA) on chemicals of concern, by disregarding four of these chemicals in the assessment tool.

3 Barriers for Effective IAQ Guidelines in Sustainable Homes

One reason for the lack of attention to IAQ may be due to the intangibility of health and the problems associated with measuring quantifiable benefits. As suggested by Dols et al (1996), references to indoor air quality by paradigms of sustainable building designs are often qualitative and general. This is supported by Bone et al (2010), who suggest that rating tools are mostly weighted towards easily definable measures of water and energy use, as opposed to health. Furthermore, Bluyssen (2010) explains the difficulty in precisely relating distinct, measurable chemical and physical parameters (in the interior environment) to impacts on health and wellbeing.

In addition, the changes in the UK building regulations towards more stringent demands on airtightness (including plans for pressure testing of new homes) will but pressure on architects and construction professionals to focus more on detailing (Ward 2008). However, as suggested by Dimitroulopoulou et al (2005), 'as dwellings become more airtight, sources of air pollution can have a greater impact on IAQ and occupants may experience adverse health effects.' Furthermore, other trade-offs between IAQ and building energy conservation (BEC) such as ventilation rates and specification of materials, may be more heavily weighted to BEC goals.

The specific emphasis on design goals by sustainability rating tools as opposed to performance goals further affects the ability of sustainable buildings to achieve targets in practice (Dols et al. 1996). Thus acclaimed sustainable buildings may, in reality, be no better than traditional building practices. This is particularly true when considering health and wellbeing criteria, due to the lack of post occupancy evaluations in this area. This is supported by Crump et al (2009), who suggest an urgent need for research into the impacts on health and wellbeing of highly energy efficient homes.

A further barrier to the successful adoption of IAQ strategies in sustainability rating tools is the lack of knowledge integration from indoor sciences (Levin 2005). The specialised nature of indoor air quality research is rarely translated into practical, comprehensive guidelines suitable to building designers and sustainable consultants. This sub-disciplinary tradition of indoor air quality research is a major problem, particularly as the building design is fundamental to the quality of the internal air.

4 Recommendations and Suggestions

These barriers result in a lack of comprehensive assessment methods which achieve environmentally friendly and healthy building design. There is an urgent need for an improvement of current systems through the development of effective IAQ criteria to counteract the trade-offs associated with energy efficient design. For instance, as suggested by Yu & Kim (2011), there is a need for criteria on the certification of materials with regards to their potential impact on the quality of indoor air. They further explain the importance of an IAQ management plan, stating; 'the plan should include IAQ monitoring to be assessed against the IAQ criteria certified by an approved body prior to occupancy' in housing projects.

This is supported by Bluyssen (2010), who suggests that existing sustainability labels do not provide sufficient information required to identify interior sources of pollution which have the potential to affect occupants' quality of life. Furthermore, Levin (2012) explains problems with low-emitting materials certification, suggesting the invalidity and unreliability of tests through variations in test atmospheres, in humidities, sample representativeness and repeatability.

5 Conclusion

Over-all, buildings should provide a healthy and safe atmosphere for living, thus the consideration of health and wellbeing issues, particularly IAQ, should be paramount in any sustainability rating tool (Yu & Kim 2011). A balance therefore is required which reiterates the fundamental triple bottom line principle of sustainable development, defined as 'an interpretation of sustainability that places equal importance on environmental, social and economic considerations in decision-making (Pope et al 2004)'. As suggested by Gibson (2001):

> 'Threats to human and ecological wellbeing are woven together in mutually reinforcing ways. So too, then, must the corrective actions be woven together- to serve multiple objectives and to seek positive feedbacks in complex systems'.

This is supported by Younger et al (2008), who refer to the analogous, complementary goals of indoor air quality and energy efficiency through the interactive relationship between climate change and air pollution. They explain that sources of anthropogenic air pollution subsequently contribute to global warming through their emission of volatile organic compounds (VOC's), carbon dioxide (CO2), and nitrous oxide (N2O). Thus BEC and IAQ goals can potentially be mutually beneficial.

6 Future Research Needs

Future research needs include the translation of existing knowledge from indoor sciences on IAQ to practical, relevant design guidelines aimed primarily at architectural and sustainable consultant professionals. The lack of knowledge on indoor air quality and associated health and wellbeing impacts in sustainable building design needs to be addressed through future research. In addition, further research is required to investigate the effectiveness of sustainability assessment methods, including emission certifications in reducing occupant exposure to indoor air pollution.

References

Bluyssen, P.M.: Towards new methods and ways to create healthy and comfortable buildings. Building and Environment 45, 808–818 (2010)

Bluyssen, P.M., Richemont, S., Crump, D., Maupetit, F., Witterseh, T., Gajdos, P.: Actions to reduce the impact of construction products on indoor air: outcomes of the European project HealthyAir. Indoor and Built Environment 19(3), 327–339 (2010)

Bone, A., Murray, V., Myers, I., Dengel, A., Crump, D.: Will drivers for home energy efficiency harm occupant health. Perspectives in Public Health 130(5), 233–238 (2010)

Boyd, D.R.: Dodging the toxic bullet: How to protect yourself from everyday environmental health hazards. D & M Publishers Inc., Canada (2010)

BREEAM, BREEAM Multi-residential 2008, Scheme Document SD 5064, BRE Global Ltd, Watford (2010)

Caroon, J.: Sustainable preservation: Greening existing buildings. John Wiley & Sons, Canada (2010)

Clausen, G., Beko, G., Corsi, R.L., et al.: Commemorating 20 years of indoor air, reflections on the state of research: indoor environmental quality. Indoor Air 21, 219–230 (2011)

Clements-Croome, D.: Intelligent Buildings Design, Management and Operation. Thomas Telford, London (2004)

Crump, D., Dengel, A., Swainson, M.: Indoor Air Quality in Highly Energy Efficient Homes – A Review, NHBC Foundation Report NF19. HIS BRE Press, Watford (2009)

Curwell, S., March, C., Venables, R.: Buildings and Health, The Rosehaugh Guide. RIBA Publications, London (1990)

DCLG (Department for Communities and Local Government), Sustainable New Homes- The Road to Zero Carbon- Consultation on the Code for Sustainable Homes and the Energy Efficiency standard for Zero Carbon Homes. DCLG Publications, London (2009)

Dimitroulopoulou, C., Crump, D., Coward, S.K.D., Brown, V., Squire, R., Mann, H., Pierce, B., Ross, D.: Ventilation, air tightness and indoor air quality in new homes. BRE Press, Watford (2005)

Dols, W.S., Persilly, A.K., Nabinger, S.J.: Indoor Air Quality in Green Buildings: A review and case study. In: Teichman, K.Y. (ed.) IAQ 1996 Conference, Paths to Better Building Environments, ASHRAE Maryland, October 6-8, pp. 139–150 (1996)

Environment and Human Health, Inc., The Green Building Debate, LEED Certification: Where Energy Efficiency Collides with Human Health, North Haven (2010)

European Collaborative Action (ECA-IAQ), Indoor air quality & its impact on man: indoor air quality and the use of energy in buildings. European Commission Joint Research Centre – Environment Institute, Report No. 17; EUR 16367 (1996)

Gibson, R.: Specification of sustainability-based environmental assessment decision criteria and implications for determining "significance" in environmental assessment (2001), http://www.sustreport.org/downloads/SustainabilityEA.doc (accessed April 6, 2012)

Levin, H.: Integrating Indoor Air and Design for Sustainability. Submitted for presentation at Indoor air 2005, Beijing, China, September 4-9 (2005)

Levin, H.: Why Green Building Rating Systems are Almost Always Wrong About IAQ. Powerpoint Presenation, ASHRAE winter meeting, Chicago, January 25 (2012)

LEED for Homes, Rating System, USGBC (Update from, version) January 1st, USA (2010)

Lee, W.L.: Benchmarking energy use of building environmental assessment schemes. Energy and Buildings 45, 326–334 (2012)

McManus, A., Gaterell, M.R., Coates, L.E.: The potential of the Code for Sustainable Homes to deliver genuine 'sustainable energy' in the UK social housing sector. Energy Policy 38, 2013–2019 (2010)

Pope, J., Annandale, D., Morrison-Saunders, A.: Conceptualising sustainability assessment. Environmental Impact Assessment Review 24, 595–616 (2004)

Ward: The potential impact of the new UK building regulations on the provision of natural ventilation in dwellings- a case study of low energy social housing. International Journal of Ventilation 7(1), 77–88 (2008)

Wasley, J.: Safe houses and green architecture: reflections on the lessons of the chemically sensitive. Journal of Architectural Education 53(4), 207–215 (2000)

Younger, M., Morrow-Almeida, H.R., Vindigni, S.M., Dannenberg, A.L.: The built environment, climate change and health, opportunities for co-benefits. American Journal of Preventive Medicine 35(5), 517–526 (2008)

Yu, C., Crump, D.: Indoor Environmental Quality, Standards for Protection of Occupants' Safety, Health and Environment. Indoor and Built Environment 19(5), 499–502 (2010)

Yu, C., Kim, J.T.: Building Environmental Assessment Schemes for Rating of IAQ in Sustainable Buildings. Indoor and Built Environment 20(1), 5–15 (2011)

Chapter 35
A Comprehensive Monitoring System to Assess the Performance of a Prototype House

Oliver Kinnane, Tom Grey, and Mark Dyer

TrinityHaus, Trinity College Dublin, Dublin 2, Ireland

Abstract. This paper presents a monitoring system and methodology designed for the evaluation of a prototype house, and proposed as an exemplar for comprehensive domestic monitoring. This system will enable assessment of a wide range of parameters for proof of concept and technology of the prototype. Consumption, occupancy patterns, indoor air quality, thermal conditions and building efficiency are monitored so as to gain a real-time understanding of building performance and occupant interaction with the house. The broad range of parameters will allow quantification of the correlation between for example; occupancy and consumption patterns, or air quality and ventilation system operation. Results of this monitoring study will inform future design iterations of this housing product.

1 Introduction

Buildings are responsible for much of all energy consumed, and dwellings account for a significant proportion of this. Although the majority of housing stock in Western Europe is built and will still exist in 2050 (The UK Low Carbon Transition Plan, 2009) and hence, much of the current focus is on upgrade of the existing stock (Spataru et al. 2010, Sinnott and Dyer, 2012), any new housing that is built should present the opportunity to achieve the highest standards of low-energy building. However, post-occupancy evaluation (POE) of low-energy buildings shows that real world building performance often doesn't match with design expectations (I.M. Pegg, A. Cripps, M. Kolokotroni 2007, Bordass et al. 2001). Also low-energy buildings can use considerably more energy than that predicted at the modeling stage (Bordass, W., Cohen, R. and Field, J. 2004).

Monitoring the actual building performance of low-energy housing can help establish the reason for this discrepancy between design or modeled expectations and real-world performance (Gill et al. 2011).

As part of a collaborative design, build and research project being undertaken by TrinityHaus of Trinity College Dublin, with an industry partner Glenbeigh Offsite Ltd., a prototype house was built using off-site, light gauge steel, construction. The house is

A. Håkansson et al. (Eds.): *Sustainability in Energy and Buildings*, SIST 22, pp. 373–379.
DOI: 10.1007/978-3-642-36645-1_35

built to assess the appropriateness of this construction method to the residential market in the Irish context and for export markets. The house was designed and built with the twin concepts of low carbon operation and adaptable form in mind (Grey et al. 2011). Significant attention was given to achieving good insulation and air-tightness so as to minimize consumption due to operational space heating. Also, considering the almost inevitability of home adaptation during its lifetime and the associated high levels of waste and energy consumption during this procedure, adaptability was planned for by enabling the easy removal of wall sections and the attachment of extension 'pods'. The house design package includes a mechanical heat recovery ventilation system, which will cover ventilation and space heating requirements.

To test this new housing-product, and the success of these concepts of low energy and adaptability an extensive monitoring study is being undertaken over the coming 24 months. During this time the house will be occupied by a family and phased extension will take place. Consumption patterns of domestic living habits in this housing type, and changes in patterns during and post extension will be monitored.

Residential energy monitoring and smart metering are becoming established in many countries, and are often driven by the electricity suppliers. Assessment of these large data sets allow for good insight into patterns of consumption of large populations (McLoughlin et al. 2012). These monitoring studies are however, limited in their scope as they are generally restricted to single or few meters.

A number of commercially available systems enable the monitoring of multiple parameters including energy, indoor air and thermal parameters, water, light, motion etc. However, no one system could be described as a fully holistic system. For an in-depth study of a prototype house an extensive monitoring system is required that allows all relevant parameters be monitored for analysis and for correlation assessment. This paper describes an extensive monitoring system appropriate for the analysis of a prototype house and applicable for future in depth home monitoring.

2 The Case Study Building

This paper describes a collaborative design, build and research project being undertaken by TrinityHaus of Trinity College Dublin, with an industry partner Glenbeigh Offsite Ltd. who specialize in offsite construction.

The so called Low Carbon Adaptable Home (LCAH) is a 3-bedroom house suitable as a family home. The house has a foot print covering 102 m^2 with a first floor area of 79.3m^2. The external envelope is 482 m^3, and contains a house volume of 500m^3.

The LCAH is built of a light gauge steel structure. The structure is wrapped in an air tightness membrane and with a layer of external insulation of 140 mm. An external render is applied to much of the house envelope. Areas that have been marked as possible future locations for extension are clad with fibre-cement rain-screen boards for easy demounting.

Fig. 1. The low carbon adaptable home, containing the described monitoring system

The air tightness of the LCAH has been measured at 0.6 air changes/hour. The external wall has been designed to achieve a U-value of 0.11 $W/m^2.K$. At a conservative estimate the energy consumption of the house is expected to approximate a maximum value of 50 kWh/ m^2/yr.

3 Research Objectives

The research objective of this project is to test this new housing product, principally from the point of view of low-carbon operation and for future adaptability. A monitoring study, of high resolution, is hence key to this assessment. This paper presents an extensive monitoring system - one approaching a holistic system - to enable assessment of a wide range of parameters for proof of concept and technology.

The designed monitoring system will enable assessment of:

- building type – its operational efficiency
- systems – efficiency of space heating and ventilation systems
- occupant consumption patterns – spatial and behavioral analysis
- indoor environment – air quality, thermal conditions
- occupancy – for the correlation of occupancy with consumption patterns

4 Research Methodology

4.1 Energy Monitoring

The electrical load represents the major proportion of the energy impact of the house, as domestic housing appliances, space heating and that portion of hot water not supplied by thermal solar panels are all to be supplied electrically. Hence, detailed electrical profiling of the house is key to developing a good understanding of its efficiency. At the design stage the electrical wiring system was designed to enable this level of resolution. The electricity monitoring strategy has a twin focus on:

1. Spatial profiling of electricity consumption
2. Behavioral / functional profiling of electricity consumption

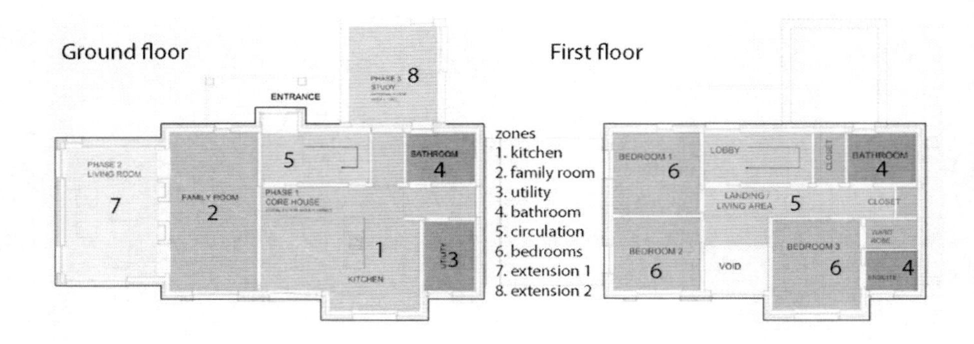

Fig. 2. Floor plans and spatial zones within the house

The spatial profiling will enable analysis within distinct house spaces, such as the kitchen, bedroom, bathroom etc. and importantly from the point of view of the LCAH prototype, also the added pods, hence, enabling analysis of the energy impact of the addition of extended spaces to the house. Distinct circuits were set up for 8 spatial zones: kitchen, family room, utility room, bathrooms, circulation space, bedrooms, extension pod 1 and extension pod 2 (Fig 2).

The reason for assessing the behavioural and functional energy profile is to enable analysis of functions (e.g. lighting and appliance usage) and behaviors (e.g. cooking and entertainment) that might be shared across multiple spatial zones. Electrical supply to the grouped common functions is monitored to capture this information.

Electricity is monitored using a total of 20 current sensors (CTs), all of which are located at the distribution board. Smart sockets are used to monitor the consumption due to individual appliances (*e.g.* fridge) and multi-smart sockets at locations where devices of common function are clustered (*e.g.*TV, DVD player, digital TV box, Playstation *etc.*).

4.2 Occupancy Monitoring

Monitoring of energy usage alone only tells a portion of the story. Relating energy usage to occupancy enables the correlation of domestic energy consumption patterns with spatial occupancy patterns in the home. Spataru et al. (2010) propose a thorough system of occupancy analysis that involves the occupants using tagging devices. In this study, occupancy patterns will be monitored using pulsed infrared sensors, which will detect motion in salient zones (Fig 4). Post processing of the data will be required to convert from motion data to sensible occupancy information.

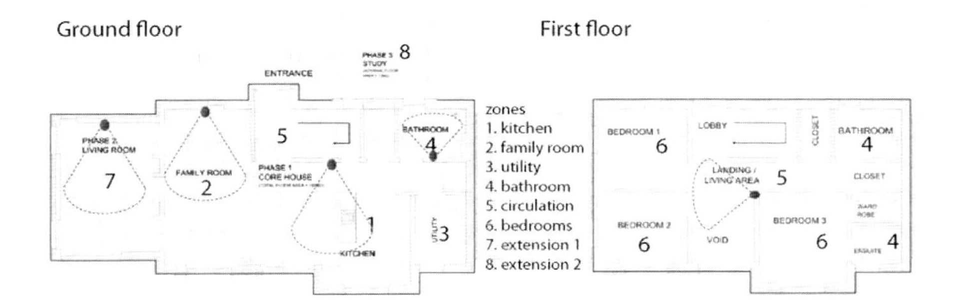

Fig. 3. Locations of monitors to assess occupancy in salient zones of the house

4.3 Indoor Environment Monitoring

As part of an overall thermal and environmental monitoring study, the indoor space is monitored for a range of parameters including temperature and humidity in multiple zones. To assess the ability of the mechanical ventilation system to maintain good air quality a host of air quality parameters including; carbon dioxide, carbon monoxide, hydrogen, ammonia and H_2S are monitored.

4.4 Other Monitoring

An array of temperature probes implanted in a horizontal line across the building envelope will provide temperature readings in the wall layers. This will enable assessment of the thermal characteristics of the designed wall configuration. The flue of the solid fuel fire is monitored with a temperature probe so that its operability can be monitored and hence, its impact on indoor thermal conditions, air quality and the efficiency of the MHRV can be assessed.

Water consumption is monitored using a pulse counter. External weather conditions, including solar incidence, wind velocity and direction are monitored. Light (lux) levels are monitored.

Consideration was given to the inclusion of contact sensors on the doors and windows, all of which are operable. However, it is presumed that the mechanical ventilation system will provide for ventilation for the vast majority of the occupancy period, and will be off when natural ventilation is chosen.

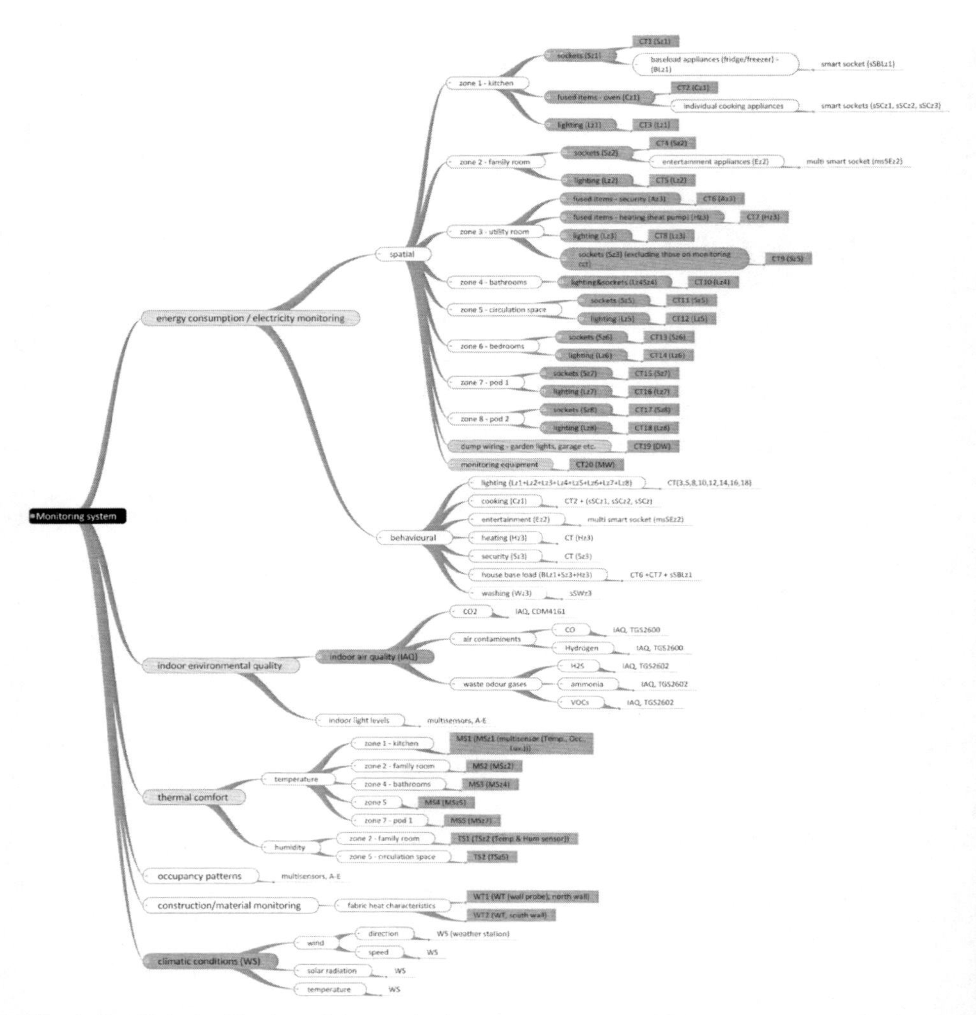

Fig. 4. Detailed specification of the monitoring system (Sensors: CT - current sensor, sS – smart socket, msS – multismart socket, MS - multisensor (of temperature, lux and PIR(occupancy)), TS – temperature and humidity sensor, IAQ – indoor air quality sensor, WT – wall temperature sensor)

5 Discussion and Results

This paper presents a comprehensive monitoring system for the thorough assessment of domestic energy consumption, environmental conditions and occupancy in a prototype house which is undergoing efficiency testing ahead of product release. This research project is ongoing and will present valuable information both for the evaluation of this prototype house and for the design of comprehensive monitoring systems for domestic properties in general. Results of this monitoring study will inform future design iterations of this housing product. A wide range of parameters are monitored to allow quantification of the correlation between consumption patterns, occupancy patterns, thermal conditions and building efficiency. This is an almost holistic monitoring system and it will enable a detailed evaluation of domestic living habits.

It is recognised that a limitation of this study is that it provides information regarding the subjective habits of a family in a single domestic building. Instead this study attempts to define a strategy for comprehensive monitoring of domestic habits, in a house-product that is due for multi-reproduction.

Occupancy of the low carbon adaptable home prototype begins this month and monitoring will continue for the coming 24 month period. A sufficient period of monitoring is required before validated findings and conclusions of this study can be published.

This paper outlines how limited monitoring equipment and budget can be optimised so as to comprehensively monitor a residential property type with a number of objectives in mind.

References

Grey, T., Dyer, M., Collins, R.: Low Carbon Adaptable Home: Seven Layers of Change. Accepted for publication in Engineering Sustainability, ICE Proceedings (2011)

Bordass, B., et al.: Assessing building performance in use 2: technical performance of the Probe buildings. Building Research & Information 29, 103–113 (2001)

Bordass, W., Cohen, R., Field, J.: Energy Performance of Non-Domestic Buildings: Closing the Credibility Gap | BuildingRating.org (2004)

Gill, Z.M., et al.: Measured energy and water performance of an aspiring low energy/carbon affordable housing site in the UK. Energy and Buildings 43(1), 117–125 (2011)

Pegg, I.M., Cripps, A., Kolokotroni, M.: Post-occupancy performance of five low-energy schools in the UK. ASHRAE Transactions 113(2) (2007)

McLoughlin, F., Duffy, A., Conlon, M.: Characterising domestic electricity consumption patterns by dwelling and occupant socio-economic variables: An Irish case study. Energy and Buildings 48(0), 240–248 (2012)

Sinnott, D., Dyer, M.: Air-tightness field data for dwellings in Ireland. Building and Environment 51(0), 269–275 (2012)

Spataru, C., Gillott, M., Hall, M.R.: Domestic energy and occupancy: a novel post-occupancy evaluation study. International Journal of Low Carbon Technologies 5(3), 148–157 (2010)

Stevenson, F., Rijal, H.B.: Developing occupancy feedback from a prototype to improve housing production. Building Research & Information 38(5), 549 (2010)

The UK Low Carbon Transition Plan: National Strategy for Climate an Energy (2009) ISBN: 9780108508394

Yohanis, Y.G., et al.: Real-life energy use in the UK: How occupancy and dwelling characteristics affect domestic electricity use. Energy and Buildings 40(6), 1053–1059 (2008)

Chapter 36
Smart Consumers, *Smart* Controls, *Smart* Grid

Catalina Spataru and Mark Barrett

UCL Energy Institute, Central House 14 Upper Woburn Place, WC1H 0NN, London, UK
{c.spataru,mark.barrett}@ucl.ac.uk

Abstract. The grid has three components: demand, transmission/distribution and generation, with the latter being mainly dispatchable, conventional power generation. A future grid based on renewable energy sources will impose serious challenges due to the variable nature of resources (wind, solar). In the transition from the current grid based on fossil and nuclear energy to a more sustainable one, based on renewable energy sources and components such as storage and with possible active participation by consumers, controls will play an important role, providing essential infrastructure for end users and system managers to monitor and control their energy usage. The uncertainty in the supply due to the integration of wind and solar energy will require intelligent control and with possible ways for shifting demand. The paper will discuss challenges, issues and advantages of demand reduction and demand shifting within a future *smart* grid with some illustrative examples.

Keywords: spatial patterns, smart grid, consumers, dynamic demand control and response, demand shifting.

1 Introduction

Across the world there are objectives and targets imposed by Governments regarding energy consumption, renewables and carbon emissions reduction. These have brought new challenges which require investigation and consideration to allow informed decision making. Current plans to achieve significant reduction in carbon emissions rely heavily on the future decarbonisation of electricity supply and the increased use of electricity to replace fossil fuels, for example, electric heat pumps to replace gas fired boilers and transport electrification with electric vehicles.

Fossil and nuclear fuels availability will reduce over the years and service demand will grow worldwide. Renewable energy sources are expected to supply a considerable share of electricity in the power system of the future, alongside conventional sources. Today's energy system has been built for conventional power generation systems, with power flow in one direction from power stations to consumers, with centralized network control and a relatively predictable demand. In the future, the grid needs to accommodate large quantities of energy coming from renewable sources which are variable and partially unpredictable. In addition, the grid will probably include integrated and active resources (such as storage for electric vehicles) and bi-directional power flows – export from consumer generation systems.

A. Håkansson et al. (Eds.): *Sustainability in Energy and Buildings*, SIST 22, pp. 381–389.
DOI: 10.1007/978-3-642-36645-1_36

Some people argue that grids are already managed *smart* and a complex system is managed effectively through the use of various technologies, such as off-peak electric heating, but that the grid is not taking advantage of its full potential [4]. In fact the term *smart* is a general term, the latest buzz word in the energy sector, which refers to a grid with more control, information and communication technology than the current grid. The *smart* grid is a continuation of the past hundred years of development, to a world where renewables and electricity services will require more dynamic control through enhanced information and communication technologies. To design *smart* grids, we first need to understand people demand services in time and space and use energy consuming technologies. Then we can look at options for controlling the supply of delivered energy for these services so as to facilitate renewable energy integration.

2 People's Demand for Services

The fundamental driver of all energy demand is the number of people and what they do. Forecasts predict an increase in ageing population along with a decrease in household size and an increased number of households [3]. An increase of 29% in number of households in England is projected for 2031 compared to 2006 [6]. The trend to more, but smaller households will cause an increase in electricity consumption per person unless balanced by efficiency gains.

The final electricity demand in UK in 2010 was 1% higher than in 2009 [7] and the domestic sector consumption was 36% of the total. As the electrification of the energy system increases to facilitate achieving carbon and renewable targets, electricity demand –both energy and power - will probably increase. According to DECC [7] in UK, without any form of demand response, peak demand could double by 2030. This will require significant investment in network reinforcement by National Grid and the Distribution Network Operators (DNOs).

To move towards a *smart* grid it is important to understand demand patterns and behaviours, and project how they might or be changed. Consumers' energy demand is complex, given the wide range of interlinking behavioural and technological factors: income, comfort desire, time clock settings; technology choice (building, heat pumps, gas boilers, microCHP) and use of technology and impact of weather. These human and technological factors combine in many different configurations. The relationship between individual consumer factors and energy consumption, and the wider context of public energy supply and society is complex on a range of different spatial and temporal scales.

3 Use of Monitoring Data and General Load Profiles Generation

Occupant behaviour has a significant role and impact on energy use and accurate quantitative data on time and location occupancy is rarely collected. With monitoring, we can better understand occupants' use of space and time in buildings, and travel.

Knowing what people actually do is an essential basis for designing *smart* energy service systems in terms of design and dynamic control. All this information can be used to understand better people's behaviour and the resultant energy flows. Of course there is a multitude of physical, psychological and sociological factors that influence human behaviour. The literature provides us with a variety of useful information and methods for occupants' use of appliances. Walker and Pokoski, [21]; Yao and Steemers [23]; Paatero and Lund [13]; Page [14] are some examples, where they looked at for how long and at what power demand appliances are likely to be in use. Richardson *et al* [15] uses Time Use Survey and derived activity profile for different occupants, creating a relationship between activity and energy use. Also, the literature is reach in information on demographic, electricity consumption, ownership level of appliances and energy consumption of certain appliances.

In this study it was used survey and monitored data from which general occupancy profiles were deduced. An example of household occupancy and their patterns and power demand consumption are shown in Fig. 1.

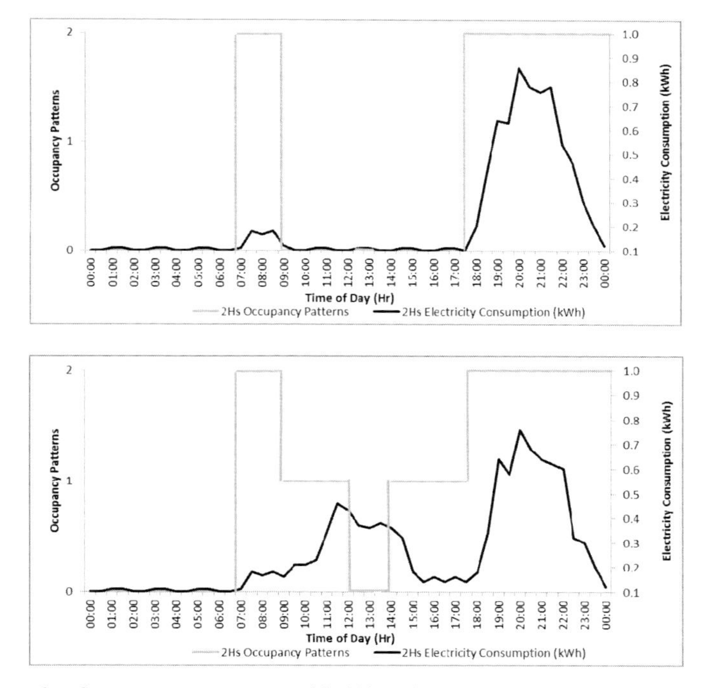

Fig. 1. Example of occupancy patterns and half hourly power demand for different occupancy

The profile represents two persons household, both working and being away from 8.30 am until 18:00 pm (first graph), but with one who works sometimes from home (second graph) in Fig. 1 and the variation in profiles for the two different situations. As it can be seen (second graph Fig.1) there is still electricity consumption between 12 and 2 o'clock, this is due to the use of washing machine and laptop left plugged in.

Across the population, these patterns will differ according to the activity of households - some will be at home during the day, others out working and so on. Of course, there are millions of households in most countries and we need to collate data on these to model them in aggregate accounting for system dynamics such as variability and peak loading. In addition there are important system dynamics problems, such as peak loading, which are often ignored or estimated simply in existing models.

Fig. 2 shows a plot of occupation where the density depicts aggregate time in zone on the floor – most time is spent in living room and bedroom.

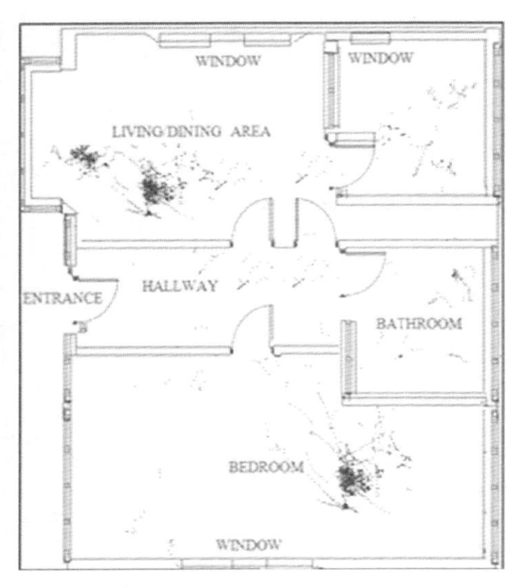

Fig. 2. Mapping occupancy density

In a previous case study [16] it has been shown that a large fraction of rooms is empty for most of the time. Also it has been shown that the densest occupied spaces were when the house was heated with electrical heaters [17].

This understanding allows us to improve design: service systems could be refined to give a more precise delivery in space and time. In the case of space heating, if rooms are thermally isolated within a dwelling, and the energy systems are designed to meet the service needs with any particular room when occupied, then energy consumption could be reduced. Concerning passive elements, it is not possible to deliver heat precisely in space, if rooms are poorly isolated or in time if the thermal mass is large. Control with person sensors for services such lighting, the television, hot water supply from taps with hand detection can all reduce energy consumption and the technologies are generally available now.

The demand for electricity varies according to end use through the day, week, month and year. For some end uses (like space and water heat, lighting) the demand depends on short and long term variations in weather. The changing composition of demand will alter the variations in total demand. The temporal pattern of demand, and particularly peak demand, determines the capacity and mix of electric power sources required. Barrett [2] covers the demand-supply balance and variable renewables integration. The least cost configuration was found given the constraints of meeting energy demands and a specified renewable energy fraction. The hourly power for each component of demand as it varies through the day, week and year was calculated. Fig. 3 shows illustrative the demand and the power management on a winter weekday for UK.

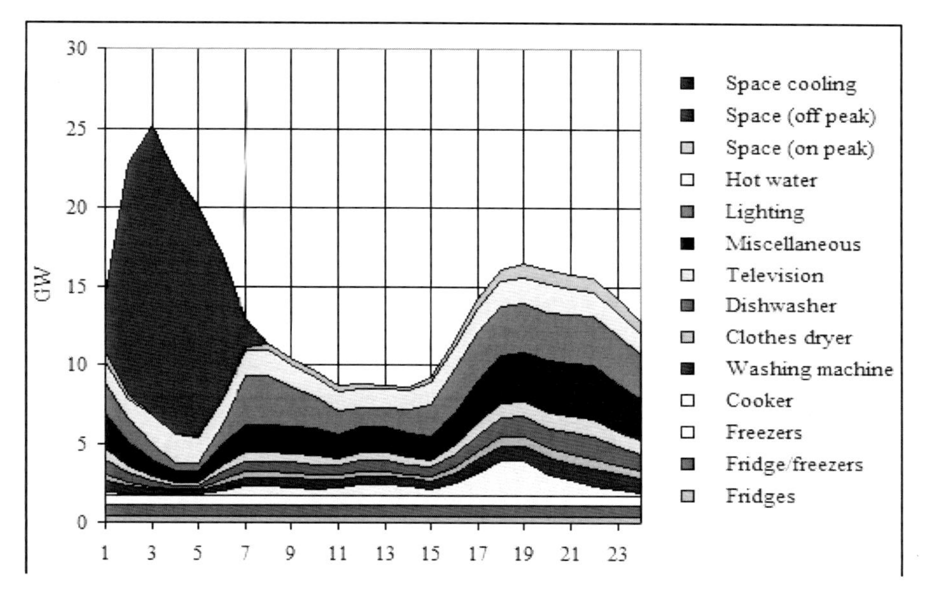

Fig. 3. Domestic demand, 24 Hrs, UK (GW)

Total demand at any time is an aggregation of millions of demands, with all of us making changes, such as every time we switch on or off a light, heater or appliance. This needs to be tested through simulations of individual consumers and aggregates of consumers in buildings, transport and industry. Moreover, of importance is how activities are distributed across time – hour of day, day of week, season etc. how they are affected by factors such as weather, and where these activities take place. This dynamic behaviour is one of the key determinants of energy load curves. These demands are then modulated by control system and the building itself to present demands to energy supply systems.

4 Loads and Demand Side Management (DSM) Potential

In the *smart* grid, DSM in the dynamic, short term is where consumers shift patterns of consumption to balance supply and demand and maximize the use of low short run marginal cost renewables. Consumers will play a role with active human control with in-home display, or through automated control. Through active control, consumers will decide how and when to use their demand-side response capacities which can be integrated as dispatchable resources ([10], [22]). Through load shifting consumers will be able to change their consumption behaviour and follow the dynamic fluctuations in the electricity prices ([1], [9], [12]). *Smart* grids will rely on domestic and other consumers' acceptance of the new systems.

The objective of *smart* management is to reduce the total cost of energy services to consumers. The total cost comprises the capital costs of the consumers' technologies, network and generators, plus the variable costs of the system for energy and operation and maintenance. Dynamic, minute to minute, management aims to minimize the short run marginal costs of the system. In a high renewables scenario this is done by managing the difference between demand and renewable supply that has to be met by dispatchable generators with higher avoidable costs (nuclear, fossil, biomass).

All connected service components and systems need to be considered together. This should be considered because the peak electricity load of a dwelling will be affected by the simultaneous operation of heat pumps, lights, dishwashers etc, Some, may be scheduled with energy storage (e.g. hot water) or service storage (e.g. clean dishes). Furthermore, optimal control strategy cannot be properly constructed without considering all service and supply systems as well as those in dwellings – the strategy should cover the whole UK demand-supply system. A control strategy for energy service systems comprises two parts: design of the human and technical system of components for controlling a system – information processing, switching etc. – and the utilization of those components to control operation in particular real time circumstances. The aim is that the two parts together achieve optimal control in the sense of minimizing or maximizing an objective function subject to constraints; for example the total cost (capital plus variable operational) of the system is minimized.

A number of possible objective functions might be pursued with control strategies, such as: minimize the total cost to dwellings; minimize the carbon of delivered energy (electricity, heat) to dwelling; minimize the heat or delivered energy to dwelling; minimize the peak power consumption of dwelling. These objectives can be conflicting and therefore the optimal control strategies may differ depending on objective functions. The optimal strategy will depend on occupancy and the design and operating characteristics of the dwelling and energy systems.

In homes, some devices can be controlled automatically, others require consumer interaction. Controllable loads include air-conditioning, space-heaters, washing machines, which can be shifted to other time or can be eliminated if not urgently needed. The aim is to calculate the optimal load shifting to minimize avoidable costs. In the UK, shiftable loads for domestic electricity usage account for about 10% ([5], [11]), but this is expected to increase with the electrification of space and water heating and any attendant heat storage. End-user types of shiftable loads and their topology are described in earlier works ([8], [18]): early shifting, late shifting, forward shifting, backward shifting, flexible, real-world. Early and late shifting types allows users to use lowest prices. Forward and backward shifting types give the possibility to postpone the consumption to a later (respectively earlier) time once the end users see a high price.

5 From *Smart* Home to *Smart* Grid Motivating Consumers

Following an analysis of service demands and the potential for load shifting, the system needs to be simulated and optimised to find how it can best be managed. Traditional optimization and simulation tools provide useful information about market operations, but they do not usually reflect the diversity of agents. An agent-based modeling approach can give us a better understanding of the behaviour diversity and over the last few years it has become an important tool for the energy sector. Some examples: Zhou *et al* [20] review agent-based simulation tools and their application to the study of energy markets; Sueyoshi and Tadiparthi [19] have developed an agent-based decision system for dynamic price changes for the U.S. whole-electricity market before and during the California energy crisis.

Energy usage needs to be optimized to maximize savings and maintain the comfort required by occupants. Intelligent agents need to be coordinated so they do not produce higher peak differences in the system than before. By combining local measurements for weather conditions and predictions from the weather forecast, the agent can look up at the local external temperature over the next 24hrs and in combination with forecast prices for electricity and with the supply energy available, control the devices so as to optimally reduce costs.

An initial attempt at developing *smart* control strategies is to look at how to maximize the utilization of low cost, low carbon supplies for a given system configuration. At the demand end, the control values can be used as one input to the control of the flows into user storage so that inputs are higher at times of high uncontrollable renewables and vice versa. On the supply side, there are system stores (e.g. pumped storage), dispatchable generators and energy trade to be controlled. If the costs of shifting demand to another time, or of disconnecting load, are less than the cost savings in delivered electricity then power management is worthwhile. Fig. 4 shows an illustrative example of a mix of CHP, renewable and dispatchable generation supplying electricity to demand, and export of surplus electricity for three days for Jan (1), April (4) and July (7) at 15min interval.

As it can be seen an improved control strategy of demand and supply components and additional storage and trade might reduce the dispatchable generation for which it is most difficult to achieve low carbon intensities.

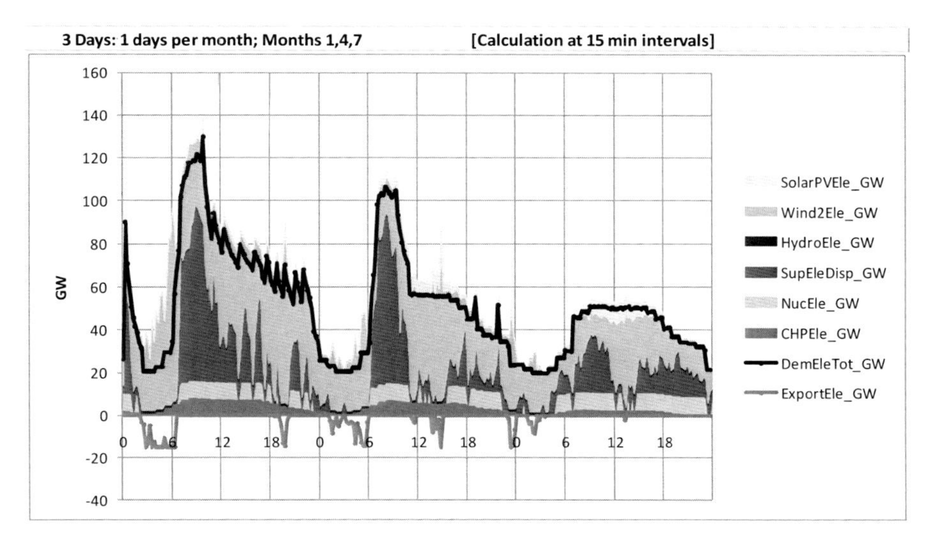

Fig. 4. Seasonal electricity demands and supplies

Intelligent agents, either technological or human, will act according to information such as the current and predicted price of electricity. They will endeavor to reduce costs by changing their energy use pattern to times when price is lower. This has to be carefully managed to avoid chaotic responses. The question is can near optimal results be obtained with a variety of consumers and suppliers operating autonomously with different stores (e.g. domestic and services heat stores, vehicle batteries) with different capacities and powers and other constraints such as temperature ranges, and in different initial states? There is the risk of controls over and under compensating because of feedback and system response times, with a chaotic outcome, or in the resultant control being far from the global optimum, because the primary customers' choice will to be shift their usage to a lower energy cost option. Alternatively, will the system have to be centrally managed? Authors will continue to research and find answers to these questions.

6 Conclusions

Good design and control of energy systems rely on accurate information about people's behaviour and their interactions with technologies; this can only be obtained through direct monitoring in conjunction with other data such as time use diaries. In a *smart* grid, households and other consumers will be more involved in the investments required for system management, and also in dynamic control.

The main objective of demand side and supply side management in a *smart* grid is to achieve low operational cost by maximizing the use of renewable energy sources. One of the biggest challenges for the future *smart* grid is the creation of an effective management structure for the system that is acceptable to consumers and suppliers. This is the essence of the *smart* grid and its potential benefits to the world for its future wellbeing. Essentially, it requires us to develop insight into the behaviour of how social structures and technologies can combine to be of benefit to all social segments, and at the same time to aid the achievement of wider energy and environmental targets essential for a sustainable society.

The battle for energy efficiency and an effective smart grid will not be won only through complex distribution networks, but mainly through end consumer participation.

References

[1] Albadi, M.H.: A summary of demand response in electricity markets. Elect. Power Syst. Res. 78(11), 1989–1996 (2008)

[2] Barrett, M.: A Renewable Electricity System for the UK. In: Boyle, G. (ed.) Renewable Energy and the Grid: The Challenge of Variability. Earthscan, London (2007) ISBN-13: 978-1-84407-418-1 (hardback)

[3] Barrett, M., Spataru, C.: Dynamic simulation of energy system. In: International Conference on Civil Engineering and Materials, Paris, France (2012)

[4] Conzelman, G., Koritarov, V.: Smart Planning for a Smart Grid, Elec-tric Light & Power (2011),
http://www.elp.com/index/display/article-display/8094560760/articles/utility-automation-engineering-td/volume-15/issue-9/features/smart-planning-for-a-smart-grid.html
[5] The Department for Communities and Local Government, The UK low carbon transition plan: National strategy for climate and energy. Technical report, UK Dept. of E & CC (2009), http://www.communities.gov.uk/documents/statistics/pdf/1172133.pdf
[6] Communities and Local Government, Household projections to 2031, England (March 11, 2009) ISBN: 978-1-4098-1285-2
[7] DECC, Planning our electricity future: a White Paper for secure, affordable and low-carbon electricity (July 2011),
http://www.decc.gov.uk/assets/decc/11/policy-legislation/emr/2176-emr-white-paper.pdf
[8] David, A.K., Li, Y.Z.: Consumer rationality assumptions in the real-time pricing of electricity. IEE Proceedings, pt. C 139(4), 315–322 (1992)
[9] Goel, L., Wu, Q., Wang, P.: Reliability enhancement and nodal price volatility reduction of restructured power systems with stochastic demand side load shift. In: IEEE PES General Meeting (2007)
[10] Ipakchi, A., Albuyeh, F.: Grid of the future. IEEE Power Energy Mag. 8(4), 52–62 (2009)
[11] MacKay, D.: Sustainable energy without the hot air UIT, Cambridge (2009)
[12] Medina, J., Muller, N.: Demand Response, Distribution grid operations: opportunities and challenges. IEEE Trans. Smart Grid 1(2), 193–198 (2010)
[13] Paatero, J.V., Lund, P.D.: A method for generating household electricity load pro-files. Int. Journal of Energy Research 30, 273–290 (2006)
[14] Page, J., Robinson, D., Morel, N., Scartezzini, J.-L.: A generalised sto-chastic model for the prediction of occupant presence. Energy and Buildings 40(2), 83–98 (2007)
[15] Richardson, I., Thomson, M., Infield, D., Clifford, C.: Domestic electricity use: A high-resolution energy demand model. Energy and Buildings 42(10), 1878–1887 (2010)
[16] Spataru, C.: Location Tracking and Energy Use – E.ON 2016 Research House. In: Ubisense User Conference, Cambridge, UK (September 2010)
[17] Spataru, C., Gillott, M., Hall, M.R.: Domestic energy and occupancy: a novel post-occupancy evaluation study. International Journal of Low-Carbon Technologies (5), 148–157 (2010)
[18] Strbac, G., Farmer, E.D., Cory, B.J.: Framework for the In-corporation of Demand-Side in a Competitive Electricity Market. IEE Proc. Genr. Transm. Distrib. (May 1996)
[19] Sueyoshi, T., Tadiparthi, G.: An agent-based decision support sys-temfor wholesale electricity market. Decision Support Systems 44(2), 425–446 (2008)
[20] Zhou, Z., Chan, W., Chow, J.: Agent-based simulation of electricity markets: a survey of tools. Artificial Intelligence Review 28(4), 305–342 (2007)
[21] Walker, C.F., Pokoski, J.L.: Residential load shape modelling based on customer behaviour. IEEE Trans. On Power Apparatus and Systems PAS-104, 7 (1985)
[22] Wang, P., Huang, J.Y., Ding, Y., Loh, P., Goel, L.: Demand side load management of smart grids using intelligent trading/metering/billing system. In: IEEE PES General Meeting (2010)
[23] Yao, R., Steemers, K.: A method of formulating energy load profile for domestic buildings in the UK. Energy and Buildings 37, 663–671 (2005)

Chapter 37
A Qualitative Comparison of Unobtrusive Domestic Occupancy Measurement Technologies

Eldar Nagijew, Mark Gillott, and Robin Wilson

University of Nottingham, University Park, Nottingham, NG7 2RD, UK

Abstract. Domestic occupancy measurement could save significant amounts of energy, either instantly via a home automation system or retrospectively via post-occupancy evaluation and feedback. However, not many localisation technologies are applicable to a domestic environment. In this paper three unobtrusive occupancy measuring technologies, i.e. Passive Infra-Red (PIR), Carbon Dioxide (CO_2) and Device-free Localisation (DfL), are compared. Their operation is explained and possible advantages and disadvantages are outlined. A qualitative experimental study then analyses the abilities of each system to detect overall occupancy, detect room level occupancy, count the number of occupants and localise them. It has been found that CO_2 and PIR sensors are very limited. The impacts of other factors, such as windows or occupants' metabolic rates, were significant on the reliability of the measured data. Device-free localisation on the other hand has great potential, but requires further research.

1 Introduction

The domestic sector consumed 30.5% of the UK's final energy in 2010 [1]. Improving the energy efficiency of a dwelling is a challenge, as a variety of factors influence its energy consumption. One attempt has been to give occupants feedback of their energy consumption and thus make them consciously change their behaviour [2]. However, even the most energy aware consumer would not be able to control on a 24 hour basis, for example, complex heating patterns. Therefore, another approach is to use an automated system, which would be able to take energy saving measures autonomously and serve as an adjunct to occupant control. Undoubtedly, there will be situations when the occupants' interests will conflict with the energy saving measures. To reduce the occurrence of such situations, the home automation system should know the number of occupants and their location. This would also allow the system to take more educated decisions and increase the resolution of information given through post-occupancy evaluation.

In a domestic environment a localisation system is required, which does not reveal the occupant's identity, is unobtrusive, needs low maintenance, is visually pleasing and is energy efficient itself. Two established systems fulfilling these requirements have been identified. They are based on Carbon Dioxide (CO_2) and Passive Infra-Red (PIR) sensors. These as well as an emerging third technology called Device-free

A. Håkansson et al. (Eds.): *Sustainability in Energy and Buildings*, SIST 22, pp. 391–400.
DOI: 10.1007/978-3-642-36645-1_37

Localisation (DfL) will be introduced and analysed in this paper. In section 2 each system will be explained in detail, along with their advantages and disadvantages. Section 3 will describe the methodology used to experimentally compare the three technologies. The results will then be described in section 4 and section 5 will conclude the paper.

2 Technology Descriptions

2.1 CO_2 Technology

Humans naturally exhale CO_2 on a constant basis, therefore the CO_2 concentration in a building can be used as an indicator of occupancy. However, Human's CO_2 generation rate depends heavily on their metabolic rate. Most research projects assume that the changes in CO_2 generation related to the metabolic rate are marginal and that an average value can be taken, as for example did Lu et al. [3] when they tried to deduce the number of occupants from CO_2 measurements in an office.

Also, CO_2 sensors have a significant reaction time. Emmerich and Persily [4] demonstrate that it is dependent on the volume of the room and the ventilation rate. However, this approach still assumes the air content to be well-mixed due to constant ventilation. In a non-ventilated room many other factors could influence the reaction time, such as the CO_2 dissipation within the room, the layout of the room, the infiltration rate, etc. This delay would be especially problematic if the residence time of occupants was short. Control strategies, which require fast switching times, could not be implemented and the efficiency of slower strategies would also be affected.

2.2 PIR Technology

Passive Infra-Red sensors detect moving objects of one temperature on a background of another temperature. Usually they are adjusted to the average human body temperature to identify occupancy more effectively. However, heat currents from HVAC systems can also trigger a PIR sensor, as mentioned by Teixeira et al. [5]. This is called false positive output. PIR sensors also suffer from false negative outputs, for example if occupants remain still. Furthermore, Akhlaghinia et al. [6] demonstrated in a domestic-like environment that PIR sensors can have difficulties to cover the desired visual area. They also point out the case in which a sensor associated with one room, is triggered by events in another room.

The PIR sensor's output is binary and can therefore not differentiate between the presences of one or several persons. A misconception is that the rate of triggering can be used to infer the number of occupants. However, PIR sensors have the advantage that they are cheap, low in energy consumption, easy to deploy and that they operate in real-time.

2.3 DfL Technology

Device-free Localisation takes advantage of the fact that the human body partially absorbs an emitted radio signal, thus decreasing the received signal strength (RSS). However, multipath propagation amongst others can also impact the quality of the signal. Filters could be employed to reduce these effects. Alternatively, the signal patterns in the environment can be learned previous to the localisation and can then be taken into account. However, the training required for such algorithms could prove problematic in a domestic environment.

Kosba et al. [7] demonstrated that the line-of-sight of the communicating radios is especially sensitive to human presence. The infrastructure should therefore be carefully designed to improve the localisation of occupants. Also, DfL has the advantage that it can be integrated into existing wireless home automation systems without any additional hardware costs. However, the estimation of the number of people present is still a challenge for this technology.

3 Methodology

The test-bed chosen to compare occupant detection strategies is a three bedroom domestic house. It is equipped with CO_2 sensors on the landing and in every living area, i.e. the living room, the dining room and all bedrooms. Those areas as well as the kitchen are also fitted PIR sensors on the ceiling. In addition, several temperature and humidity (TH) sensors are fitted throughout the house. All the CO_2, PIR and TH sensors send their recorded data approximately every four minutes via routers to a data logging system. The system stores these values and simultaneously records the RSS associated to each of the sensors.

The house is also equipped with a tag based ultra-wideband (UWB) tracking system, called Ubisense. This will be used as a reference during the tests to measure the actual location of the participants. The layout of the CO_2 and PIR sensors is illustrated in figure 1. The TH sensors are not represented, as the relation between their positions and the occupants' positions might not be linear due to the routers. The UWB Ubisense sensors are also shown in figure 1.

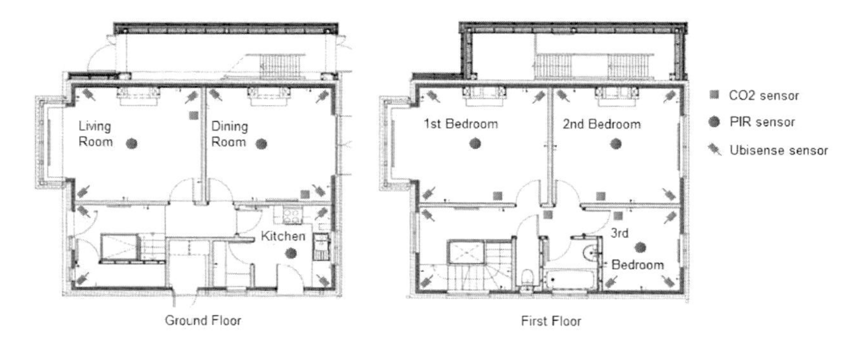

Fig. 1. Floor plans of the house with CO_2, PIR and Ubisense sensors

Each of the three systems mentioned will be tested to determine whether it can detect occupancy, differentiate the number of people present and has the potential to localise an occupant within a room.

4 Results

The house was occupied between 10:45 and 18:00 by occupants A, B and C. Each occupant had been given a UWB tag with an individual identification number. Table 1 shows the actual location of the occupants transcribed from the Ubisense recordings along with circumstances that could affect CO_2 or PIR results.

Table 1. Timeline of actual room occupation and added circumstances

	Occupied Rooms	Occupant	Additional Circumstances
10:45 – 11:00	Living Room	A	
	Dining Room	B	
11:00 – 12:00	Living Room	A	
	Dining Room	B	
	2nd Bedroom	C	
12:00 – 13:00	Living Room	A	
	Dining Room	B	All interior doors were closed
	2nd Bedroom	C	
13:00 – 14:00	Dining Room	B	
	1st Bedroom	A	
	2nd Bedroom	C (until 13:15) –> 0	
14:00 – 15:00	Dining Room	B (until 14:20) –> A+B	Cooking food
	Kitchen	A (until 14:20) –> 0	
15:00 – 16:00	Dining Room	A+B	MVHR on
16:00 – 17:00	Living Room	A	
	Dining Room	B	Window in Living Room was opened
	2nd Bedroom	0 (until 16:25) –> C	
17:00 – 18:00	Living Room	A	
	Dining Room	B	Occupant A exercised until 17:30 All interior doors were closed
	2nd Bedroom	C	

4.1 CO_2 Measurements

The house was unoccupied between approximately 18:00 the previous day and the start of the experiment. The data of all the CO_2 sensors is plotted in figure 2 to give an impression of their relative responses before discussing each in detail. The initial average CO_2 content measured was between 400 and 500ppm. These values correspond to

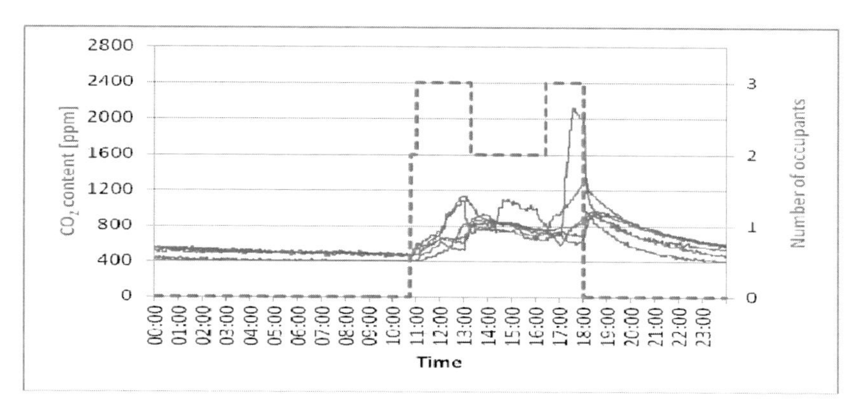

Fig. 2. CO2 content in all rooms versus overall human presence

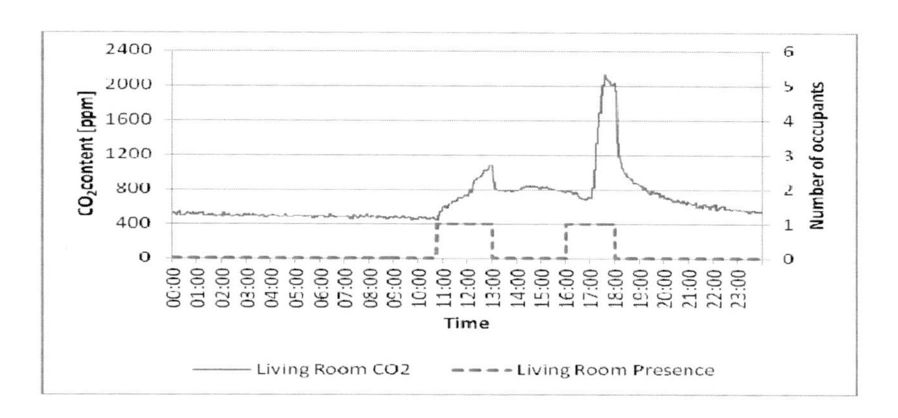

Fig. 3. CO2 content versus human presence in the living room

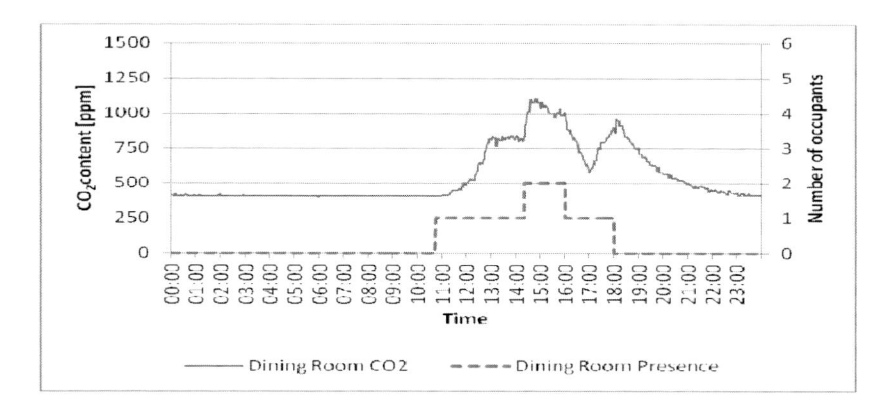

Fig. 4. CO2 content versus human presence in the dining room

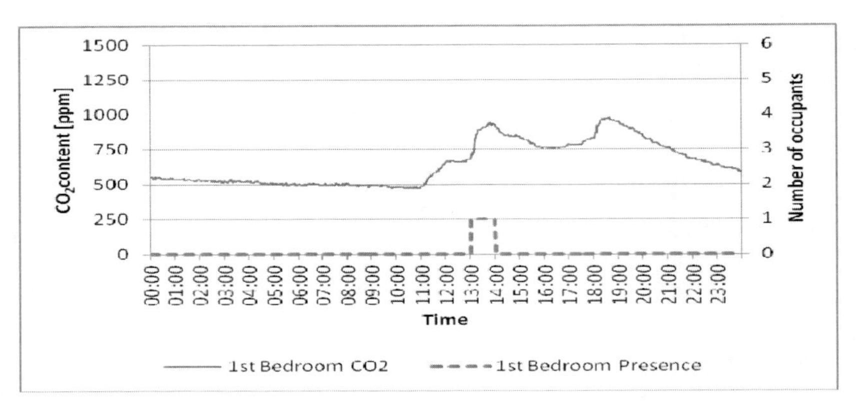

Fig. 5. CO2 content versus human presence in the 1st bedroom

typical outdoor CO_2 concentration and can therefore be taken as baseline reference. Figures 3 to 5 show the individual room's CO_2 measurements against recorded occupancy.

In order to explore the effects of variable space separation, the interior doors have been closed from 12:00 to 13:00 and from 17:00 to 18:00 as shown in table 1. It can be seen that before 12:00 the CO_2 content increased in individual rooms within the house, even unoccupied ones. Once the doors were closed, the content in the occupied rooms accumulated faster whilst it slightly decreased in the unoccupied rooms. Very important is also the moment when the doors were opened again. The air with the higher CO_2 content mixed with the air of the unoccupied rooms. This could lead to misinterpretation of occupation or mask the actual change of location of an occupant, as happened in the experiment. As shown in table 1 occupant A, who occupied the living room, moved to the 1st bedroom when all doors were opened at 13:00. These results suggest that a CO_2 sensor's output is highly dependent on the air circulation within its space.

The ability of the CO_2 system to detect the number of people has been examined. Two persons were present in the dining room between 14:20 and 16:00. In figure 4 the CO_2 levels shown are distinctively higher during that period. On the other hand, the very sharp rise in the living room between 17:00 and 17:30 is not related to an increased number of occupants, but only to the increased metabolic rate of a single occupant due to physical exercise as mentioned in table 1. Therefore, conclusions cannot be drawn reliably on the number of occupants based on CO_2 content.

4.2 PIR Measurements

The measurements represented in figure 6 are the times each PIR sensor has been triggered within four minute intervals. It shows that overall occupancy has been detected correctly.

Presence has also consistently been detected within the single rooms, as shown in the figures 7 to 9.

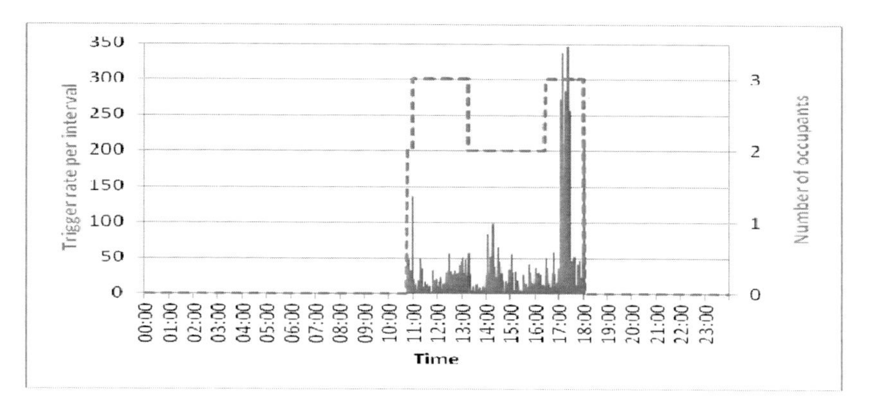

Fig. 6. PIR measurements in all rooms versus overall human presence

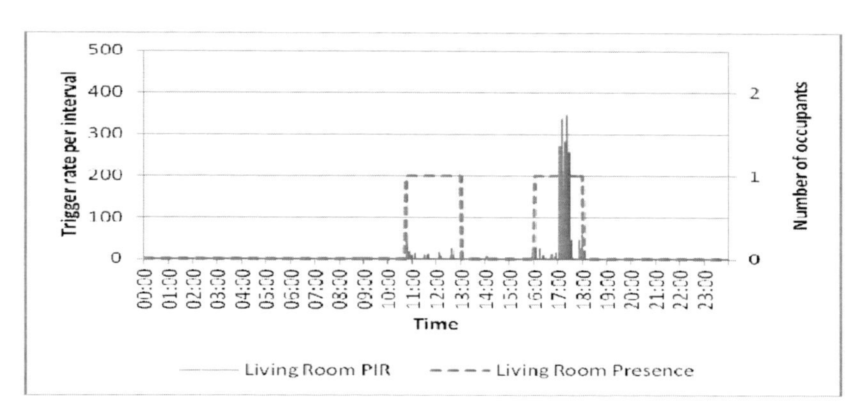

Fig. 7. PIR values versus human presence in the living room

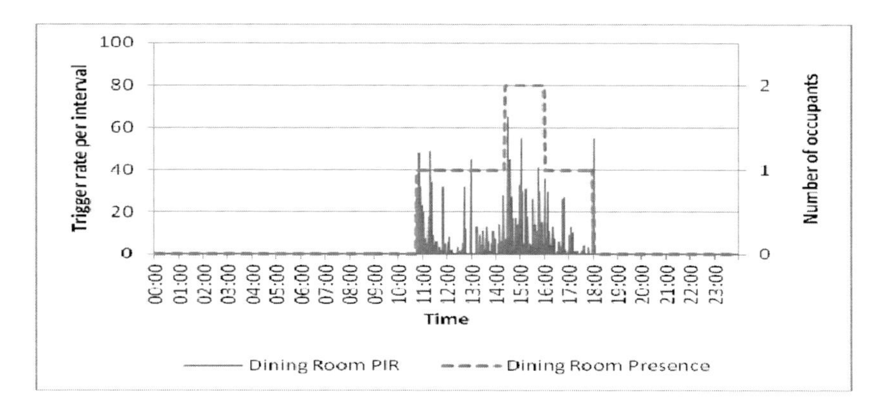

Fig. 8. PIR values versus human presence in the dining room

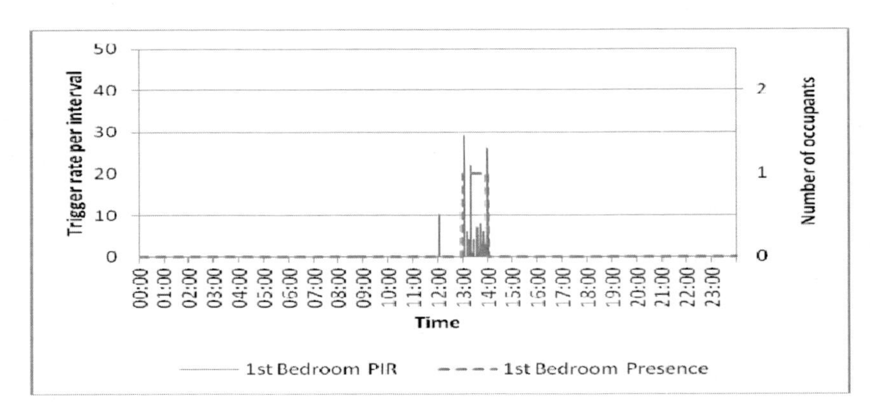

Fig. 9. PIR values versus human presence in the 1st bedroom

However, the PIR sensors were prone to negative false outputs. Several time intervals had no PIR measurements although the room was occupied. This occurred in all rooms and was probably caused by a reduced amount of movement of the occupant. The longest period of negative false measurement was in the living room for half an hour between 11:38 and 12:07. Positive false outputs on the other hand have not been detected. Figure 9 shows a value outside the occupied periods, however, it is related to the closing of the 1st bedroom's door.

Also, having two people in the same room did not result in greater PIR peaks as can be seen in figure 8. The average might be slightly higher over that period, but the average was also high around 11:00 when just a single person was present. The only real information that can be deducted from PIR values is the amount of movement that is taking place in the sensor's visual area. The assumption that greater movement equals more people present cannot always be applied. Over the whole day the biggest PIR values were measured in the living room between 17:00 and 17:30, which were caused by the physical activity of occupant A.

4.3 DfL Measurements

The signal strength of all the PIR sensors, CO_2 sensors, TH sensors and routers is shown in figure 10.

The signal strength for the occupied period from 10:45 to 18:00 has a clearly distinguishable pattern compared to the rest of the day. The fluctuations mainly before that period are due to the RSS varying between two gain stages.

A signal "sniffer" has been used during the experiment to monitor the network traffic. It showed that the connections were constantly rerouted, which made it difficult to associate a location or room to the signal strength of a specific radio.

Figure 11 represents the RSS values of two sensors, whose signal clearly responded to the combined human presence over the entire measurement period. It is

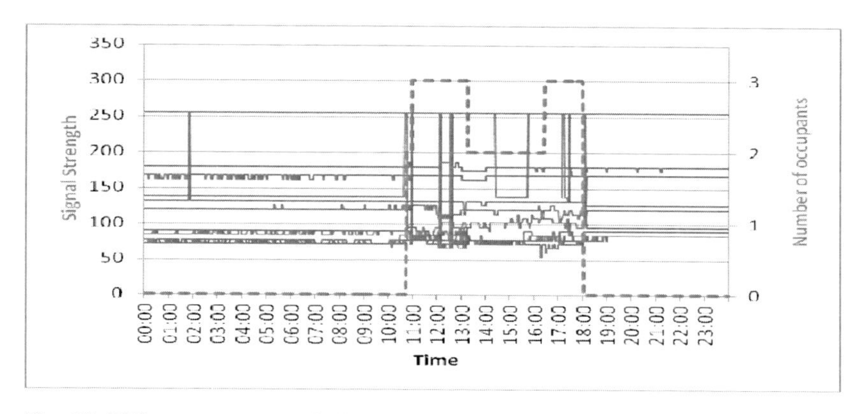

Fig. 10. RSS measurements of all sensors and routers versus overall human presence

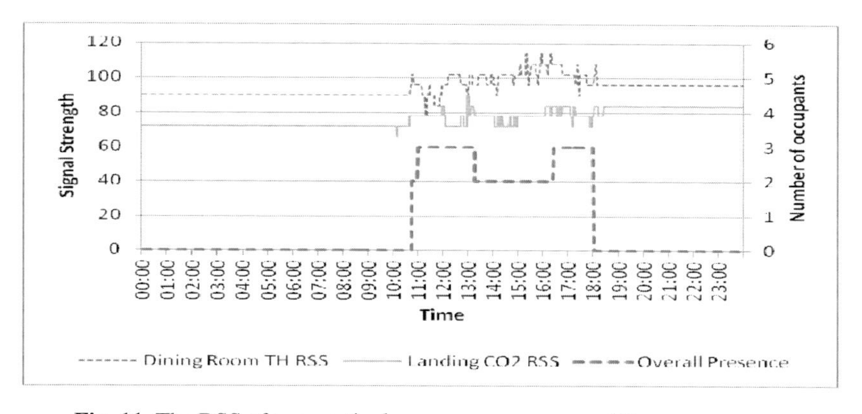

Fig. 11. The RSS of two particular sensors versus overall human presence

noticeable that the signals started varying before the occupants entered the house and also after they left it. As radio waves travel through walls, DfL purely aimed at indoor applications could be prone to give feedback related to events that take place outside the desired area.

Table 2. Findings of qualitative comparison

	CO2	PIR	DfL
Overall occupancy	Yes	Yes	Yes
Room occupancy	Partially	Partially	Unknown
Number of people	Partially	No	Unknown
Potential to localise	No	No	Unknown
Additional information	CO_2 sensors have a slow reaction time and are highly dependent on air circulation patterns and occupants' metabolic rate	PIR sensors operate in real time but are prone to negative false outputs	DfL operates in real time, but the relation between RSS and human presence is not linear and it is prone to events outside the building

5 Conclusion

Three unobtrusive occupancy measuring technologies that could be applied to domestic environments have been outlined and compared. The findings are summarised in table 2.

All three technologies have intrinsic restrictions; however DfL seems to have the greatest potential as the research literature suggests that it should be able to localise occupants. The exploitation of the data as well as the optimal setup would require further research work.

References

[1] Macleay, I., Harris, K., Annut, A.: Digest of United Kingdom Energy Statistics (DUKES) 2011, London (2011)

[2] Darby, S.: The Effectiveness of Feedback on Energy Consumption: A Review for DEFRA of Literature on Metering, Billing and Direct Displays. Environmental Change Institute, University of Oxford (2006)

[3] Lu, T., Knuutila, A., Viljanen, M., Lu, X.: A novel methodology for estimating space air change rates and occupant CO_2 generation rates from measurements in mechanically-ventilated buildings. Building and Environment 45, 1161–1172 (2010)

[4] Emmerich, S.J., Persily, A.K.: State-of-the-Art Review of CO_2 Demand Controlled Ventilation Technology and Application. National Institute of Standards and Technology NISTIR 6729 (2001)

[5] Teixeira, T., Dublon, G., Savvides, A.: A Survey of Human-Sensing: Methods for Detecting Presence, Count, Location, Track, and Identity. ACM Computing Surveys (2010)

[6] Akhlaghinia, M.J., Lotfi, A., Langensiepen, C., Sherkat, N.: Occupancy Monitoring in Intelligent Environment through Integrated Wireless Localization Agents. In: Proceedings of IEEE Symposium on Intelligent Agents, Nashville, USA, March 30-April 2, pp. 70–76 (2009)

[7] Kosba, A.E., Abdelkader, A., Youssef, M.: Analysis of a Device-free Passive Tracking System in Typical Wireless Environments. In: Proceedings of 3rd International Conference on New Technologies, Mobility and Security, Egypt, Cairo, December 20-23, pp. 1–5 (2009)

Chapter 38
Review of Methods to Map People's Daily Activity – Application for Smart Homes

Stephanie Gauthier and David Shipworth

UCL Energy Institute, 14 Upper Woburn Place, London, UK
s.gauthier@ucl.ac.uk

Abstract. People's daily activity in their home has widespread implications, including health and energy consumption, yet in most environmental studies people's activity is only estimated by using screening or structured observation. This paper reviews the current protocols and standards, and then identifies a mixed-method approach to measure people's activity levels in free-living environments. One of the key issues is to gather accurate measurements while using 'discreet' observatory methods to have minimum impact on their behaviour. With the recent emergence and advancement of more accurate and affordable sensing technologies, this problem might be overcome. Drawn from physiological research, heart-rate monitoring, accelerometry, and automated visual diary, were used in a field study, which monitored a small sample of UK households during the winter of 2012. Within a smart home, these methods could potentially be used to forecast energy demand for heating and to manage power distribution peaks.

Keywords: activity monitoring, ubiquitous sensor technologies, thermal comfort.

1 Introduction

People's daily activity in their home has been associated with health and energy consumption. In the UK, almost 32% of the total energy consumed is attributed to dwellings, of which 60% is used for space heating (DECC, 2011). This demand for heat varies greatly between dwellings. The main reason for these disparities has been attributed to different building fabrics and types of heating systems, but also to occupants' behaviour. A Danish study investigating the heat consumption in 290 identical homes found that the highest heat consumption was up to twenty times higher than lowest (Fabi et al., 2012). This study shows that occupant's behaviour plays a key role in the amount of energy consumed for heating. Moreover, it raises the question on the significant disparity in heat demand between occupants. Why does one occupant heat his/her home more than their neighbours? What are their expectation toward thermal comfort and their response toward thermal discomfort?

The adaptive thermal comfort model states behavioural, physiological and psychological adaptive processes (Auliciems, 1981, de Dear et al., 1998, Humphreys and Nicol, 1998). One common element of these three processes is the type and level

A. Håkansson et al. (Eds.): *Sustainability in Energy and Buildings*, SIST 22, pp. 401–411.
DOI: 10.1007/978-3-642-36645-1_38

of activity that a resident carries out. To investigate the physiological adaptive processes based on the heat balance, climate chamber experiments were implemented by Fanger (1970). However in this experimental context, it is difficult to look at behavioural and psychological adaptive processes. This is especially due to the lack of behavioural opportunities. To overcome this problem, studies have been carried out in free-living environments (de Dear et al., 1998; Parsons, 2001; Hong, et al., 2009). However current methods only estimate resident's activity through screening or structured observation. As reducing energy consumption as become a priority - Climate Change Act 2008, it has become important to gather a comprehensive evaluation of people's daily activity in their home. Recently, wearable instruments have been developed to objectively capture physical activity (Chen and Bassett, 2005). Used extensively in physiology, heart-rate monitors and accelerometers are able to assess activity by determining its type, duration and intensity. Also these tools have been able to measure a person's energy expenditure, which is a key variable when trying to determine the physiological response to thermal discomfort (Gauthier, Shipworth, 2012).

The majority of research on activity level estimation has been focusing on the use of pre-established metabolic rate values for a given workload, posture or activity (ISO 8996). These are implemented through questionnaires, direct observation or written diaries as self-reported activity type, duration, and intensity. While these try to depict activity in free-living environment, it is uncertain whether the actual activity has been recorded. The aim of this study was to develop a new set of methods to accurately estimate the activity level and duration in daily life. This approach was implemented in a field study during the winter of 2012. This paper presents and reviews the findings of this study.

2 Current Methods to Determine Activity Levels

Arisen from laboratory experiments in climate chambers, the current standards combine knowledge of the human body's physiology and of heat transfer theories. They form parts of the International Standard Organisation in BS EN ISO 7730:2005, and BS EN ISO 8996:2004. A person's physical activity is characterised by its type, intensity, duration, and frequency (Chen and Bassett, 2005). In the standards, methods have been developed to estimate activity level and to analyse the relationship between activity level and thermal comfort. As described in ISO 8996, metabolic rate (M) is a measure of activity level and is defined as the rate, at which the human body utilises oxygen, food, and other sources to produce energy, per surface area of the body. In summary, it refers to the rate of production of energy in time per surface area, and is expressed in watts per squared meter (W/m^2), or in metabolic unit (met), where 1 met = 58.2 W/m^2. Based on this definition, a person's metabolic rate consists of two components:

- Body surface area (BSA), which is assumed to be $1.8m^2$ for a man of 70kg, and $1.6m^2$ for a woman of $1.6m^2$ (Parsons, 2001).

- Energy expenditure (EE), which refers to the energy used per unit of time to produce power, and is expressed in watts (W) or more often in mega-joules per day (MJ/day) (Jeukendrup & Gleeson, 2004).

Metabolic rate (M) can estimated by using the following equation:

$$M = EE / BSA \ (W/m^2)$$

Where: M: metabolic rate; EE: energy expenditure; BSA: body surface area.

Methods to estimate or to measure human energy expenditure range from direct to indirect methods with associated level of complexity and cost. ISO 8996 provides the methodological framework to estimate this metabolic rate. It includes four levels, screening (1), observation (2), analysis (3) and expertise (4), summarised below:

- Level 1, screening: Metabolic rate is estimated by reviewing the subject mean workload for a given occupation (level 1A) or for a given activity (level 1B). Each activity-intensity corresponds to a metabolic rate range, as follow: resting (55 to 70 W/m^2), low level of activity intensity (70 to 130 W/m^2), moderate (130 to 200 W/m^2), and high (200 to 260W/m^2). This estimation method only provides rough information and is associated with a great risk of error.
- Level 2, observation: Metabolic rate is estimated by observing the subject-working situation at a specific time. Information such as time and motion are required for this type of study, including body posture, type of work, body motion related to work speed. The accuracy of the results is estimated to be within ± 20% (ISO 8996, Table 1).
- Level 3, analysis: Metabolic rate is determined from the subject heart rate recordings over a representative period. This method uses an indirect determination, based on the relationship between oxygen uptake and heart rate under defined conditions. This method holds an accuracy of ± 10% (ISO 8996, Table 1).
- Level 4, expertise: Experts determine metabolic rate using three different methods: (1) oxygen consumption measured over short periods, (2) doubly labelled water (DWL) method over longer periods - 1 to 2 weeks, (3) direct calorimetry method in laboratory. Each one of these methods requires specific measurements, which undermine their application in field studies, over longer periods of time. These methods have an accuracy of ± 5% (ISO 8996, Table 1).

In summary, these four methods can be divided into subjective and objective type. The subjective methods as level 1 and 2 apply questionnaire, observation, diaries, and/or activity log. They are used in most studies, as relatively low cost. However these methods often provide a biased assessment and are associated with a great risk of error. This is an issue when incorporating their estimated results within the predictive thermal comfort model (ISO 7730), as metabolic rate has been proven to be its most influential variable (Gauthier and Shipworth, 2012). High inaccuracy of metabolic rate estimation will undoubtedly undermine prediction of thermal comfort levels to design new home or intervention strategies for reducing energy demand in homes. To overcome these limitations, objective methods in level 3 and 4 measure physiological mechanism such as heart-rate, body temperature and metabolic effect.

These provide a reliable assessment of activity level, but their applicability in free-living environment may be limited. However with the recent advancements in more accessible, accurate and affordable sensing technologies, this may be overcome. The review of current methods used to estimate activity level reveal some limitations in terms of accuracy and of usability in fieldwork. To address those issues, this paper then explores alternative methods to measure, to observe and to analyse people's activities in households.

3 Field Study Methods

As sensing technologies have developed over the past few years, many significant advances have taken place in the area of people's activity assessment (Trost, 2005). One of the most noticeable has been the rapid uptake of accelerometry, which measures movement as bio-mechanical effect. This objective technique enables the estimation of activity level in a free-living environment, for period of time representative of a person daily activity level, with minimal impact and discomfort. Supported by other methods such as heart-rate monitoring and automatic visual diary, this study's mixed-method approach aim to collect objective and accurate measurements of daily activity level in home. This paper addresses this objective with a view toward establishing an evidence-based protocol to implement activity measurement.

3.1 Accelerometry

To estimate energy expenditure, accelerometers quantify activity level by measuring acceleration of a person in movement. Acceleration is defined as the change in speed over time, it is expressed in units of gravitational acceleration (g), with 1 g = 9.8 m/s^2. It is influenced by the frequency, the duration and the intensity of the body movement. If the acceleration value is equal to zero, then the subject might be static or has a constant speed (Chen, 2005). When a body is in movement, then energy expenditure is related to the acceleration of the body mass (BM). As small portable devices, accelerometers are of two types; piezoelectric or piezoresistive sensors (Bonomi, 2010).

The first type, piezoelectric accelerometers consist of a piezoelectric element with a seismic mass. When acceleration occurs, the mass causes the piezoelectric element to bend, which displace charge to build-up on one side of the sensor. This results in variation in the output voltage signal (Godfrey, 2008). To measure acceleration in three axes, several unidirectional sensors are assemble into one instrument (Bouten, et al. 1997). These accelerometers are relatively small and lightweight. Their outputs are the amplitude and the frequency of acceleration signals; these are rectified and integrated in a time interval to determine the activity counts (Bouten, et al. 1997). An 'activity count' is an arbitrary unit varying across devices (Rothney, 2008).

The second type of device, piezoresistive accelerometer consists of polysilicon structure with springs (Bao, 2000). As the human body accelerates, it causes displacement of the silicon structure, resulting in a change in capacitance. This change is processed into an analog output voltage, which is proportional to the acceleration. The outputs are raw acceleration signals, which are often analysed using recognition techniques to identify different types of activity.

The main limitation of accelerometer lies with its underestimation of metabolic rate due to the confounding effect of several factors, including temperature (Jeukendrup and Gleeson, 2004). For example, if a participant was to stay seated in a cold room, the accelerometer will indicate low level of energy expenditure, which might be misleading. In this instance, other methods such as heart-rate monitoring or observation should support the evaluation of metabolic rate.

3.1.1 Instrument

The accelerometer used in this field study is one of the sensors housed in the SenseCam (Vicon Motion Systems, Microsoft, UK). It comprises of a tri-axial piezoresistive accelerometer, Kionix KXP84 (refer to Table 1). This instrument captures body movement in three orthogonal directions. The raw acceleration signal was sampled and stored on a SD card within the SenseCam and later transferred to a computer for analysis. The device is lightweight; of a similar size to a badge, it was worn on the chest without discomfort.

Table 1. Accelerometer performance specification

Parameters	KXP84
Range	± 2 g
Sensitivity	819 counts/g ± 25
Resolution	1.22 mg
Power supply	3.3 V
Operating temperature	- 40 to + 85 °C

3.1.2 Analysis Method

Accelerometer output was derived from raw acceleration signals, followed by the calculation of the vector magnitude as $\left(\left(\sqrt{x^2+y^2}\right)+|z|\right)$, movement intensity (MI). Combining the signals of the three axes makes the MI value insensitive to orientation of the accelerometer with regards to the body. The acceleration in the 'z' axis (vertical) was isolated from the 'x' (mediolateral) and 'y' (anteroposterior) axes, as acceleration in 'z' differs from the rest of the dimensions because of gravity (Chen and Sun, 1997). The resulting signal was then averaged over epoch of ten seconds. All processing was done in R (http://cran.r-project.org).

To determine activity level from motion sensors, most analysis methods are based on controlled experiments in laboratory, often using indirect calorimetry or the DWL techniques. The accelerometer's outputs are compared to the results of physiological metabolic rate measurements. Combining these two sets of results, regression analyses are carried out to determine activity level from the accelerometer readings (Bouten, et al., 1996, Chen and Sun, 1997, Freedson, et al., 1998). Validation studies have reported correlation values from 0.58 to 0.92 (Chen and Bassett, 2005). It should be noted that most of the studies have associated activity counts rather than gravitational values (g) with energy expenditure. Activity counts are difficult to interpret as each device has proprietary data processing methods and assumptions. The device used in this study gives raw accelerometry values expressed in (g), which are easier to interpret. However only a limited number of studies have reviewed the outputs of piezoresistive accelerometers (Van Hees, et al., 2011). This study will only present the signal output; future calibration work is required to determine activity level from these results. Calibration could be completed through experimental study, from which traditional a regression model could be derived, or through more advance processing approaches, such as activity pattern recognition (Godfrey, et al, 2008). For example, patterns in the accelerometer's output signals can be recognised with an activity detection algorithm. These can detect types of activity, for example: lying, sitting, standing and walking (Van Hees, et al., 2009). Combined these results with controlled experiments in laboratory; estimation of energy expenditure can be even more accurate (Gyllensten and Bonomi, 2010).

3.2 Heart-Rate Monitoring

Heart-rate (HR) monitors have become more accessible and reliable in recent years as the demand for training tools in endurance sports increased (Achten and Jeukendrup, 2003). To follow ISO 8996 level-3 approaches, this field study included the monitoring of HR to estimate energy expenditure (EE). As reviewed in this standard, HR levels show a significant relationship with oxygen uptake for heat rates above 120 beats per minutes (bpm) (Parsons, 2001). The estimation of metabolic heat production can be determined using hear rate counts as summarised in the following equation (ISO 8996):

$$HR = HR_0 + RM\ (M - BMR)\ \mathit{(bpm)}$$

Where: HR: heart rate; M: metabolic rate; BMR: basal metabolic rate; RM: increased in heart rate per unit of metabolic rate; HR_0*: heart rate at rest under thermo-neutral conditions.*

This equation has been developed in a set of equations (Annex C, ISO 8996), where metabolic rate is estimated from heart rate recordings as a function of the subject gender, age and weight. This method holds some limitations. At rest, small movements can increase HR, while EE remains almost the same; also emotions could increase HR, while EE remains almost the same (Jeukendrup and Gleeson, 2004). In those instances, other methods such as accelerometry and observation should support the evaluation of metabolic rate.

3.2.1 Instrument

In the field study, 2 devices manufactured by Kalenji were used to monitor HR:

- Sensors and transmitter, Kalenji CW 300 coded; fitted in a chest strap belt, it records the heart electric activity using electrocardiography.
- Receiver and cataloguer, Kalenji Cardio Connect; fitted in an independent device, it could be attached to the belt or kept in the participant's pocket.

During the experiment, continuous recordings, with a 2 seconds sensing interval, were taken. The datalogger memory capacity allowed over 35 hours of recording time and was able to store information from multiple sessions. Data was transferred with the proprietary Geonaute software, and gathered as raw HR values in beats per minutes (bpm).

3.2.2 Analysis Method

The output from the heart-rate monitor was analysed with the ISO 8996, level-3 approach and the associated set of equations, to estimate metabolic rate. The result was then averaged over intervals of ten seconds.

3.3 Automatic Visual Diary

To validate the recordings of accelerometers and heart-rate monitors, each participant wear a SenseCam. This device generates an individual photographic diary, which was concurrent to the other recording methods.

3.3.1 Instrument

In this study, SenseCam is used as a data-logger and an automatic diary. Primarily used in the field of cognitive psychology, this tool has been used as an external memory aid for patients with neurodegenerative disease and brain injury (Hodges, et al., 2006). The device is worn around the neck and placed on the chest. Of similar size to a badge, the SenseCam takes photographs when triggered manually and/or automatically, by timer or by changes in sensor readings. The sensors include temperature, light level, PIR, accelerometer and magnetometer (Gauthier, 2012). The SenseCam provides two types of outputs: (1) a record of measurements taken by each sensor and (2) a visual diary of the participants activity in their home. As an automatic diary, SenseCam aim is to validate the evaluation of metabolic rate in field experiments over continuous periods of time.

3.3.2 Analysis Method

The aim of the automatic diary is to support the evaluation of metabolic rate in field experiments over longer and continuous periods of time - 3 to 10 days. As a validation tool, the recorded images were compared with the acceleration and the heart-rate readings. For each sequence, the activity was attributed to one of the five classes defined by Aminian (1999), as (1) dynamic or static; (2) lying, sitting or standing. Even if the use of visual diary does not prevent bias from the observer; it offers a rich picture of the participants' activity level.

3.4 Study Design and Participants

The field study was conducted between January and March 2012. Using a case-study approach, a purposive sample of nine participants was chosen based on their gender, age and weight as described in ISO 8996, table C.1. All participants were living in

London, in different dwellings dating from 1850 to 2008. Some incorporated features such as retrofitted central or communal heating systems. The participants were asked to wear at the same time the SenseCam and the heart-rate monitor in their home during ten consecutive days. In addition, physical parameters of the indoor environment, including ambient air temperature and relative humidity were measured continuously at five minute-intervals. To record these two parameters, three sets of four dataloggers were placed in the living room and in the bedroom. At the end of the monitoring period, a semi-structured interview was carried out. Participants were asked to indicate how the monitoring went, and if the tools used affected their activity.

4 Field Study Results

Drawn from literature on wearable ubiquitous sensor technologies, this field study used two instruments to monitor activity level: the SenseCam and the Kalenji heart-rate monitor. Initial results were analysed to estimate the movement intensity (MI) and estimate metabolic rate (M). Figure 1 and figure 2 illustrate an example of a 1.5-hour sequence over lunchtime for one participant.

Over this sequence, the average MI was 1.06 g. No acceleration data was retrieved for a period of 15 minutes. According to the visual diary, the participant might have chosen to switch the 'privacy mode' on, so no data was recorded. Over the same 1.5-hour sequence, average metabolic rate was 83.1 W/m^2, or resting state. Following ISO 8996, Table A.2, classification of activity, the participant was in resting position 87% of the time, carrying out light manual work 25% of the time and sustaining moderate work 2% of the time.

In four instances MI rose above 1.5 g; these peaks of activity were also recorded by the HR monitor, but only in three instances. The visual diary provided an explanation to the discrepancy between MI and HR readings. In this instance the SenseCam was dropped to the floor; hence the high level of acceleration and the low HR reading. The three peaks highlighted in Figure 2 correspond to the highest levels of HR in this sequence. The activity levels were above 165 W/m^2, and correspond to sustained moderate work (ISO 8996, Table A.2). The review of the visual diary did confirm these levels of activity. The first peak corresponds to the participant setting up lunch, the second peak was attributed to climbing up the stairs to the first floor and getting ready to leave the house, finally the third peak was accredited to leaving the house.

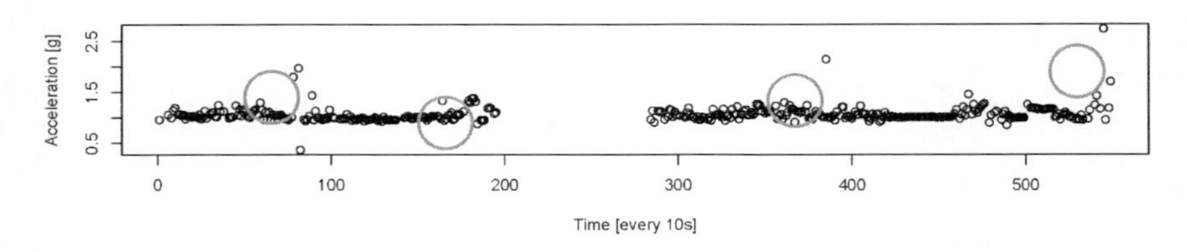

Fig. 1. Movement intensity in g, derived from accelerometer recording as $((\sqrt{x^2+y^2})+|z|)$, every 10 seconds over 1.5-hour monitoring period

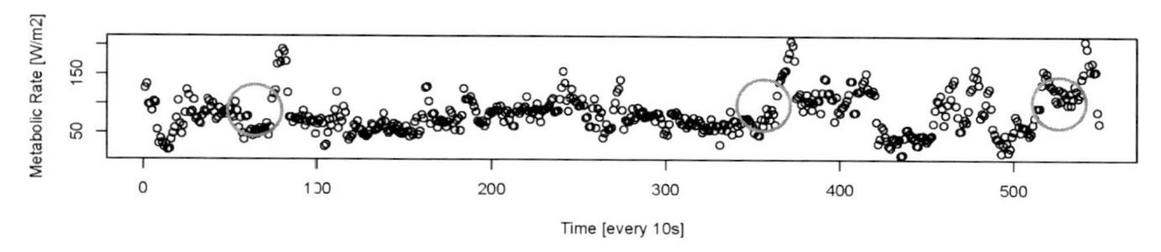

Fig. 2. Metabolic rate in W/m^2, derived from heart rate recording, every 10 seconds over 1.5-hour monitoring period. Three peaks over 165 W/m^2.

5 Conclusion

Methods to estimate or to measure human energy expenditure range from direct to indirect methods with associated level of complexity and cost. This paper focuses on objective and automatic methods, assessing energy expenditure in free-living conditions. Recent advances in activity monitoring instruments and data logging have assisted the development of activity estimation techniques, including heart-rate monitoring and accelerometery. Supported by visual diary, this mixed-method offers number of advantages, including:

- A rich picture of the participants' activity pattern;
- A measured value for metabolic rate, with an accuracy of ± 10%;
- Longer and continuous periods of monitoring, recording variability of a person activity in time;
- A collection method, which has minimum impact on the participants' activity through the use of. automated tools.

Although significant advances in the technology have occurred over the past decade, much remains to be learned about the processing of the accelerometers outputs in field-based research. Future studies may address the following:

- The impact of epoch length should be investigated, in particular how this parameter may vary with the participant's characteristics.
- Calibration of piezoresistive accelerometer should be carried out in laboratory and free-living environments, to evaluate the relationship between MI and M, by developing thresholds for each type of activity and associated prediction equations.
- New approaches should be developed to determine the relationship between the accelerometer and HR monitor data. As these methods are complementary to each other, a set of rules could be developed to establish a 'tree-decision' structure, similar to Aminian (1999).

In conclusion, the proposed mixed-method contributes to estimate pattern in daily activity, which could be used in smart homes to forecast energy demand for heating and to manage power distribution peaks. For the purposes of this research, this mixed-method was carried-out in dwellings over the winter period, however it may be applied to different settings and seasons.

Acknowledgments. The author would like to thank the reviewers and to acknowledge UK EPSRC support of this current PhD research, part of the CDT in energy demand reduction at UCL Energy Institute, UK.

References

Achten, J., Jeukendrup, A.: Heart-rate Monitoring Applications and Limitations. Sports Medicine 33(7) (2003)

Aminian, K.: Physical activity monitoring based on accelerometry: validation and comparison with video observation. Medical & Biological Engineering & Computing 37 (1999)

Auliciems, A.: Towards a psychophysiological model of thermal perception. International Journal of Biometeorology 25 (1981)

Bao, M.: Micro mechanical transducers: pressure sensors, accelerometers, and gyroscopes. Elsevier, Amsterdam (2000)

Bonomi, A.: Physical Activity Recognition Using a Wearable Accelerometer, Sensing Emotions. Philips Research Book Series 12. Springer (2010)

Bouten, C., et al.: Daily physical activity assessment: comparison between movement registration and doubly labeled water. J. Appl. Physiol. 81 (1996)

Bouten, C., et al.: A triaxial accelerometer and portable data processing unit for the assessment of daily physical activity. IEEE Transactions on Biomedical Engineering 44 (1997)

Chen, K., Bassett, D.: The technology of accelerometer-based activity monitors: Current and future. Med. Sci. Sports Exerc. 37 (2005)

Chen, K., Sun, M.: Improving energy expenditure estimation by using a triaxial accelerometer. J. Appl. Physiol. 83 (1997)

de Dear, R., et al.: Developing an Adaptive Model of Thermal Comfort and Preference. Final Report on ASHRAE Research Project 884. ASHRAE, Atlanta, USA (1998)

Department of Energy and Climate Change (DECC). Energy consumption in the United Kingdom (2011), http://www.decc.gov.uk/ (accessed April 30, 2012)

Fabi, V., et al.: Window opening behaviour: simulations of occupant behaviour in residential buildings using models based on a field survey. Presentation of conference: The Changing Context of Comfort in an Unpredictable World, Cumberland Lodge, Windsor, UK, April 12-15. Network for Comfort and Energy Use in Buildings, London (2012)

Fanger, P.O.: Thermal Comfort. Danish Technical Press, Copenhagen (1970)

Freedson, P., et al.: Calibration of the Computer Science and Applications, Inc. accelerometer. Med. Sci. Sports Exerc. 30 (1998)

Gauthier, S.: Mapping Occupants Thermal Discomfort Responses in Households Using SenseCam. In: M'Sirdi, N., et al. (eds.) Sustainability in Energy and Buildings. SIST, vol. 12, pp. 437–445. Springer, Heidelberg (2012)

Gauthier, S., Shipworth, D.: Predictive thermal comfort model: Are current field studies measuring the most influential variables? In: Conference proceedings: The Changing Context of Comfort in an Unpredictable World, Cumberland Lodge, Windsor, UK, April 12-15. Network for Comfort and Energy Use in Buildings, London (2012)

Godfrey, A., et al.: Direct measurement of human movement by accelerometry. Medical Engineering & Physics 30, 1364–1386 (2008)

Gyllensten, I., Bonomi, A.: Identifying types of physical activity with a single accelerometer: evaluating laboratory-trained algorithms in daily life. IEEE Trans Biomed Eng. 58(9) (2010)

Hodges, S., et al.: SenseCam: A Retrospective Memory Aid. In: Dourish, P., Friday, A. (eds.), Springer, Heidelberg (2006)

Hong, S., et al.: A field study of thermal comfort in low-income dwellings in England before and after energy efficient refurbishment. Building and Environment 44(6) (2009)

Humphreys, M., Nicol, J.: Understanding the adaptive approach to thermal comfort. ASHRAE Transaction 104(1) (1998)

ISO, Ergonomics of the thermal environment - Analytical determination and interpretation of thermal comfort using calculation of the PMV and PPD indices and local thermal comfort criteria. BS EN ISO 7730:2005 (2005)

ISO, Ergonomics of the thermal environment - Determination of metabolic rate. BS EN ISO 8996:2004 (2004)

Jeukendrup, A., Gleeson, M.: Sport Nutrition: An Introduction to Energy Production and Performance, 2nd edn. Human Kinetics, Champaign (2004)

Parsons, K.: The estimation of metabolic heat for use in the assessment of thermal comfort. In: Moving Thermal Comfort Standards into the 21st Century, Windsor, UK (2001)

Rothney, M., et al.: Comparing the performance of three generations of ActiGraph accelerometers. J. Appl. Physiol. 105 (2008)

Trost, S., et al.: Conducting Accelerometer-Based Activity Assessments in Field-Based Research. Med. Sci. Sports Exerc. 37 (2005)

Van Hees V., et al.: Estimating Activityrelated Energy Expenditure Under Sedentary Conditions Using a Tri-axial Seismic Accelerometer. Obesity (Silver Spring) (2009)

Van Hees, V., et al.: Estimation of Daily Energy Expenditure in Pregnant and Non-Pregnant Women Using a Wrist-Worn Tri-Axial Accelerometer. PLoS ONE 6(7), e22922 (2011)

Chapter 39
Optimizing Building Energy Systems and Controls for Energy and Environment Policy

Mark Barrett and Catalina Spataru

UCL Energy Institute, Central House 14 Upper Woburn Place, WC1H 0NN, London, UK
{mark.barrett,c.spataru}@ucl.ac.uk

Abstract. This is an informal introduction to some aspects of energy system optimisation to provide sustainable services to people in dwellings. This paper advances data, methods and results of optimising building energy systems and controls for energy and environment policy using quantitative techniques. Optimisation can aid the design of systems to meet policy objectives efficiently and at low cost. Three optimisation methods were applied: genetic algorithm (GA), particle swarm optimisation (PSO) and steepest decent (SD). It was concluded that the higher the energy price, the greater the efficiency of the dwelling envelope and heating system to achieve least cost. Ultimately, optimisation should be done across all systems and stock, and simultaneously for configuration, size and controls.

Keywords: Building Efficiency, Energy System, Dynamic Simulation, Optimisation.

1 Introduction

The UK Government's objectives include to cut 1990 greenhouse-gas emissions by 80% by 2050 [6] and for 20% of energy consumption in 2020 to be from renewable sources as specified by European Parliament and Council. Residential buildings are estimated to constitute around 17% of the UK's total CO_2 emissions in 2010 [2]. The government has a number of policies to encourage energy efficiency and low carbon energy systems for buildings; for retrofit including Feed-in-Tariffs, the Renewable Heat Incentive, and the Green Deal, and for new build building regulations. These policies aim to save energy and carbon using currently available technologies, at relatively low cost. However, building energy efficiency and supply systems are complex with many interlinking factors that affect final performance. The challenge is to design whole systems that realise energy and carbon savings at least. The building systems will affect network and national scale planning; for example on the peak load of gas or electricity networks. Internal and external conditions and systems will affect performance: technology options and characteristics (heat pumps, gas boilers, microCHP, storage), costs (capital, fuel and operational costs), feasible system configurations (such as combinations of devices installed and connection to distribution

A. Håkansson et al. (Eds.): *Sustainability in Energy and Buildings*, SIST 22, pp. 413–423.
DOI: 10.1007/978-3-642-36645-1_39

systems), selected dwelling archetypes (building size, building fabric), wider system (network) constraints (electricity price, variable renewable supply), control strategies for occupancy profiles (heating, electricity export, energy storage). The relationship between building energy flows and the local and national system is complex and can be influenced by a range of interlinking factors at different scales.

The simulation and optimisation model presented here allows flexible configurations in terms of building component set, sizing and connection, and the setting of different operational control strategies. This enables testing of different technologies and combinations of them manually and through optimisation and thereby the discovery of low cost configurations or packages. These packages are costed and may be used to construct building energy programmes with lowered risk of regret because optimal packages are installed in the first place without requiring a second, costly installation.

The paper will address the problem of simulating and optimizing key aspects of package design: component set, component connection and control. In so doing it will contribute towards energy policy and programme planning. There are two main challenges: first, to simulate the real time operation of systems with temporally varying demands driven by social activities and weather and ensure the system will deliver energy services reliably across hours, days and months. There are many building simulation packages, some sophisticated, such as the public EnergyPlus [10].

Second, the problem is to optimise the system so as to find a combination of system components (building efficiency, energy conversion, storage) and operational control that will deliver least cost given a delivered energy cost whilst meeting objectives such as energy services and constraints such as carbon emission. Building energy optimization applications tend to focus on either the control of building energy systems [8], or determining the optimum building efficiency (insulation, ventilation control) level given energy prices, such as using BeOpt [1]. However, for optimal designs all of these are interdependent – energy efficiency, energy system, energy system control – and so must be simultaneously optimized, but to cover all these aspects means less detail on each.

In this paper, we will confine our attention to dwelling efficiency and energy system sizing and control. The dwelling is a single zone with a detached envelope of given height, width and depth and glazing fraction, and an electric heat pump provides space and water heating. The design variables in our optimization include those determining the sizing and efficiency of building and energy system components, or the control of the heating system, or both. The system can then be optimized to find the minimum total annual cost for different delivered energy prices.

2 Simulation

The dwelling-energy system simulated is designed to provide comfort to occupants when they are active in the dwelling, i.e. not asleep. The focus is on heating so water heating is included but other services, such as lighting, are not considered in detail. The simulation is coded in Excel and Visual Basic for Applications. The heating system has to produce heat such that the comfort temperature is maintained during the occupied period which is set, in this case between 8 am and 10 pm. The time taken for the dwelling to heat up depends on the current dwelling temperature (*Tinternal* °C), the comfort temperature (*Ttarget* °C), the ambient temperature (*Tambient* °C), the maximum output of the heater (in this case, *HPmax* W, a heat pump), and the specific loss (W/°C) and the thermal capacity (Wh/ °C) of the dwelling. The simulation accounts for these and includes solar gain (*SolarGain k*W) and hot water load (*Hot-Water k*W). The gross space heat loss (*SpLossGross* kW) and net heat loss (*SpLossNet k*W) are calculated; the hot water load is added to this and then the electricity input to the heat pump (*HPin k*W) is calculated assuming the heat pump operates at 30% of the ideal Carnot cycle efficiency between ambient temperature and a hot water temperature of 55°C. Figure 1 shows the temperatures and energy flows for a highly efficient dwelling for six sample days for months 1,3,5,7,9 and 11 across the year. We see the dynamic heating and cooling of the dwelling; the pulses of heat from the non-modulating heat pump heating up the dwelling in the winter and maintaining its temperature. We also see that dwelling temperatures are high in the summer without any heating – this flags the risk that highly efficient buildings can pose a risk of overheating without careful design, especially given predicted increasing ambient temperatures due to global warming. This may lead to increased air conditioning and accompanying electricity demand. The simulation calculates a penalty function which is the integral of underheating – that is the number of hours underheating occurs times the number of degrees underheated.

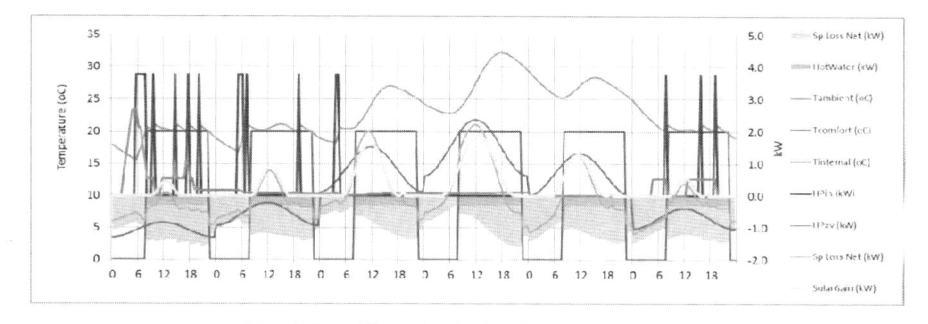

Fig. 1. Dwelling simulation for 6 sample days

3 Optimisation

Correctly selected and applied, optimization aids the planning of, regulatory and market frameworks that will facilitate the approach to least cost solutions. The objective (called the *objective function* in optimisation) here is to minimise the total annual cost of building and operating the dwelling and energy system within constraints including achieved comfort. Capital costs comprise the annuitized costs for insulation, ventilation control, and the heating system annuitised at 3%/a real interest rate and using the different lifetimes of the components; and the running costs of the energy system – mainly the fuel cost. The costs are minimised by applying optimisation to a set of decision variables (DVs). The U-value, ventilation and heater DVs determine the capital costs of the building elements; for example, the lower the U-value, the greater the cost per unit area (m^2) of that element. The capital costs of energy efficiency and heating systems may be divided into materials or equipment costs and installation costs and these fixed and variable manufacturing and installation costs, both of which have scale economies. The unit costs of energy efficiency and heating systems are taken from manufacturers' data. In the Figure 2 are shown envelope insulation costs per m^2 of envelope area; for infiltration and ventilation control costs given in £/m^3 per air change per hour (ACH). We see a varying fixed cost per m^2 according to the measure (e.g. wall or glazing) and increasing marginal costs as efficiency increases.

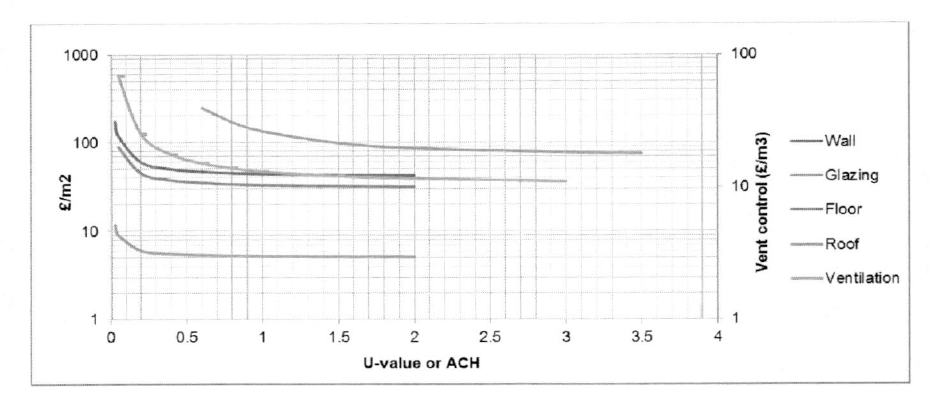

Fig. 2. Energy efficiency costs

In this exploration a heating system comprising an air source heat pump directly supplying space and water heat loads is assumed, with no storage. Figure 3 shows the cost of purchasing heat pumps and boilers of different sizes. Here we see unit costs (£/kW) declining with installed capacity (kW) as fixed installation and manufacturing costs become a smaller fraction of total cost.

Using these capital cost curves and the fuel cost, the optimiser runs the simulation testing out different combinations of DVs until a near least cost is found. Different sets of decision variables (DVs) can be used in optimization. The current possibilities

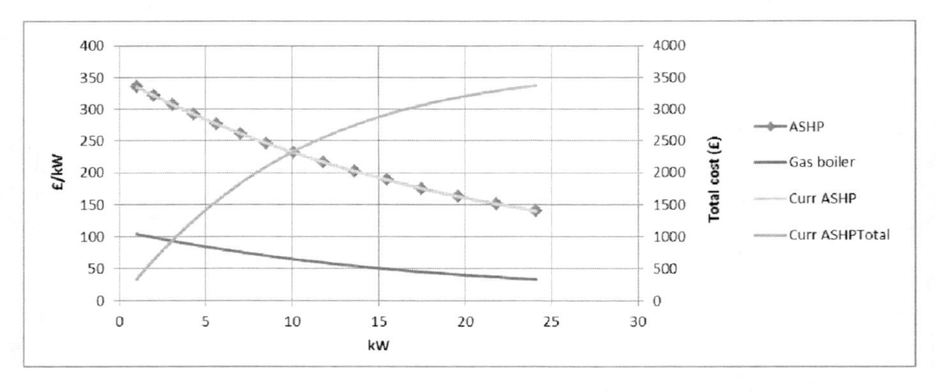

Fig. 3. Heater conversion equipment

are shown in the Table: decision variables current values in yellow, minimum and maximum values in red. The first four DVs relating to building shape and glazing ratio are not used in this paper. The next four DVs are the U-values of the envelop elements and the next DV is the ventilation rate. The Heater DV is the maximum heat output of the heater and the last two DVs are the times when heating is turned on in order to warm up the building to meet target comfort temperature when occupancy commences, and the time to turn the heater off.

Table 1. Decision variables

		DVs	Curr	Min	Max	
Specifications	Building volume	m3	252	250		
Decision variables	**Length**	**m**	**7.0**	3.0	20.0	CN
	Width	**m**	**6.0**	3.0	20.0	CN
	Height	**m**	**6.0**	3.0	20.0	CN
	Glazing	**%**	**15%**	15%	100%	CN
	Wall	**U-value**	**0.13**	0.01	1.60	CN
	Glazing	**U-value**	**0.84**	0.70	5.60	CN
	Floor	**U-value**	**0.10**	0.01	2.00	CN
	Roof	**U-value**	**0.03**	0.01	2.00	CN
	Ventilation	**ACH**	**0.26**	0.05	2.00	CN
	Heater	**kW**	**3.5**	0.10	30	CN
	Heating	**On**	**5.54**	1.00	24.00	D
		Off	**19.75**	1.00	24.00	D

A good optimisation technique accurately finds a global optimum or best solution with a small number of calculations so as to minimise solution time. We have to select optimisation techniques appropriate to the system and DVs modelled. Common optimisation techniques used for energy systems are: linear programming (LP) for continuous linear systems; hill climb/steepest descent (SD) for continuous linear or non-linear systems; genetic algorithms (GA), particle swarm (PSO) optimization and other biological algorithms such as ant colony optimization (ACO) for most types of problem. In this case we have a system with continuous variables with combine non-linear (marked CN in the table above) and discontinuous variables (marked D). This means we cannot use linear programming (LP) as it requires continuous linear functions. For speed and accuracy, multiple techniques can be used iteratively. Here we will use a combination of GA, PSO and SD methods.

Genetic algorithm (GA) is a technique based on the mechanism of natural selection. Genetic algorithms (GAs) can operate on discrete and non-linear systems and can therefore be applied to a wide range of systems and decision variables, whether concerning technologies or people. The advantage of this method is that it does not required the property of continuous differentiability and convexity of the objective function; and it can search broadly thereby improving the chance of converging on a global rather than a local minimum.

The particle swarm optimisation (PSO) method was first proposed in 1995, and is a fast, efficient and accurate when applied to a diverse set of optimization problems [7]; [9]; Eberhart and Shi, 2001). The potential solutions (particles) move through the problem space with a route determined by current speed and direction, neighbours and the current optimum particle. As for GA, the function can be non-linear and non-continuous. As a number of particles are spread over the function 'landscape', PSO is capable of broad searches and finding global optima.

The hill climb or steepest descent (SD) method works with any order dimensional space and can be applied to continuous linear or non-linear functions. It uses a geometric construct to find the line of steepest descent to the optimum (minimum). The method is fast compared with GA or PSO with the other optimization methods and is a robust method that does not rely on derivatives to provide function minimization. However, if the objective function has several minima, the SD method will find the local minimum depending on its starting coordinates and it may that this is not the global minimum.

For this work, a hybrid optimizer incorporating SD, PSO and GA was written in Excel VBA with which the different techniques can be used separately or in combination sequentially and iteratively. The idea is that in combination these methods are more likely to find the global optimum, and the SD method can rapidly and accurately find at least the local minimum. By running the optimizer with different starting points, the probability of finding the global optimum is increased.

Optimisation for developing energy technologies and policy requires assumptions about the future. In order to improve confidence in the robustness of optima, sensitivity analysis can be used, but this generally relies on subjectively assumed ranges. Very often it isn't known if there are several local optima, or certainty that the global optimum has been found. In using more optimisation evaluations, marginal improvements in the objective function often rapidly decrease as shown in the next Figure where we see the reduction in cost after 1000 function evaluations is small.

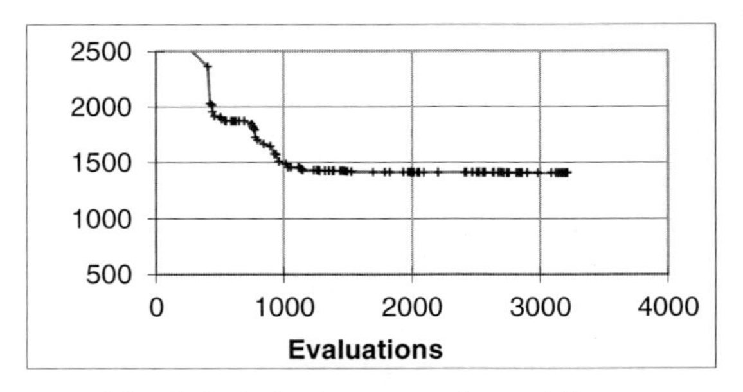

Fig. 4. Optimization approach to minimum

In real world systems sharp optima are rare. Optima are generally broad allowing decisions to take account of decision variables, factors and criteria not included because reliable or not quantifiable in simulation or optimisation modelling. For example, there are dwelling variable limits to the thickness of insulation that can be installed.

4 Results

The hybrid optimiser determines DVs giving near to least cost. Of most importance is the heat loss of the dwelling. The next figure shows the U-values and heat losses of the different elements of the dwelling for a sample optimisation. Of particular note is the dominance of losses from windows, because of the difficulty in reducing their U-value, and ventilation because of sharply increasing costs due to mechanical ventilation at low air change rates.

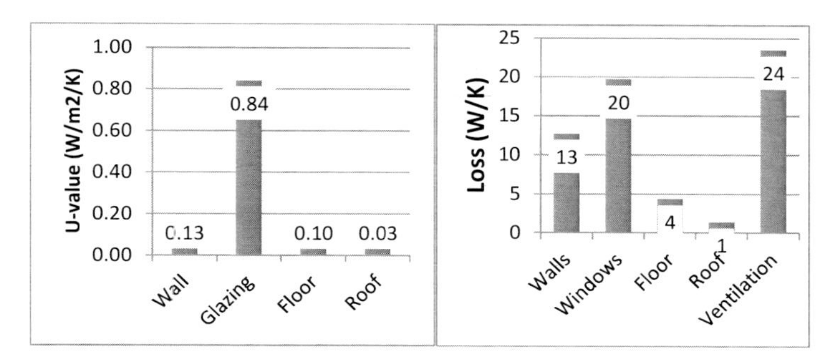

Fig. 5. U-values and loss factors

The optimisation was run with electricity costs ranging from 10 to 80 £/GJ (3.6 to 29 p/kWh). The cost of off-shore wind power, the UK's largest renewable resource, delivered to consumers might be in the range 35 to 50 £/GJ (11 to 22 p/kWh). For the UK, this represents the indefinitely sustainable cost of electricity. The trend in optima for energy and heat losses is shown in the next Figure.

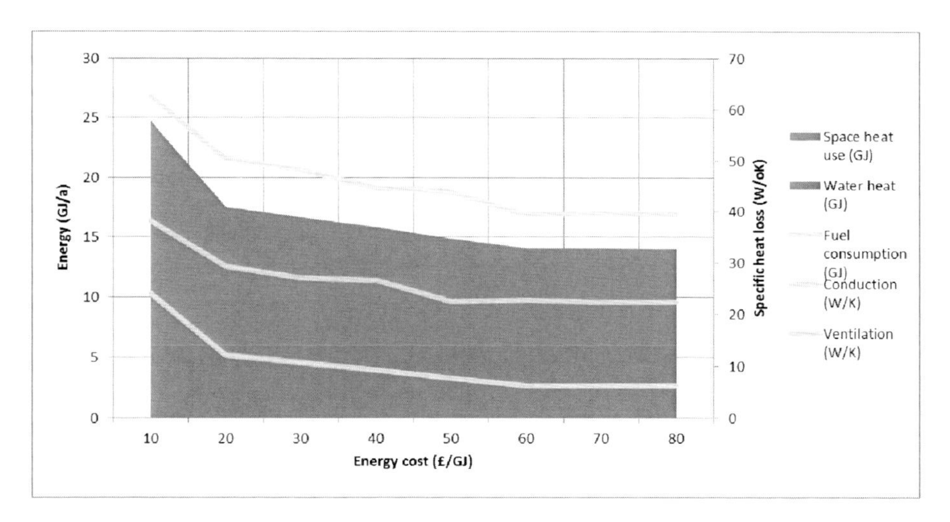

Fig. 6. Optimisation results – U-values, losses and heater power

The general trends are predictable; as energy costs increase, there is more capital investment in energy efficiency, lower U-values and ventilation rates, so the specific loss (W/°C) of the dwelling decreases. Not so predictable is that the heater capacity (kW) also reduces with increasing energy cost. This is a balance because the lower specific loss means less power is needed generally and because the losses incurred by having a higher average temperature because of a low powered heater are smaller, so the heater can start some time before occupied periods.

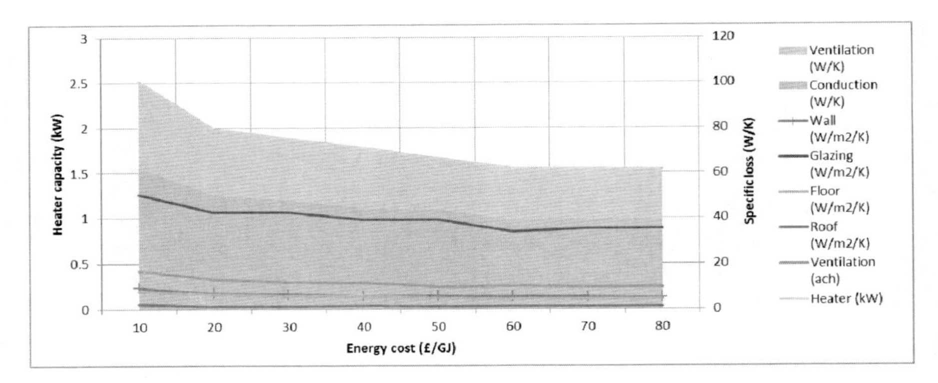

Fig. 7. Optimisation results – U-values, losses and heat power

Figure 8 shows the capital and annuitised costs of the dwelling envelope and heating system, and the annual fuel costs. As the fuel costs rise across the range, there is increased capital investment in energy efficiency of about 30% and a reduction of about 10% in the heating system cost so total investment increases by 20%.

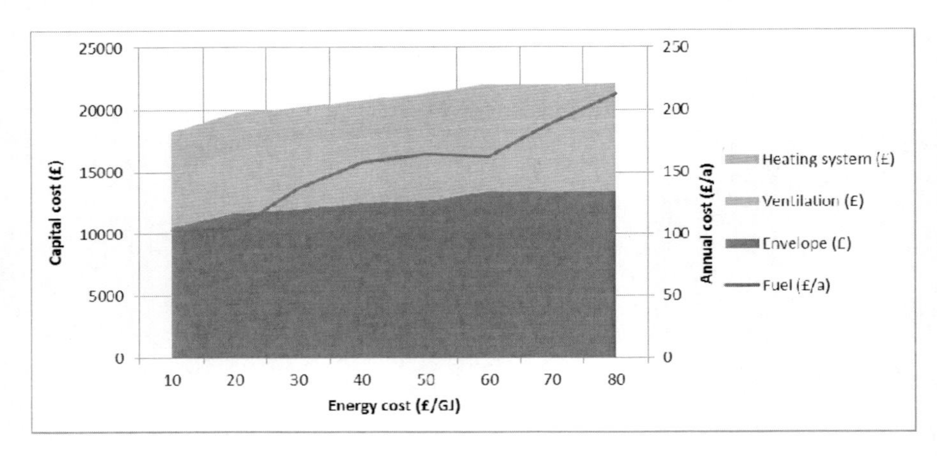

Fig. 8. Optimisation results – capital and energy costs

The next figure is interesting in that the total annual cost of heat services to the dwelling increases by 20% from 1200 £/a to 1440 £/a; whilst the energy unit cost increases by 700%, with the total energy cost rising 100% from 100 £/a to 210 £/a. This underlines how insulation insulates households against the vagaries of energy

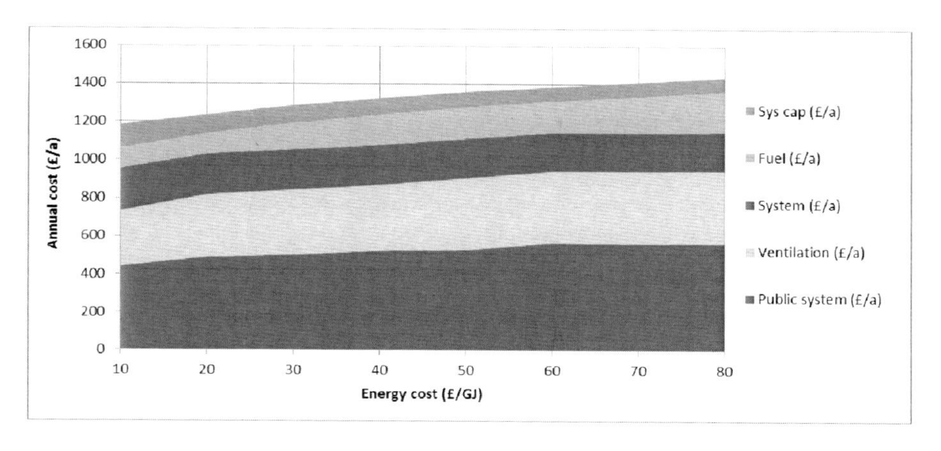

Fig. 9. Optimisation results – total annual costs

cost variations – an eightfold energy price rise increases the cost of space heating by just 20%.

5 Discussion and Policy Implications

It may be seen that the interactions between the costs of the components and energy flows are complex, and this is without accounting for any external system issues. In general, the higher the delivered energy price, the more energy efficiency, but the marginal savings in heat reduce with increasing insulation thickness. The lower the heater power (kW), the less it will cost, but the slower the dwelling will heat up to comfort temperature so the heater will have to be turned on earlier, and therefore the average dwelling temperature and total heat loss will be higher. The efficiency cost curves are such that even the lower electricity costs drive efficiency in terms of U-values and ventilation control towards high levels as shown in Table 1 for DV optimum values for an electricity cost of 40 £/GJ.

This has important policy implications. In the UK, to meet policy objectives, between now and 2050, some 15 to 20 million existing dwellings should be refurbished and re-equipped and 10 million new dwellings built, and similar efforts will be required for non-domestic buildings. At the same time of the order of 100 GW to 300 GW of new energy supply transformation and distribution infrastructure will be needed. Typically, buildings have lifetimes of decades or centuries and energy systems of 20 to 40 years, so it is important that the right package is installed initially. The cost of returning to a building to further increase efficiency is much higher than doing it in the first place because of repeated fixed installation costs, quite apart from the social installation capacity required. The optimisation shows that the efficiency levels should approach that, which is the maximum technically feasible given constraints such as maximum insulation thickness.

6 Further Work

This work could be improved in several ways.

- The house energy systems should efficiently cater for occupancy that varies on long and short timescales, and for weather. For example, average occupancy and comfort temperatures will change as the UK population ages and household structure change – probably such that space heat demand will increase with longer occupancy and higher temperatures. Buildings and systems should be designed for a wide range of occupants. The heating system control setting should be optimised to change across days and the year accounting for predicted occupancy and weather so as to reduce energy costs.
- Rather than assuming a constant fuel price, a price which varies across the hours of the year could be used to reflect cost changes for to the varying level of demand summer to winter and day to night, and the varying generation by renewables. Changes to electricity demand because of heat pumps and energy efficiency will incur different capital and running costs on the electricity supply system. An account of this would likely increase the value of peak winter demand. Demand is higher in the winter but so is wind output and so the effects of these on prices are in opposite directions. In conjunction with this, heat storage could be included as this allows better utilisation of renewables which will be at a lower short run marginal cost and carbon content than conventional supplies. To include this, the system simulation and optimisation would have to be extended to storage sizing and control.
- The analysis could be applied to a range of domestic and non-domestic buildings and thence to the whole UK building stock. It is not anticipated that the results would be very different from those presented here because the cost curves for insulation and so forth will be similar, except that there will generally be economies of scale for larger buildings and mass efficiency programmes.

References

[1] Anderson, R., Christensen, C., Horowitz, S.: Program Design Analysis using BEopt Building Energy Optimization Software: Defining a Technology Pathway Leading to New Homes with Zero Peak Cooling Demand. Presented at the 2006 ACEEE Summer Study on Energy Efficiency in Buildings, Pacific Grove, California, August 13-18 (2006)

[2] DECC, Statistical Release: Provisional Figures for 2010, National Statistics Information on (2011), http://www.communities.gov.uk (last accessed July 2012)

[3] DTI, Meeting the Energy Challenge, A White Paper on Energy, CM 714, The Stationery Office (2007), http://www.decc.gov.uk/assets/decc/publications/white_paper_07/file39387.pdf (last accessed July 2012)

[4] Eberhart, R.C., Shi, Y.: Particle swarm optimization: developments, applications and resources. In: Proceedings of IEEE International Conference on Evolutionary Computation, vol. 1, pp. 81–86 (2001)

[5] European Parliament and Council, 2009/28/EC on the promotion of the use of energy from renewable sources and amending and subsequently repealing Directives 2001/77/EC and 2003/30/EC (2009)
[6] HM Government, The Carbon Plan: Delivering our low carbon future. Amended 2nd December 2011 from the version laid before Parliament on (December 1, 2011), http://www.official-publications.gov.uk (accessed July 2012)
[7] Kennedy, J., Eberhart, R.C.: Particle Swarm Optimization. In: Proceedings of the IEEE International Joint Conference on Neural Networks, vol. 4, pp. 1942–1948 (1995)
[8] Klein, L., Kavulya, G., Jazizadeh, F., Kwak, J., Becerik-Gerber, B., Varakantham, P., Tambe, M.: Towards Optimization Of Building Energy And Occupant Comfort Using Multi-Agent Simulation. Automation in Construction 22, 525–536 (2012)
[9] Shi, Y., Eberhart, R.: A Modified Particle Swarm Optimizer. In: Proceedings of IEEE International Conference on Evolutionary Computation Anchorage, vol. 6, pp. 9–73 (1998)
[10] US Department of Energy (USDOE), EnergyPlus Energy Simulation Software (2012), http://apps1.eere.energy.gov/buildings/energyplus/ (accessed July 2012)

Chapter 40
Towards a Self-managing Tool for Optimizing Energy Usage in Buildings

Naveed Arshad, Fahad Javed, and Muhammad Dawood Liaqat

Department of Computer Science, LUMS School of Science and Engineering, Lahore, Pakistan
{naveedarshad,fahadjaved,dawoodliaqat}@lums.edu.pk

Abstract. Smart grid is the next generation of electricity generation, transmission and distribution technology. A major component of smart grid is an overlay communication network for two-way communication between the power providers and the customers. With this feature smart grid provides exciting new ways of energy management and conservation. One of ways to conserve energy using a smart grid is to control and optimize energy usage in buildings. Buildings consume more than one third of the energy produced in the world. Therefore, conserving energy in buildings is cited as the "most important fuel" in energy generation. To this end, we have developed a self-managing approach to optimize energy usage in buildings. We have evaluated our approach using a software tool called Power Conservation Analysis Tool (PCAT). Our initial results using PCAT show upto 38% savings in the energy bills of customers that could directly translates into reduction in energy production costs for power producers.

Keywords: optimization, energy usage, self-managing tool.

1 Introduction

With global warming and impending scarcity of fossil fuel, cleaner sources and better utilization of energy has been considered as a major goal for future technology advancements and research. Energy conservation has been cited as the 'most important fuel' in energy generation [1], [10]. Correspondingly, the smart grid provides a number of ways to conserve energy. One of the ways in which a smart grid provides energy savings is through Advanced Metering Infrastructure (AMI). Through an AMI the energy usage in buildings is monitored through smart meters that are capable of communicating with the smart grid in real time. Electric devices in buildings are connected through a home area network (HAN) using standards such as Zigbee [18], wi-fi, radio or other communication protocols, thus making it possible for all the devices on the network to be managed remotely. The control of these devices through home area networks provide newer ways of conserving energy.

Buildings are the biggest consumers of energy. According to a study, out of all consumers of energy in EU, 37% of energy is consumed by buildings including both residential and commercial. This is ahead of energy consumption in industrial sector which consumes 28% and transportation sector which consumes 32% [13].

A. Håkansson et al. (Eds.): *Sustainability in Energy and Buildings*, SIST 22, pp. 425–434.
DOI: 10.1007/978-3-642-36645-1_40

One of the ways of energy conservation in buildings is if the consumers control their device usage during peak periods. However, traditionally most energy users in buildings are not interested in manually managing their electric devices to save energy. Rathnayaka and colleagues identified several challenges to of energy management in smart homes [5]. Some of these challenges include a lack of intelligence to handle uncertainty, making passive decisions and others. Research has shown that consumers are willing to engage with the smart grid provided that its interface with the consumers is simple, accessible and in no way interfere with the normal day-to-day life of the consumers [2], [6] . In addition, consumers are also interested in finding that how much can they save in energy bills. This is especially true in case of countries where energy is a scarce and expensive.

Our hypothesis in this research is that if the consumers plan their energy usage according to the supply of energy then just by shifting some energy load to off-peak hours can save energy. To this end, we propose a mechanism where the consumers are not only aware of the energy supply position but their devices are also turned on or off in a self-managed way through employing the priorities and conditions set by the consumers.

To realize this goal, a collaborative technique of energy usage with a constant monitoring of energy devices inputs is required. These inputs are then used in creating a plan to reduce the cost while satisfying conditions and priorities of the customer. To our knowledge, most such systems that provide some sort of interface to the consumer to plan and save power are very few in number and are very rudimentary from a software perspective [8], [14] and [15]. In essence, what is required is a self-managing system that takes user defined goals and produces energy usage plans for the user. Additionally, if the user behavior changes such system must also be able to replan based on the new information.

2 Scope and Assumptions

Smart grid vision foresees that the purchasing of electricity will be available directly to the end users rather than through a distribution company. Thus in future, consumers will be charged different prices at different times of the day i.e. hour-ahead pricing, and day-ahead pricing. In this scenarios, where there are time varying prices of electricity, consumers can reduce their electric devices usage expenses by using electric devices at those times when the price of electricity is lower instead of those times when the price of electricity is higher.

Types of Electric Devices

Electric devices have different consumption profiles, therefore, it is imperative that we classify the devices according to their power profiles for controlling them to save energy costs. The devices in almost all types of buildings i.e. residential, commercial, etc. can be divided into four categories: 1) Low Power Low Usage (LPL) 2) Low Power High Usage 3) High Power Low Usage (HPL) 4) High Power High Usage (HPH). Since the bulk of energy in any given household is used by the HPL and HPH devices, we believe that if we can manage these devices more intelligently especially in the HPH category then significant cost savings is possible. The devices in this category include air conditioners, water heaters, electric cars, water pumps, etc. Note that our system is mostly focused on residential buildings. For commercial buildings this categorization of devices may not hold.

Device Usage
Amongst electric devices, there are many for which consumer's acceptable usage time ranges can be defined. For example, a consumer might wish to run the water pump for the duration of 45 minutes anywhere between 12:00 AM to 6:00 AM. If one schedule the run of the water pump to a particular 45 minutes time-slot when the price of electricity is the lowest then one can certainly reduce electric power consumption expenses. In such a manner, if one tries to save power consumption expenses on multiple devices in a home on daily basis then the accumulated savings can be significant for the whole billing cycle duration.

There are trade offs between consumer preferences regarding the usage timings of the electric devices and the times when the price of electricity is lower. If electricity consumers use their electric devices at those times when the price of electricity is higher even though the use of those electric devices could be delayed to the times when the price of electricity will be lower, then certainly the consumers will be having relatively higher electric power consumption expenses. So, if we have consumer preferences regarding the electric devices usage timings then we can reduce the electric devices usage expenses by optimally switching electric devices ON/OFF in a time varying electricity prices-aware manner while satisfying the preferences, at the same time. Given a set of electric devices, user preferences regarding the usage timings for each of the devices and the time varying prices of electricity, we describe an approach to develop electric devices usage plan for consumer.

3 Approach

In our approach we assume that "hour-ahead" pricing is available from the energy market. This means that the energy prices are known one hour before.

Priorities and Constraints Specification
A typical building may have a number of devices on the HAN. PCAT has an editable list of devices present on the HAN. Using PCAT a consumer selects the devices that are flexible in their usage. The consumer provides a set of priorities and constraints for each device. These are specified at a high-level by using a very simple user interface. Each device can have multiple set of priorities and goals. For this paper, we assume that these goals are not conflicting with one another.

Other than providing this the consumer also specifies the maximum total cost of electricity that he or she is willing to pay at the end of a billing cycle. This is especially beneficial for usage in countries where the price of per unit electricity increases with more usage. These pieces of information are used by our optimization algorithm to generate a plan for energy management and consequently energy savings.

Optimization Strategy
We consider a set of electric devices for which a home consumer wish to generate the optimal usage plan such that the power consumption expenditure is minimized by

taking advantage of the time varying prices of electricity. For each of the devices to be controlled on the HAN, average hourly energy consumption profile is available. Each device can have multiple consumer preferences regarding its usage timing limits and the duration to keep the device in a certain state. For simplicity purposes, in this paper, we have only considered devices whose states can be turned on or off only. However, in reality multiple ways of energy conservation are possible. This includes controlling the thermostat, reducing the voltage supplied to a certain device, providing energy from multiple source i.e. a solar cell and the electricity from the power company, etc.

We would like to stress here that our optimization strategy does not turn the devices on or off without implicit permission by the user. The devices can only be turned on or off in the given time windows specified by the user.

Let us discuss the energy conservation strategy for an electric dishwasher. An automatic electric dishwasher can have a preference that it should be scheduled to run anywhere between 9:30 AM to 12:30 PM for 45 minutes. Similarly for the same dishwasher another preference can be that it should be scheduled to run anywhere between 4:00 PM to 7:00 PM for 45 minutes. Consider the later preference, we call 4:00 PM as the lower time limit and 7:00 PM as the upper time limit during which the dishwasher must be switched on. Our objective is to find that particular 45 minutes time slot anywhere between the lower and upper time limits at which the price of electricity is the minimum.

However, the energy price is only known before the start of the hour. Therefore, to develop a 24 hour plan ahead of of its start, we use historical average of prices. If the price of electricity for each hour is between a bound of the actual electricity prices then the originally created plan will be followed. These bounds are determined by the maximum bill amount a consumer is willing to pay. However, if the difference of the actual energy price and expected energy price is more than a given bound then a replanning process will take place to calculate another plan based on the new information about the energy prices.

Optimization Algorithm

We have used the concept of timing windows to implement our algorithm. The detailed implementation is described as follows:

Let $D = \{d_1, ..., d_n\}$ is the set of n electric devices. The electric power consumption of the n devices is represented by the set $K = \{k_1, ..., k_n\}$ where each element k_i denotes the electric power consumption for the device $d_i \in D$. The preferences for an electric device $d_i \in D$ are represented using the set P^i where each element p^i_j is itself a tuple defined as $p^i_j =< l_{p^i_j}, u_{p^i_j}, o_{p^i_j} >$. Among the elements of the tuple, $l_{p^i_j}$ is the lower time limit, $u_{p^i_j}$ is the upper time limit and $o_{p^i_j}$ is the ON duration for j^{th} preference of ith device. Time varying prices of electricity are given as a set $C = \{c_1, ..., c_m\}$ where each element c_t denotes the price of per unit electricity for the time duration between time instances t (**inclusive**) and $t + 1$ (**exclusive**).

Our objective is to find a set G^i corresponding to each set of preferences P^i such that for each preference $p^i_j \in P^i$ the set G^i has one element g^i_j which is an ordered pair of the form $g^i_j =< r_{g^i_j}, s_{g^i_j} >$. Inside the ordered pair, $r_{g^i_j}$ and $s_{g^i_j}$ denote the most suitable times to turn ON and OFF the ith device, respectively for satisfying j^{th}

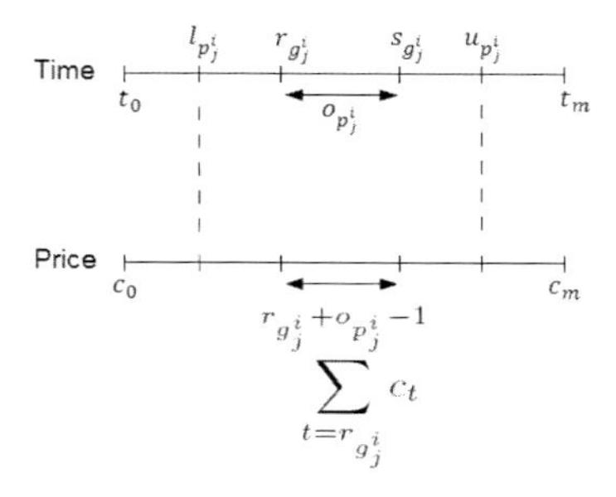

Fig. 1. Finding optimum switch-ON time

preference. Each $s_{g_j^i}$ can be calculated as $s_{g_j^i} = r_{g_j^i} + o_{p_j^i}$ that is, the time at which to turn ON the device **plus** the duration to keep it ON. The overall electricity consumption expense of the devices should be the minimum if the generated plan consisting of all G^i is followed.

We have to find each $r_{g_j^i}$ for which $f(r_{g_j^i})$ is the minimum where $f(r_{g_j^i})$ denotes the expense of switching ON the ith device at the time $r_{g_j^i}$ and keeping it ON for the duration determined by $o_{p_j^i}$. This is according to j^{th} preference of the same device. The function $f(r_{g_j^i})$ can be defined iteratively as:

$$f(r_{g_j^i}) = \sum_{t=r_{g_j^i}}^{r_{g_j^i}+o_{p_j^i}-1} c_t$$

The function $f(r_{g_j^i})$ can also be defined recursively. For recursive case when $r_{g_j^i} > l_{p_j^i}$ then

$$f(r_{g_j^i}) = f(r_{g_j^i} - 1) - c_{(r_{g_j^i}-1)} + c_{(r_{g_j^i}+o_{p_j^i}-1)}$$

For base case when $r_{g_j^i} = l_{p_j^i}$ then again

$$f(r_{g_j^i}) = \sum_{t=r_{g_j^i}}^{r_{g_j^i}+o_{p_j^i}-1} c_t$$

Recursive definition is cheaper in terms of calculation because it avoids repetitive calculations. In either of iterative or recursive case, we have to minimize $f(r_{g_j^i})$ subject to following constraints:

$$r_{g_j^i} \geq l_{p_j^i}$$

$$r_{g_j^i} \leq u_{p_j^i} - o_{p_j^i}$$

$$\forall i, j$$

Figure 1 shows how the time window determined by each $< r_{g_j^i}, s_{g_j^i} >$ pair can slide within the time limits determined by the preference p_j^i. We have to find those points $r_{g_j^i}$ and $s_{g_j^i}$ on the timeline where the expenditure value of the sliding window comes out to be the minimum.

4 Architecture and Implementation

PCAT is a web application developed for displaying electricity consumption statistics and graphs to electricity consumers. PCAT is based on the model that electricity consumers have smart electricity meters installed at their homes. These meters are capable of transmitting periodic electricity consumption statistics of the consumption database of the PCAT. Consumers can then logon to PCAT to view their power consumption statistics and various types of charts on daily, weekly, monthly, yearly and on billing cycle duration basis. Consumers can check their tariff based on their chosen tariff package and the consumption made during the billing cycle. Consumers are also able to plan the consumption of their electric devices during the billing cycle while staying within a consumer chosen desired bill amount.

The most interesting component of PCAT is the Optimization Manager (OM). The OM interfaces with the HAN to access electric device information and updates its database accordingly. OM also interfaces with the smart grid interface to get upto date energy prices and also to get information from the AMI. Based on the customer preferences, it invokes a planning module to generate a plan for energy consumption based on historical price data. The planning module gets the estimated price information from an estimation module that uses a moving averages algorithm to calculate the estimated electricity prices. The OM also triggers a replanning process, if the energy prices are not in between certain bounds. The replanning is performed for the rest of the hours in a 24 hour cycle.

Other than the OM, the layers architecture has a Presentation Layer (PL) and a Data Layer (DL). The PL is used as an interface with the consumer and the DL is used to connect with the database that holds all the necessary data to run the application.

PCAT is developed as a web application using Microsoft ASP.NET, Visual C\# and Microsoft SQL Server.

5 Evaluation

This section reports preliminary experimental results using PCAT for optimizing power utilization in a typical household scenario.

We run our experiments using profiles of electric devices in an average-sized house. This include air conditioners, washing machine, dishwasher, water pump, etc. In all, we considered 7 different devices in our sample household. Table 1 lists the devices and respective energy profiles. We considered the general use of these devices and set the preferences accordingly. The 15 preferences we used for experiments are shown in figure 3. In this figure the preferred time of usage is shown as an interval on the time line. Some devices, such as ACs are used multiple times in a day whereas some devices

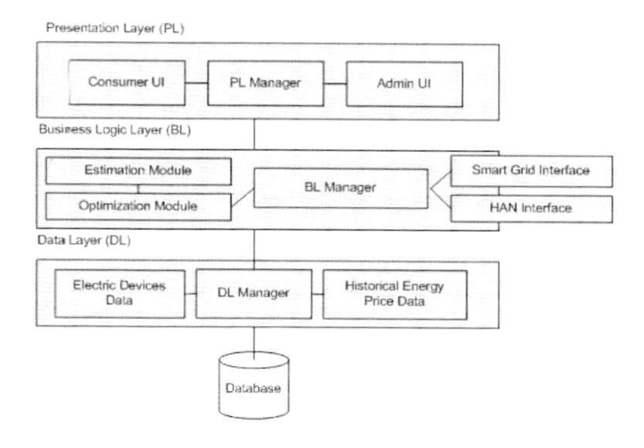

Fig. 2. PCAT Architecture

are only used once. It can be noted that different types of devices are being used at night so that the services they provide are available in the morning. This include the water pump, dish washer and battery charging for electric vehicles. Our third source of data is the power tariff that is applicable for a given 30 minute period. We used the data from the New York Independent System Operator's [1] website for first 7 days in the month of September 2009.

Table 1. Devices

Device ID	Device Name	Average Hourly Consumption (kWh)
1	Air Conditioner 1	2
2	Air Conditioner 2	1.5
3	Air Conditioner 3	2
4	Washing Machine	1.5
5	Dishwasher	1
6	Electric Car Battery Charger	2.5
7	Water Pump	1.5

A user oblivious to the price variation will use the device any time within the limits of his preferences. Using this hypothesis we run two types of simulation. For both of these simulations we fixed the amount of energy that is used the house.

Our first simulation uses a pseudo-random selection of device usage. This is to simulate a typical energy usage behavior in a household. The second simulation is performed using the Optimization Manager (OM) of PCAT. We executed multiple runs of both the scenarios and averaged out the results. The results of this comparison are shown in figure 4.

[1] http://www.nyiso.com/

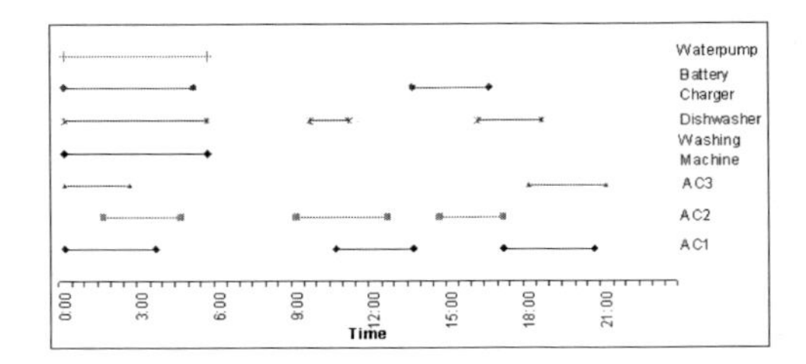

Fig. 3. Preferences for the 7 devices in the household

As we can see the optimized plan costs less than the typical scenario on each of the seven days. The typical plan is at an average 19% more costly than the optimal plan and on some days, such as on Day 6 it is more than 38% more costly than the optimized plan. It was noted during the multiple runs of the typical plan that not even once did the typical plan match the performance of the optimized plan.

The benefits from optimization is due to the variations in price in a day. We observed that on a given day, the price of power varied from \$356/MWh to \$5.64/MWh with standard deviation as high as \$65 over a period of a single day. This variation in price is used by our optimizer to deliver the results we see in our experiments.

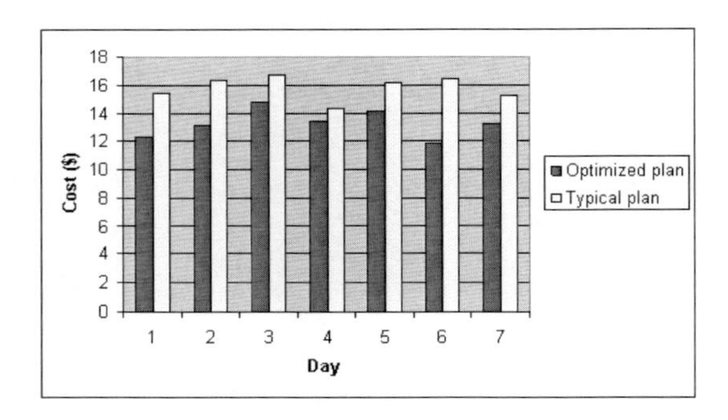

Fig. 4. Comparison of optimized plan and average of typical plan oblivious of price

6 Related Work

A variety of self-managing power management techniques have been developed which provide power management support to large scale computing systems such as server farms etc. Works such as by Milenkovic and colleagues use the power-awareness in management of cloud data centers [11]. Nathuji and colleagues proposed mechanisms to manage data centers by using the heterogeneity of machines [12] and and Zhu and colleagues proposed a framework for hierarchical optimization scheme for managing large scale, distributed data centers [17]. Works of Ha and Abras provide methods to optimize power consumption in smart-homes [3,9] . However, these systems are made with the scope that the user does not have the capability to provide the preferences of usage thus these systems have a heavy component to predict the usage of electricity and to calculate the effectiveness of plan through secondary measures. Through its user interface, PCAT is able to bypass these procedures by gathering the usage and effectiveness of plan data directly from the user. This not only simplifies the optimization methods but also make it efficient enough in terms of speed for large scale deployment.

There are some supporting works that can better serve users in conjunction with PCAT. These include Genio by Gï¿œrate et al. which uses Ambient Intelligence for human interaction with home environment where user can control home appliances by talking in a natural way [7]. Alkar et al. [4] developed a low cost, secure Internet based wireless system for home automation which can control a wide variety of devices. And Yuksekkaya et al. [16] developed a low cost, user friendly, wireless interactive home automation system which can be controlled by GSM, Internet and speech.

7 Conclusions and Future Work

In this paper, we have showed that home electricity consumers can save money on electricity bills by taking advantage of the time varying prices of electricity together with saving electricity production costs. The basic rational is that when we have time varying prices of electricity available then we can reduce electric devices usage expenses by using electric devices at those times when the price of electricity is lower.

Many future opportunities ahead in the area of smart grid for the research community. Other than device on/off there are various ways of intelligent energy usage. For example, through reducing the voltage to a device, through adding multiple sources of energy i.e. solar panel on the rooftop etc. Using these various ways of conserving energy has a potential for very complex, interesting and meaningful problems for the research community. Moreover, using an extension of HAN the self-managing system could also be applied to a Neighborhood Area Network (NAN) where a neighborhood becomes an energy island and has its own ways of producing and consuming electricity.

We further plan to extend PCAT to handle some of the challenges of energy management in smart homes as identified by Rathnayaka and colleagues [5]. These include handling conflicting multiple user priorities, multiuser buildings and so on and so forth.

References

1. Energy Use in the New Millennium – Trends in IEA Countries. International Energy Agency (2007)
2. The smart grid: An introduction. US Department of Energy (2009)
3. Abras, S., Pesty, S., Ploix, S., Jacomino, M.: An anticipation mechanism for power management in a smart home using multi-agent systems. In: 3rd International Conference on Information and Communication Technologies: From Theory to Applications, ICTTA 2008, pp. 1–6 (2008), doi:10.1109/ICTTA.2008.4530305
4. Alkar, A., Buhur, U.: An internet based wireless home automation system for multifunctional devices. IEEE Transactions on Consumer Electronics 51(4), 1169–1174 (2005), doi:10.1109/TCE.2005.1561840

5. Dinusha Rathnayaka, A.J., Potdar, V.M., Kuruppu, S.J.: Energy Resource Management in Smart Home: State of the Art and Challenges Ahead. In: MSirdi, N., Namaane, A., Howlett, R.J., Jain, L.C., Howlett, R.J., Jain, L.C. (eds.) Sustainability in Energy and Buildings. SIST, vol. 12, pp. 403–411. Springer, Heidelberg (2012)
6. Feinberg, R.: Achieving customer acceptance of the smart grid. The Intelligent Project (2009)
7. Gárate, A., Herrasti, N., López, A.: Genio: an ambient intelligence application in home automation and entertainment environment. In: sOc-EUSAI 2005: Proceedings of the 2005 Joint Conference on Smart Objects and Ambient Intelligence, pp. 241–245. ACM, New York (2005), http://doi.acm.org/10.1145/1107548.1107609
8. Gill, K., Yang, S.H.Y.S.H., Yao, F.Y.F., Lu, X.L.X.: A zigbee-based home automation system (2009), http://ieeexplore.ieee.org/lpdocs/epic03/wrapper.htm?arnumber=5174403
9. Ha, D.L., de Lamotte, F., Huynh, Q.H.: Real-time dynamic multilevel optimization for demand-side load management. In: 2007 IEEE International Conference on Industrial Engineering and Engineering Management, pp. 945–949 (2007), doi:10.1109/IEEM.2007.4419331
10. Javed, F., Arshad, N.: A penny saved is a penny earned: Applying optimization techniques to power management. In: 16th IEEE International Conference on the Engineering of Computer- Based Systems (ECBS 2009), San Francisco, CA, USA, April 13-16 (2009)
11. Milenkovic, M., Castro-Leon, E., Blakley, J.R.: Power-Aware Management in Cloud Data Centers. In: Jaatun, M.G., Zhao, G., Rong, C. (eds.) Cloud Computing. LNCS, vol. 5931, pp. 668–673. Springer, Heidelberg (2009)
12. Nathuji, R., Isci, C., Gorbatov, E.: Exploiting platform heterogeneity for power efficient data centers. In: Fourth International Conference on Autonomic Computing, ICAC 2007, June 11-15, p. 5 (2007), doi:10.1109/ICAC.2007.16
13. Pérez-Lombard, L., Ortiz, J., Pout, C.: A review on buildings energy consumption information. Energy and Buildings 40(3), 394–398 (2008), http://www.sciencedirect.com/science/article/B6V2V-4N8BMX9-2/2/5d59d6827f1f8d36a2a6480a0c95650e, doi:10.1016/j.enbuild.2007.03.007
14. Wallin, F., Bartusch, C., Thorin, E., Bdckstrom, T., Dahlquist, E.: The use of automatic meter readings for a demand-based tariff, pp. 1–6 (2005), doi:10.1109/TDC.2005.1547125
15. Wallin, F., Dotzauer, E., Thorin, E., Dahlquist, E.: Automatic meter reading provides opportunities for new prognosis and simulation methods, pp. 2006–2011 (2007), doi:10.1109/PCT.2007.4538626
16. Yuksekkaya, B., Kayalar, A., Tosun, M., Ozcan, M., Alkar, A.: A gsm, internet and speech controlled wireless interactive home automation system. IEEE Transactions on Consumer Electronics 52(3), 837–843 (2006), doi:10.1109/TCE.2006.1706478
17. Zhu, X., Young, D., Watson, B., Wang, Z., Rolia, J., Singhal, S., McKee, B., Hyser, C., Gmach, D., Gardner, R., Christian, T., Cherkasova, L.: 1000 islands: Integratedcapacity and workload management for the next generation data center. In: International Conference on Autonomic Computing, ICAC 2008, pp. 172–181 (2008), doi:10.1109/ICAC.2008.32
18. Zigbee, http://www.zigbee.org/ (accessed March 20, 2012)

Chapter 41
A Library of Energy Efficiency Functions for Home Appliances

Hamid Abdi*, Michael Fielding, James Mullins, and Saeid Nahavandi

Center for Intelligent Systems Research, Deakin University,
Geelong Waurn Ponds Campus, Victoria 3217, Australia
hamid.abdi@deakin.edu.au

Abstract. Emerging home automation technologies have the potential to help householders save energy, reduce their energy expenses and contribute to the climate change by decreasing the net emitted greenhouse gas by reduced energy consumption. The present paper introduces a distributed energy saving method that is developed for a home automation system. The proposed method consists of two levels functional elements; a central controller and distributed smart power points. The smart power points consist of a proposed a library of energy saving functions that are specifically developed for different home appliances or devices. Establishing a library of high performing, energy saving functions can speed up development of a buildings control system and maximum the potential reduction in energy consumption.

Keywords: Home automation, distributed control, energy saving, energy efficiency, function library.

1 Introduction

Over the past three decades, there has been an increasing impact and consequences of climate change worldwide with a documented increase in environmental pollution and extreme weather conditions. The current profile of fossil fuels usage needed to support the modern-human lifestyle is said to be one of the main reasons of climate change. Energy use in residential and commercial buildings accounts for between 20% and 40% of the total energy consumption of developed countries, which corresponds to a similar percentage of greenhouse gas emissions worldwide [1]. Todays, the smart technology is widely discussed by researchers and manufacturers with the intention to introduce new services, or improve the quality and efficiency of the existing services, in electrical grids [2], buildings [3], and home appliances [4]. This technology is beneficial for energy saving and environmental sustainability of homes or residential places [5]. Smart home technologies are increasing in popularity thanks to more efficient and lower costs sensors, controllers and home automation networks [3]. The smart home technologies offer a more convenient lifestyle by using

* Corresponding author.

A. Håkansson et al. (Eds.): *Sustainability in Energy and Buildings*, SIST 22, pp. 435–444.
DOI: 10.1007/978-3-642-36645-1_41

intelligent/programmable controllers and new or advanced functions for home equipment [6]. A smart-home technology provides additional features including fire and smoke detection and alarm systems, HVAC control, remote supervisory options, and security systems [3, 4]. The open structure of smart-home technology has also been used to develop specific and novel features such as voice control for homes and new ideas related to iPhone and iPad apps, for example [7].

Smart-home technology is centrally based on Home Automation Networks (HANs), a sub-category of Personal Area Networks (PANs). There are number of PANs specifically designed for home and building automation and management systems including the well-known protocols of C-Bus, LonWorks, EIB, BACnet, HomePlug, UPnP, X10, IEEE 802.15.4 [8], HAVI, Jini, and LnCP [9]. A comparative study of some of the HAN protocols has been presented in [6]. There are some emerging home protocols based on wireless ZigBee [6, 10] and wireless Bluetooth [11] but which have not currently commercially available. Using current commercial technology the security of the wireless HANs and interoperability of home appliances and electronic devices are bottlenecks of the HANs systems. For example, the diversity of home appliances, electronic devices, and networks resulted in interoperability problems such as not all home devices being able to communicate with all HAN protocols. This difficulty has been addressed by proposing a universal middleware bridge (UMB) in [12] and it was shown that the UMB is able to "solve interoperability problems caused by the heterogeneity of several kinds of home network middleware". Although an improvement, the UMB is still unable to support all protocols and communicate between all devices.

Among the literature of HANs, Z. Ye et al in [13] developed a low cost and practicable plug and play (PnP) HAN, however they focused mainly on the network protocols layers of the system. Other researchers have also worked on different HAN protocols, hardware and implementation of the networks, but there is gap in the development of specific functions for these networks specifically for the energy saving purposes. Literature is available that discusses using home automation systems to develop appropriate control strategies for different appliances but they are not developed for HANs. For example, the usage pattern of appliances has been discussed by Edwin et al [14] and the energy history has been employed in a data mining method to discover significant patterns of device usage.

The present paper proposes a control strategy of energy saving for HANs through the establishment of an energy saving function library. Although energy saving using HANs has been discussed in the literature, no specific function library has been developed thus far. In this paper we propose a distributed energy saving method for home automation systems based using a combination of a modified conventional HAN and smart power points. The method is applicable for different HANs, however we use the application layer of the HAN suggested by Z. Ye et al in [13] to develop the function library.

The paper is organized as follows. In Section 2, home automation systems and their potential for energy saving is discussed. In Section 3, the concept for distributed energy saving systems based on smart power points and home automation networks is introduced. Then in Section 4, two levels of energy saving functions are presented that are for the central controller and smart power points, and which includes the proposal of a library of functions for smart power points for different appliances. Finally, the concluding remarks are presented in Section 5.

2 Home Automation Systems

Home automation system include all the hardware, wiring, and software to automatically perform some of the home services. Generally, a home automation systems consist of conventional 1) power points, 2) a series of actuators such as motors, pumps, heaters, valves, and relays 3) various sensors and switches such as thermal and humidity sensors, motion, fire and smoke detectors, 4) home entertainment equipment such home audio and video systems, 5) security cameras, 6) controllers, 7) interface panels or screens, 8) a software and a computer interface system and 9) network interface units. Figure 1 shows a commercially available model for a home automation system based C-Bus system developed by Clipsal, one of the largest manufacturers of components for a buildings electrical system. The Clipsal systems offers various features thanks to the programmability of the digital controllers and interface screens. The system consists of a series of controllers and interface panels (LCD touch screens), wall switches, and a bank of output modules which can include simple on/off control relays, speed controllers or dimmable outputs for varying light brightness. The network of the system is based on a wired C-Bus network. The C-Bus controller connects different relays, switches, controllers, control panels, sensors, and other devices so they can communicate over C-Bus.

3 Distributed Systems Based on Smart Power Points

Using the conventional system shown in Figure 1, the power points are unable to provide any detailed information of the power use by connected devices. In this paper, we propose a modification to the conventional system by adding a power metering feature to individual power points. This can be achieved by metering the power at the relay modules in Figure 1 or by using smart power points (SPP). The SPPs are assumed as a modified version of the conventional power points. The SPPs can measure the power usage at the power point and can communicate with the network. In order to have this feature, the SPPs should employ an onboard microcontroller system to perform the readings and package the data to be sent. Additionally, the microcontroller allows for SPPs to perform local processing and control. The programmability of the power points can be used to program the SPP for any specific appliance or device that is attached to the power point. The power metering in each SPP can perform a precise analysis of the energy consumption and in turn can be used to develop appliance-specific energy saving strategies.

Z. Ye et al in [13] has introduced a Communication Unit (CU), whereby the appliances are connected to the network by the CUs as shown in Figure 2. The the proposed CU has a digital input, a digital output, an RS232 port, and the authors state that "application software is also needed which may be implemented either in the CU or in the appliance". Such claims for CUs does not suggest local energy saving functions and power metering. We use the base of their definition of CUs and we propose SPPs as a more advanced version of the CUs. Later in this chapter we focus on energy saving at SPPs, which contributes to further development of the application layer of the CUs.

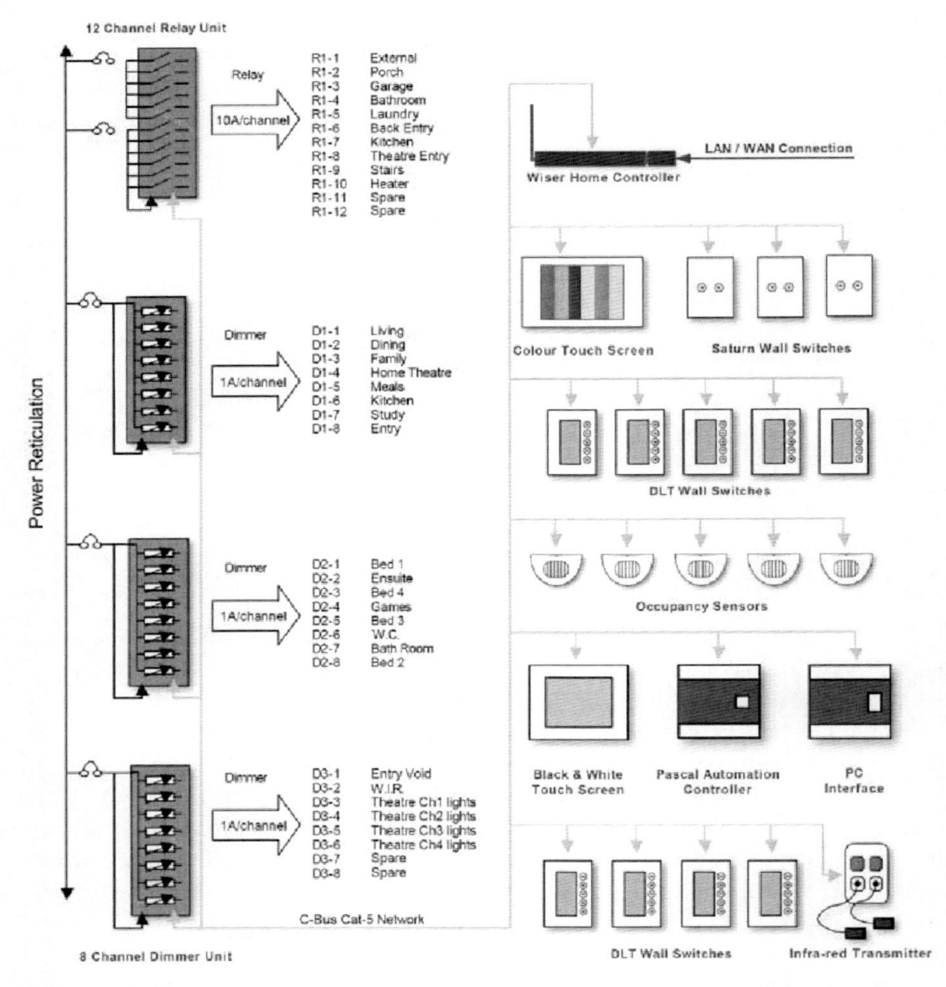

Fig. 1. Example of the commercially available model for home automation systems based on Clipsal products [15]

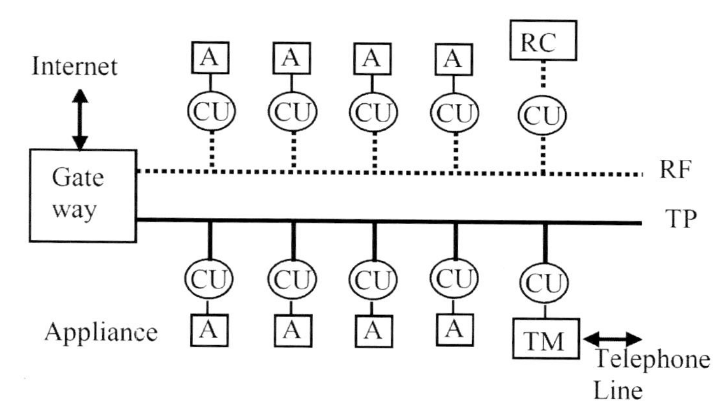

Fig. 2. Home automation network proposed by Z. Ye et al in [13], the CUs are attached to different appliances and home devices and can communicate wirelessly (RF line)or on twisted pair (TP line)

3.1 Smart Power Points

The proposed HAN network below is similar to the network in Figure 2, but we use SPPs in place of the CUs. The wireless communication is between the central controller and the SPPs as shown in Figure 3.

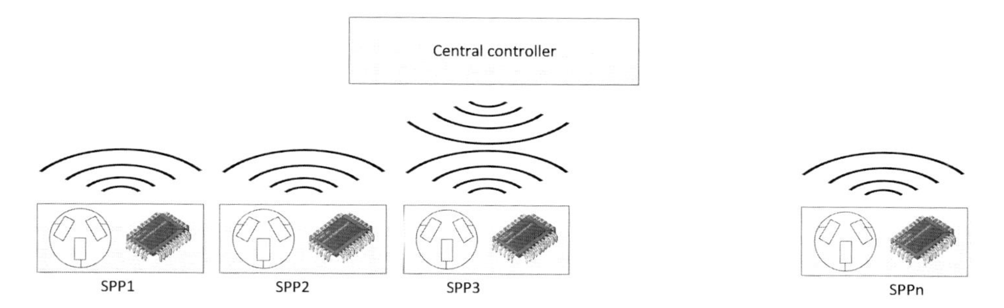

Fig. 3. Proposed network based on SPPs and wireless communication

3.2 Central Controller

The Central Controller in Figure 3 has a significantly reduced processing and communications requirement in comparison to the controllers in Figure 1 and 2. This is because the SPPs have local computational ability. This also simplifies the application layer of the Central Controller as it is only required to perform high-level home management functions. Therefore, there is much a reduced requirement for high speed hardware specifications, as required in [16]. The energy saving functions of the central controller is introduced later.

3.3 SPPs and Local Calculations

The application layer of the SPPs is based on the application layer of CUs proposed by Z. Ye et al in [13]. They proposed an object oriented methodology for this layer to describe different appliances and their operation. However, their focus has been on the network and they have not offered a detailed study of the application layer. In the present paper, we extended the application layer to include energy saving functionality.

The objects of the CUs application layer are shown in Table 1 and consists of an Appliance Object (AO), Appliance Identification Object (AID) and an Appliance Operation Object (AOO). We use a similar structure for the application layer of SPPs and provide further development for the objects. For this purpose, the structure of the objects remains very similar, with only minor changes to some of the parameters. This is achieved by modifying the objects of CUs and adding a new object for the SPPs as shown in the second column of the Table. Using the modified objects, the SPPs are able to measure local power usage and perform local control or calculations by the Appliance Calculation Object (ACO). The ACO is tasked to perform general local calculations as shown in figure 4. The ACO function allows various calculations at the SPPs. It is possible to assign the type of energy saving by the Energy_Save_Fuc variable, therefore, an SPP can employ specific energy saving functions developed for a specific type of appliance.

The calculation can be a linear/ non-linear combination of the input values and parameters with arithmetic or logical operations. This includes various computational ability including addition, subtraction, comparison, multiplication, or other mathematic or boolean operations. It is also possible to implement various functions such as exponential, trigonometric functions and filter functions, though is limited by the speed and functionality of the selected micro controller and size of available local memory.

4 Distributed Energy Saving for Home Automation Systems

Energy saving control strategies can be performed both locally on the SPPs and centrally on the central controller. Such a control system is considered as distributed control system. This system has several advantages including, 1) improved local energy saving ability due to local processing and decision making, 2) reduced requirement for high speed processor to be used within the central controller, 3) simplified energy saving control strategies in the central controller, 4) improved reliability of overall total system and energy saving, 5) lower communication overhead, and 6) improved data security.

4.1 Energy Saving Function by Central Controller

The central controller receives wireless information from SPPs that include an SPPs power reading, the outputs of the calculation object, and various status bits. The information will be only sent to the central controller if they are required for the

Table 1. The objects of the application layer of CUs form [13] and modified objects for the proposed SPPs in this paper, a new ACO object and Energy_Save_Fuc have been added to perform local calculation and energy saving, Description of the object parameters is shown in Appendix 1.

Proposed system by Z. Ye et al in [13]	Modified objects for SPPs
AO{ char Appliance_Name[16]; long int Appliance_ID; int OO_Number; AOO Aoo1; AOO Aoo2; }	AO{char Appliance_Name[16]; long int Appliance_ID; int OO_Number; int Power_use; char Energy_Save_Fuc; ACO Calc1, Calc2; AOO Aoo1, Aoo2 }
AOO{int R/W; int Priority; int Group_permission; char OO_name[16]; int Parameter_type; }	AOO{int R/W; int Priority; int Group_permission; char OO_name[16]; int Parameter_type; int Power_use }
No calculation	ACO{int in1, in2 boolean bin1,bin2; int par1, par2 int out1, out2 boolean bout1,bout2 char function_ID }

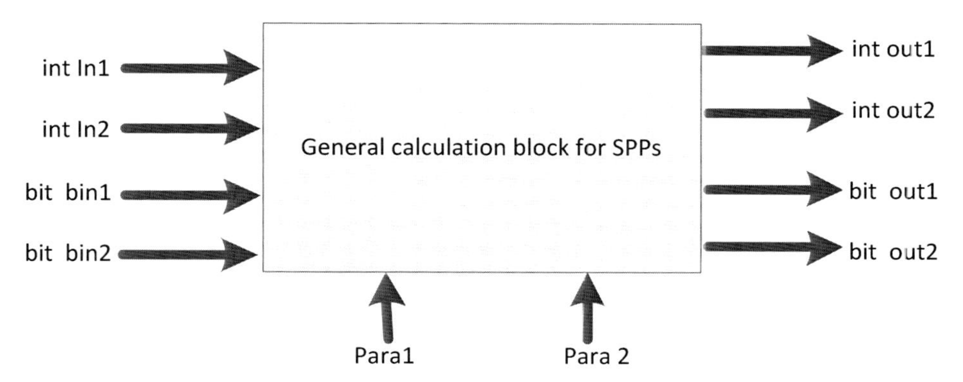

Fig. 4. Local calculation at SPPs, accepts two integer values and two binary values as an input and includes two adjustable parameters for each object, the output comprises two integer values and two binary values

control routine of the central controller or periodically for data logging purposes. The controller also receives information from different sensors, motion detectors, smoke detectors, and wall switches. The user can communicate with the controller via an interface PC software or an interface screen similar to that in Figure 1. Upon instruction from an operator or automatic routine, the controller performs a high-level energy saving controls such as total shutdown, partial shutdown, occupancy analysis for active and passive energy saving, standby power elimination for PCs, TVs and other entertainment devices. For example, based on the information for the motion detectors, when tenants leave the home then the central controller sends a command to all SPPs to turn off the appliances.

4.2 Energy Saving Functions for SPPs

SPPs have a built in microcontroller that has an on-chip RAM and non-volatile memory. The chip also supports USB connections, an analogue input channel for measuring power and several digital input and output channels. The microcontroller can perform on/off control of the power at the SPP via a relay mechanism.

Each electric devices or appliance connected to a SPP require a specific energy saving routine. For example, the energy saving for TVs are different from Fridges. Therefore, we define a library of energy saving functions that are for different home devices and appliances. We assume 256 different energy saving functions and we propose 50 functions in the current library. The library can be extended similarly for up to 256 different types of home devices or appliances.

In this library, specific group of functions are assigned to the kitchen SPPs that are for different appliances. The kitchen energy saving category is Energy_Save_Fuc=50-69. For example microwave has an energy saving code of Energy_Save_Fuc=50-52, fridge has Energy_Save_Fuc=53-55, oven has Energy_Save_Fuc=56-58, and fan Energy_Save_Fuc=59-61. The sitting room general-purpose power points category is Energy_Save_Fuc=100-119, bed rooms Energy_Save_Fuc=120-159, and similarly for outdoor Energy_Save_Fuc=180-199, laundry Energy_Save_Fuc=200-204, garage Energy_Save_Fuc=205-209 and other places. For the entertainment devices such TV and related equipment, the function type of (Energy_Save_Fuc=80-99) is assigned. For computers, laptops and printers the function of (Energy_Save_Fuc=20-39) is assigned.

For each of these function types, the related energy saving strategy has been defined. A simple function can be based on time or logical operations. However, the programmability of the SPPs allows us for further development of more efficient energy saving functions.

4.3 Energy Saving Interface

The energy saving in the central controller and the SPPs requires some inputs from the user side. This requires an interface to the user to take the inputs. Such interface has been developed for the central controller and individual SPPs by a software package that connects to the SPPs by USB port.

5 Conclusions

Energy consumption in residential and office buildings is responsible for 20-40% of the greenhouse gas emissions for different countries. Home automation systems have a great potential for energy saving for the buildings using smart technology. The present paper introduced distributed control system for home automation systems with a focus on energy saving. The proposed system consisted of smart power points, a wireless home automation network, and a wireless central controller. We proposed a distributed method for energy saving that allowed lower computation load for the central controller, lower communication overhead and higher security as well as local power saving features. The method provided two levels of energy saving for the central controller and distributed smart power points. An extendable function library for up to 256 energy saving functions for the smart power points were introduced for different types of home devices or appliances. Having such library of high performing energy saving functions can speed up development of homes' control systems and reduce the energy consumption.

Appendix

Table 2. Definition of parameters of objects for the application layer of CUs in [13]

Parameter	Type/Size	Comment
Appliance_Name	16 character	The name of the attached appliance
Appliance_ID	48 bit	A unique for the appliance , 2 bytes for the manufacturer, 2 bytes for appliance type and 2 bytes for sequence number
OO_Number	8 bit	Number of operation object
R/W	bit	Status bit of the AOO object
Priority	8 bit	Three levels of priority for the AOO objects
Group_permission	16 bit	Belonging to a specific group of appliances
Parameter_type	16 bit	The type of the parameter of each AOO

References

[1] Perez-Lombard, L., Ortiz, J., Pout, C.: A review on buildings energy consumption information. Energy and Buildings 40, 394–398 (2008)

[2] Farhangi, H.: The path of the smart grid. IEEE Power and Energy Magazine 8, 18–28 (2010)

[3] Snoonian, D.: Smart buildings. IEEE Spectrum 40, 18–23 (2003)

[4] Jiang, L., Liu, D.Y., Yang, B.: Smart Home Research 2, 659–663 (2004)

[5] Fox-Penner, P.: Smart power: climate change, the smart grid, and the future of electric utilities: Island Pr (2010)

[6] Osipov, M.: Home automation with ZigBee. Next Generation Teletraffic and Wired/Wireless Advanced Networking, 263–270 (2008)

[7] Petkov, P., Köbler, F., Foth, M., Medland, R., Krcmar, H.: Engaging energy saving through motivation-specific social comparison, 1945–1950 (2011)

[8] Callaway, E., Gorday, P., Hester, L., Gutierrez, J.A., Naeve, M., Heile, B., Bahl, V.: Home networking with IEEE 802.15. 4: a developing standard for low-rate wireless personal area networks. IEEE Communications Magazine 40, 70–77 (2002)
[9] Lee, K.S., Choi, H.J., Kim, C.H., Baek, S.M.: A new control protocol for home appliances-LnCP 1, 286–291 (2001)
[10] Gill, K., Yang, S.H., Yao, F., Lu, X.: A ZigBee-based home automation system. IEEE Transactions on Consumer Electronics 55, 422–430 (2009)
[11] Sriskanthan, N., Tan, F., Karande, A.: Bluetooth based home automation system. Microprocessors and Microsystems 26, 281–289 (2002)
[12] Moon, K.D., Lee, Y.H., Lee, C.E., Son, Y.S.: Design of a universal middleware bridge for device interoperability in heterogeneous home network middleware. IEEE Transactions on Consumer Electronics 51, 314–318 (2005)
[13] Ye, Z., Ji, Y., Yang, S.: Home automation network supporting plug-and-play. IEEE Transactions on Consumer Electronics 50, 173–179 (2004)
[14] Heierman III, E.O., Cook, D.J.: Improving home automation by discovering regularly occurring device usage patterns, 537–540 (2003)
[15] http://www.clipsal.com/homeowner/home
[16] Han, I., Park, H.S., Jeong, Y.K., Park, K.R.: An integrated home server for communication, broadcast reception, and home automation. IEEE Transactions on Consumer Electronics 52, 104–109 (2006)

Chapter 42
Smart Energy Façade for Building Comfort to Optimize Interaction with the Smart Grid

Wim Zeiler[1,2], Rinus van Houten[2], Gert Boxem[1], and Joep van der Velden[1]

[1] TU/e, Technical University Eindhoven, Faculty of Architecture, Building and Planning
[2] Kropman Building Services Contracting

Abstract. An Intelligent Electrical Energy supply Grid, a Smart Grid, is being developed to cope with fluctuations in energy generation from the different renewable energy sources. Energy demand and energy need to be better balanced to achieve improved overall efficiency. The process control of the energy flows in the buildings in relation to the outside environment and the user behavior also needs to become smart, intelligent and capable of adaptation to changing conditions. Otherwise you get the combination of a smart infrastructure but a dumb client, which is not good for the business of the client. Especially is it of great importance to take in account the goal of the energy use: human comfort. There is need for dynamic individual local comfort control instead of only process control at room level. Especially with these new process control possibilities, the interaction becomes essential of the outdoor active and passive energy processes with indoor through the façade. The façade is as such passive and active energy source on the one hand and a critical factor in relation to the perceived thermal comfort. The façade can be seen as an energy interface that should be optimized to perceived comfort of the occupants and their energy consumption.

Keywords: Software Agents, Building Automation, Thermal Comfort.

1 Introduction

The environmental impact of the built environment needs to be reduced. Energy is not a goal but a mean to achieve something. At the moment energy use in the built environment accounts for nearly 40% of the total energy use in the Netherlands. Most of this energy (nearly 87% for non-residential and 72% for residential buildings) is used for building systems or room heating with the goal of providing comfort of the occupants of the buildings. Optimizing all energy flows in connection to comfort is not carried out in practice as yet. In addition, the common methods used to predict the amount of energy needed to generate thermal comfort are far from optimal and could be tremendously improved by using a more precise and detailed approach.

The instable supply of some of the applied renewable energy sources, such as wind and sun, and the variation in local generation, makes it necessary to improve the current stability of the centralized infrastructure of energy supply. An Intelligent

A. Håkansson et al. (Eds.): *Sustainability in Energy and Buildings*, SIST 22, pp. 445–452.
DOI: 10.1007/978-3-642-36645-1_42

Electrical Energy supply Grid is being developed to cope with fluctuations in energy generation from the different energy sources. To better match energy demand and energy need to achieve improved overall efficiency, the process control of the energy infrastructure in the buildings also needs to become smart, intelligent and capable of adaptable behaviour in changing conditions. Normally only simple approaches are applied to incorporate the comfort demand of occupants or their behaviour and use of appliances. Often only on the level of house or building and only sometimes on room level, see Fig. 1 and 2.

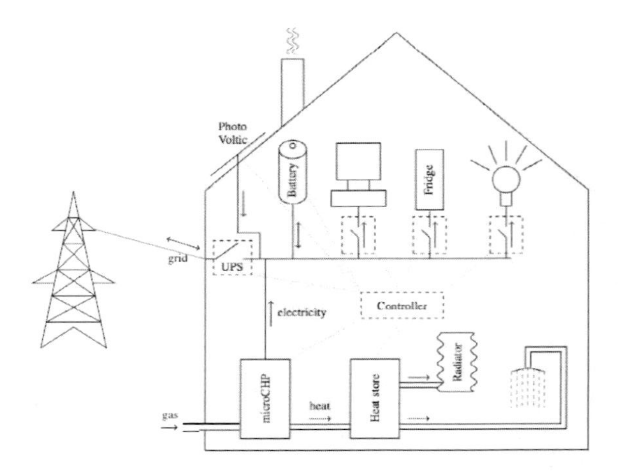

Fig. 1. Model of domestic energy streams of a dwelling [Molderink et al 2010]

Besides the smart application of smart grind technology it is of great relevance integrating a more advanced comfort controls strategy for energy management of buildings not only for the energy consumption but for the comfort management.

New energy management systems should be developed, which do not only optimize the energy consumption, but also optimize the comfort for the users by applying new insights about comfort and apply predictive process control based on short term and long term weather forecasts. Multi agent systems will be used to cope with all the dynamic changing influences and be coupled onto existing BEMS platforms.

The developed systems can be used to optimize new buildings but far more important also used for optimizing existing buildings. The viability of the solutions will be demonstrated in a real office situation in cooperation with Kropman Building Services who developed their own BEMS SCADA (Supervisory Control And Data Acquisition) system, InsiteView. The functionality of the InsiteView will be extended and developments are ongoing for a new generation of BEMS that incorporate Agent technology for predictive environmental and user adaptive comfort-energy process control.

In former research on predictive environmental-adaptive and user adaptive energy management systems, TU/e and Kropman investigated the application of multi agent software. These projects represent the starting point for further development and improvement. Especially the following projects are relevant:

- SMART and IIGO: In the SMART, Smart Multi Agent inteRnet Technology) [Akkermans et al 2002, Kamphuis et al 2002] and IIGO, Intelligent Internet mediated control in the built environment) projects, which were partly financially supported by SenterNovem. field test were done in the Kropman offices in Rijswijk, Utrecht and Nijmegen [Jelsma et al 2003, Hommelberg 2005, Kamphuis et al 2005, Zeiler et al 2005, Zeiler et al 2006]. These field tests showed the potential of the combination of Building Energy Management systems and Multi Agent System applications.

- EBOB: Within the EU-FP5 project EBOB (Energy Efficient Behaviour in Office Buildings) [Claeson-Jonsson 2005, Opstelten et al 2007] Kropman worked the aspects of personalizing and presenting data of the energy management system to generate energy awareness.

- Flexergy: The project focuses on the integral optimization of energy flows within the built environment when fitting in decentralized sustainable energy concepts. The research outcomes are tested in an existing office of Kropman [Zeiler et al 2008, Zeiler et al 2009, Pruissen and Kamphuis 2010]. This design methodology should lead to solutions that offer more flexibility to the energy infrastructure; Flex(ible)en)ergy. However in the Flexergy project the user was still represented by a comfort level day profile based on the room temperature setting. Field tests were held at Kropman Utrecht [Pruisen and Kamphuis 2010, Zeiler et al 2010].

We now want to look more closely to the individuals on working space and personal level. So we do not look only to room temperatures and thermostat settings of hot water taps but really look into the most important dynamic parameters related to the individual thermal comfort, the actual occupancy and the actual parameters of the facade. Especially the focus will be on the dynamic characteristics of the interaction between façade, user and outdoor environment. The energy supply to a building must be related to actual dynamic changing comfort needs, behaviour of the occupants of the building and the behaviour of the building itself due the weather conditions. Therefore, more actual information is needed.

The central research question is to determine whether it is possible to come to a comfort – sustainability optimization of the smart indoor building energy grid. This optimization is strongly influenced by the effective use of the façade as a decentralized renewable energy source of passive and active energy. This leads to the research questions how to design a process model to predict and to control the necessary energy flows within a building in relation to the changing outside conditions and based on the individual momentary comfort demand of occupants.

Research Method

The goal and intended result is to design, build and test an intelligent energy grid within buildings with the actual individual human need as leading principle. Therefore, the first step is to apply an appropriate design approach. A hierarchical functional decomposition approach is used to structure the energy infrastructure of a building

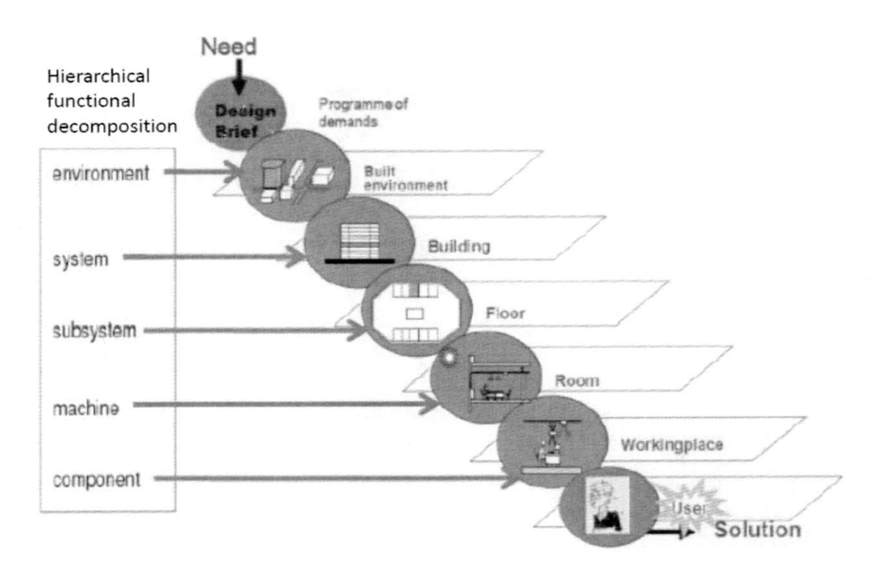

Fig. 2. Hierarchical functional decomposition of the built environment

[Zeiler & Quanjel 2007]. This method approach makes it possible to study the energy flows connected to heating, cooling, ventilation, lighting, and power demand, within a building on the different levels of hierarchical functional abstraction, see Fig.2.

Compared the common approaches our approach offers the possibility to focus on the level of workplace and the level of the individual. This enables us to look more closely on the comfort and energy demands of individuals and to built a more detailed process representation. The individual user has become leading in the whole process to optimize the necessary use of energy to supply the occupants with their own preferred comfort environment and energy for their activated appliances. On different levels abstraction a functional representation of the building and its occupants will be made. First simple representations of the process will be made to look into the interrelations between the different levels within a building system, see Fig. 3: building level (possible energy supply from the grid and renewable energy generation within the building related to the weather forecast and current weather conditions), room level (energy exchange depending on the outside environmental conditions and internal heat load), workplace level (workplace conditions and energy need from appliances) and human level (defining the different comfort needs of individuals and the resulting energy demands).

Based on the abstract representation of occupant, workplace and room, the influences of different characteristic parameters for comfort and energy consumption will be measured. In a rapid prototyping approach, process representations of some of the different levels will be built. This will form the basis of defining different agents within the multi agent process control system. Using data from an existing building and its users it is possible to fit the model immediately with a real live situation. By making a coupling with the Building Management system of the building all the necessary data will be made available to fit and to investigate the behavior of the model compared to real historical data of before the intervention. The insights gained from the modeling of the local personal comfort on personal level, workspace level and room level, leads to a concept for monitoring and management of the comfort and the energy flows in a real building in a more detailed and accurate way.

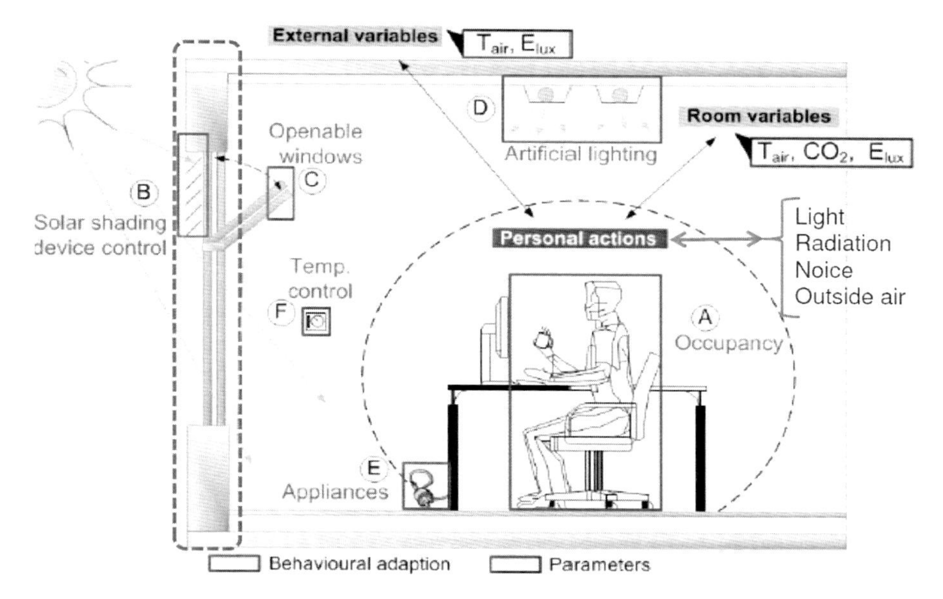

Fig. 3. Detailed representation of the most important aspects on room level and workspace level

Significance of the Proposed Research

Instead of applying new cooling, heating or ventilation devices the research is focussed on the optimal process control of comfort demand and the necessary energy for that. This means that the developed process control strategy could be applied in new as well as in existing buildings. It could become the leading technology for realising the technical savings potential of advanced control systems in the Dutch built environment which is no less than 19% of the total energy usage of the Dutch built environment [Kester and Zondag 2006]. Sensing, monitoring and actuating systems in relation to the user perception and preferences play the key role in reducing overall energy consumptions in buildings. Therefore we start with looking more closely to the perceived comfort.

Traditionally, calculations of human comfort have been based on the theory of Fanger [1970]. Basically, this theory makes it possible to calculate the thermal sensation indicator representing the perceived comfort with 6 parameters: Temperature, Relative Humidity, Air Velocity and Radiant Temperature. However, activity level and clothing insulation of occupants have a strong effect on thermal comfort, but they are variable and usually not measurable. As a result, in practice the comfort control is simply done on the room level by controlling the room temperature. The variations of the other parameters of influence on the individual comfort demand are not taken into account. This omission results in differences depending on for example the place of the workspace in the room, e.g. close to a window or more in the back of a room. As a result the comfort is only controlled within a broad range, resulting in more complaints and more energy use than necessary. In theory, 95% of the users should be satisfied if all conditions stay within the specific ranges, however in field studies there is a much smaller satisfaction range of between 80% to 50% (Zimmerman 2008).

Since the human sensation of thermal comfort is a subjective evaluation that changes according to personal preferences, the development of an HVAC control system on the basis of the PMV model has proven to be impossible [Mirinejad et al 2008]. Despite all the recent attempts to find correlations, significant dependencies, standards, and optimal set points there is a growing understanding that every person has its own personal temperature preference and tolerance ranges [Noom 2008, Zimmermann 2008] and those might change even a little during the day depending on emotional state and fatigue. Therefore, it should be possible for occupants to adapt their individual comfort preference comfort profiles. Furthermore, it might be possible with the help of low cost sensors to register the changes of individual parameters connected to the global and also to the local comfort and respond to these dynamic changes. By optimizing the responses to the individual human comfort differences energy conservations of up to 25% are possible [Buitenhuis and Drissen 2007, van Oeffelen et al 2010]. Thermal comfort for all can only be achieved when occupants have effective control over their own thermal environment [van Hoof 2008]. This led to the development of Individually Controlled Systems (ICS) with different local heating/cooling options [Filippini 2009, Wanatabe et al. 2010]. Our intention is to design and built an experimental workplace with an individual controlled heating/cooling panel above the workplace to test our specific approach to comfort and energy management. The implementation of such detailed dynamic approach to individual comfort control is new.

Based on the experiences with multi-agent system projects and a literature review on the latest developments concerning human comfort a concept is being developed for the optimization of individual comfort and energy consumption by the use of an intelligent building energy grid with a combination of low cost wireless sensors. The application of low cost wireless sensors offers new practical applicable possibilities [Neudecker 2010, Gameiro Da Silva et al 2010]. If so, then energy demand and energy supply could become more balanced and less energy wasted. A promising technology to achieve the necessary dynamic process control is by using Multi Agent System technology [Qiao et al. 2006, Dounis and Caraiscos 2009, Lee 2010]. Agent technology in combination with low cost sensor networks can be implemented at different levels of building automation. Individual agents for individual climate control for each user of the building in combination with feedback on the energy consumption (costs/ sustainability) leads to better acceptance of the individual comfort and a reduction of the energy consumption [Jelsma et al 2003, Kamphuis et al 2005].

References

Akkermans, H., Kamphuis, R., Warmer, C., Jelsma, J., Wortel, W., Zeiler, W.: Advances and Experiences in Linking ICT and Power: Innovatieve e-Services for Smart Buildings. In: Proceedings CRIS 2002: Power Systems and Communications Infrastructures for the Future, Bejing (September 2002)

Buitenhuis, H., Drissen, R.: Installatieoptimalisatie Academie voor Engineering, Vraaggestuurd regelen

Claeson-Jonsson, C.: Final publishable report, Energy Efficient Behaviour in Office Buildings (2005)

Dounis, A.I., Caraiscos, C.: Advanced control systems engineering for energy and comfort management in a building environment – a review. Renwable and Sustainable Energy Reviews 13, 1246–1261 (2009)

Fanger, P.O.: Thermal Comfort. Danish Technical Press, Copenhagen (1970)

Filippini, G.J.A.: De mens centraal bij het ontwerp van het binnenklimaat: ontwerp en ontwikkeling van een duurzaam lokaal klimatiseringsysteem, MSc thesis Technische Universiteit Eindhoven (2009)

Gameiro Da Silva, M., Paulino, A., Brito, A., Valetim, R., Gomes, J., Fonseca, R.: A wireless network web based monitoring solution for indoor environmental quality and energy performance assessment of buildings. In: Proceedings Clima 2010, Antalya, May 10-12 (2010)

Hommelberg, M.P.F.: 2005, Careful Buildings, software agents voor een gebouwbeheersysteem, Master thesis Building Services TU/e (January 2005) (in dutch)

van Hoof, J.: Forty years of Fanger's model of thermal comfort: comfort for all? Indoor Air 18, 182–201 (2008)

Jelsma, J., Kamphuis, R., Zeiler, W.: Learning about smart systems for comfort management and energy use in office buildings. In: Proceedings ECEEE Summer Study (2003) ISBN: 91-631-4001-2

Kamphuis, I.G., Warmer, C.J., Zeiler, W., Wortel, W., Akkermans, J.M., Jelsma, J.: SMART: Experiences with e-services for Smart Buildings. In: ISPLC Conference, Athens, March 27-29 (2002)

Kamphuis, I.G., Warmer, C.J., Jong, M.J.M., Wortel, W.: IIGO: Intelligent Internet mediated control in the built environment: Description of a large-scale experiment in a utility building setting, ECN rapport ECN-C—05-084 (October 2005)

Kester, J.C.P., Zondag, H.A.: Demand Side Management achter de meter, Raming verduurzamingspotentiëlen, ECN rapport E—06-037 (November 2006)

Lee, J.: Conflict resolution in multi-agent intelligentenvironments. Building and Environment 45(2010), 574–585 (2010)

Mirinejad, H., Sadati, S.H., Ghasemian, M., Trab, H.: Control Techniques in Heating, Ventilating and Air Conditioning (HVAC) Systems. Journal of Computer Science 4(9), 777–783 (2008)

Molderink, A., Bakker, V., Bosman, M.G.C., Hurink, J.L., Smit, G.J.M.: A Three-Step Methodology to Improve Domestic Energy Efficiency. In: Proceedings of the 2010 IEEE Innovative Smart Grid Technologies Conference, Gathersburg, USA, January 19-21 (2010)

Neudecker, F.: Eliminating wires and batteries in building management the new standard: Energy harvesting wireless sensors. In: Proceedings Clima 2010, Antalya, May 10-12 (2010)

Noom, P.: Het individu leidend; Een omgekeerde benadering van het thermisch comfort ten behoeve van de gebruiker, [Dutch], MSc thesis TU Eindhoven (2008)

Oeffelen, E.C.M., Van Zundert, K., Jacobs, P.: Persoonlijke verwarming in kantoorgebouwen. TVVL Magazine 1, 6–11 (2010)

Opstelten, J., Bakker, J.E., Kester, J., Borsboom, W., van Elkhuizen, B.: Bringing an energy neutral built environment in the Netherlands under control, ECN-M-07-062, Petten (2007)

van Pruissen, O.P., Kamphuis, R.: Multi agent building study on the control of the energy balance of an aquifer. In: Proceedings Improving Energy Efficiency in Cmmercial Buildings, Frankfurt, April 13-14 (2010)

Qiao, B., Liu, K., Guy, C.: A multi-agent system for building control. In: Proceedings IEEE/ACM International Conference on Agent Technology, Hong Kong, December 18-22 (2006)

Watanabe, S., Melikov, A.K., Knudsen, G.L.: Design of an individually controlled system for an optimal thermal microenvironment. Building and Environment 45, 549–558 (2010)

Zeiler, W., Wortel, W., Kamphuis, R., Akkermans, H., Jelsma, J., Bakker, L.: FACT; Forgiving Agent Comfort Technology. In: Proceedings 8th REHVA World Congress, CLIMA 2005, Lausanne (2005)

Zeiler, W., van Houten, R., Kamphuis, R., Hommelberg, M.: Agent Technology to Improve Building Energy Efficiency and Occupant Comfort. In: 6th International Conference for Enhanced Building Operations ICEBO 2006. Shenzhen, China, CD, vol. I-I-5, 8 pages (2006)

Zeiler, W., Quanjel, E.M.C.J.: Integral design methodology within industrial collaboration. In: Proceedings of the 4th Symposium on International Design and Design Education, DEC 2007, Las Vegas, September 4-7 (2007)

Zeiler, W., Noom, P., Boxem, G., van Houten, M.A., Velden, J.A.J., van der Haan, J.F.B.C., Wortel, W., Hommelberg, M.P.F., Kamphuis, I.G., Broekhuizen, H.J.: Agents to improve Individual Comfort and save Energy. In: Kodama, Y., Yoshino, H. (eds.) Proceedings AIVC 2008, Kyoto (2008)

Zeiler, W., Boxem, G., Houten, M.A., van Savanovic, P., Velden, J.A.J., van der Haan, J.F.B.C., Wortel, W., Noom, P., Kamphuis, I.G.: Multi-agent system process control ontology. In: Sharma, D. (ed.) KES AMSTA 2009, pp. 1–10. Uppsala University, Uppsala (2009)

Zeiler, W., van Houten, M.A., Boxem, G.: The user as leading factor in comfort control technology: the Forgiving agent technology approach. In: Proceedings Clima 2010, Antalya, May 1-12 (2010)

Zimmermann, G.: Individual comfort in open-plan offices. In: Proceedings DDSS 2008, Eindhoven (2008)

Chapter 43
Building for Future Climate Resilience

A Comparative Study of the Thermal Performance of Eight Constructive Methods

Lucelia Rodrigues and Mark Gillott

Department of Architecture and Built Environment,
University of Nottingham, Nottingham, UK
{Lucelia.Rodrigues,Mark.Gillott}@nottingham.ac.uk

Abstract. A great deal of literature has been published in recent years around the need to mitigate climate change and the building industry is already working to make buildings more energy efficient. However, some changes to our climate cannot be avoided so we will need to change the way we design, construct, refurbish and use buildings to adapt to the likely increases in temperature. A great proportion of British housing is now being built using Modern Methods of Construction (MMC) systems, and this number is expected to rise significantly over the next decade. All systems are potentially able to deliver good buildings, so how to choose? Sustainability should be the order, but it is only achievable if future climate resilience is considered. Otherwise, the use of MMC to build dwellings that use less energy for heating today could result in a future undesirable scenario when energy for cooling is also needed. In this work, the occurrence of overheating today and in the future in a highly insulated $100m^2$ space built using eight different walls constructions has been investigated in a parametric study. The building was dynamically simulated with few parameters to allow easy comparison of the performance of each constructive system. It was found that there is a high risk of overheating in houses and this risk will not be mitigated by one solution alone. Although this not a comprehensive study by any means, it is the start of a discussion to instigate further research that could inform design decisions that address future climate resilience.

Keywords: Climate Resilience, Buildings Energy Efficiency, Modern Methods of Construction, Thermal Mass.

1 Introduction

Due to ongoing shortages in UK housing supply, the Government has increased the rate of housebuilding to a new target of 240,000 additional homes a year [1]. Simultaneously, since 2007 an ambitious target has been set for all new houses to meet net carbon dioxide emissions (zero carbon) from 2016 in an attempt to tackle climate change and meet the targets set by the Kyoto Protocol that came into force in 2005 [2].

A. Håkansson et al. (Eds.): *Sustainability in Energy and Buildings*, SIST 22, pp. 453–462.
DOI: 10.1007/978-3-642-36645-1_43

In order to deliver more houses of better quality at a faster rate, the Government and its agencies are prioritising the modernisation of the housebuilding sector through the promotion of Modern Methods of Construction (MMC). MMC refers to a number of innovative methods and products, and the innovative use of traditional materials, mostly through off-site construction. This has already resulted in a growth of 20% in MMC permanent buildings between 2000 and 2005 and a growth of 24% between 2005 and 2008 [3]. According to the organisation Buildoffsite the UK market for off-site solutions was worth £2 billion per annum in 2005 and had grown to more than £6 billion by 2008 [4]. It aims to achieve a tenfold increase by 2020.

Stricter building regulations and new building standards (such as the Code for Sustainable Homes) have also been implemented to support more energy efficient housing construction. These changes in building regulations have resulted in an increase in insulation levels to reduce the heating season and save energy. However, they have also resulted in buildings that are much more sensitive to any alteration in energy inputs, especially if they are built using certain common MMC configurations that incorporate low thermal mass materials [5]. Generally, thermal mass refers to a material's capacity to absorb, store and release heat and, well applied, can help control the indoor temperature of a building. A highly insulated building with low levels of thermal mass tends to be more thermally responsive in shorter periods of time (i.e. quickly getting too hot or too cold). Various researchers have speculated that, in future climate scenarios, well insulated houses with low levels of thermal mass could result in substantially higher and uncomfortable room temperatures [6-10].

With the temperatures becoming warmer [11, 12] there is a greater risk of overheating inside these buildings and in the UK the houses are particularly vulnerable to the impact of this [6, 9, 13-15] as most of them rely on natural ventilation alone to overcome the issue. It is likely that as a response to warmer indoor temperatures more home owners in the UK will seek to install air-conditioning as it is generally now within economic reach, and its use is already rising [16]. This could easily negate the energy savings intended through their design.

In this work the occurrence of overheating in highly insulated buildings in the UK and the influence of the thermal mass of the walls in regard to this issue has been investigated. Eight different wall construction types were selected offering various degrees of heat storage capacity. As it seemed unreasonable to categorise each type with regards to its 'weight' (i.e. lightweight or heavyweight construction), they were characterised by their admittance, decrement factor and time constant which ways to identify the quantity of thermal mass. A special- purpose simple model was built in TAS by EDSL, a modelling and simulation tool capable of performing dynamic thermal simulation of buildings. The advantages of producing a simpler model include fewer inputs allowing efficient application of the principle of superposition and the study of the importance of each input. Tas was selected for having a good workflow methodology, for having full flexibility to model complex systems and to be as accurate as any of the other competitors [17].

The aim was to firstly clarify if thermal mass is still essential in the UK when U-Values are reduced to a point when almost no conductance happens through the fabric. Low U-Values already mean large wall thicknesses and so the addition of a layer just for its heat storage capacity might be undesirable. Secondly, this work investigated if occupancy would have an influence on the effectiveness of thermal mass. With full time occupants there is no large temperature fluctuation but with part time occupancy people may have to wait for the house to warm up, which might use even more energy than keeping the house warm continuously.

2 Scope and Method

The aim of this work was to determine the difference in the performance of the different building fabrics under the same conditions. The model and assumptions were kept the same for all the simulations. The only change was the wall construction type whilst floor, roof and windows were kept the same.

A Base Case with no shading or occupancy, and minimum infiltration, was simulated only to allow comparison. Next, the simulations were divided in 2 sets:

1. Cases 0 to 3: Model in today's climate with a parameter changed in each case (Table 1)
2. Cases 4 to 7: Model is future climate scenarios, years 2020, 2050 and 2080 (Table 2)

The starting point to decide the build up of the walls assessed was to use the most common construction methods (traditional and modern) and achieve a U-Value of $0.12W/m^2K$ in order to comply with the higher UK standards (Code for Sustainable Homes level 6) and international standards such as Passivhaus. With that in mind eight different wall construction types were selected:

1. Brick and Block full fill cavity wall (BB)
2. Timber frame part fill cavity wall (TF)
3. Insulated concrete formwork wall (ICF)
4. Steel frame wall (SF)
5. Structural insulated panel wall (SIPs)
6. Cross laminated timber wall (CLT)
7. Solid Concrete Block wall (SB)
8. Precast concrete panel wall (PCP)

Each one was characterised through relevant values such as admittance, decrement factor and time constant (Figure 1). Service voids or air gaps were considered where appropriate. The constructions that do not have brickwork externally received a 5mm external surface finishing with similar absorptance to the brick used in the other ones (i.e. 0.7). Internally all the walls have the same surface finishes, either of lightweight plaster or of plasterboard. The location of the extra insulation to achieve the desirable U-Value took into account practicality and best positioning to maintain the thermal mass characteristics if any. The same floor, roof, windows and shading were used for all cases (concrete floor with timber finishing and roof tiles on timber structure for the roof). Full details including thermal properties layer by layer can be found at Rodrigues [18].

In Figure 1 it can be observed that decrement factor and time constant do not follow a pattern that relates in any way to admittance. In addition, construction types such as SIPs and CLT are in the middle of the chart even though they do not have any material that would traditionally characterise thermal mass. The performance of each wall type will be investigated dynamically in the next section in order to understand if these figures are really meaningful to characterise thermal storage capacity.

2.1 Assumptions

A simple model was built composed of one zone of $100m^2$ (10x10m) and $270m^3$ of volume. All the walls are external, made up of the same building material and of an area of $30m^2$. Each layer was assumed to be isotropic. The south wall contains a $10m^2$ (10% of the floor area) window with a 50mm frame. The model was the same throughout the simulations as it was not the scope of this work to experiment with different dimensions and configurations.

All the work considered the climate of the city of Nottingham, latitude 53°N, longitude 1.25°W and altitude 117m, in the East Midlands of England. The weather data Design Summer Year Weather Data (DSY) Nottingham, made of hourly collected data on the year 2002 and developed by the Chattered Institute of Building Services Engineers [19], was used. This is the recommended data for the design of buildings focused on summer performance and overheating assessment, and considering a year with a hot, but not extreme summer. Nottingham represents well the UK climate in average, presenting a temperate climate, with prevailing low temperatures throughout the year and an annual average dry bulb temperature around 10°C. The highest and the lowest dry bulb temperatures recorded in that year were 29.1°C (August) and -6.7°C (January).

This weather data was used to simulate the climate in the future using the UK Climate Impact Programme (UKCIP) scenarios for building environmental design. The programme provides information on how the UK's climate is likely to change until 2080 as a response to rising levels of greenhouse gases in the atmosphere. It is based on probabilistic projections at a national level for the years 2020s, 2050s and 2080s under the high, medium and low emission scenarios and at 10, 50 and 90% probability levels [11]. As the low emission scenarios represent a future where there is a commitment to a large reduction of greenhouse gas emissions on a global scale, which seems unrealistic, the worst case scenario (emissions continue to increase until the middle of the century) was selected.

The data was morphed using the Climate Change World Weather File Generator for World-Wide Weather Data (CCWorldWeatherGen [20]) and the results show that the average temperature in Nottingham is expected to rise by over 4°C while relative humidity is expected to fall by about 5%. Temperatures may go as high as 36°C while relative humidity may fall below 30%. Night time temperatures also rise making the use of strategies such as night time ventilation more difficult.

The results were extracted as temperature ranges compared against thermal comfort benchmarks. according to the CIBSE criteria. The CIBSE benchmark for overheating in living rooms is 26°C, which should not be exceeded; if that benchmark is exceeded, it should not be for longer than 1% of the time above 28°C. In bedrooms the desirable temperature is 23°C and the temperatures should not exceed 25°C but if they do they should not stay above 26°C for more than 1% of the time [21].

The general model settings are summarised below [18]:

- Occupants: 2 sedentary people in the space resulting in 1.4W/m^2 sensible gains and 0.8W/m^2 latent gain. Case 1 and 2 have full time occupancy and in Case 3 occupants were out during the day and at home from 8pm to 8am (12h).
- Lighting and equipment: no gains were assumed.
- Infiltration: was assumed to be sufficient to provide occupants with their need for fresh air (8l/s per person), which adds to approximately 0.2ACH at atmospheric pressure in this model
- Ventilation: introduced in Case 7 by means of windows opening in summer only. It was assumed that the windows would start to open when temperature in the house reached 22°C and be fully opened at 28°C just when there was occupancy. If external temperatures exceed the internal temperature, the window will close. No ther means of cooling were assumed.
- Heating: was assumed for Cases 2 to 7 during the heating period (1st of October to the 30th of April) when the house was occupied (i.e. full time in Case 2 and part time in cases after that). The thermostat was set to a lower limit of 19°C and an upper limit of 21°C and radiators were used as emitters.

Table 1. Summary of Cases 0, 1, 2 and 3

			Case 0 Weather data 2002, shading, infiltration 0.05ACH	Case 1 As Case 0 + 2 full time occupants, infiltration 0.2ACH	Case 2 as Case 1 + winter constant heating	Case 3 as Case 2 but with intermittent occupancy and heating
			C0	**C1**	**C2**	**C3**
Construction type	1. Brick and Block full fill cavity wall	**BB**	C0 BB	C1 BB	C2 BB	C3 BB
	2. Timber frame part fill cavity wall	**TF**	C0 TF	C1 TF	C2 TF	C3 TF
	3. Insulated concrete formwork wall	**ICF**	C0 ICF	C1 ICF	C2 ICF	C3 ICF
	4. Steel frame wall	**SF**	C0 SF	C1 SF	C2 SF	C3 SF
	5. Structural insulated panel wall	**SIPs**	C0 SIPs	C1 SIPs	C2 SIPSs	C3 SIPs
	6. Cross laminated timber wall	**CLT**	C0 CLT	C1 CLT	C2 CLT	C3 CLT
	7. Solid Concrete Block wall	**MT**	C0 MT	C1 MT	C2 MT	C3 MT
	8. Precast concrete panel wall	**PCP**	C0 PCP	C1 PCP	C2 PCP	C3 PCP

Table 2. Summary of Cases 4, 5, 6 and 7

			Case 4 As Case 3 but with weather data high-emissions 2020	Case 5 As Case 4 but with weather data high-emissions 2050	Case 6 As Case 5 but with weather data high-emissions 2080	Case 7 As Case 6 but with ventilation
			C4	C5	C6	C7
Construction type	1. Brick and Block full fill cavity wall	BB	C4 BB	C5 BB	C6 BB	C7 BB
	2. Timber frame part fill cavity wall	TF	C4 TF	C5 TF	C6 TF	C7 TF
	3. Insulated concrete formwork wall	ICF	C4 ICF	C5 ICF	C6 ICF	C7 ICF
	4. Steel frame wall	SF	C4 SF	C5 SF	C6 SF	C7 SF
	5. Structural insulated panel wall	SIPs	C4 SIPs	C5 SIPs	C6 SIPs	C7 SIPs
	6. Cross laminated timber wall	CLT	C4 CLT	C5 CLT	C6 CLT	C7 CLT
	7. Solid Concrete Block wall	MT	C4 MT	C5 MT	C6 MT	C7 MT
	8. Precast concrete panel wall	PCP	C4 PCP	C5 PCP	C6 PCP	C7 PCP

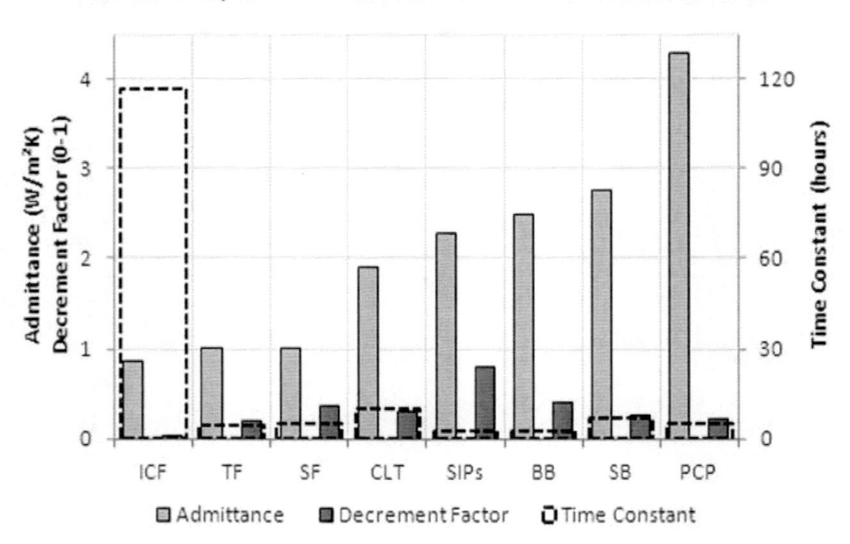

Fig. 1. Admittance and Decrement Factor against Time constant of each of the wall types

3 Results and Discussion

It was found that the addition of shading (and consequent elimination of solar gains in summer) diminished overheating for all wall types in Case 0 if compared to Base Case. Temperatures go above 25°C for 1% or less of the time in the case of SB and PCP and between 1 and 2% in the case of BB, ICF and CLT. TF, SF and SIPs presented temperatures above 25°C for more than 2% of the time.

Once occupancy is introduced, in Case 1, overheating is observed. If the 'above 25°C' criterion is used then all material types present overheating to a certain degree (Figure 2). However, if a higher temperature is acceptable then PCP, SB and ICF are the top performers reaching above 26°C for less than or around 1% of the time followed by CLT and BB with less than 2%. TF, SF and SIPs were the worst cases reaching above 28°C.

Case 3 has reduced overheating due to part time occupancy. In the heating season, Case 3 is just being heated up for half of the time (12h) while Case 2 house is permanently heated. However, the difference in heating demand is actually quite small suggesting that constant heating might not mean significantly higher energy bills in a highly insulated house (Figure 3). In the case of the higher mass wall types (PCP, SB) the difference is even smaller (less than $2kWh/m^2$). PCP had the lowest heating demand of all wall types in Case 2 and 3, although in Case 3 the difference between wall types practically disappeared.

In Cases 4, 5, 6 and 7 (future climate scenarios), if 25°C is considered as a limiting temperature, than all cases presented some degree of overheating regardless of the wall construction (Figure 4). In Case 4 PCP and ICF maintained 25°C just above 5% of the time while BB, TF, SF and SIPs all exceeded 25°C for more than 6% of the time. In Case 6, in 2080, 25°C was the temperature for almost 30% of the time in all cases. It is clear that another means of cooling should be introduced, in this case ventilation in Case 7. The most accentuated differences occurred when peak temperatures are considered with PCP being always around 2°C lower than TF and SF and up to 4°C lower than the peak external temperature (Figure 5). The second best performer is SB always around 1.5°C below TF and SF.

In summary, Case 7 comprises full summer shading (which has been used since Case 0), part time occupancy (which has been applied since Case 3), winter heating (since Case 3) and summer natural ventilation (the window starts to open at 22°C and is 100% open at 28°C). Case 7 assumes the high-emissions 2080 climate change scenarios.

As it can be seen in Figures 5 and 6, even in Case 7 with all the mitigation strategies described, summer overheating has not been completely eliminated even though ventilation improved greatly the situation. Just PCP and SB are within acceptable levels of overheating while BB and CLT are just below the border line (1% above 28°C) and ICF and SIPs just above it. SF and TF presented unsatisfactory levels of overheating. Peak temperatures reach 28°C in all cases, although PCP stayed around 2°C below SF and TF and BB and CLT around 1°C below.

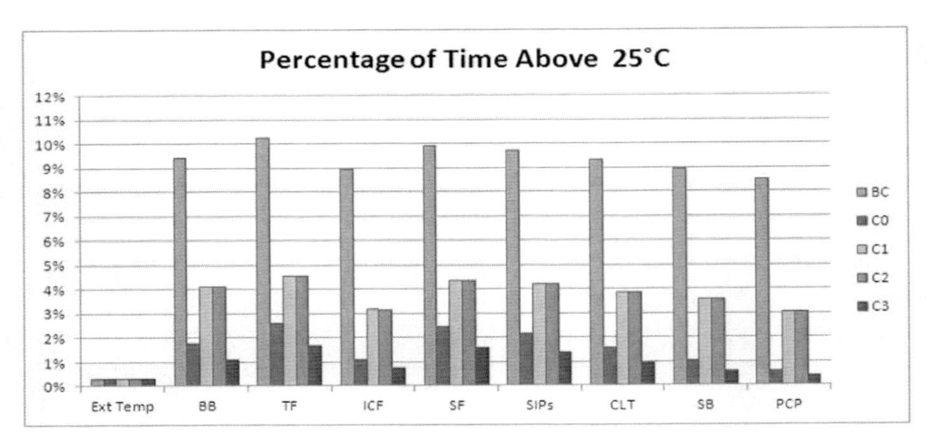

Fig. 2. Comparison between Cases 0 to 3 - percentage of time with temperatures above 25°C

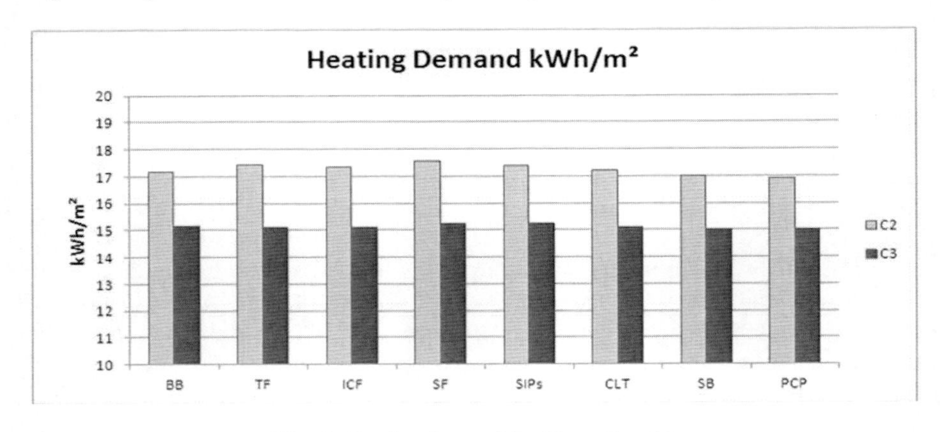

Fig. 3. Heating demand for Cases 2 and 3

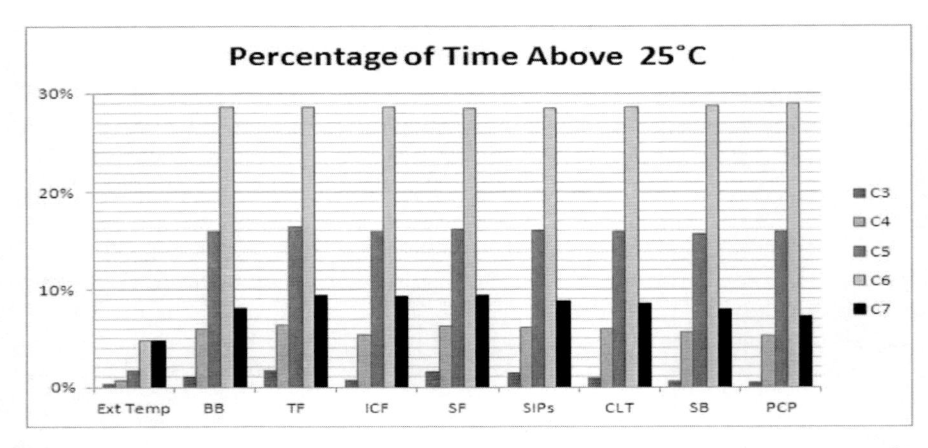

Fig. 4. Comparison between Cases 3 to 7 - percentage of time with temperatures above 25°C

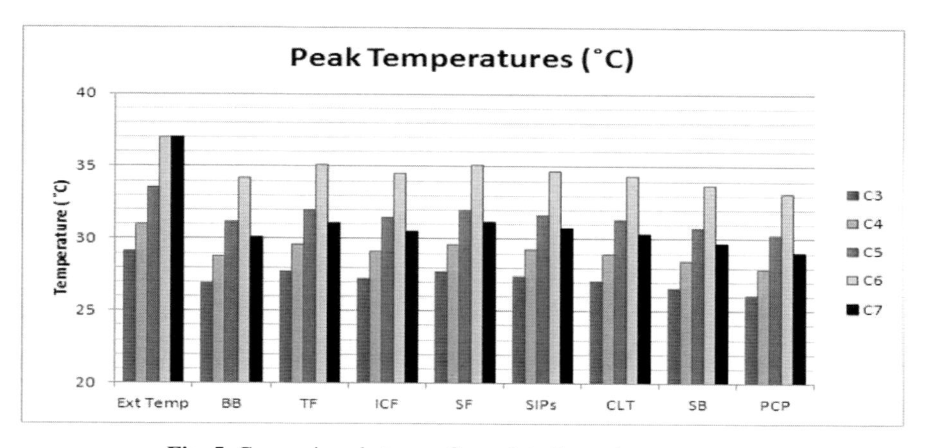

Fig. 5. Comparison between Cases 3 to 7 - peak temperatures

4 Conclusions

Eight different wall construction types were selected representing the most common types used in housing construction in the UK. These are distinguished by differing quantities of thermal mass and diverse thermal properties. The only fixed parameter across all wall types was a U-Value of 0.12W/m^{2}K. The parameters admittance, decrement factor and time constant do not necessarily characterise a construction type with regards to thermal mass. For example, SIPs presented a reasonably high admittance but had a significantly poorer performance than BB and CLT (both of which had similar admittance to SIPs) when summer overheating is considered. Decrement factor and time constant values also do not match the results of the simulations in view of best and worst performers.

A simple model was used to investigate the performance of each wall type. Suggestions for further work include testing with various room and window sizes. Further work could also include detailed assessment of each wall type in laboratory conditions, to determine the impact of the quantity of thermal mass in their performances. Liaising with suppliers is recommended in order to understand the full limitations of each wall component and investigate new options.

As the results have shown, high thermal mass may reduce summer overheating even in highly insulated buildings. In the case of the model used here, the difference in the peak temperature between the top performer (PCP) and the worst performers (TF and SF) was always at least 2°C, despite the change in other inputs such as occupancy and ventilation. However, it is clear that thermal mass alone offers limited benefits and should always be considered with other passive strategies. The fact that conductive heat transfer is very low due to the low U-Values of the envelope means that the building will be less susceptible to external temperature changes and more reliant on good design that considers passive mitigation strategies. Although not a comprehensive study, these results suggest that further work in needed to inform design decisions that consider future climate resilience.

References

1. Department for Communities and Local Government, Household Projections, 2008 to 2033, England in Housing Statistical Release C.a.L. Government. National Statistics, London (2010)
2. Department for Communities and Local Government, Definition of Zero Carbon Homes and Non-Domestic Buildings: Consultation, London (2008)
3. Department for Communities and Local Government, Innovative Construction Products and Techniques, London (2008)
4. Buildoffsite, Yearbook. Buildoffsite, London (2008)
5. Gillott, M., Rodrigues, L.T., Spataru, C.: Low-carbon housing design informed by research. Proceedings of the ICE - Engineering Sustainability 163(2), 77–87 (2010)
6. Arup, Your home in a changing climate. Three Regions Climate Change Group, London (2008)
7. Hacker, J.N., Saulles, T.P.D., Minson, A.J., Holmes, M.J.: Embodied and operational carbon dioxide emissions from housing: A case study on the effects of thermal mass and climate change. Energy and Buildings 40, 375–384 (2007)
8. Dunster, B.: UK Housing and Climate Change: Heavyweight vs. lightweight construction, ArupResearch+Development, London, p. 42 (2005)
9. Orme, M., Palmer, J.: Control of Overheating in Future Housing- Design Guidance for Low Energy Strategies, DTI Partners in Innovation Programme, St Albans (2003)
10. Zero Carbon Hub, Carbon compliance for tomorrows new homes - Topic 3 Future Climate Change, in Carbon compliance for tomorrows new homes, NHBC Foundation, London (2010)
11. Jenkins, G., Murphy, J., Sexton, D., Lowe, J., Jones, P., Kilsby, C.: UK Climate Projections: Briefing report, in UKCP09 scientific reports, M.O.H. Centre, Editor, Climatic Research Unit, University of East Anglia, University of Newcastle: Exeter, UK. p. 60 (2009)
12. Hacker, J., Capon, R., Mylona, A.: Use of climate change scenarios for building simulation: the CIBSE future weather years, in Technical Memoranda TM 48, CIBSE, Editor, Chartered Institution of Building Services Engineers (CIBSE), London (2009)
13. Orme, M., Palmer, J., Irving, S.: Control of Overheating in Well-insulated Housing. FaberMaunsell Ltd. (2002)
14. Goodier, C.I., Wood, G.A., Shao, L.: Briefing: Community resilience to extreme weather. Proceedings of the ICE - Urban Design and Planning 161(3), 97–99 (2008)
15. Rodrigues, L., Gillott, M.: The Summer Performance of the BASF House. In: PLEA 2011 - 27th Conference on Passive and Low Energy Architecture. Louvain-la-Neuve, Presses universitaires de Louvain, Belgium (2011)
16. Littlefair, P.: Avoiding Air-conditioning. Constructing the Future 24, 11 (2005)
17. Gilbert, B.: Building Performance Simulation Software - A Review for Kingspan Century. B.G. Associates, London (2006)
18. Rodrigues, L.: An Investigation into the Use of Thermal Mass to Improve Comfort in British Housing, in Department of Architecture and Built Environment, University of Nottingham, Nottingham (2009)
19. CIBSE, Guide J: Weather, solar and illuminance data. Chartered Institution of Building Services Engineers London (2002)
20. Sustainable Energy Research Group, Climate Change World Weather File Generator for World-Wide Weather Data (CCWorldWeatherGen), University of Southampton, faculty of Engineering and the Environment, Southhampton (2009)
21. CIBSE, Guide A: Environmental Design. 7th edn. Chartered Institution of Building Services Engineers, London (2006)

Chapter 44
Exploring Indoor Climate and Comfort Effects in Refurbished Multi-family Dwellings with Improved Energy Performance

Linn Liu[1] and Josefin Thoresson[2]

[1] Division of Energy Systems, Department of Management and Engineering, Linköping University, Sweden
linn.liu@liu.se
[2] Department of Thematic Studies – Technology and Social Change, Linköping University, Sweden

Abstract. The building stock in Sweden includes many older residential dwellings often with inadequate building envelopes and poor insulation resulting in high energy use and uncomfortable indoor climate. Improving energy performance in multi-family dwellings by refurbishment processes is the key factor to success in order to meet national and European energy goals to reduce energy use in the building sector by 50% through 2050. How is indoor environment affected when dwellings are refurbished to become low-energy dwellings? This paper aims to explore parameters for indoor climate and comfort in refurbished dwellings transformed into low-energy dwellings from an inter-disciplinary perspective, taking into account both quantitative and qualitative aspects of indoor climate using technical measurements, a questionnaire survey, and qualitative interviews. Based on a combination of methods, the results show that the indoor climate has largely been improved and user satisfaction was high in the refurbished dwellings. Results also showed that however indoor temperatures were too high during summer, resulting in dissatisfaction from residents. Overheating can be prevented by providing information to the residents about the functionality of the heating system and by adding shade in front of the windows.

1 Introduction

There is considerable pressure to improve the energy system and to reduce the energy demand in the building sector (Nässén & Holmberg, 2005). In Sweden, buildings are responsible for almost 40% of primary energy utilization. Reducing energy use in residential buildings is the key factor to success in order to meet national energy targets to reduce energy demand and energy use in the building sector by 50% through 2050 (Swedish Energy Agency, 2012). The European Commission directive on Energy Performance of Buildings also takes up this issue and claims that the housing sector has the largest potential to decrease the demand side for energy (Directive 2010/31/EU).

A. Håkansson et al. (Eds.): *Sustainability in Energy and Buildings*, SIST 22, pp. 463–478.
DOI: 10.1007/978-3-642-36645-1_44

The building stock in Sweden includes many older multi-family dwellings from 1950-1985. According to statistics from the Swedish National Board of Housing, Building and Planning (2012) and SCB (2012), there are about 165,000 multi-family buildings of a total of 2,100,000 buildings of various types in Sweden. 57% of multi-family buildings were built from 1961-1985 and in virtually the same way from an architectural point of view. The most common facade material for these multi-family buildings is brick. The roofs are often covered with concrete roof tile. Exhaust ventilation systems are the dominant ventilation system. These buildings often have inadequate building envelopes and poor insulation and facade U values were in general 0.38-0.4 W/m^2K and 2.0-2.2 W/m^2K for windows of multi-family buildings from 1961-1985. Refurbishments in order to improve these deficits are necessary in order to lower energy demand. However, refurbishments result in both improved energy performance and thermal comfort in the dwellings. Refurbishments that convert old residential buildings into low-energy buildings but still provide a good living environment for the tenants are an important target (Power, 2008). So far refurbishments of older residential buildings have only been done on a small scale in Sweden. The function of dwellings in low-energy refurbished residential buildings has hardly been studied. A study of a refurbished building into passive house standard has however indicated deficits of uneven temperatures between floors of a three-story house (Jansson, 2010). However, what influence the outcome on the indoor environment in low-energy buildings has mainly been studied in newly built low-energy buildings (Hauge et al., 2010; Isaksson & Karlsson, 2006).

This paper aims to explore parameters affecting indoor climate and comfort in dwellings refurbished to become low-energy dwellings with an improved energy performance by about 50%. An interdisciplinary socio-technical approach combining several data collection methods has been used for the study. This is based on the fact that the building forms an energy system composed of both technical elements and social aspects such as the members of the construction team and the residents living in the building. They are key actors for the performance of residential buildings (c.f. Rohracher, 2001). The following section will present perspectives on indoor comfort and climate in buildings and the study object. Then the different methods are presented, followed by a presentation of the results and a discussion and comparison of the methods used.

2 Indoor Comfort and Environment in Low-Energy Buildings

Indoor comfort and environment is more than temperature alone and can be defined by temperature, humidity, light, and sound. Thermal environment is one part of the indoor climate (Isaksson & Karlsson, 2006). To get a good indoor climate, the Swedish National Board of Health and Welfare recommends that the indoor temperature should not exceed 24°C during wintertime and 26°C during summertime and not go beyond 18°C (Socialstyrelsen, 2005), while the Forum for Energy Efficient Buildings (FEBY) does not recommend indoor temperatures that exceed 26°C for more than 10% of the time during summer (FEBY, 2009). However, it is important to discuss the meaning of a good indoor climate and how to create one. A distinction can be made between the concept of good comfort as something universal where there is an optimum temperature, and the opinion that the perception of good comfort is shaped by history and society, meaning that a desirable temperature involves the socio-cultural context and predominant conventions (Chappels & Shove, 2005; Shove, 2003). That can explain why there are always some people who are dissatisfied with their indoor temperature, even if it is seemingly at a recommended level. Studies of satisfaction with indoor climate in new energy-efficient buildings have also shown a discrepancy between measurements of indoor temperature and the residents' opinion about the temperature. Measured temperatures were often higher than was experienced by the residents, which points toward subjectivity in experiences between different residents (Isaksson & Karlsson, 2006; Hauge et al., 2010).

3 The Study Objects

The objects for the study are buildings located in a residential area centrally located in the city of Linköping, Sweden. The area consists of six individual multi-family buildings with approximately 100 dwellings and a large multi-family building with 186 dwellings. All buildings in the area were designed by the same architect and constructed during 1979-80 in similar ways. Due to mold damage and poorly functioning technical systems, one tower-block building, Föreningsgatan 23 (F 23) was refurbished into a low-energy building during 2008-2009. F 23 contains 19 dwellings on six floors. Two dwellings in the building have been used as study objects for technical measurements. Before the refurbishment F 23 was a par with the other buildings in the area with similar features, like most multi-family buildings built from 1961-1985. The buildings used district heating for heating and a mechanical exhaust air ventilation system for ventilating. The dwellings are flats for rent and the residents do not purchase or own their dwellings.

The refurbishment resulted in the construction of a new facade and the roof was insulated. The windows were replaced to triple glazed windows with U-value of 1.1 W/m^2K. A heat recovering exchanger (HRX) ventilation system was installed with a heat recovery efficiency of about 80%. The building still uses district heating for heating. The new exterior wall includes a drain slot which makes it possible to add about 30 cm of insulation. The seal is located 8 cm outwards from the inside. Table 1 shows a comparison between the construction of the building before and after renovation.

Table 1. Comparison of U values of F23 before and after renovation

	Before renovation	After renovation
Facade	U was 0.21 W/m^2K	U_{ave} is 0.15 W/m^2K
Floor	0.2m cement U=3.4 W/m^2K	0.2m cement U=3.4 W/m^2K
Roof	U_{tot} 0.13 W/m^2K	U_{tot} 0.17 W/m^2K
Window	Dubble-glazed, U: 1.8 W/m^2K	Triple-glazed U:1.1W/m^2K
Thermal bridges	ψ_{ave} 0.13 W/m K	ψ_{ave} 0.1 W/m K

F 23 had no individual energy meter before the refurbishment but energy use was estimated by dividing the total heating demand of all the buildings in the area. The heating area of F 23 is 2192 m^2. The building's annual heating demand before refurbishment was estimated at about 245 MWh, which is equal to 131 kWh/m^2. In 2011 the annual heating demand was reduced to 147 MWh which is equal to 67 kWh/m^2 as shown by the energy meter.

4 Method Design

The paper originates from a case study. Within the case study methodology a number of data collection techniques can be used (Yin, 1994). Indoor comfort and climate in buildings are depending on both the social context and technical potentials which are inseparable and can provide a better understanding of parameters affecting indoor climate and comfort (Rohracher, 2001). A characteristic combination of interdisciplinary methodologies in studying indoor comfort and climate has been used for this study combining a quantitative and qualitative approach from both a technical and a social point of view (cf. Isaksson & Karlsson, 2006; Karlsson & Moshfegh, 2007). Measurements of the thermal comfort from two dwellings have been used as input in BES software in order to simulate and calculate the energy use and indoor climate of the study object. Similar parameters which have influence on the indoor climate have been investigated by using a survey with the title "My indoor climate." The results of the survey do not have to be the same as the results from measurements and simulations (cf. Hauge et al., 2010; Isaksson & Karlsson, 2006; Jansson, 2010). Good indoor comfort and how to achieve it can largely be related to interests and concepts in the construction process (Chappells & Shove, 2005). Qualitative interviews have been conducted with the project team in the refurbishment process in order to study the planning and construction process. The interviews answer questions about reasons for the outcome affecting the results of measurements and residents' experience of the indoor climate.

The interdisciplinary approach in this study has several advantages. First, a similar approach for studying indoor environment in low-energy buildings has seldom been used. The users of the building and their perceptions and activities in the building are important for energy use and the indoor environment in low-energy buildings (Karlsson et al., 2007), but there are few studies focusing on end-user perspectives and experiences of indoor climate. Studies of indoor comfort often only have a technical perspective, focusing on technologies for improved energy performance in buildings (Rohdin et al., 2011). This study used questionnaires as a method for studying resident experiences of their indoor environment. There are studies using questionnaire surveys study indoor environment (Frontczak et al. 2012; Yoshino et al., 2012; Farrja et al., 2010), but few papers have used physical measurements combined with questionnaire methods (Tiberiu & Vlad, 2012; Dahlan et al., 2011) where the focus is indoor climate. There are a few examples of studies studying users' satisfaction with the indoor climate combining technical measurements with resident interviews (Isaksson & Karlsson, 2006; Karlsson & Moshfegh, 2007). This study also comprises interviews but with project team participants. Research following organization of construction projects is rare (Rohdin et al., 2011), even if studies exploring construction teams' intentions are considered important (Hamza & Greenwood, 2009). A second advantage is that combining and comparing these methods can reveal other discoveries, problems and consequences than if the methods were used and analyzed separately. The following presentation gives a closer description of each method used.

4.1 Technical Measurements and BES

Technical measurements of parameters affecting the indoor climate have been conducted from two dwellings in the studied building F 23. These measurements were carried out during May and June 2011. Each measurement period lasted for 11 days. The measurements included indoor climate measurements (indoor temperature, CO_2 concentration, air flow, humidity) and electricity consumption on a building level and on a household level. Predicted mean vote (PMV), and predicted percentage of dissatisfied (PPD) as two important indoor climate factors have been calculated from collected data from the measurements. The collected data have also been used as input in the simulation software IDA ICE 4.0, which has mainly been used for testing the indoor climate in this project. A Building Energy Simulation (BES) tool models the buildings' heating, cooling, ventilation, lighting, and other energy systems. There are several other energy simulation programs available, for example DEROB-LTH, Energy Plus, ESP-r and TRNSYS. IDA Indoor Climate and Energy 4.0 (IDA ICE 4.0) is one type of building energy simulation tool developed especially for indoor climate and energy design tasks. It accurately models the building construction and its control systems. Local weather climate profile, position of the building, material data, thermal bridges and air change rate are the basic inputs which are required by the program. A modeled building is based on many different zones. There is also more required input data such as indoor temperature, material construction, internal energy from the tenants and electrical equipment, and air flow if the building uses a mechanical ventilation system. By means of IDA ICE 4.0, heating, ventilating and air conditioning (HVAC) systems in the building can be modeled using mass flow networks or plant networks. (Thollander & Rohdin, 2011). The output data includes detailed lowest possible energy consumption and best possible occupant comfort. With help of a detailed and dynamic multi-zone simulation the study of thermal indoor climate and energy use of the entire building can be approached (EQUA, 2012).

The two dwellings were modeled and simulated in IDA ICE 4.0. Apartment A is on the second floor and heating area is $76m^2$. Apartment B is a double-story apartment which is on the top two floors; the heating area is 120 m^2. The supply air flows and exhaust air flows of the measured apartments are in the range of 5-16 l/s and 10-20 l/s. A high value of exhaust air flow is required on account of the laundry rooms in each apartment. Heat losses φ from thermal bridges which have been used as input data are different for different types of thermal bridges and vary from 0.05-0.15 W/m K. All the other input data such as internal heat releases from the electrical appliances and human activities have been obtained from measurements and standardized values. A measurement instrument Innova[1] was placed in the living room of each apartment.

[1] Innova: A instrument which measures two indoor climate indexes in particular: PMV and PPD. PMV and PPD are based on indoor temperature, air velocity, and air humidity values which are collected by Innova.

Clo-and met-value are two input data for Innova which is supposed to represent real people. Clo[2]-value should be set to 1.0 and met[3]-value should be set to 1.2 which represents a person wearing ordinary indoor clothing, and is in dormant form (Warfvinge, 2000). All the parameters measured by Innova such as indoor temperature, air flow and indoor humidity, are used as input in calculations of PMV(Percentage Mean Value) and PPD (Predicted Percentage Dissatisfied), which are two indexes normally used together to calculate the percentage of a large number of people who are dissatisfied with the thermal indoor climate.

Two climate files of Linköping have been used for the simulation, one for 2011 and the other for a normal year. The average temperature of Linköping during 2011 was 7.3 °C and for a normal year it is 7 °C.

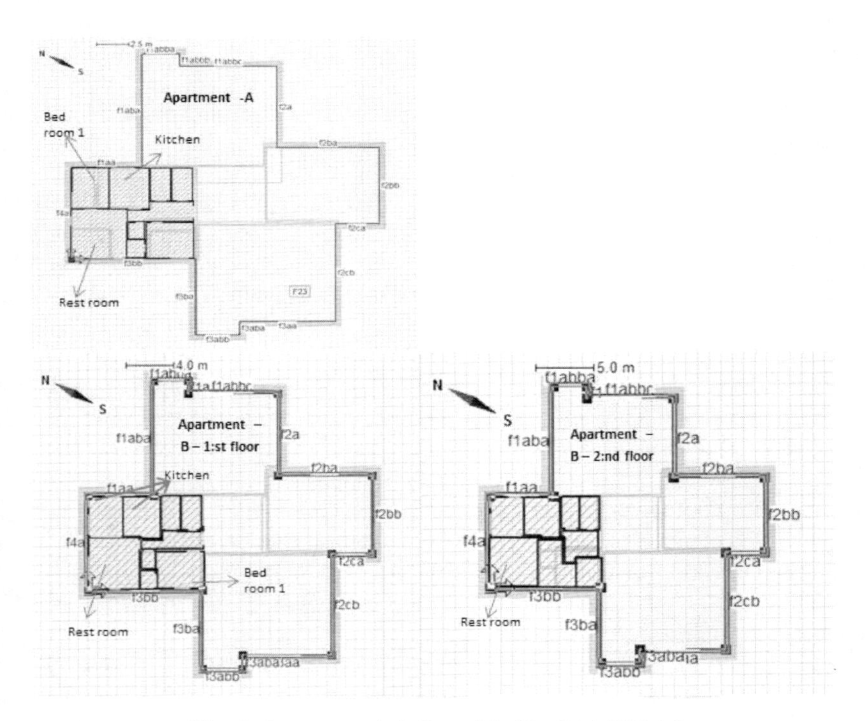

Fig. 1. Apartment A & B modeled by IDA ICE 4.0

Figure 1 shows 2D pictures of apartment A and B by using of the simulation software. Figure 3 shows the results of temperature variations in the simulated apartments.

During the measurement, some of the Tinny loggers which are used for temperature measurements stopped when their memories were full. Therefore the time steps on the x-axis are different in picture b and e from other pictures below. Since the

[2] Clo: cloth thermal resistance. 1 clo corresponds 0.155oCm2 / W.

[3] Met: heat generating value. 1 met=60 W/skin area.

measurements started at different times, for apartment A the measurements started at 12:00 a.m. and for apartment B it started at 2:00 p.m., the x-axis therefore started with different times in the pictures in figure 2 below. Since there are no registered values of the heat demand by individual apartment, the validation could only be carried out by comparing the measured indoor temperatures with the simulated indoor temperatures in those rooms where the residents spend most of their time.

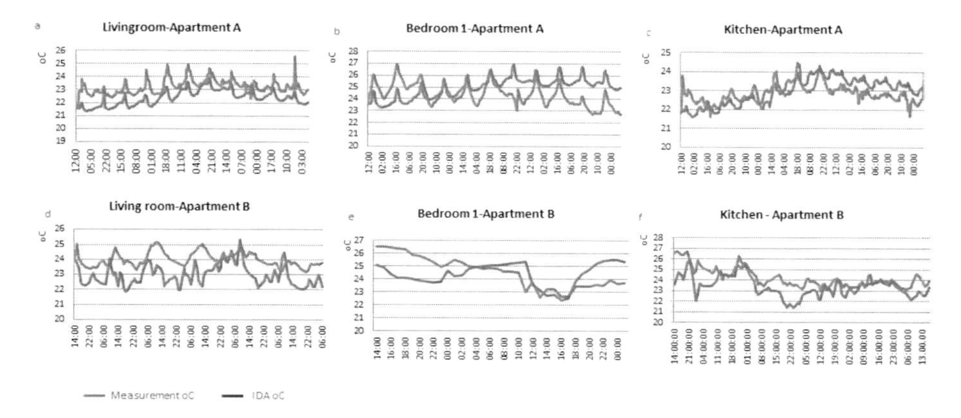

Fig. 2. Validation of temperatures in different rooms from apartment A

A mean deviation of the validation is about 2°C. The deviations in the validation of the building model are due to several factors. The first one is the that climate file of the year 2011 used as input data during validation was created manually. Solar radiation was one of the six terms that were included in the climate file. The values of solar radiation were gathered from a STRANG model[4]. The STRANG model allows a deviation of solar radiation up to 12.9%. The second deviation concerns measured temperatures. During the measurement period, temperature loggers may have inadvertently been moved, which would make the recorded temperature data insufficiently comprehensive. The last factor is human behavior and activities which may affect the simulation results to differ from reality. People with different habits regarding energy use can lead to higher or lower internal loads which in turn will affect the indoor temperature and the heat demand. Since the model is based on only two apartments, the simulation results cannot be generated as the situation of the whole building's indoor climate.

[4] STRANG model system measures include global radiation, photosynthetic active radiation, ultraviolet radiation (CIE weighted), direct radiation along with the sun duration of a horizontal resolution of approximately 11 x 11 km in one hour. STRANG covers Scandinavia's geographic area and operated at SMHI. (SMHI, 2012) Information on direct solar radiation and global radiation from the radiation network of SMHI has been used for the validation of our model.

4.2 Questionnaire Survey

A questionnaire has been used and sent out to residents in the whole area in order to get more generalized results of how residents from both refurbished and non-refurbished dwellings experience their indoor climate. The questionnaire method does not require actual contact with the respondent and is a widely used method to collect data from a large number of respondents, as in this case (Bryman, 2004). A pre-printed, standardized questionnaire was used called "My indoor climate." which is based on the so-called "Örebro model" (Örebro model, 2012) to study opinions about the indoor climate. The questionnaire includes five categories: the environment, air quality, noise situation, indoor temperature and residents' complaints. The questions were mainly closed questions where the residents could rank their experience of their indoor environment and did not include matters of why and how. If the research design does not demand too many open questions and has few follow-up questions, or includes "how" and "why" questions, a questionnaire is a suitable choice of method for studying user experiences (Thollander, 2011). In total, 80 questionnaires were distributed to the tenants in F 23 and to tenants living in non-renovated buildings in the neighborhood, of which 42 tenants chose to respond to the survey including 11 from F 23. This represents a total response rate of 53%, which seems acceptable. The data collected from the questionnaire are used in order to analyze the data qualitatively and no statistical analyses using a software tool have been done from the data (Bryman, 2004).

4.3 Qualitative Interviews

Ten qualitative in-depth interviews with project members on the refurbishment team including employees of the property owner and the construction company were conducted for the study. The employees were construction engineers, plant engineers, energy and environmental managers and contact persons for the residents. The interviewees were all connected to the refurbishment project. The aim of the interviews was to explore experiences and thoughts affecting the construction process and the outcome of the refurbishment. Conducted interviews lasted about one hour and were recorded and transcribed. The questions followed an interview template but took a semi-structured form. The questions were based on issues as to why the refurbishment was done in certain ways, which provided the opportunity to ask follow-up questions. That made the interview method useful in this case (Yin, 2007). The interviews have been interpreted and analyzed by searching for how events and processes are described concerning the refurbishment process.

5 Results from the Measurements and Simulations

The results from the measurements collected by the output from Innova which are the PMV and PPD indexes as shown in Table 2.

Table 2. Indoor temperature and PMV PPD as measured data

	Temperature variation °C	PMV	PPD
Apartment A	21.60~25.40	-0.6~-0.4	12~9%
Apartment B	23.10~25.30	-0.5~-0.4	11~9%

In Sweden ISO 7730[5] is used as an indoor comfort standard. According to ISO 7730 an acceptable indoor climate is when the PMV is between -0.5 to 0.5, which corresponds to a PPD below 10% (ISO 7730, 2005). The measurement results of PMV and PPD in Table 2 also meet ISO 7730 requirements (ISO 7730, 2005) for a standard and acceptable indoor climate. Table 3 shows the simulated results of three selected rooms in each apartment where the households spent most of their time.

Table 3. Indoor temperature and PMV and PPD as simulated data

Föreningsgatan 23	Temperature variation	PMV	PPD
Apartment A-Bedroom2	21 °C ~ 25°C	0.3~0.6	7~12%
Apartment B-Bedroom1	22.5 °C ~ 26.5 °C	0.5~0.7	10~17%
Apartment A-Living room	21 °C ~ 25 °C	0.3~0.5	7~10%
Apartment B- Living room	23.5 °C ~25 °C	0.4~0.5	9~10%
Apartment A-Kitchen	21 °C ~27 °C	0.3~0.8	7~20%
Apartment B-Kitchen	22.8 °C ~26.8 °C	0.6~0.8	12~20%

The PMV and PPD from the simulated results of each apartment's restroom match with the measurements. The minus sign from Table 2 is due to the errors made by calibrating too low values of the clo- and met-value in Innova. This gave a wrong perception that the residents experienced a colder climate than they actually do.

Table 3 also shows that apartment B has higher indoor temperatures than apartment A. When the indoor temperature in bedroom 1 of apartment B achieves almost 27°C the corresponding PMV is equal to 0.7, which represents a warm indoor temperature. This is because the bedroom receives afternoon sunlight as it is situated at the west side of the building. Another reason is that apartment B is on the top floor and receives more sunlight than apartment A. However, simulation results indicate that there is a good indoor environment and they fulfill ISO 7730's requirements for a good indoor climate.

[5] ISO 7730: Ergonomics of the thermal environment -- Analytical determination and interpretation of thermal comfort using calculation of the PMV and PPD indices and local thermal comfort criteria.

6 Questionnaire Survey Results

The answers from the questionnaire survey are presented and divided between residents in the refurbished building F23, and residents living in non-refurbished dwellings in the same area. Figure 4 presents residents' experience of problems related to their indoor air quality.

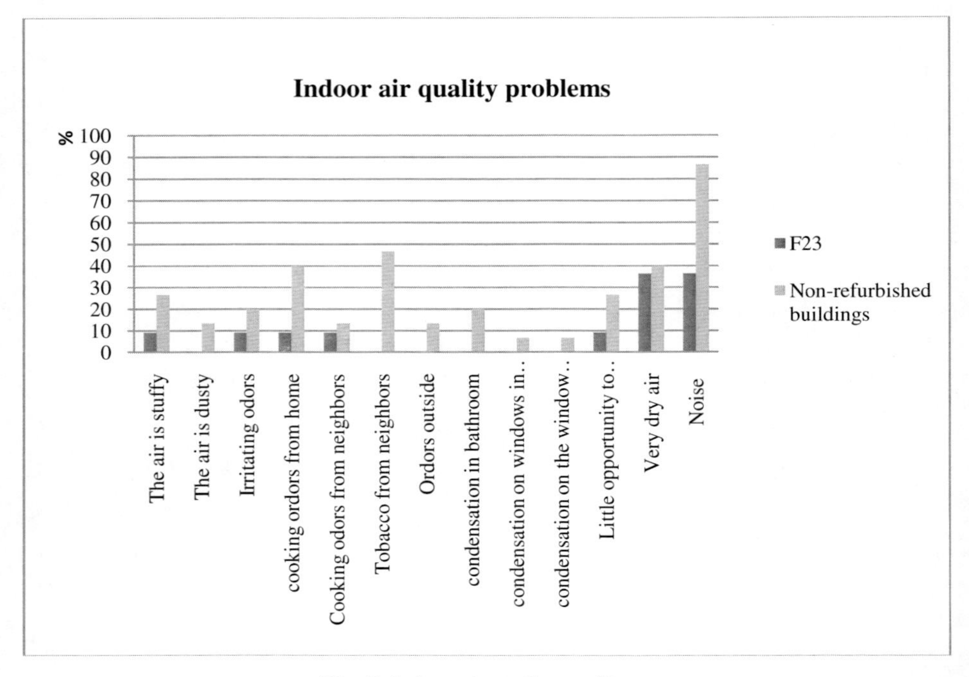

Fig. 3. Indoor air quality problems

Residents from F 23 are more satisfied with the indoor air quality than residents from non-refurbished buildings in all aspects surveyed. Few residents in F 23 complained about unpleasant smells from the ventilation channel compared to 20-45 % of the residents from the non-refurbished buildings. 20 % of the residents in non-refurbished buildings complained about residual moisture in the bathroom after showers. That problem was not experienced in F23. Condensation is a common problem in old buildings and a sign of the buildings poor air tightness. As the non-renovated buildings use mechanical air ventilation system, the supply air temperature is the same as the outdoor air temperature and the air is not preheated or filtered, which might explain the dissatisfaction about stuffy and dusty air in non-refurbished dwellings and not in F 23. However, the problem of dry air is comparable in F 23 and in non-refurbished dwellings. Regular maintenance of ducts, fans and registers are necessary in order to get functioning air ventilation and efficient fans (Swedish Energy Agency, 2012). Only one resident in F 23 complained about cooking odor or other unpleasant smells from their neighbors compared to 40 % in non-refurbished buildings. Odor problem had been a problem in F23 after renovation but was solved when the property owner installed a filter in the ventilation channel. Unpleasant noise was experienced as a problem in both F 23 and non-refurbished dwellings but was considered a bigger problem in non-refurbished dwellings. A well-insulated building can also keep out a higher degree of unpleasant noise.

Figure 5 presents experiences concerning indoor air temperature problems. According to the simulation results, the indoor temperatures in the two dwellings can vary from 21 °C to above 27 °C (see Table 3) which exceeds the recommended indoor temperature (Socialstyrelsen 2005). 18% of the residents in F 23 also experience problems with a high indoor temperature during summer compared to only 9% in non-refurbished dwellings. A well-insulated building is in need of less heating and the problem with overheating can be caused by the high temperature of the supply air from the ventilation system.

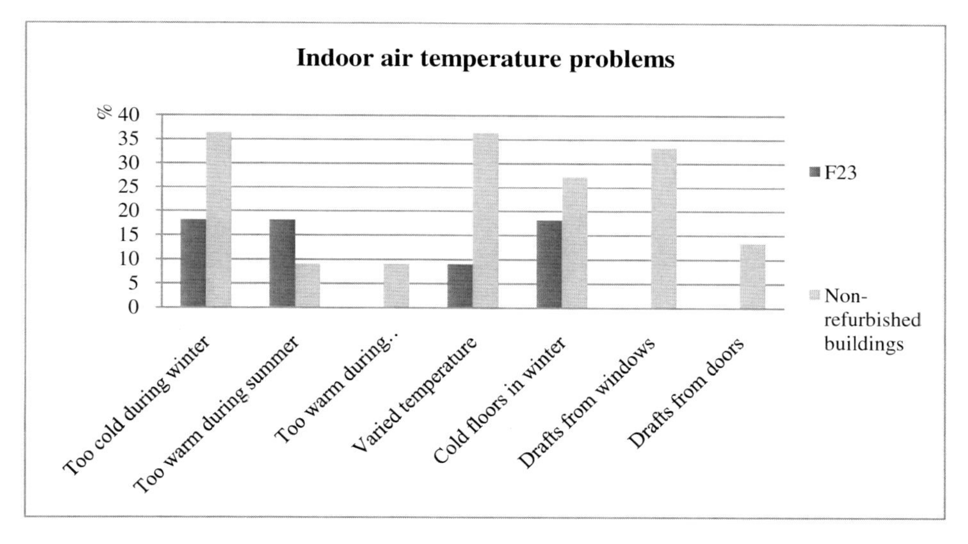

Fig. 4. Indoor air temperature problems

The heating elements in F23 turn off automatically when the indoor temperature drops to 14°C, otherwise the internal heat and the HRX system will cover the temperature difference in order to keep the indoor temperature at 21°C. Except for high indoor temperature during summer, residents in F 23 were more satisfied with their indoor temperature compared to residents in non-refurbished dwellings. A uniform indoor temperature is related to well-insulated walls, windows with a low U-value, an HRX system and no cold bridges removed during the refurbishment. Buildings with thermal bridges and poor insulation in walls and in windows can lose much heat through leakages, cool down indoor air and cause cold drafts. These problems were also experienced in non-refurbished dwellings. No extra insulation was added to the floors in F 23 during the refurbishment which might explain why 18% of the residents experiences cold floors during wintertime.

7 Results from the Interviews

The members of the refurbishment project team all had the opinion that the refurbishment of F 23 was an energy project. The well-known goal among all interviewees was to improve energy performance by around 50% to meet new energy goals and standards for buildings, and to meet perceived societal pressure to reduce energy use in buildings. The impression in the project team was that the residents were mostly positive about the refurbishment. Some residents questioned the extent of measures taken to reduce energy demand. A side benefit with building airtight and installing a ventilation system with a heat recovering exchanger (HRX) in order to reduce energy demand might be improved indoor comfort and climate according to some project team members. However that was seldom discussed in the interviews nor was residents informed of this by the project team. The indoor environment was mentioned by one interviewee as a change from the refurbishment that affects the residents in a positive way.

> "The residents are hopefully affected in a positive way because the indoor environment should be more uniform. Less draft and in particular with the ventilation. They will now get preheated fresh air that will result in less draft and a better indoor environment."

Indoor comfort and climate issues were however only discussed by the interviewees to a small extent, and there were no measureable goals regarding indoor comfort or climate in the refurbishment process.

The energy goal should be reached by installing energy saving technologies that the project team had already tested in earlier projects. The choice of well-known technologies was due to the fact that the project team did not want to take any risks in testing new technologies for heating and ventilation. They only implemented measures that they believed the management team could handle. An example was the choice of windows. There were affordable windows on the market with even better energy performance. These windows were not installed since there had been condensation on the outside of the windows in earlier projects that resulted in complaints from residents. The construction team was also given a day of training where they learned the importance of building airtight and installing the technologies in order to get a low-energy demand for heat in the building. The aim was to reduce air leakages resulting in a lower use of energy for heating. The residents in refurbished dwellings were not given any information about installed technologies or how they should act in order for the system to work properly and function in an optimal way. The idea was that installed technologies in connection with the refurbishment were meant to be invisible and simple for the residents, and not affect their daily life.

The installed energy-saving technologies were supposed to be unnoticed by the residents, who were not expected to contribute to lower the energy demand by changing their energy use. There should not be any changes for the residents to live in a refurbished energy-efficient dwelling from before the refurbishment according to the project members. However, one project member admitted that the installed thermostat was not adjustable to the same extent by the residents and turned off the heat more often after the refurbishment. That could result in residents experiencing a lower indoor temperature than actual temperature, depending on the resident. The interviewee also expressed apprehensions about differences in residents' perceptions of their indoor climate.

> "There is... how shall I put it... a bit of psychology in how you perceive the indoor temperature."

The thermostat was not adjustable by the residents either. This was acknowledged by a couple of team members who thought some residents might not be accustomed with the fact that they cannot regulate the thermostat by themselves according to their needs, as they could do to some extent before the refurbishment.

> "Before you get used to it you might perceive it as... very strange and you can perceive it as very hard to not be able to directly affect the indoor temperature."

Another example of trials for invisible technologies was the seal in the insulation that was located 8 cm outwards from the inside. The residents can then drill up to 8 cm in the wall before they damage the seal, according to a member of the project team. The residents were however provided with very limited information about the seal and how far they could drill in the walls.

8 Discussion

There were no major differences between the technical measurements of the indoor climate, the residents' experience of the indoor climate, and the expectations on improvements to the indoor climate by several members in the project team. Overall the measured parameters of the indoor climate indicated that it should be an acceptable indoor climate that fulfills ISO 7730's requirements for a good indoor climate. The simulated PMV and PPD values from the measurements in the two dwellings indicated that the residents should be satisfied with the indoor temperature most of the time. The questionnaire also confirmed that residents in F23 in general were more pleased with their indoor climate than residents in non-refurbished buildings. The residents were however overall more satisfied with the temperature in their dwellings in the refurbished building, except for the indoor temperature during summertime. Rooms with windows directed to the south also had an indoor temperature exceeding recommended levels during summer. The measured values indicated a warm indoor temperature but not to the extent that the questionnaire results indicated the point of discrepancy between measured parameters and the residents' experience (cf. Hauge et al., 2010; Isaksson & Karlsson, 2006; Jansson, 2010). High indoor temperatures have been a problem in buildings refurbished to low-energy buildings (cf. Jansson, 2010). There might be problems other than unsatisfied residents that go along with a high indoor temperature in low-energy buildings. A fast-growing trend is the demand for air-conditioning (cf. Chappels & Shove, 2005). Even if it still is far from being a normal standard in dwellings in Sweden, it is more and more becoming a domestic standard in for example the UK and may become a demanded standard also in Sweden. A potential risk with low-energy buildings, with the aim to reduce energy demand, is that a high indoor temperature during summer can increase the demands for air-conditioning, which can eat up the energy saving potential with the energy-efficient building.

An interesting note is that the project team did not plan for or choose technologies for improving the indoor climate. The intention from the project team was not to deliver any specific comfort conditions. The idea was that technical measures, such as building airtight and providing mechanical heat exchange ventilation systems, should increase the energy efficiency. The meaning of good comfort and how that could be delivered was not specified in the project. However, a few team members anticipated that a low-energy refurbishment should result in a better indoor comfort but without a specification of what that would entail. Parts of the project team were aware that not all residents have the same opinion about what a good indoor climate is, but this knowledge was not used in the design of the refurbishment. The installed technical systems were chosen only with the intention to be invisible and not adjustable to the needs and understanding of the individual resident. That was based on an understanding that comfort for the residents is simplicity. The residents should not have to deal with the installed technical equipment temperature and ventilation, or know how to handle it. The residents however expressed a higher satisfaction about their indoor climate in the refurbished dwellings, but that does not answer whether they are pleased with the supposed invisible technical systems in their dwellings or not. The dwellings in the refurbished building overall got higher ratings in the questionnaire regarding the indoor climate than the dwellings in similar non-refurbished dwellings. That can also depend on other factors as there were mold problems in the non-refurbished buildings which might have had an effect on the questionnaire results.

The project team carefully chose technologies they knew how to handle. They also made sure of that by providing training to the construction team in order to install the technologies properly, because some technologies might be new to them. To provide both written and oral information and instructions to the residents about the functionality of the technical systems in low-energy buildings and how they work, has been deemed crucial in order for the residents to know how to avoid indoor comfort problems (Mlecnik et al., 2012; Isaksson & Karlsson 2006). The residents were however not given any information from the project team about the functionality of the ventilation system and their indoor temperature, even if the technologies also were new to them. Information about the function of the technical systems can potentially avoid problems with non-functioning systems such as high indoor temperature. To inform the residents that it is possible to call maintenance to lower the supply curve of heat when they perceive the indoor temperature as too warm instead of simply opening the windows and letting out heat may produce a more uniform indoor climate and save energy.

9 Conclusions

This study combines results from technical measurements and simulations from two dwellings, qualitative interviews with the project team, and a questionnaire survey covering the residents' experiences of their indoor climate. This socio-technical approach provides a larger picture of the thermal comfort in a low-energy refurbished building with dwellings with an improved energy performance by 50%. Further studies need to include measurements and simulations on the building level too and more in-depth studies of resident experiences. Continuing refurbishments into low-energy dwellings need to take several aspects into account. The outcomes from this study indicate that the residents were more satisfied with their refurbished dwelling compared to non-refurbished dwellings, but information about the functionality of the technical system should be improved in order to keep a low-energy demand and to avoid high indoor temperatures. Information that a low-energy refurbishment can also result in a better indoor environment should also be provided to the residents in order to avoid complaints about the extent of a low-energy refurbishment. The results show that it is important to be aware of the high indoor temperatures during summertime. A suggestion of a way to avoid overheating and maintain low-energy demand is to add shade in front of the windows in order to obtain a lower indoor temperature during summertime. Another suggestion is to lower the indoor temperature during summertime by adjusting the predetermined supply air's temperature in the HRX system from 21°C to 20°C during summer.

References

Bryman, A.: Social Research Methods. Oxford University Press, Oxford (2001)

Swedish National Board of Housing, Building and Planning.: Så mår våra hus, `http://www.boverket.se/Bygga-forvalta/sa-mar-vara-hus/` (accessed June 20, 2012)

Catalina, T., Iordache, V.: IEQ assessment on schools in the design stage. Building and Environment 49, 129–140 (2012)

Chappells, H., Shove, E.: Debating the future of comfort: environmental sustainability, energy consumption and the indoor environment. Building, Research & Information 33(1), 32–40 (2005)

Dahlan, N.D., Jones, P.J., Alexander, D.K.: Operative temperature and thermal sensation assessments in non-air-conditioned multi-story hostels in Malaysia. Building and Environment 46, 457–467 (2011)

Directive 2010/31/EU of the European Parliament and of the Council of 19 May 2010 on the energy performance of buildings, 18.6.2010. Official Journal of the European Union 153(13) (2010)

Al-ajmi, D.F., Loveday, D.L.: Indoor thermal conditions and thermal comfort in air-conditioned domestic buildings in the dry desert climate of Kuwait. Building and Environment 45(3), 704–710 (2010)

FEBY: Kravspecifikation för minienergihus. LTH rapport EBD-R–09/26, IVL rapport nr A1593, ATON rapport 0903 (2009), `http://www.energieffektivabyggnader.se/download/18.712fb31f12497ed09a58000141/Kravspecifikation_Minienergihus_version_2009_oktober.pdf` (accessed February 25, 2012)

Frontczak, M., Andersen, R.V., Wargocki, P.: Questionnaire survey on factors influencing comfort with indoor environmental quality in Danish housing. Building and Environment 50, 56–64 (2012)

Hamza, N., Greenwood, D.: Energy conservation regulations: Impacts on design and procurement of low-energy buildings. Building and Environment 44(5), 929–936 (2009)

Hauge, Å.L., Thomsen, J., Berker, T.: User evaluations of energy efficient buildings: Literature review and further research. In: Proceedings of Renewable Energy Conference 2010, Trondheim, Norway, pp. 97–108 (2010)

Isaksson, C., Karlsson, F.: Indoor climate in low-energy houses - an interdisciplinary investigation. Building and Environment 41(12), 1678–1690 (2006)

Jansson, U.: Passive houses in Sweden. From design to evaluation of four demonstration projects. Department of Architecture and Built Environment. Lund, EBD-T–10/12 (2010)

Karlsson, F., Moshfegh, B.: Energy demand and indoor climate in a low-energy building-changed control strategies and boundary conditions. Energy and Buildings 38(4), 315–326 (2006)

Karlsson, F., Moshfegh, B.: A comprehensive investigation of a low-energy building in Sweden. Renewable Energy 32(11), 1830–1841 (2007)

Karlsson, F., Rohdin, P., Persson, M.-L.: Measured and predicted energy demand of a low-energy building: Important aspects when using building energy simulation. Building Services Engineering Research and Technology 28(3), 223–235 (2007)

Mlecnik, E., Schütze, T., Jansen, S.J.T., de Vries, G., Visscher, H.J., van Hal, A.: End-user experiences in nearly zero-energy houses. Energy and Buildings 49, 471–478 (2012)

Nässén, J., Holmberg, J.: Energy efficiency — a forgotten goal in the Swedish Building sector? Energy Policy 33(8), 1037–1051 (2005)

Power, A.: Does demolition or refurbishment of old and inefficient homes help to increase our environmental, social and economic viability? Energy Policy 36(12), 4487–4501 (2008)

Rohdin, P., Glad, W., Palm, J.: Low-energy buildings – scientific trends and developments. In: Palm, J. (ed.) Energy Efficiency, pp. 103–124. InTech (2011) ISBN: 978-953-307-137-4

Rohracher, H.: Managing the technological transition to sustainable construction of buildings: A socio-technical perspective. Technology Analysis & Strategic Management 13(1), 137–150 (2001)

SMHI: Sweden's Meteorological and Hydrological Institute, `http://strang.smhi.se/` (accessed June 26, 2012)

Socialstyrelsen.: Temperaturer inomhus (2005) ISBN: 91-7201-972-7

Shove, E.: Comfort, Cleanliness and Convenience: The social organization of normality, Berg, Oxford (2003)

Statistics Sweden (SCB).: Statistics of multi-family buildings 2010 (2010)

Swedish Energy Agency.: Energy Mode 2012. Swedish Energy Agency, Eskilstuna (2012)

Thollander, P.: Questionnaires. In: Karlsson, et al. (eds.) Interdisciplinary Energy System Methodology, pp. 85–88. Division of Energy Systems, Linköping University (2011)

Thollander, P., Rohdin, P.: Case study research. In: Karlsson, et al. (eds.) Interdisciplinary Energy System Methodology, pp. 12–16. Division of Energy Systems, Linköping University (2011)

Warfvinge, C.: Installationsteknik AK för V. Studentlitteratur, Lund (2000)

Yin, R.K.: Fallstudier: design och genomförande. Liber, Malmö (2007)

Yoshino, H., Guan, S., Lun, Y.F., Mochida, A., Shigeno, T., Yoshino, Y., Zhang, Q.Y.: Indoor thermal environment of urban residential buildings in China: winter investigation in five major cities. Energy and Buildings 36, 1227–1233 (2004)

Örebro model: Indoor climate problems, `http://www.inomhusklimatproblem.se/index.html` (accessed June 27, 2012)

Chapter 45
Occupancy-Driven Supervisory Control Strategies to Minimise Energy Consumption of Airport Terminal Building

D. Abdulhameed Mambo[1], Mahroo Efthekhari[1], and Thomas Steffen[2]

[1] School of the Civil & Building Engineering, Loughborough University, Le11 3tu, UK
[2] School of Aeronautical and Automotive Engineering, Loughborough University, Le11 3tu, UK

Abstract. The most cost-effective way to improve the energy efficiency of a building is often achieved through efficient control strategy. Such strategies may include shutting down plant or setting back/up setpoints of indoor environment systems as the case may be during the period that the building is not occupied and providing optimal setpoints for comfort during occupancy. In most cases, airport terminal indoor environment systems run on designed conditions and do not have fine control based on detailed passenger flow information. While opportunities for complete shut-down of HVAC and lighting systems are limited in busy airport terminals due to round-the-clock operations, this paper uses a professional building software to examined the potentials of applying appropriate setpoints during occupancy conditions and setback operation during inoccupancy conditions as an energy saving strategy for the indoor spaces of airport terminal. Based on some acquired site information, existing HVAC and lighting control system, a thermal model of a real UK airport terminal building was constructed. This base model was upgraded to a more energy efficient model based on real-time passenger flow. Results showing improved energy and CO_2 savings are presented.

Keywords: Building Control, Indoor Comfort, Airport Terminal's CO_2 emission savings, Airport Terminal's Energy Consumption.

1 Introduction

HVAC and lighting systems in buildings must be augmented with a good control scheme to provide comfort under any varying load conditions. Efficient control is often the most cost-effective way to improve the energy efficiency of a building. Airport buildings contain many spaces that are different in function and structure and the operations within these buildings are round-the-clock. These leads to a complicated building system such as heating, ventilation, air-conditioning, electric lighting and hot water systems that is difficult to predict. This complexity is further compounded by the non-linear and time-varying nature of the variables inside and

A. Håkansson et al. (Eds.): *Sustainability in Energy and Buildings*, SIST 22, pp. 479–489.
DOI: 10.1007/978-3-642-36645-1_45

outside of the building affecting these systems. As a result, the HVAC and Lighting systems are run on full schedules thereby leading to a substantial waste in energy. This paper examines the potentials, in terms of energy savings, of applying Chartered Institution of Building Services Engineers (CIBSE) recommended setpoints for visual and thermal comfort with setback operation in a real UK airport terminal.

2 Comparison and Selection of Simulation Tools

Computer based building design and development is beneficial in studying complex buildings such as the airports but the fragmentations within the building industry has reflected in the development of these tools, such that whole-building simulation is still an open issue (Salsbury 2005). For example, simulating advanced controller is still limited in most state-of-art building simulation tools. Some are better at specifying local controllers such as TRNSYS and ESP-r while EnergyPlus offer ease in specifying supervisory control (Pan et al 2011). Although domain independent simulation platforms such as MATLAB/SIMULINK, LABVIEW, SIMBAD and Dymola are efficient in design and testing of controllers but they do not have all the models to accurately simulate buildings forms and systems (Trčka & Hensen, 2010).

The complex nature of airport terminal building and systems has caused the trial with several building modelling tools in order to develop an accurate model. EnergyPlus, a new generation building-energy-analysis tool, that was suitable for analysing building performances with unusual building systems (Yiqun et al 2011) such as airport was selected. Indeed, Griffith et al (2003) used the earliest form of EnergyPlus (Version 1.0.3) to study the influence of advanced building technologies such as optimised envelop system and schedules for a proposed Air Rescue and Fire Fighting Administration Building at Teterboro Airport and find that the results obtained compare well with those obtained using DOE-2.1E. Ellis and Torcellini (2005) confirmed the reliability and accuracy of EnergyPlus in simulating tall buildings.

Standard control tools within EnergyPlus includes low level control, high level control and the Energy Management System (EMS) based on the EnergyPlus runtime language (Ellis et al 2007). The Low-Level Control simulates a particular closed-loop hardware controls that has a specific task to accomplish. They are usually found in the input of an EnergyPlus object. High-Level (Supervisory Control) operates at a higher level than the local loop in control hierarchy. This type of control affects the operation of local control and can jump across system boundaries and can be used to manage and control the running of other component objects, part of or the entire system.

The major shortcoming of EnergyPlus was that it does not have a friendly user interface. To overcome this problem, DesignBuilder was used for the modelling process. DesignBuilder was the first and most comprehensive user interface to the EnergyPlus dynamic thermal simulation engine. It combines rapid building geometry, HVAC and lighting modelling and ease of use with state-of-the-art dynamic energy simulation based on EnergyPlus. Through the DesignBuilder (DB 2011) and for the first time, the advanced HVAC and Dayligthing features in EnergyPlus are now accessible in a user-friendly graphical environment. The latest DesignBuilder v3 provides a powerful and flexible new way to model both air and water sides together in full detail with a good range of components including all ASHRAE 90.1 baseline HVAC systems.

3 Results of HVAC Probe

The indoor temperature of an airport terminal was monitored from 26th October to 2nd November 2011. Fig. 1 show results for the baggage reclaim area of the arrival concourse. It can be seen that the indoor temperature for this area hovers between 20-22 degree Celsius throughout the week under review as against the 12 – 19 degree Celsius recommended by Chartered Institution of Building Services Engineers (CIBSE) for such spaces. The same situation was observed for all the other spaces monitored in the terminal.

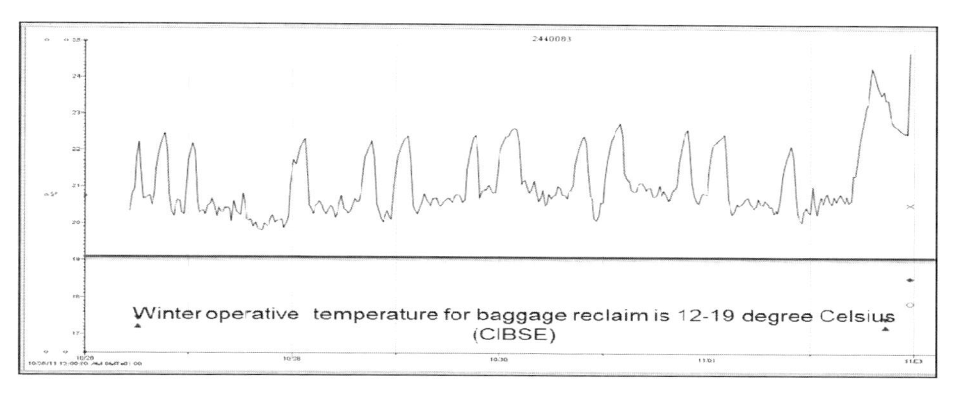

Fig. 1. Temperature and lighting setpoint for the baggage reclaim

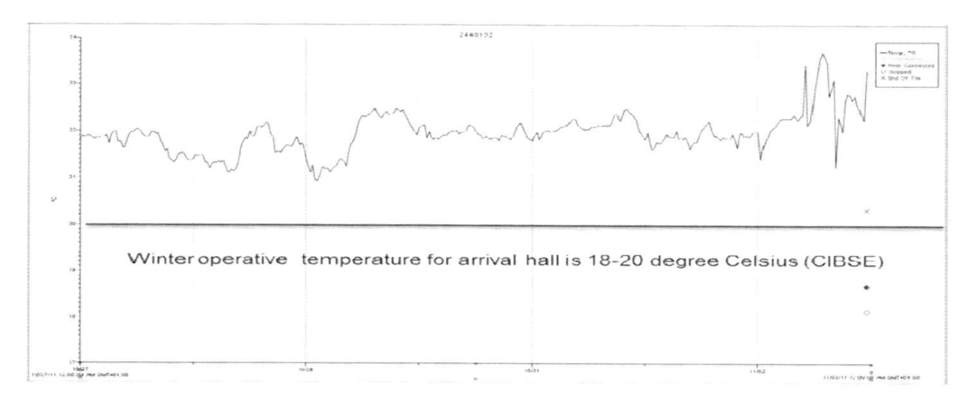

Fig. 2. Temperature and lighting setpoint for the Departure hall (Airside)

For example, in fig. 2 for the departure hall on the airside the temperature band is between 21-24 degree Celsius as against the CIBSE recommended 19-20 degree Celsius and the temperature swings for all the spaces monitored does not vary in consonance with passenger flow information for the period under review.

4 Real-Time Flight Schedules

In addition, fig. 3 below shows real time plane arrival times plotted against the time-interval between any two consecutive arrivals for the period 26th October to 3rd November 2011. Here, it was assumed that it took two hours to complete processing of arriving passenger to accommodate any delays, although the actual time recommended by International Civil Aviation Organisation (ICAO) was 45 minutes for international arrival passenger processing from disembarkation to completion of last clearance process (ICAO 2005). For domestic passengers, it is much less. Using the very conservative 2 hours benchmark, Up to 40 hours opportunity exist for the week under review to implement setback operation. When this is extrapolated across the airport terminals and for a whole year, the savings in energy will be significant.

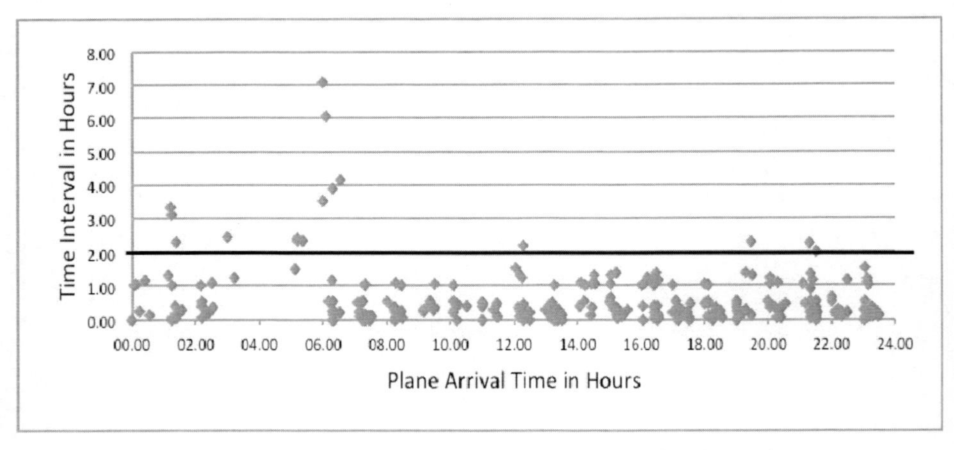

Fig. 3. A plot of plane arrival time versus arrival time intervals

In addition, Fig. 4 shows that similar opportunity to save energy using efficient controls exists in the departure areas of the terminal especially in the airside where only boarding passengers were allowed. Four hours minimum was selected to accommodate the up to three hours check-in time allowed for international flight and any delays that might occur. Although, ICAO recommends only one hour from presentation at first processing point to the scheduled time of flight departure. Even by this very conservative minimum time, about fifty hours (two days of the week) opportunity exists to implement energy saving strategy. It can also be clearly seen that there were only two departures flights between 22.00 hours to 6.00 hours and no flight at all between 0.00 hours and 6.00 hours for the entire week.

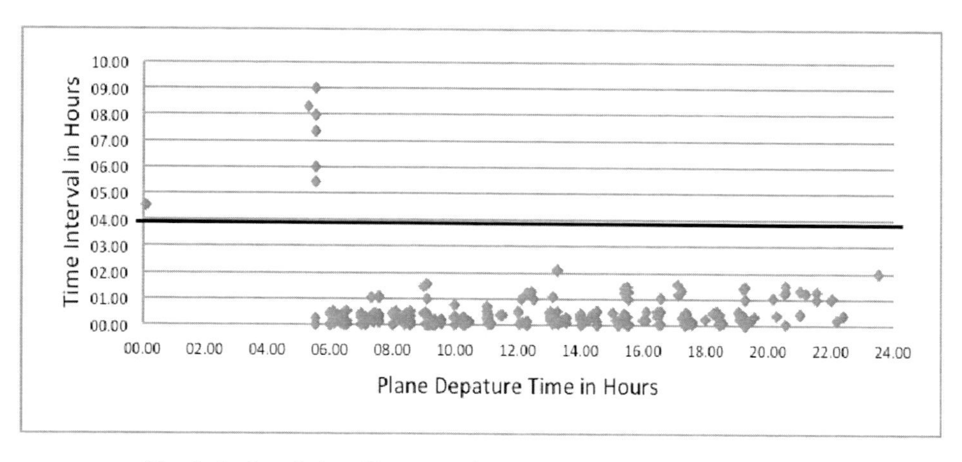

Fig. 4. A plot of plane Departure time versus Departure time intervals

5 Case Study – Building Layout

This UK airport is composed of three terminals (Terminal 1, 2 and 3). Our case study is Terminal 2. This terminal was constructed in 1992 on the North-West part of the airport site. The terminal is made up of five-floor central building covering a gross floor area of about 18,000 m^2 and has two piers of four floor levels measuring about 5,400 m^2 spanning to the left and right direction of the central building. The ground and the first floor contain the arrival halls, the third floor, the departure halls, and the fourth floor is made up of lounges, offices and the control room on the central building it mainly housed the plant rooms on the piers. The fifth floor is mainly plant rooms.

The terminal is heated by gas boilers located in the central and eastside of the terminal. There are air-cooled chillers externally located on steelwork frames in the main plant rooms. The air handling units comprises of Inlet damper, mixing box, HPHW Frost Coil, Panel Filter, Bag Filter, Carbon Filter, Cooling Coil, HPHW Re-heat Coil, Supply Fan, Extract Fan. The building has no lighting and Dayligthing control but the luminaries are currently being upgraded and Introduction of lighting control is also being considered. For the purpose of this study, lighting control will be introduced into the energy efficiency model.

6 Modelling of Building Geometry and HVAC Systems

The building geometry was modelled in DesignBuilder by importing the 2D AutoCAD drawings in the dxf format and tracing the external walls and defining the zones based on the functions and type of the HVAC system in the indoor space for each of the floors. Fig. 5 provides 3D geometric form of the building.

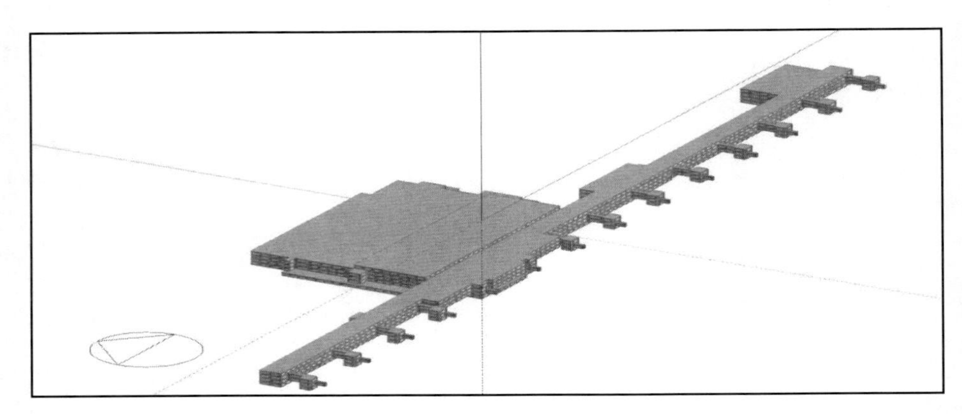

Fig. 5. 3D view of the designed model

The HVAC and Dayligthing modelling was done using a recently approved Version 3 which allows access to a wide range of EnergyPlus HVAC systems through an easy to use diagrammatic interface and calculations with integrated graphical daylight distribution contour plots and reports for LEED, BREEAM and Green Star.

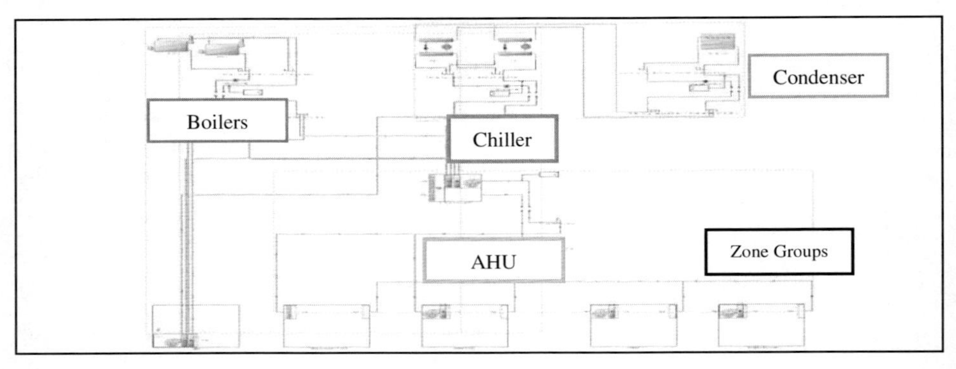

Fig. 6. Schematics of the HVAC system

For this case study, there are twenty-two thermal zones in the building. However, these zones are further sub-grouped into six zone groups according to the HVAC system type (see fig. 6). The building model was zoned according to passenger flow such that the areas accessible to the public were separated from the areas that were restricted to only passengers and staff. Occupancy in the restricted areas such as the Check-in, Customs, Security, passport control and baggage reclaim areas can easily be linked to arriving/departing passenger planes. However, in the public spaces such as the booking hall, some retail areas and some offices, the flow of people needs to be estimated and therefore more complicated to control. Generally, terminal arrival process is also less complicated as passengers are mostly interested in picking their baggage and checking-out quickly (see fig. 7).

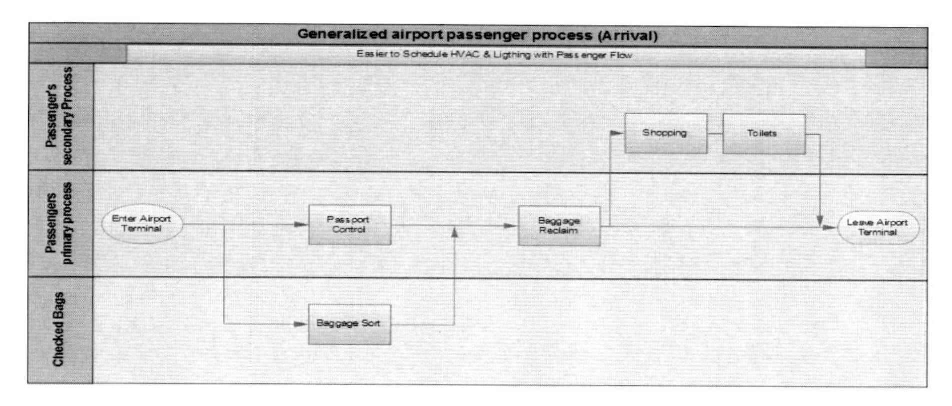

Fig. 7. A generalised airport passengers' arrival process

The departing process takes longer time since passengers spend more time at airport terminal. A typical passenger flow for departure is shown in fig. 8 below.

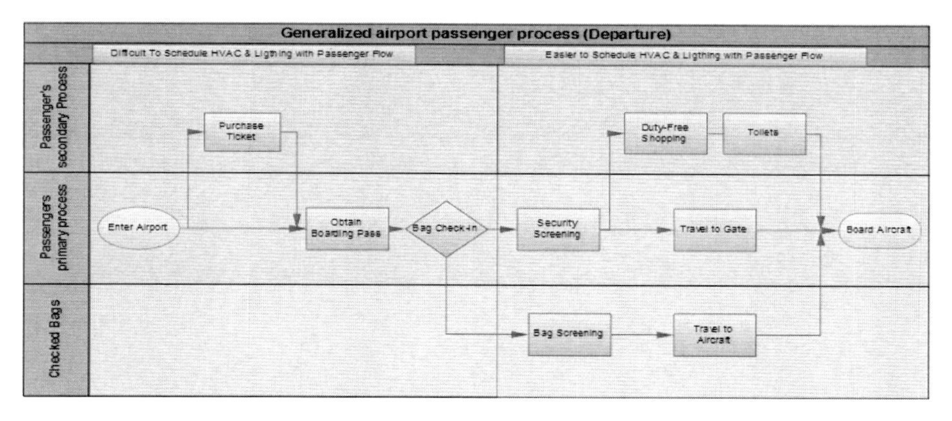

Fig. 8. A generalised airport passenger's arrival process

The model was checked by ensuring that occupancy data was inherited correctly so that changes at block and building level produce the needed effect.

7 Simulations

The base model of the terminal was constructed as described above. For the entire week under review, HVAC and lighting systems were scheduled to run for 24 hours and a temperature setpoint of between 21 - 23 degree Celsius was applied to all the indoor spaces of the terminal building to simulate an average condition of what was observed from the indoor monitoring results as shown in fig. 1 and 2. For the energy saving scenario, CIBSE recommended setpoints were applied to the various indoor spaces. HVAC and lighting systems were scheduled to vary with arrival and departure flight time in the restricted areas of the terminal building while the public areas were scheduled to run for 24 hours. When passengers vacate an area, the heating energy was reduced and indoor temperature is allowed to fall back to 12 degree Celsius and general indoor lights are in energy saving mode. *Fig. 9* shows how the internal gains vary correspondingly with passenger flow.

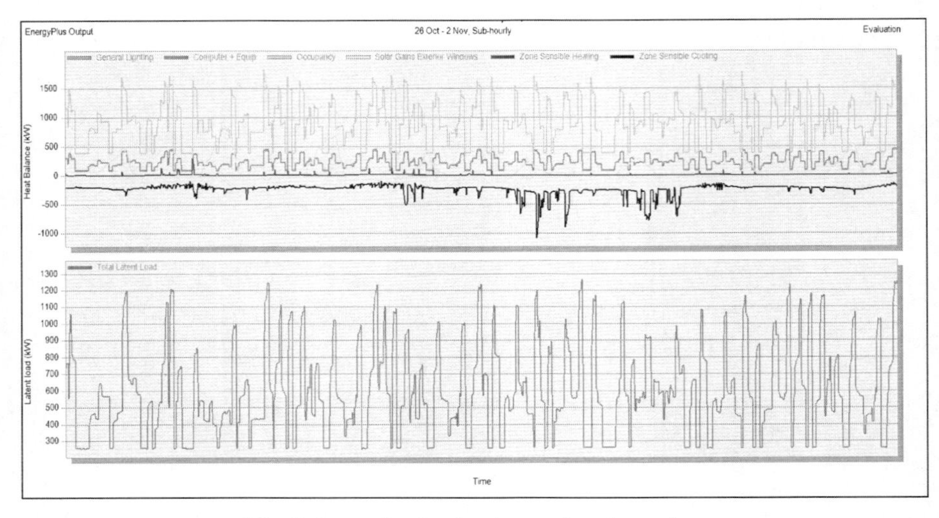

Fig. 9. Internal gains for the week under review

8 Results

The results are summarised in figure 10, 11, 12 and 13. It can be seen that selectively relieving HVAC and lighting setpoints to energy saving mode during passenger

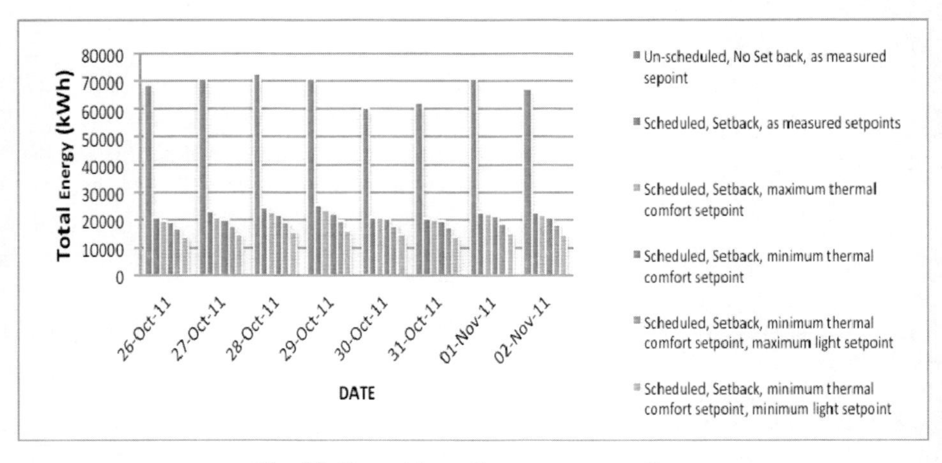

Fig. 10. Comparison of energy consumptions

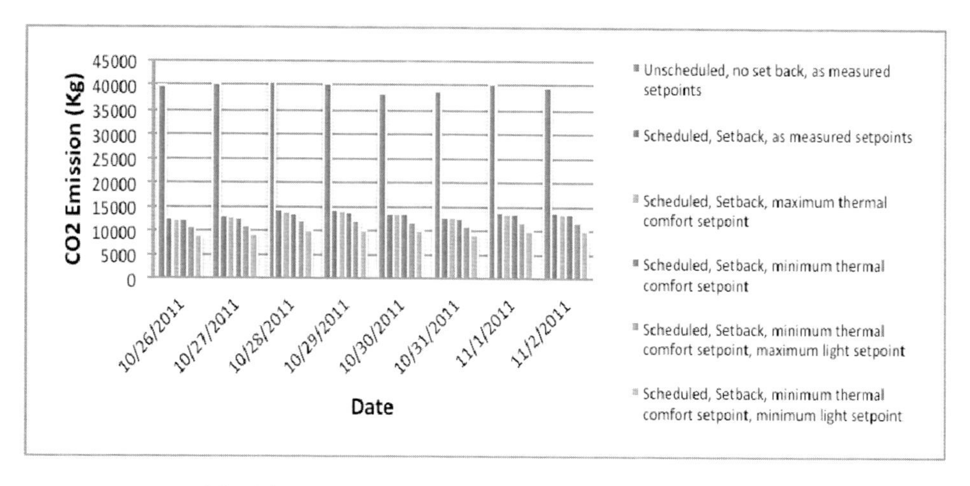

Fig. 11. Comparisons of Co2 Emissions from Energy Use

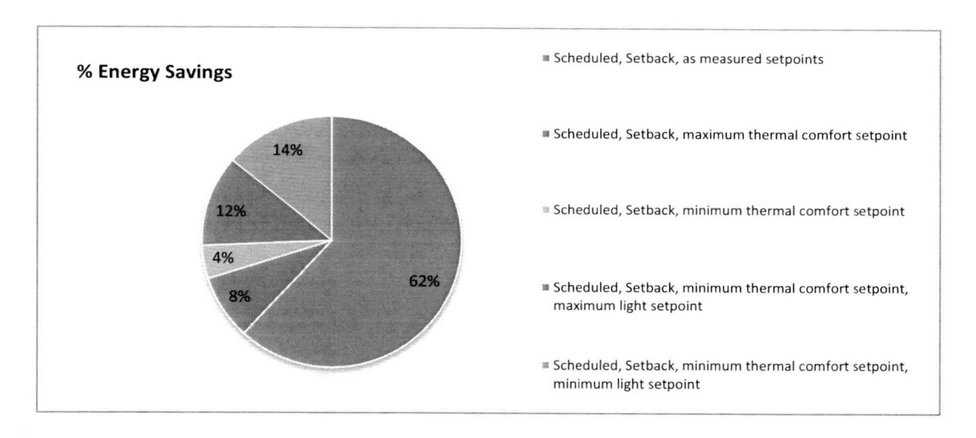

Fig. 12. energy saving potentials of retrofit options

inoccupancy period has great potentials in saving energy and reducing carbon emission in airport buildings. From *fig. 10 and 11*, Up to 60% energy savings and from figure 12 and 13, about 70% carbon emission savings results was achieved for our case study in the period under review. Providing the right setpoints as recommended by CIBSE for the various indoor spaces in the terminal is responsible for about 40% energy savings and 30% CO_2 emission savings.

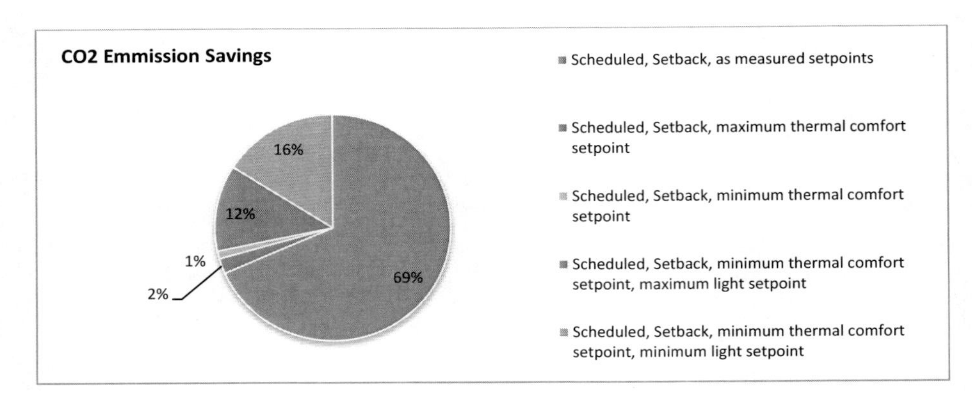

Fig. 13. CO2 emission saving potentials of retrofit options

9 Conclusions

This paper presented a case study of an existing airport terminal building aimed at developing HVAC and lighting control strategies that ensures sufficient comfort and optimal energy use. With professional building software, various supervisory control retrofit options were investigated. These options include; setback operation based on real time flight schedule and minimum comfort setpoint application for both HVAC and lighting in airport terminal building. Through integrated simulation, the building HVAC and lighting control systems setpoints were optimised and rated in terms of energy and CO2 emission savings. The result shows that setback operations based on realtime passengers' occupancy profile has a huge potential in reducing energy used and carbon emission from the airport terminal building investigated. This investigation is a precursor proof for the design of an intelligent indoor environment control system for airport building, which is currently under further investigation.

Acknowledgement. The authors gratefully acknowledge the support of Tim Walmsley, Andy Sheridan and Chris Paling of the Environment Group, Manchester Airport for facilitating the effort on data collection. We also acknowledge the financial support provided by the Engineering and Physical Sciences Research Council, UK in its Airport Sandpits Programme and Petroleum Technology Development fund (PTDF) Nigeria.

References

Chattered Institution of Building Services Engineers (CIBSE). CIBSE Guide A: Environmental Design (2006)

DesignBuilder Simulation and CFD Training Guide (2011), `http://www.designbuilder.co.uk/downloadsv1/doc/DesignBuilder-Simulation-Training-Manual.pdf` (retrieved)

Ellis, P.G., Torcellini, P.A., Crawley, D.B.: Simulation of energy management systems in EnergyPlus, Building Simulation 2009. In: 11th International Building Performance Simulation Association Conference and Exhibition, pp. 1346–1353 (2007), http://www.ibpsa.org/proceedings/BS2007/p189_final.pdf (retrieved)

Griffith, B., Pless, S., Talbert, B., Deru, M., Torcellini, P.: Energy Design Analysis and Evaluation of a Proposed Air Rescue and Fire Fighting Administration Building for Teterboro Airport Energy Design Analysis and Evaluation of a Proposed Air Rescue and Fire Fighting Administration Building for Teterboro Airport. Technical report, NREL/TP-550-33294 (2003)

ICAO, International Standards and Recommended Practice, Annex 9 to the convention of International Civil Aviation, 12th edn. (July 2005), http://www.icao.int

Mathews, E.: HVAC control strategies to enhance comfort and minimise energy usage. Energy and Buildings 33(8), 853–863 (2001)

Pan, Y., Zuo, M., Wu, G.: Whole building energy simulation and energy saving potential analysis of a large public building. Journal of Building Performance Simulation 4(1), 37–47 (2011)

Salsbury, T.I.: A Survey of Control Technologies in the Building Automation Industry Proc IFAC World Congress (2005), http://www.nt.ntnu.no/users/skoge/prost/proceedings/ifac2005/Fullpapers/02117.pdf (retrieved)

Trčka, M., Hensen, J.L.M.: Overview of HVAC system simulation. Automation in Construction 19(2), 93–99 (2010), http://www.bwk.tue.nl/bps/hensen/publications/10_autcon_trcka.pdf

Yiqun, P., Mingming, Z., Gang, W.: Whole building energy simulation and energy saving potential analysis of a large public building. Journal of Building Performance Simulation 4(1), 37–47 (2011)

Chapter 46
An Investigation into the Practical Application of Residential Energy Certificates

Alan Abela[1], Mike Hoxley[1], Paddy McGrath[1], and Steve Goodhew[2]

[1] School of Architecture Design and the Built Environment, Nottingham Trent University
[2] School of Architecture Design and Environment, Plyymouth University

Abstract. The Energy Performance of Buildings Directive (EPBD) 2002/91/EC introduced various obligatory requirements intended to achieve the reduction of use of energy resources in buildings and consequentially the reduction of the impact of energy use in buildings. Article 7 of the directive formally specified the current European requirement for the energy certification of buildings. In order to implement this requirement, a general framework for establishing a methodology of calculation of the total energy performance of buildings became necessary. The Maltese methodology for the issuance of energy performance certificates for residential property was developed and introduced by the Ministry of Resources and Rural Affairs in 2010. This methodology differs from that of most other European countries since the energy used for cooling in summer is taken into consideration when carrying out the calculation. Most states only consider the energy for heating in winter for residential energy certificates. A study of the results produced by the Maltese certification process is being used to identify whether the methodology implemented is an accurate tool for environmental monitoring of energy use in Maltese residential property. The analysis is utilised to establish a benchmark for energy use in different residential property typologies. This analysis is developed further to highlight the strengths and weaknesses of the certification procedure as a design tool, and to understand whether the procedure can be effectively applied in the cost optimisation of residential construction or refurbishment projects.

1 Introduction

Malta covers just over 300 km^2 in land area, and is the smallest and most densely populated country in the European Union. It is also one of the southernmost states in the European Union. Possibly as a result of the mild Mediterranean climate, traditional building practices were not as formally regulated as in other European Union states, and until the implementation of the EPBD in 2006, there were no energy related building regulations (Buhagiar 2007).

Residential property in Malta is generally constructed with a flat concrete roof, with walls in either limestone or concrete brick. The use of insulation in walls is not common, although traditional construction consists of a double leaf limestone wall with a central air gap. The application of insulation on roofs is increasingly more

A. Håkansson et al. (Eds.): *Sustainability in Energy and Buildings*, SIST 22, pp. 491–499.
DOI: 10.1007/978-3-642-36645-1_46

widespread although this is a practice which has become established over the past ten years. The introduction of the Minimum Performance of Buildings Regulations in 2006 stipulated maximum U-values for walls at 1.57 W/m^2 K, and for roofs at 0.58 W/m^2 K (Building Regulations Office Malta 2006). The maximum value for windows is 5.8 W/m^2 K but these are limited to a maximum of 20% of the wall area, and lower U-values are required for any increase in the glazed area above the 20% maximum. These U-values are significantly higher than those stipulated in other EU states.

The national calculation tool for the Energy Performance Rating of Dwellings in Malta (EPRDM) is the basis for the Maltese official procedure for calculating the energy performance of dwellings. The procedure takes account of the net energy required for space heating and cooling, water heating, lighting, and ventilation, after subtracting any savings from energy generation technologies. It calculates the annual values of delivered energy consumption (energy use), primary energy consumption, and carbon dioxide (CO_2) emissions, both as totals and per square metre of total useful floor area of the dwelling per annum (Ministry for Resources and Rural Affairs Malta 2011).

The procedure consists of a monthly calculation within a series of individual modules. The individual modules contain equations or algorithms representing the relationships between various factors which contribute to the annual energy demand of the dwelling.

The procedure was developed locally and is based on ISO EN 13790:2008 *Energy performance of buildings – energy use for space heating and cooling*, using a monthly calculation step.

The calculation does not differentiate between new and existing buildings and to date there are no benchmark values established. Some countries have had considerable experience with building certification but these are in North and Central Europe (Poel et al 2007) and their results cannot be applied to a Mediterranean climatic conditions and building types. Registration of Energy Performance Certificates (EPCs) in Malta commenced in January 2011.

After the methodology had been implemented for twelve months, the certificates registered were analysed in order to obtain an understanding of the calculated energy performance of residential property in Malta.

2 Data Analysis

2.1 Data Collection

A total of 249 EPCs were registered with the Ministry for Resources and Rural Affairs during 2011, and this analysis is based on the data extracted from these certificates. The certificate data was obtained from the data registry where all certificates are lodged electronically. The certificates were first analysed on the basis of property type and EPRDM values. Further analysis was carried out on the properties of the building envelope, the properties of the heating, cooling, hot water and lighting systems, and any alternative energy installations.

Forty three certificates were asset type assessments, i.e. based on actual as-built properties, whilst 206 certificates were design type assessments, i.e. assessments based on the plans of a proposed construction. Table 1 shows the distribution of certificates by property type. The majority of certificates are for single storey dwellings in multiple dwelling units, namely flats, maisonettes, and upper floor flats, accounting for just over two thirds (68%) of certificates issued. Terraced houses and duplex flats account for approximately one quarter (26%) of certificates, whilst the quantities of bungalows, villas, and identical units are too low to be statistically significant. The distribution of certificates issued reflects the predominance of flats and maisonettes in the Maltese housing market (National Statistics Office Malta 2005).

Table 1. Distribution of EPCs issued in Malta 2011

Type of Dwelling	No of Certificates	Average EPRDM kWh/m^2yr
Bungalow	1	125.3
Duplex Flat	19	116.2
Flat	107	136.8
Fully Detached Villa	8	98.7
Semi Detached Villa	2	303.5
Maisonette	40	137.8
Identical Units	4	113.1
Terraced House	46	117.2
Upper Floor	22	168.6

The Maltese EPC does not define the energy performance in terms of a band as is the case with many other housing energy certificates, and indeed for white goods also. The certificate denotes the energy performance of the property numerically using the EPRDM (Energy Performance Rating of Dwellings in Malta) indicator. The EPRDM is the calculated value of the primary energy requirement of the dwelling for heating, cooling, lighting and hot water per square metre per annum, net of any alternative energy produced on site. The certificate values of the EPRDM range from 24.0 to 395.7 kWh/m^2yr, with an average value of 133.0 kWh/m^2yr. This is equivalent to an average Dwelling CO_2 Emission Rate (DCER) of 33.2 $kgCO_2/m^2yr$. The calculated energy demand for heating, cooling, lighting and hot water actually varies between 6.95 and 114.69 kWh/m^2yr, with an average value of 37.89 kWh/m^2yr. In order to place these values in context, the requirements for a Passive House in Central Europe are a maximum of 120 kWh/m^2yr of primary energy for heating, hot water and household electricity (Passive House Institute 2010).

Table 1 demonstrates that the average EPRDM values for flats and maisonettes (single storey dwellings) are remarkably comparable at 136.8 and 137.8 kWh/m^2yr respectively. Similarly, terraced houses, which have two or more storeys, and duplex flats have average EPRDM values of 117.2 and 116.2 kWh/m^{2}yr respectively. As expected, properties on the upper floor have a higher average EPRDM of 168.6 kWh/m^{2}yr.

The subsequent analysis presented hereunder is based on the extraction of the relevant figures by the lead author of this paper from the 249 certificates registed in Malta during 2011.

2.2 Alternative Energy

Examination of the makeup of the calculated energy values shows that the contribution of photovoltaic and wind turbine installations accounts for just 0.5% of domestic primary energy demand, with the total contribution from alternative energy rising to 1% due to an energy benefit in the methodology for the use of second class water, claimed by 63 properties (25%). Just five properties, all at the design stage, have included a photovoltaic installation and two of these five also included a wind turbine. The contribution of solar water heating is intrinsic to the domestic hot water calculation within the methodology and is not identified separately as 'alternative energy'. This also applies to the contribution from heat pumps for space heating.

2.3 Heating

The main component of the primary energy requirement calculated for Maltese dwellings is heating, which accounts for over 40% of the demand, followed by cooling at 30%, and domestic hot water at 18%. Lighting contributes to 10% of the primary energy demand.

The relatively high proportion of primary energy demand arising from heating can be attributed to the fact that Maltese homes typically do not have a heating system installed, and hence the default heating system, electric heating, is applied. Out of the certificates submitted, 109 (44%) had electric heating, 133 (53%) had heat pumps with an average coefficient of performance of 3.67, and just 7 (3%) had heating systems using gas or wood. The average EPRDM for all properties with heat pumps installed reduces to 91.15 kWh/m^{2}yr whilst the average EPRDM for properties with the default electric heating installation is practically double at 184.44 kWh/m^{2}yr.

2.4 Domestic Hot Water

The second largest component of the primary energy load is domestic hot water. Only 60 (24%) of the certificates include solar water heaters. The average EPRDM for properties with solar water heaters is 102.2 kWh/m^{2}yr, with an average primary energy requirement for water heating of 7.44 kWh/m^{2}yr whilst the average EPRDM for the 189 properties (76%) without solar heating installed is 142.7 kWh/m^{2}yr with a primary energy requirement of 49.23 kWh/m^{2}yr. Sixteen properties (6.4%) use gas as a fuel for water heating whilst all the others use electricity.

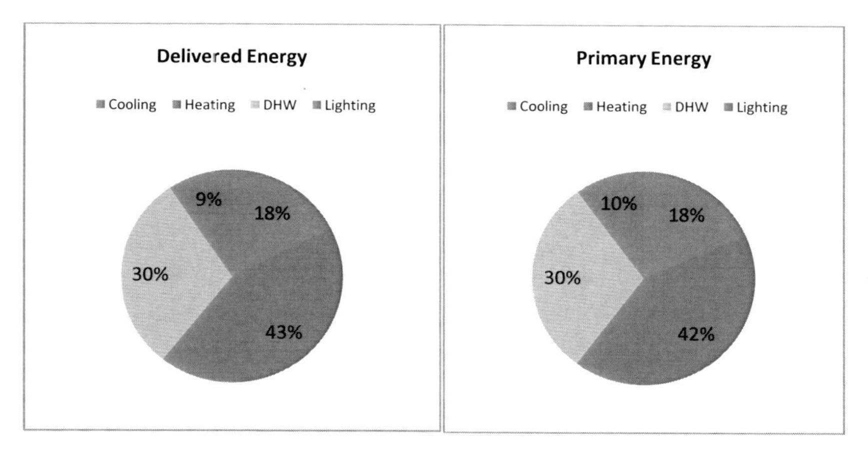

Fig. 1. Breakdown of Residential Delivered Energy and Primary Energy by Component

2.5 Cooling

Whilst cooling accounts for 18% of the primary energy in residences, this is generally provided by a heat pump for both cooling and heating, having a positive effect on the overall energy performance. The default coefficient of performance for cooling was applied in 56 of the certificates, implying that these properties either did not have a cooling system installed or planned, or that data on the cooling system was not available. The overall average coefficient of performance for cooling was 3.02 which is marginally higher than the default value of 2.8. The average primary energy for cooling is 24.64 kWh/m^2yr.

2.6 Lighting

The lighting load is 10% of the overall primary energy with an average value of 13.45 kWh/m^2yr. This value corresponds to an average of 80% of all light fittings installed indicated as being fitted with energy saving bulbs.

2.7 Building Envelope

In order to investigate the effect of the opaque building envelope, the sum of the product of the U-Values (U) and areas (A) for the envelope elements was plotted against the sum of the heating and cooling loads. Both parameters were divided by the total floor area (TFA) of the property. Figure 2 does not show any direct correlation between the properties of the opaque building envelope (UA) and the heating and cooling demand. This could be due to the effect of solar radiation which contributes considerably to reducing the heating load in winter, as well as driving the cooling load up in summer. Investigation of the effect of the glazing properties and orientation is an area for further analysis by the authors of this paper.

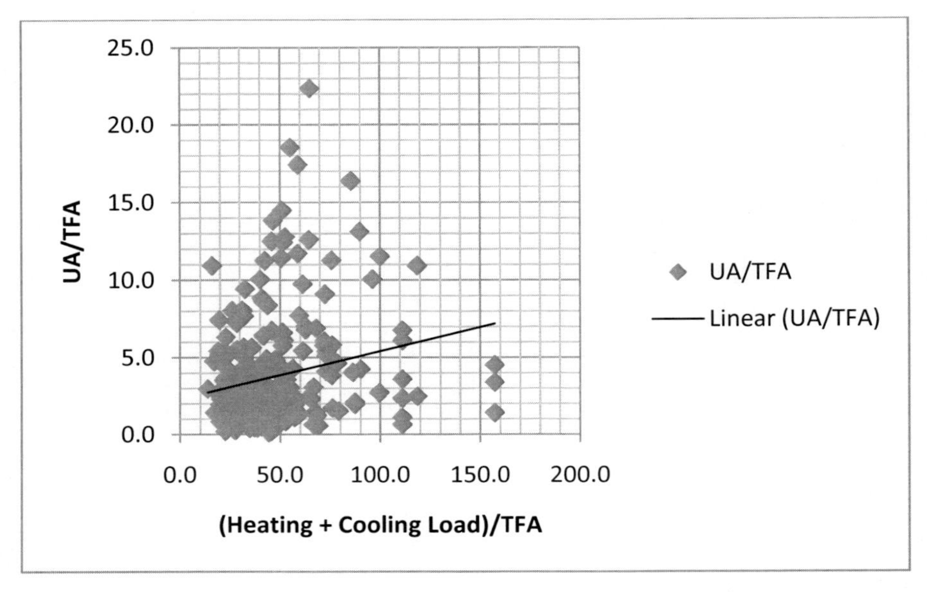

Fig. 2. UA/TFA plotted against Heating and Cooling Load/TFA

3 Development of Benchmarks

Prior to the introduction of the EPRDM methodology, no formal assessment of the energy use in Maltese dwellings was available Although the directive indicates that the certificate should indicate typical values for comparison purposes, these values have not been provided. On the basis of this analysis, carried out on the first set of 249 certificates registered at the Ministry of Resources and Rural Affairs during 2011, the following benchmark values are being proposed for different Maltese dwelling types.

Flats and Maisonettes	137 kWh/m^{2}yr
Terraced Houses and Duplex Apartments	117 kWh/m^{2}yr
Upper Floor Properties	168 kWh/m^{2}yr

The data collected was insufficient to propose values for bungalows, semi-detached and fully detached villas. However it is clear that the properties of the building envelope are not the most significant variables affecting the EPRDM. It is therefore proposed that bungalows can be categorised together with upper floor properties, whilst semi- and fully-detached villas be considered under the same category as terraced houses and duplex apartments.

3.1 Comparative Values

From the certificate data collected, the average floor area for Malta is calculated at 136m^2, and the average delivered energy is 5,175 kWh/yr. This is the value of the energy delivered to the residence for heating, cooling, lighting, and domestic hot water excluding appliances. The primary source of energy to Maltese dwellings is electricity, with LPG also used for some heating and cooking. In a previous study on energy use in Maltese dwellings (Abela 2011), the actual average delivered energy to Maltese homes was estimated at approximately 6,000 kWh/yr. This figure is for all consumption including appliances.

Table 2. Comparison of statutory limitations for primary energy consumption in dwellings

Category	Primary Energy	Constituent Components
Passive house	120	heating, cooling, hot water, household electricity
France RT 2005 Region H3 Electric Heating including heat pumps	130	heating, cooling, hot water
France RT 2005 Region H3 Heating using fossil fuels	80	heating, cooling, hot water
France RT 2012 Region H3	40	heating, cooling, hot water, lighting, auxiliaries
Spain maisonette/apartment Almeria mainland	52	heating, cooling, hot water
Spain single family dwelling Almeria mainland	79	heating, cooling, hot water
Malta maisonette/apartment Proposed current benchmark	137	heating, cooling, hot water, lighting, auxiliaries
Malta all typologies Proposed future benchmark	70	heating, cooling, hot water, lighting, auxiliaries
Italy Region A and B	40	cooling only (not yet implemented in methodology)

4 Discussion

Analysis of the certificates submitted shows that the most effective measures in reducing the calculated primary energy requirements of Maltese dwellings are the use of heat pumps for heating and the use of solar water heaters. Just over half of certified properties have heat pumps installed whilst just under a quarter have solar water heating installed. The average EPRDM for properties with both heat pumps and solar water heating installed calculated on the basis of the existing data is 57.2 kWh/m^{2}yr. In view of the fact that not all properties have access to sufficient roof space to install a solar water heater, the target EPRDM value for proposed benchmarking is 70 kWh/m^{2}yr as an immediate short-term goal for 2016, representing a reduction in calculated energy use of nearly 50% over the current average value.

The methodology used for calculation of primary energy use in Maltese dwellings suggests that passive energy saving measures are not as effective as active measures. The mild climate with relatively low temperature differences between indoors and outdoors results in the benefits of reducing U-values not being as pronounced as in more northern climates. Shading techniques could be used to reduce cooling loads, but the methodology indicates a corresponding increase in the winter heating loads. This could be due to the fact that the methodology does not allow for the use of a monthly shading factor but only permits the input of an annual value. Both shading and ventilation are areas where further development of the methodology could result in the production of more accurate results.

5 Conclusions

The certification methodology applied in Malta provides a reasonable approximation of the actual pattern of energy use in dwellings. The results obtained from the first group of certificates both match actual data and fall within the bands indicated by other regions with a similar climate (CSTB 2006) (Salmerón et al 2011). The calculation procedure is based on EN 1SO I3790:2008, and it is therefore logical to expect that although the calculation gives correct results on an annual basis, the results for individual months close to the beginning and the end of the heating and cooling season can have large relative errors (BSI 2008), although this has not been verified for Mediterranean climates. In the Mediterranean region the relationships between energy demand calculation and building design and operation become more complex (Tronchin and Fabbri 2008).

The data analysis indicates benchmark values that can be considered representative of average Maltese properties. These benchmarks can be applied to provide a ranking for certified properties, allowing owners and developers to place their properties on a scale. The benchmark values identified are comparable to those established by other Mediterranean countries with more mature certification systems (see Table 2).

The results of this analysis indicate that the most effective interventions for improving the energy performance of Maltese dwellings are the use of heat pumps for heating and solar water heating for domestic hot water. Whilst these will not bring residential energy consumption down to the 'Nearly Zero' value indicated as a target in the recast directive 2010/31/EC, these two measures together can result in a reduction of the order of 50%. The recast also places an emphasis on cost optimization, and the introduction of 'active' measures for energy efficiency is expected to be more cost effective than passive measures for existing housing.

Whilst research in other countries indicates that discrepancies are expected between the calculated energy consumption of homes and what actually happens when people live in them, it is acknowledged that these discrepancies are confusing for a household that needs to use the EPC when applying for subsidies, for example (Sunikka-Blank and Galvin 2012). Further investigation into the data is required to examine the effect of energy saving measures in greater detail, specifically on the operation of the property on a monthly basis, and consequently to understand whether these are being handled accurately by the calculation methodology. In most circumstances, the Energy Performance Certificate is the only available tool for the public to gauge the energy efficiency of a residence, and hence the accuracy of the certificate has economic and social implications.

References

Abela, A., Hoxley, M., McGrath, P., Goodhew, S.: A comparative study of the implementation of the energy certification of residential properties in Malta in compliance with the Energy Performance of Buildings Directive. In: Passive House Conference, Malta (2011)

Buhagiar, V.: Technical Improvement of Housing Envelopes in Malta. In: Braganca, L., et al. (eds.) COST C16 Improving the Quality of Existing Urban Building Envelopes: Facades and Roofs. Delft University Press, Netherlands (2007)

British Standards Institute, Energy performance of buildings - Calculation of energy use for space heating and cooling (ISO 13790:2008). British Standards Institute (2008)

Building Regulations Office Malta, Technical Guide F: Conservation of Fuel Energy and Natural Resources. Valletta, Malta (2006)

Centre Scientifique et Technique du Bâtiment (CSTB). Réglementation Thermique 2005 Th-CE (2006)

Centre Scientifique et Technique du Bâtiment (CSTB). Réglementation Thermique 2012 Th-CE (2010)

European Parliament and Council . Energy Performance of Buildings Directive 2002/91/EC. Brussels, Belgium (2002)

European Parliament and Council. Energy performance of buildings (recast) 2010/31/EC. Brussels, Belgium (2010)

Ministry for Resources and Rural Affairs Malta Implementation of the EPBD in Malta Status in November 2010 In: Maldonado, E. (ed.) Implementation of the Energy Performance of Buildings Directive, European Union, Brussels, Belgium (2011)

National Statistics Office. Census of Population and Housing 2005 vol. 2 Dwellings. Government Press, Malta (2005)

Passive House Institute. Active for more comfort: The Passive House. Darmstadt, Germany (2010)

Poel, B., van Cruchten, G., Balaras, C.A.: Energy performance assessment of existing dwellings. Energy and Buildings 39, 393–403 (2007)

Salmerón, J.M., Cerezuela, A., Salmerón, R.: Escala de califación energética para edificios existentes. Instituto para la Diversificacion y Ahorro de la Energía, Madrid Spain (2011)

Sunikka-Blank, M., Galvin, R.: Introducing the prebound effect: the gap between performance and actual energy consumption. Building Research & Information 40(3), 260–273 (2012)

Tronchin, L., Fabbri, K.: Energy performance building evaluation in Mediterranean countries: Comparison between software simulations and operating rating simulation. Energy and Buildings 40, 1176–1187 (2007)

Chapter 47
Post-Occupancy Evaluation of a Mixed-Use Academic Office Building

Katharine Wall and Andy Shea

BRE Centre for Innovative Construction Materials (CICM),
Department of Architecture and Civil Engineering, University of Bath, Bath, BA2 7AY
{k.wall,a.shea}@bath.ac.uk

Abstract. The paper presents the results from a Building User Study (BUS) survey undertaken as part of the Technology Strategy Board's (TSB) Building Performance Evaluation project. The results are from a mixed-used academic building on the University of Bath campus and are presented in relation to the design strategies used within the two distinct parts of the building, the new build and the refurbished, joined by an atrium. The summary indices from the BUS survey for both parts of the building are presented along with the results from twelve overall variables. The variables are then discussed in terms of the use and operation of the building in six sections: air quality; lighting characteristics; sources of noise; satisfaction with temperature; comfort, productivity and perceived health; and design and image to visitors. The paper concludes by highlighting potential changes to the redevelopment of buildings of this type in the future, many of which are more widely applicable.

1 Introduction

In depth post-occupancy evaluation is being undertaken on a mixed-use academic building on the University of Bath campus as part of an £8 million Technology Strategy Board (TSB) project, Building Performance Evaluation (BPE) [1]. The aim of the project is to learn lessons from a wide range of new build and refurbished buildings across the UK that can be used to inform the construction industry to enable lower carbon buildings to be delivered. This paper focuses on presentation of the results of a Building User Studies (BUS) [2] survey for the case study building.

2 Research Methodology

This paper focuses on the performance of a case study building in terms of the occupants' perception of the building, however, data from a number of sources used as part of the post-occupancy evaluation are presented. The methodologies utilised include: site visits; interviews and meetings with external design team members and internal stakeholders; a detailed survey of the building; a design process review; technical review; review of controls; Building User Studies (BUS) survey; analysis of energy demand and

A. Håkansson et al. (Eds.): *Sustainability in Energy and Buildings*, SIST 22, pp. 501–510.
DOI: 10.1007/978-3-642-36645-1_47

building occupancy profiles; heat flux measurements; thermal imaging survey; computer modelling; and TM22. Occupants' perceptions are focused on as conditions within a building have an effect on their productivity as well as wellbeing and can impact on business costs and result in failure to reach efficiency targets [3].

The case study used in this paper is a 5200 sq.m mixed-use academic building on the University of Bath campus, referred to as 4 West. The building comprises two sections joined by an atrium, the new building, built on part of the footprint of the existing 1960s CLASP (Consortium of Local Authorities Special Programme) building, and the refurbished part of this same building, which fronts onto the university's main walkway (the parade), shown in Figure.1. The building is predominantly concrete frame with concrete floors and the main block of the new section contains an innovative cast-in ducted cooling system. The new section consists of clad brick walls, cavity, metal frame and plasterboard, shown in Figure.2. The new section of the building houses a five-storey academic office block and two-storey office block, three lecture theatres, a three-storey atrium connecting the two sections, as well as plant rooms, server rooms and showering facilities. The refurbished section houses a one-storey office area, a student information centre, a café, plus plant room and laboratories.

Fig. 1. Refurbished façade and parade

Fig. 2. New-build section

Quantitative and qualitative data are collected as part of the BUS survey on occupant feedback of buildings in relation to 65 variables [2], including: design; image; needs; thermal comfort; lighting; noise; air quality; control; health; and productivity. The survey uses a benchmarking database of approximately 80 buildings within the UK for comparison. The BUS survey was undertaken in November 2011, approximately 20 months after the initial occupation of the building. A total of 126 occupants completed the survey, 86 from the new building and 40 from the refurbished areas, which represents over 80% of occupants on site.

3 Design Strategy

The 1960s CLASP system buildings make up a significant amount of the building stock on the University of Bath campus, all of which are in need of refurbishment. The 4 West project was the first of these building to be tackled and it was envisaged that this would be an exemplar for redevelopment. This process started in 2001 with the removal of asbestos within the building to be demolished and the treatment of the asbestos in the refurbished part of the building. A feasibility study was undertaken to establish if the building should be dismantled and at the time it was seen as both more economical and efficient to demolish a large proportion of the building and create a new 'statement' building in the centre of the campus. The redevelopment of the building was delayed and it is now thought that redevelopment of similar buildings in this way may well be unachievable. The lessons learnt from this process have been incorporated into the procedure and practices at the university and will be vital to the refurbishment of similar aged building.

The superstructure of the building is reinforced concrete and the floor/ceiling slabs for the new office blocks include a Kiefer natural cooling and ventilation system, which comprises ducts cast into the slab. The Kiefer cooling system [4] delivers air through finned ductwork in the slab where heat exchange takes place between the slab, space and supply air. The majority of the concrete soffits are exposed to work more effectively with the ventilation and cooling strategy. Services are concealed in corridor areas using a metal panel ceiling system, whereas in the seminar rooms suspended ceilings are used to aid acoustics, in other rooms acoustic absorption is provided by the room fittings. Raised access floors are also provided throughout the new building, except the atrium and the main stairwell space. Ventilation is mixed mode, with the Kiefer cooling system providing mechanical ventilation and cooling to the majority of the new building. The areas within the building that are served by the Kiefer cooling system also have openable windows, providing occupants with natural ventilation. These are operated by the occupants and are not integrated with the mechanical ventilation system. Openable windows and louvres are also available on the second floor of the refurbished part of the building and on the south façade of the parade level, which includes the café and student services. The lecture theatres have air handling units (AHUs) to regulate temperature within these areas, which is achieved through floor distribution. Heating is provided through a traditional wet system that is serviced from the district heating system on the campus.

The annual carbon emissions for 4 West were 77.7 $kgCO_2/m^2/year$ for the period 1st December 2010 to 30th November 2011. This figure was calculated using a modified version of CIBSE's TM22 [5] developed for the TSB Building Performance Evaluation studies, and is based on a Gross Internal Area (GIA) of 5571 m^2 and carbon factors of 0.550 $kgCO_2/kWh$ for electricity and 0.194 $kgCO_2/kWh$ for gas.

Table 1. Carbon emissions, in $kgC0_2/m^2/year$, for 4 West compared to good practice and typical buildings outlined in Energy Consumption Guide 19 [7]

Building	4 West	Naturally ventilated, cellular		Air-conditioned, standard		Air-conditioned, prestige	
Carbon Emissions		Good practice	Typical	Good practice	Typical	Good practice	Typical
$kgCO_2/m^2/yr$	90	32	57	85	151	143	226

Carbon Buzz's benchmark for University Campus' is 89.6 kg CO_2/m^2/year with a database average of 131.0 kg CO_2/m^2/year [6]. Using Energy Consumption Guide 19 [7] and the carbon factors (0.52 kgCO_2/kWh for electricity, 0.19 kgCO_2/kWh) and area (treated floor area; 95% of GIA, 5292m^2), provides the figure of 90.24 kgCO_2/m^2/year for comparison, shown in table 1.

4 Results

4.1 Summary of Occupants' Perception

BUS survey response data from the case study building are compared to the benchmarked mean as well as the scale midpoint and their respective 95% confidence upper and lower intervals [2]. Each one of the 65 variables is also given a traffic light (Red, Amber, Green) colour code in relation to how the case study building has performed compared to the benchmarked data and the scale midpoint. Red represents worse performance, amber the same performance and green better, these are also represented by different shapes on the scale, diamond, circle and square respectively [2].

To compare building performing in relation to benchmarked data a number of indices are created using the BUS data. The summary index is the average of the comfort and the satisfaction indices. The comfort index is the average of the standard or z-scores, which present variables on a common scale, with mean=0 and standard deviation=1 [8], for overall comfort, lighting, noise, temperature and air quality in summer and winter. The satisfaction index is the average of the z-scores for design, needs and productivity [8]. Figure 3 shows the summary index for the new building (1.12) which lies in the 93rd percentile and the top quintile. Figure 4 shows the same index for the refurbished building (0.64) which lies in the 85th percentile and the top quintile. These figures show graphically these values and highlight how well they perform overall compared to the other buildings used in the benchmark (shown by the rings on the chart) as well as each other.

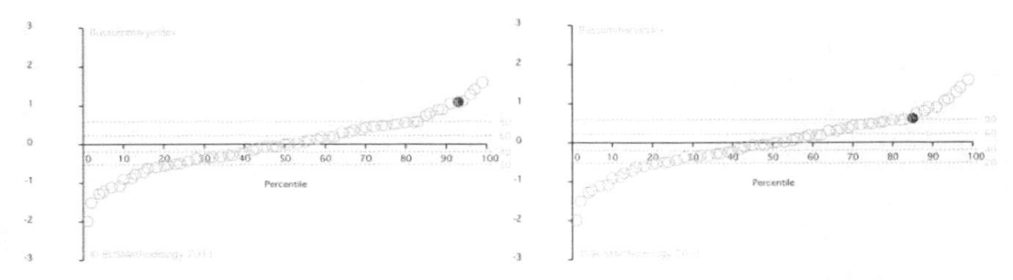

Fig. 3. Summary index, new [8]

Fig. 4. Summary index, refurbished [8]

4.2 Occupants' Perception in Relation to Aspects of the Design Strategy

There are a number of overall variables that give an impression of how the two sections of the building are performing. These overall variables are presented for the new part of the building in Figure 5 and represent an easy way to show that the new section of the building performs better than the benchmark in all these variables, with only perceived health performing in the same range as the benchmark. The performance of the refurbished section of the building, shown in Figure 6, shows the same trend as that for the new section, although the majority of the scores are lower, with air in summer and winter, perceived health, productivity and temperature in winter all in the same range as the benchmark. This provides a simple way of indicating how this section of the building is performing.

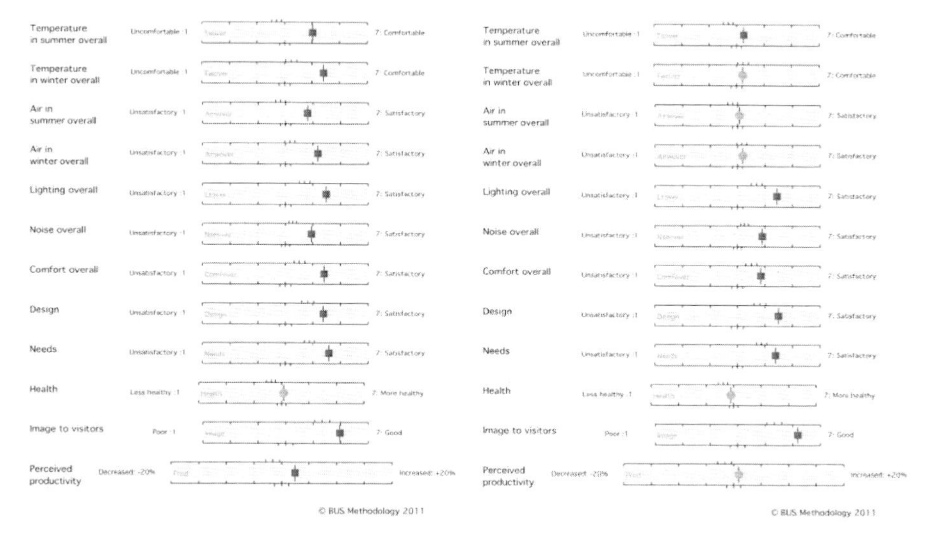

Fig. 5. Overall variables for new section [8] **Fig. 6.** Overall variables for refurbished [8]

This section discusses a selection of these overall variables in relation the design strategy, highlighting issues that influence occupants' experience in the building.

Air Quality

Occupants were asked about specific elements of the air in both summer and winter, including whether it was dry, humid, fresh, stuffy, odourless, smelly, still or draughty. In the majority of the new building ventilation is mixed mode with the Kiefer cooling system providing mechanical ventilation and cooling and restricted openable windows available for natural ventilation. The majority of the variables for air were seen as satisfactory by the participants in the new part of the building, however, two variables performed below the benchmark with air is summer seen as too dry by 36% of participants and air in winter too still by 50%. The first of these could relate to the comments in relation to the Kiefer cooling system, which was perceived by many as full air conditioning and was thought to "dry my skin and give me headaches" as well as being "dry and making my eyes sting". There were a number of comments seeking fresher air into the offices, with one respondent finding themselves "needing to get some fresh air a couple of times a day" and another stating that "it would be better if we could open windows at night to change the air". As occupants appear to want to open their windows to get fresh air, like one respondent, they find that "in winter it is too cold to open the windows while in the office" this may result in the feeling of still air in winter. There were also comments about cooking and cigarette smoke smells coming through the ventilation system, which may be related to the location of air intakes for the AHUs, which, for many areas, are cited in an under croft below the parade, where pollutants such as vehicle exhaust and cigarette smoke could infiltrate the system.

Lighting Characteristics

Lighting in both parts of the building is controlled by a mix of user operated switches and motion detector sensors. As well as the overall lighting question there were four on other lighting variables: artificial light; glare from lights; natural light; and glare from sun and sky. The respondents within the new part of the building rated these above (glare) and at the benchmark (natural and artificial lights). A similar trend was seen in the refurbished part of the building, but the artificial lighting was rated below the benchmark, 31% of respondents felt there was too much artificial lighting. Comments from occupants highlighted that the main lights were often too bright, with one saying that the "lighting in my office was so bright it hurt my eyes so I requested it to be made dimmer". The majority of offices within the building are relatively large individual rooms with two sets of twin fluorescent strip lighting and although the controls for these had, in some cases, been modified, they were still too bright for some and there was a desire for better individual task lighting, one occupant stated that there is "no individual lighting of each desk in shared offices; some colleagues (away from windows) need lighting but then it's too glaring for those sitting by a window".

Sources of Noise

Specific questions about noise related to colleagues and other people, noise from inside and outside, and unwanted interruptions. Overall the new part of the building performed slightly better than the refurbished part, with a score of 4.95 compared to 4.9. However, the results highlight two areas where the respondents felt that there was too much noise in the new part of the building, with only one area in the refurbished part. In both parts of the building, occupants reported that there was too much 'other noise from inside', with 38% and 33% of occupants reporting this in the new and refurbished parts respectively. This seem to originate mainly from events taking place in the atrium, as there are internal windows that open into this space from both parts of the building, this disturbance was described by one respondent; "if I have my window open it is really noisy as it opens into the atrium, I hear all the people going in and out, conversations etc. Far worse is when the atrium is booked for events". The atrium has developed to be used for functions and was not designed for this and so these issues were not considered when the building was designed. The ventilation system was also a popular complaint from occupants in the new part of the building, with one respondent highlighting this and giving a reason why this might be the case, "the building is generally very quiet, which makes the hum of the ceiling fans/colleagues headphones etc. all the more distracting". In the new building, hand dryers, toilet flushing and the talking lifts were also mentioned as being disturbances, again, probably more noticeable as the academic block is often very quiet. In the refurbished part the majority of noise complaints came from those working in the cafe, which related to the coffee making equipment used, with one respondent stating that "because 4 West Café is very busy, lots of people gathered makes a noise that is uncomfortable, as well as the equipment that we use in the café (especially the coffee grinders)". 'Noise from outside' was seen as being too much by 33% of respondents in the new building, which when reviewing the written responses tended to come from the atrium when events were held as well as the student bar and one respondent thought that "Samba bands should be banned from practicing anywhere near the building". There seems to be some discrepancy as to whether the atrium is inside or outside space by different users of the building, as events in this space were mentioned in relation to both of these categories.

Satisfaction with Temperature

Temperature in summer and winter was assessed through two further variables, hot/cold and stable/varies. The results from these show that in the new part of the building only 'Temperature in winter: hot/cold' was on the benchmark, the others were above. In the refurbished part of the building the temperature in winter was deemed as too cold by 60% of respondents, with 24% rating it as '7', the lowest score on the scale. This did not translate to a poor score on the overall winter variable, which was 5.15, for this part of the building, which is in the 97th percentile of the benchmarked results. There is no comments section specifically for temperature, but the cold environment was mentioned within the following section of the questionnaire (with typical quotes from respondents provided after each): Behaviour change ("have to bring extra clothes to work"); Comfort ("in winter it is freezing"); Health ("Apart from the coldness in winter I don't feel any influence for my health"); Hinder ("the front door, during cooler days and in winter lets lots of cold air in and draught takes it directly to our working area"); Perceived productivity ("Ability to concentrate is lowered when it's cold"); and Requests for changes ("to have heating checked as it can be quite cold"). These mainly relate to the cafe, where the automatic door opens when people walk past and from queues for service, and from an office in student services which is mainly glass.

Comfort, Productivity and Perceived Health

The new part of the building obtained an average score of 5.41 (89th percentile) for overall comfort with 82% of respondents being satisfied with the building, this compares with 4.85 (64th percentile) for the refurbished part of the building, where 63% were satisfied with the building. The average perceived productivity increase was

3.09% for the new part and 1.11% for the refurbished part, both above or at the benchmark. Comments dealt with locality of facilities, which increased productivity in the new part, due to "cups of tea being readily available" for one respondent, as kitchens are provided at regular intervals, especially in the academic block where there is one per floor. This was seen to decrease it in the refurbished part with "staff spending a great deal of time coming and going from counter space to 4 West kitchen", cold was also cited as an issue. Perceived health within the new part of the building averaged at 4.07, just about the mean score with 18% of respondents perceiving their health to have improved and 14% declined. The score for the refurbished building of 3.9 for perceived health, is just below the mean score, but still within the benchmark, 26% of respondents felt less healthy within the building, whereas 19% felt more healthy. The comments related to health in the new building mentioned general aspects such as, "my work involves sitting down at a desk all day, this does not make me feel healthy". In the refurbished part of the building the comments focused mostly on cold.

Design and Image to Visitors

The design of both building was seen as satisfactory by the respondents, with both scores above the benchmark. The score for the new part of the building was 5.38, slightly below the 5.49 average score for the refurbished building. Again, both buildings were seen as having a good image to visitors and performed better than the benchmark and again the refurbished building scored slightly higher, 6.18, than the new building, 5.97. Comments made by respondents in relation to design of the new part of the building focus on the feeling that there is "little communication across floors" in the main five-storey block, which seems somewhat to be blamed on the provision of kitchens on each floor. There were also several complaints about wasted space and observations that the individual offices could have been a bit smaller to increase the number of occupants. There was also a few people who commented that the design was "a bit sterile" and that the "design is general purpose and hence soulless", many others did appreciate the design with comments such as "I like the height and space". The comments for the refurbished part of the building seemed to also incorporate parts of the new building, as the only entrance for the second floor offices is through this space, focusing on the feeling of space, light and the modern design, present in the atrium.

5 Conclusions

In the new part of the building there seemed to be three related issues, the Kiefer cooling system being perceived as an air-conditioning system that meant respondents wanted to open windows in order to obtain fresh air. This causes noise issues from events in the atrium, summarised by one respondent; "If left open the internal window generates a lot of noise from a large number of people using the atrium for example, if closed this is not a problem but it gets very stuffy". Several respondents also attributed some aliments to the "air-conditioning", with people complaining of increasing headaches, dry skin and sore eyes. To be certain of the air quality within the building an air quality test could be undertaken as well as checks on the air flow of the ventilation fans.

In the refurbished part of the building the issues highlighted related to the bright lights and cold temperatures, these caused both lack of concentration and lower productivity than the new building and increased health issues, such as colds. These issues seemed to be concentrated around the café and student services office areas, surrounded on two sides by a glass curtain wall. The former has an automatic door that opens when people are queuing or walking past.

The refurbished part of the building scored slightly higher than the new part of the building in terms of both design and image to visitors, although as the two parts are attached via the atrium and those on the second floor of the refurbished part have to access their offices through it this might have affected their response. This does, however, suggest that as well as being much more expensive than originally thought to dismantle the old building, the image of the new building is perceived by occupants as not quite as good as the refurbished part. This might have consequences on future redevelopments of these types of buildings, in terms of user satisfaction and economy, as well as embodied carbon.

There are a number of lessons that can be learnt from these findings that are applicable to both further developments on campus as well as redevelopment of similar buildings around the UK. The main lessons identified are listed below:

- The interaction between openable windows and mechanically ventilation systems needs to be considered.
- Users need to be provided with more information about how mechanical ventilation systems work, so they are aware of what is being provided to them and how it's set up to operate. This may enable the system to work more effectively as well as the users to feel happier with their environment.
- Location of intake units for air supply should be carefully considered, with actions taken to ensure that unwanted odours do not infiltrate the system.
- The lighting system for the building should have been thoroughly assessed to ensure that spaces were not over-lit and that lighting can be switched in sets to allow more flexibility of use. Task lighting should also be considered in shared offices, especially if flexibility of the main lighting is low.
- The use of spaces should be thought through carefully and communicated by the client to the design team, so that items such as acoustic buffering can be added to atrium spaces during design, if appropriate.
- If buildings are likely to have a lower than average background noise, such as academic departments, then the specification of products and additional sound proofing to bathrooms should be considered.
- Possible buffer spaces should be taken advantage of or designed into buildings or ideas generated as to how to avoid doors being open for considerable amounts of time during winter, especially in café areas with lots of footfall.

References

1. Technology Strategy Board (TSB) (2011) Building Performance Evaluation, Non-Domestic Buildings: Technical Guidance. Version 4 (February 2011) (unpublished)
2. Usable Building Trust. BUS Methodology (2009), http://www.usablebuildings.co.uk/WebGuideOSM/index.html (accessed February 22, 2012)
3. Thomas, L.: Evaluating design strategies, performance and occupant satisfaction: a low carbon office refurbishment. Building Research and Information 38(6), 610–624 (2010)
4. Core Group. University of Bath, 4 West Project: Concrete Cooling. Report. CGW/10/079 (2010) (unpublished)
5. Chartered Institute of Building Service Engineers (CIBSE) TM22 Energy Assessment and Reporting Methodology, 2nd edn. CIBSE, London (2006) ISBN 190328760X
6. Carbon Buzz. CIBSE Benchmark Category: University Campus (2012), http://www.carbonbuzz.org/sectorbreakdown.jsp?id=3 (accessed March 14, 2012)
7. Action Energy. Energy Consumption Guide 19: Energy use in offices. 2nd edn. Action Energy, London (2003)
8. Arup. 4 West BUS Results (2011) (unpublished)

Chapter 48
The Human as Key Element in the Assessment and Monitoring of the Environmental Performance of Buildings

Wim Zeiler, Rik Maaijen, and Gert Boxem

TU/e, Technical University Eindhoven, Faculty of the Built Environment, Netherlands

Abstract. To further reduce the environmental load future buildings must be much more sustainable than the existing buildings. Currently most decisions about the building sustainability are made by applying sustainability assessment tools. However these tools are not really suited for monitoring the environmental performance of buildings during its whole life cycle. New methods and approaches are necessary to asses and monitor the environmental performance of buildings. Optimizing comfort for occupants and its related energy use is becoming more important for facility managers. Presently however HVAC installations often do not operate effectively and efficiently in practice, because the behaviour of occupants is not included. This result in comfort complains as well as unnecessary high energy consumption. As the end-user influence becomes even more important for the resulting energy consumption of sustainable buildings, the focus should be how to integrate the occupants in the building's performance control loop. This leads to new approaches which enable the inclusion of occupant's behaviour in the process control of the building's performance to help facilities managers operate and maintain their sustainable buildings more efficiently. In an experiment in a real in-use office building a wireless sensor network was applied to describe user behaviour. The results showed that it is possible to capture individual user behaviour and to use this to further optimize comfort in relation to energy consumption. Based on our experiments we could determine the influence of occupants' behaviour on energy use and determine possible energy reduction by implementing the human-in-the-loop process control strategy.

Keywords: Sustainable assessment, monitoring, human behaviour, energy management.

1 Introduction

Buildings represent a significant contribution to energy use and consequent green house emissions (1). Following the Kyoto summit, and all the other summits on sustainable development, it is clear that one of the driving forces in the design and refurbishment of the building stock is determined by sustainability factors (2). The

A. Håkansson et al. (Eds.): *Sustainability in Energy and Buildings*, SIST 22, pp. 511–520.
DOI: 10.1007/978-3-642-36645-1_48

European Union and its Member States have a large number of on-going policy initiatives directly aimed at supporting of sustainability of the built environment. The climate and energy strategies are aimed, so that by 2020 renewable energy will represent 20% of energy production; a reduction of greenhouse gas emissions by 20% (base 2005) and achieving energy savings of 20%. The targets go even further: to reduce CO2 emissions by 80-90% (Nearly Zero) by 2050. In addition, Directive 2006/32 EC requires Facility Managers to reduce energy consumption and operational costs of existing buildings. A recent report of the Pacific Northwest National Laboratory (PNNL) gave the results of a post-occupancy evaluation of 22 'green' federal buildings from across the United States. PNNL found that, on average, green buildings, compared to commercial buildings in general use 25% less energy, emit 34% less carbon dioxide, cost 19% less to maintain and have 27% more satisfied occupants (3). So sustainability is a way to reduce energy consumption and reduce operational cost as well. Therefore conducting (sustainability) performance based assessments of buildings operation is of great importance.

There are many sustainable assessment tools available to support design teams in their quest for green effective buildings (4), which makes it difficult to choose, which tool should be used to implement the new business strategy most effectively. There is a pressing need for practical tools for sustainable facilities management (5). The priority for the near future is to provide insight into the consequence of building design decisions on building sustainability performance. Facilities managers need decision support tools to make their (future) building more resilient to risk, cost-effective to maintain and run, use less energy and other resources and are more comfortable and better places to work. Only then, progress can be made towards more effective, productive as well as more sustainable buildings.

Some of the current sustainability assessment tools like BREEAM recognize the importance of such an operational analysis, by granting points when such analysis is performed and the results are implemented into the design. This way must become possible to carry out a operational analysis-study fulfilling the requirements of BREEAM-NL (BRE Environmental Assessment Method for the Netherlands) credit MAN 12 (6).

2 Methodology: Facility Management and User Behaviour

According to the CEN definition facilities management is the integration of processes within an organisation to maintain and develop the agreed services which support and improve the effectiveness of its primary activities (European Committee for Standardisation). There is more attention to the importance of people as part of its remit, as can be seen in the definition of facility management by the International Facility Management Association: Facilities management is a profession that encompasses multi disciplines to ensure functionality of the built environment by integrating people, place, process and technology (7). Finch takes even a step further by stating that 'care' of people should be preeminent in any definition of facility management. By putting the user first, organizational efficiency will follow (8).

The energy management within buildings can improve by applying the latest developments from ICT technology (9). The potential savings of energy due to better use of ICT technology is well documented by Røpke (10), however, in most of the research focusing on improved ICT often overlooks the role of user in reducing the energy consumption. Overall the role of the occupant in relation to the energy consumption is important (11): occupant presence and user behaviour have a large impact on space heating, cooling and ventilation demand, energy consumption of lighting and room appliances (12) and thus on the energy performance of a building (13). An analysis of occupant behaviour on the energy consumption (14), shows that conservation oriented behaviour of occupants can reduce energy consumption by one-third in normal buildings, while in more efficient buildings, by nearly half (47%). Reduction of or optimizing of energy use is often done without really taking in to account the goal of the energy consumption, human comfort. However, trying to optimize energy efficiency, without addressing occupant comfort, is not going to work (15). Several models have been developed to describe human behaviour and to include it in building performance analyses (12, 16, 17, 18). However, only a few studies successfully demonstrate energy reduction from real occupancy behavioural patterns that have been determined (19).

Still, as until now user behaviour has not been part of the comfort system control strategy in offices. As there are not many specific research results of the effect of user behaviour in existing office buildings, first a user-actions analysis was performed in cooperation with Royal Haskoning, one of the major Dutch HVAC engineering consulting companies.

3 Analysis of Human Behaviour on Energy Consumption

In the 3th floor of one of their office was chosen as it is a characteristic and representative example of their office working space. Fig. 1 shows the floor of the building and Fig. 2 illustrates the parameters which might have an influence on the personal actions. For the calculation of the effects of the user behaviour on the energy consumption of the building, the latest version of the VABI Elements heat/cooling load calculation tool was used. VABI (Vereniging voor Automatisering Bouw en Installaties, Society for Automating Building Construction and Building Services) is the most important Dutch software developer of tools for building systems, with emphasis on HVAC systems, thermal aspects, electricity and solar energy. The 3rd floor of the case study office was modeled in the VABI model, see Fig. 1, this made it possible to calculate the effects caused by actions of the occupants.

To determine the importance of these behavioural actions on the energetic building performance, the spread in outcomes resulting from the behaviour interactions were determined by basic calculations, see Fig. 3 and Fig. 4. The input parameters were based on observations of the occupants during a week. To test the sensitivity of the process outcome, in relation to specific user actions, input parameters were changed within an acceptable and realistic bandwidth based on the observations. The output results from the VABI model for the office space 3.20 – 3.22 are shown in Fig. 3 and represent the total sum of the heating and cooling demand for a year. A high bandwidth means that the parameter is an interesting factor of the occupants' behaviour as it has a major impact on building performance.

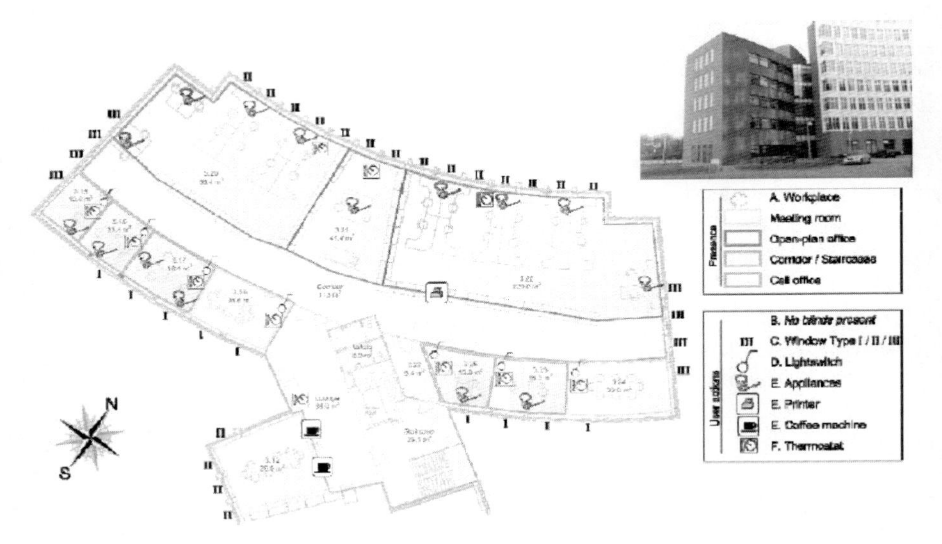

Fig. 1. Test case 3rd floor of an existing office building

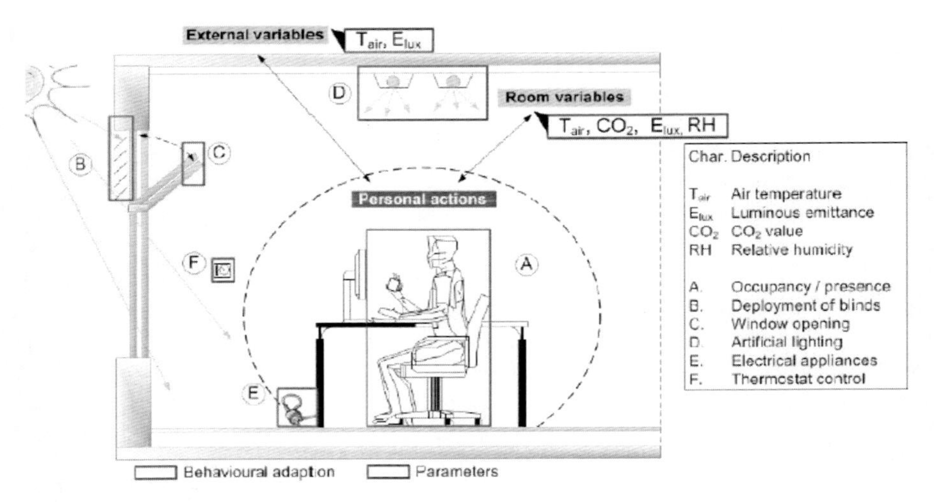

Fig. 2. Personal actions and parameters in an example office

Based on the above results in Fig. 3, it is concluded that some of the parameters related to user behaviour (occupancy, lighting, electrical appliances and temperature setting) have a clear and high influence (up to plus or minus 30%) on building performance. The results of the first measurement period are used for estimating the energy saving potential applying the new proposed bottom-up approach. Therefore the building is modelled in HAMBase, making use of the Matlab Simulink software environment:

- Energy demand using input parameters as assumed in the design phase of the building systems
- Applying real data obtained from the measurements in the case study building, using the gained temperature set points, and profiles of electrical appliances energy use;
- Implementing the new approach where energy is sent to those spots where needed, e.g. the positions of the building occupant.

The measurements are during the winter, when there was only a heating demand. The acquired profiles for electrical appliances use and occupancy patterns are also applied in the summer situation. The applied values in the simulation are presented in Table 1.

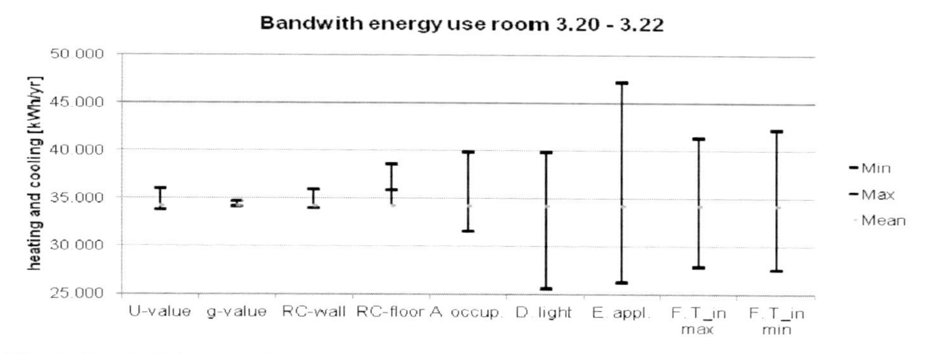

Fig. 3. Bandwidth of results from VABI elements for the total energy demand room 3.17 as caused by changing the specific input parameter

Table 1. Simulation input data with three different reference data

Simulation input		A. Design	B. Measured	C. New approach
Appliances		10W/m²	Measured profiles	Measured profiles
Lighting	Power	10W/m²	Installed power/zone	Installed power/zone
	Schedule	8-18hr	Measured profiles	Measured profiles
Metabolism	Power	10W/m²	1 Met/prs	1 Met/prs
	Schedule	8-18hr	Measured profiles	Measured profiles
T [°C] (heating)	Day	22 (8-17hr)	22 (8-19hr)	If present 22 else 19
	Night	19	19	19
T [°C] (cooling)	Day	24	23	If present 23 else 25
	Night	25	25	25

Fig. 4 shows the simulation results of heating and cooling of the building as designed, the actual profiles as measured in the building and the new individual controlled conditioning based on occupancy. For all the three different situations \ both heating and cooling situation. The actual measured energy demands are higher than designed in both situations, while in the new approach the heating and cooling demand is lower. This illustrates that the actual energy use is higher than designed.

Mainly the cooling demand shows an increase (+43%) compared to the designed situation. Based on the first data it can be concluded that energy savings can be gained, especially for the cooling demand.

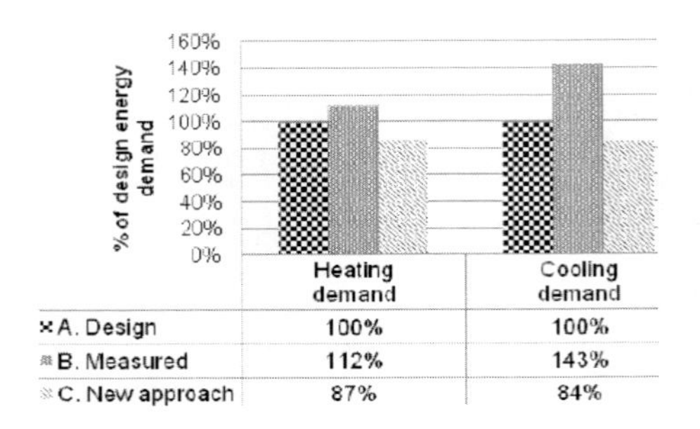

Fig. 4. Results of the HAM Base simulation

4 Approaches to Include the User in the Control Loop

This underlines the importance for focusing within facilities management on the inclusion of human behaviour to improve building process control performances. Therefore it was necessary to think about a way to integrate the human in the process control loop on the level of room/floor and even building. The concept of how the position of the building occupant could be leading in the system control is presented in Fig. 5.

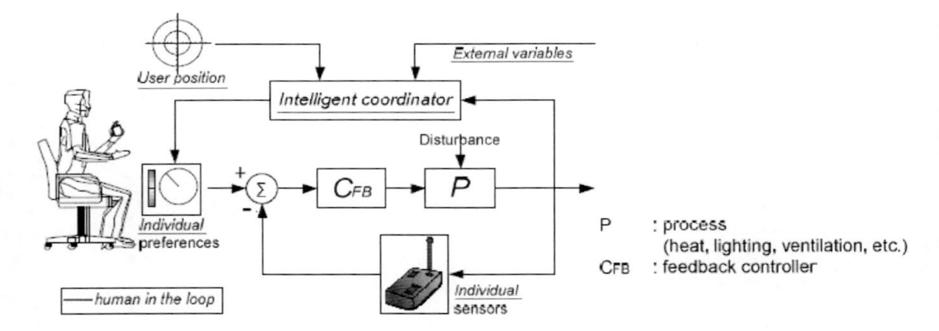

Fig. 5. Proposed block diagram of the controlled system and the intelligent coordinator for taking the human in the control loop of building systems

This makes it possible to apply the individual preferences for HVAC process control by determining the occupant position and behaviour. In order to control the operation of building systems, local intelligent field controllers send their signal to the intelligent supervisor. The input signals are diverse, including the user position but also other external variables like weather data. The intelligent coordinator makes its decision and sends acknowledge signals to the individual building systems. To apply the individual preferences while maintaining the comfort level, individual controlled systems with local HVAC options show high potential, as developed in recent research. The human in the control loop of building services systems, can only be done if users can be located within the building. Low-budget wireless sensor networks with portable nodes show high potential for real-time localization and monitoring of building occupants. Therefore static wireless sensor nodes were mounted on the walls and communicate with mobile nodes (or in the future smart phones) carried by the occupant to determine the position of the occupant on workplace level. The measurement set-up is shown in Fig. 6.

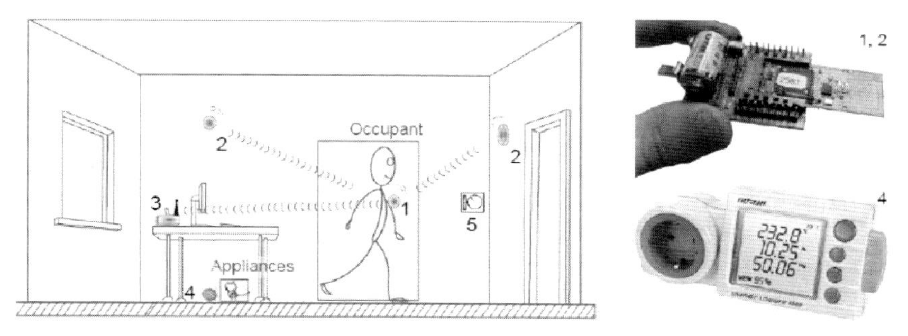

Fig. 6. Schematic representation of the measurement set-up with the mobile node (1) determining his position based on the signal strength from the static nodes (2), where the mobile node updates its position to the server via the receiver (3). The most important user influence on building performance, electrical appliances (4), is measured and the thermostat control (5) to be applied in future building energy simulations

The measurements were performed on the fourth floor of Royal Haskoning, an international engineering company in The Netherlands, Rotterdam. The wireless static nodes for position tracking of the occupants were placed on points of interest e.g. the workplaces, printer, coffee machine and toilet (Fig.7). Based on the signal strength the nodes identify in which zone the occupant is located, Fig. 8.

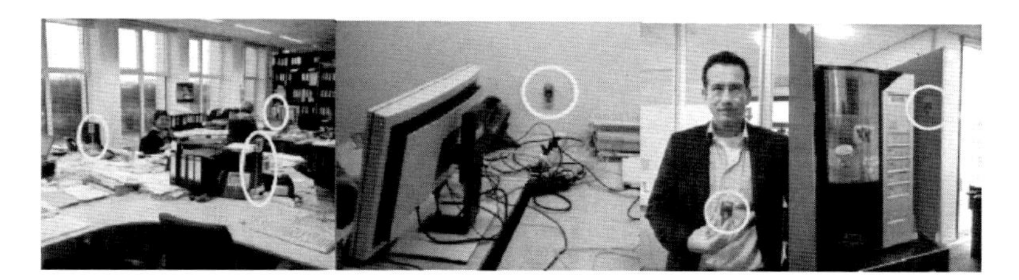

Fig. 7. Wireless sensor nodes used as well as a user with mobile sensor

Using MATLAB the data is put into usable information. Figure 9 shows the mean occupancy level (i.e., presence on the office floor or at workplace) with standard deviation for the case study over the course of a reference day (representing the entire observation period). There can be considerable differences amongst offices, though this occupancy shows a comparable pattern with occupancy patterns found in literature (20). The mean occupancy level is low as it never exceeded 50%. It is likely that occupancy patterns vary from one day to another day during a week. Figure 9 shows the occupancy level between 7AM and 7PM, with the highest occupancy on Tuesday and the lowest on Thursday.

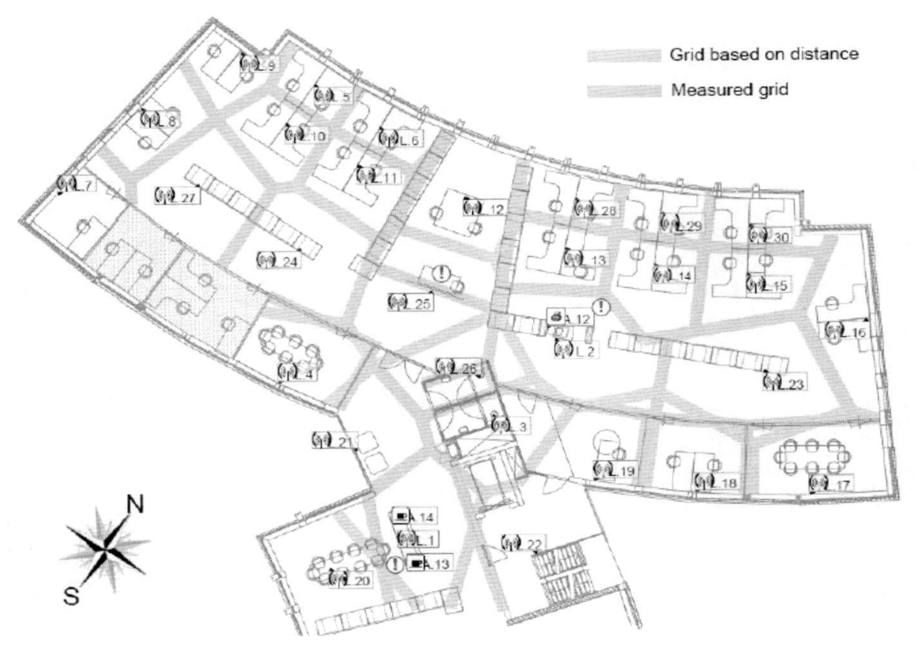

Fig. 8. Floor plan with calculated and measured grid formed by 30 static nodes

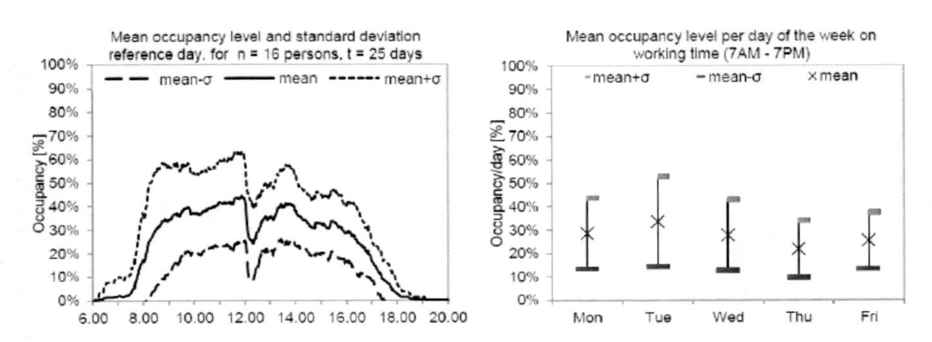

Fig. 9. Mean occupancy level and standard deviation over a week

5 Conclusion

Big steps need to be made to reach future targets regarding the reduction of energy consumption and maintaining the requested comfort level in the built environment.

With increasing energy performances, the influence of the occupant becomes significant and should be looked into. This is still not sufficiently done within current assessment tools. In the used case study the human influence is 3-5 times higher than variations in building parameters. From measurements of 20 employees during 6 weeks on an office floor it is clear that individual occupant's behaviour can be distinguished. Further research towards integrating the actual effects of human behaviour on the energy performance of a building is necessary to determine the actual performance (energy/comfort). This could form the basis for new insights that could lead to new ways for energy management leading to further reduction of the energy consumption as well as adjustments to the present assessment tools.

Acknowledgements. Royal Haskoning Consultants supported this research by allowing the experiments in their Rotterdam office. Sense Observational systems enabled the use of their wireless sensor appliances.

References

1. Morrissey, J., Horne, R.E.: Life cycle cost implications of energy efficiency measures in new residential buildings. Energy and Buildings 43(4), 915–924 (2011)
2. John, G., Liyanage, C., Clements-Croome, D.J.: Supportability in the Built Environment: Enhancing the Life Cycle Performance of Building Facilities. In: Proceedings 18th CIB World Conference W070 (2010)
3. Fowler, K.M., Rauch, E.M., Henderson, J.W., Kora, A.R.: Re-Assing Green Building Performance: A Post Occupancy Evaluation of 22 GSA Buildings. Pacific Northwest National Laboratory PNNL-19369 (2010)
4. Sanuik I.: Review of global environmental assessment methods (2011), `http://www.bsria.co.uk/news/global-env-assess/` (accessed October 27, 2011)
5. Elmualim, A., Shockley, D., Valle, R., Ludlow, G., Shah, S.: Barriers and commitment of facilities management profession to the sustainability agenda. Building and Environment 45(1), 58–64 (2010)
6. Dutch Green Building Council, BREEAM-NL 2010 - Keurmerk voor duurzame vastgoedobjecten. Versie 1.11 maart 2010, 53–56 (Dutch)
7. IFMA, What is FM?, International Facility Management Association (2012), `http://www.ifma.org`
8. Finch, E.: Florence Nightingale: Pioneer of Facility Management. In: Proceedings W070, 18th CIB World Building Congres, Salford (2010)
9. Webb, M.: SMART 2020: Enabling the low carbon economy in the information age, report of The Climate Group Annex 47 Cost-Effective Commissioning for Existing and Low Energy Buildings (2008)
10. Røpke, I., Christensen, T.H., Jensen, J.O.: Information and communication technologies – A new round of household electrification. Energy Policy 38(4), 1764–1773 (2010)
11. Groot de, E., Spiekman, M., Opstelten, I.: Dutch Research into User Behaviour in relation to Energy Use of Residences. In: Proceedings PLEA 2008, Dublin (2008)
12. Page, J., Robinson, D., Morel, N., Scartezzini, J.L.: A generalized stochastic model for the simulation of occupant presence. Energy and Buildings 40(2), 83–98 (2007)

13. Hoes, P., Hensen, J.L.M., Loomans, M.G.L.C., de Vries, B., Bourgeois, D.: User behavior in whole building simulation. Energy and Buildings 41(3), 295–302 (2009)
14. Brahme, R., O'Neill, Z., Sisson, W., Otto, K.: Using existing whole building energy tools for designing net-zero energy buildings – challenges and work arounds. In: Proceedings IBPSA Conference, Glasgow (2009)
15. Nicol, J.F.: Comfort and energy use in buildings – Getting them right. Energy and Buildings 39(7), 737–739 (2007)
16. Akhlaghinia, M.J., Lofti, A., Langensieppen, C., Sherkat, N.: Occupant Behaviour Prediction in Ambient Intelligence Computing Environment. Journal of Uncertain Systems 2(2), 85–100 (2008)
17. Tabak, V., de Vries, B.: Methods for the prediction of intermediate activities by office occupants. Building and Environment 45(6), 1366–1372 (2010)
18. Kwok, S.S.K., Yuen, R.K.K., Lee, E.W.M.: An intelligent approach to assessing the effect of building occupancy on building cooling load prediction. Building and Environment 46(8), 1681–1690 (2011)
19. Dong, B., Andrews, B.: Sensor-based occupancy behavioral pattern recognition for energy and comfort management in intelligent buildings. In: Proceedings IBPSA Conference, Glasgow (2009)
20. Mahdavi, A.: The human dimension of building performance simulation. In: Twelfth international IBPSA Building Simulation Conference, Sydney (2011)

Chapter 49
The Effects of Weather Conditions on Domestic Ground-Source Heat Pump Performance in the UK

Anne Stafford

Centre for the Built Environment, Leeds Metropolitan University, Northern Terrace, Queen Square Court, Leeds LS2 8AJ

Abstract. Unpredictable and variable weather is often cited as one of the factors which may contribute to the underperformance of heat pumps in the UK, compared with other European countries. In this study, 10 similar ground-source heat pump systems, installed in existing social housing in North Yorkshire, were monitored intensively over a period of almost two years. A weather station, closely co-located with six of the ten dwellings, was also established giving data on local external temperatures and other parameters. Differences in the performance characteristics of the heat pump systems over 2010 and 2011 are assessed with particular reference to differences in local weather conditions.

Keywords: Ground Source Heat Pumps, Seasonal Performance Factor, Energy Monitoring.

1 Introduction

Evidence is emerging that heat pumps in the UK may be tending to exhibit slightly poorer in-situ system performance, than is generally the case in Europe (Huchtemann and Müller 2012), (EST, 2010). Differences have been attributed variously to occupant behaviour, design and sizing issues, poor installation practice, uncertainties with respect to building fabric, and last but not least, the rapidly variable and unpredictable nature of UK weather conditions compared to conditions found in other parts of Europe. All of these factors may have a part to play, but this paper focusses primarily on the effects of weather conditions.

Data is analysed from 10 similar single-dwelling heat pump systems, over a period of almost two full calendar years (2010 and 2011). The winter periods of 2010, and the early part of 2011 in the UK encompassed some particularly severe cold weather conditions, while the latter months of 2011 were relatively mild. The details of the monitoring systems have been published elsewhere (Boait et al. 2011) but a brief overview is given below for convenience. Weather data was obtained from a weather station co-located with six of the ten dwellings monitored, and located less than 15 miles (24 km) from three of the remaining four, and less than 25 miles (40 km) from the fourth.

A. Håkansson et al. (Eds.): *Sustainability in Energy and Buildings*, SIST 22, pp. 521–530.
DOI: 10.1007/978-3-642-36645-1_49

System performance, expressed as a monthly performance factor (SPF) according to the equation given below, was calculated for each dwelling. This calculation method is in accordance with the definition of SPF4 given by Nordman et. al. (2010), in that the system boundary includes the energy consumed by a supplementary back-up electric heater which is integral to the heat pump, but does not include the distribution pump energy.

$$SPF_{month} = Q_{month}/E_{month}$$

where Q is the total heat output of the heat pump over the period (in this case a given calendar month), and E is the electricity consumption over the same period.

The system performances are assessed with reference to both average external temperatures, and degree days (to base 15.5).

2 Brief Description of Systems and Monitoring Protocols

2.1 Dwellings and Heat Pump Systems

The systems studied were all installed during the winter of 2007-2008 in small social housing bungalows near Harrogate in North Yorkshire. All but one of the dwellings were off gas-grid. All were of similar size and construction, dating from between 1967 and 1980, and had received fabric upgrades to at least the UK "Decent Homes" standard (i.e. 300mm loft insulation, cavity wall insulation and double-glazing) (DCLG 2006).

The heat pumps supplied both space-heating (SH) and domestic hot water (DHW) and were connected to a conventional wet radiator heating system, with radiators oversized by about 30% to compensate for the lower output temperatures associated with heat pumps. The DHW production of the heat pumps was, in all cases but one, supplemented by the presence of a separate electric shower. Secondary space-heating in the form of an electric fire was also present in all cases, but occupants reported that these were rarely or never used on grounds of cost. The 6kW IVT Greenline C6 heat pump systems (IVT 2012) were capable of providing the required internal demand temperatures, even during severe weather, and may in fact have been somewhat oversized for the dwellings.

2.2 Monitoring Protocols

The electricity consumption of the heat pumps, disaggregated into various components, was monitored, together with the total heat output and the heat output to the SH circuit. (Thus heat output to DHW could be obtained by subtraction). In addition, system flow and return temperatures, DHW storage temperatures and cold water flow into the storage tank (equivalent to hot water volume usage) were monitored. Inside the dwellings, data collected included room temperatures, relative humidity and CO_2 levels.

All data was recorded at 10 minute intervals via radio transmission to a logger located in the dwelling or in a neighbouring dwelling. Data from the logger could be downloaded via GSM modem, thus minimising disruption to the occupants.

2.3 Review of Weather Conditions over the Monitoring Period

The monitoring period for the heat pumps extends over most of the two calendar years of 2010 and 2011. Monitoring equipment was installed in 9 of the 10 dwellings during December 2009, and in the final dwelling in February 2010. At this time a weather station was also established, on the site of six of the ten dwellings. This location corresponds approximately to latitude 54.16°N, longitude 1.65°W with an elevation of about 140m above sea level. Data was recorded on external air temperature, humidity, pressure, wind-speed, wind-direction rainfall and insolation. The weather station was dismantled on 13th December 2011.

Figure 1 shows the variation in daily average and monthly average air temperature over the monitoring period.

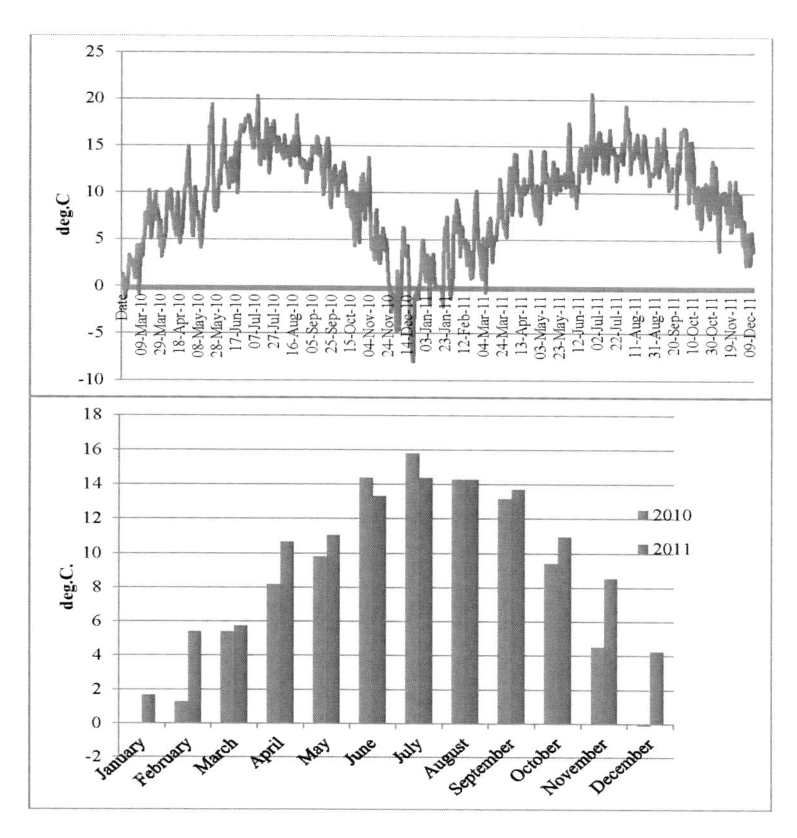

Fig. 1. Daily and Monthly Average External Air Temperature

The plot of monthly average temperatures gives a clearer picture of the differences between the two years. 2011 tended to be somewhat warmer in the late winter and spring compared with 2010, and considerably warmer in the latter part of the year, but slightly cooler in June and July. In August, the average monthly temperature was virtually identical in both years. Note that the average temperature shown for December 2011 relates only to the period 1st- 13th, as the weather station was dismantled after this date, and so is less reliable than the other figures. However, the Weather Underground station at Jennyfields, Harrogate (Weather Underground 2011) around 20 km distance from the project weather station, reports a monthly average value of 4.8 °C for December 2011, which is reasonably close to the value shown in Figure 1.

3 Results

3.1 Relationship of System Performance to External Air Temperature

Heat pump performance depends to a large extent on temperature difference. The heat pumps studied were all weather compensated systems controlled via a target radiator return temperature which varied with external temperature in order to keep internal demand temperatures constant. Temperature lifts tend to be larger when the system is delivering domestic hot water (DHW), as this typically has a higher demand temperature compared with space-heating output temperatures. Thus the variation of SPF throughout the year is bi-modal with the highest SPFs occurring in the spring and autumn, when space heating dominates but temperature lift is lower than in mid-winter, as shown in Figure 2.

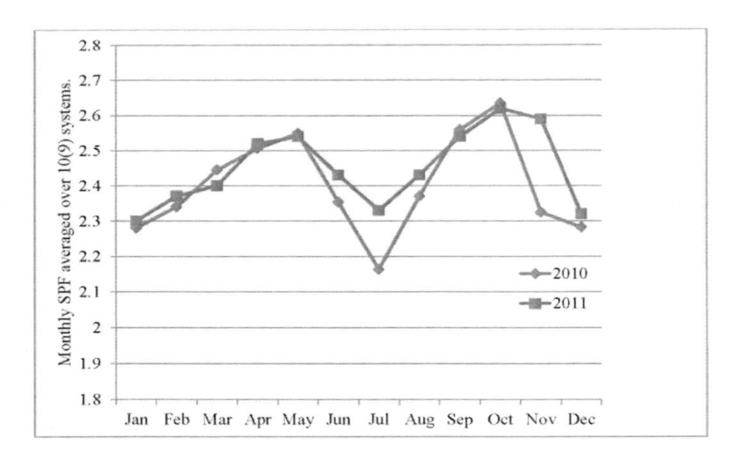

Fig. 2. Monthly Seasonal Performance Factor averaged over all systems

For 2011, only 9 of the systems are included in this calculation, since it was not possible to disaggregate the distribution pump energy in the case of the tenth system. Slightly better performance was obtained in all months of 2011, compared with the same month in 2010, except for March, September and October. Considerably better performance was obtained in the summer months and in November.

The marginally poorer performances recorded in March, September and October 2011 may be partially attributable to factors such as reduced reliability of weather data due to weather station issues (September, both years), a heat pump system breakdown leading to loss of data for one system (September and October 2011), and to the fact that although average air temperature for March 2011 was slightly higher than for 2010, the degree days value (to base 15.5) was also slightly higher. However the performance differences in these months are very small.

Over the summer months, the heat pump output is dominated by DHW production. This project was conceived as an action research project, and as the project progressed, average DHW tank storage temperatures (set-point temperatures) tended to be reduced in a number of systems as Harrogate Borough Council and participating tenants acted upon feedback from the project regarding the influence of this parameter on performance. Figure 3 shows the difference in average storage temperatures over the two years.

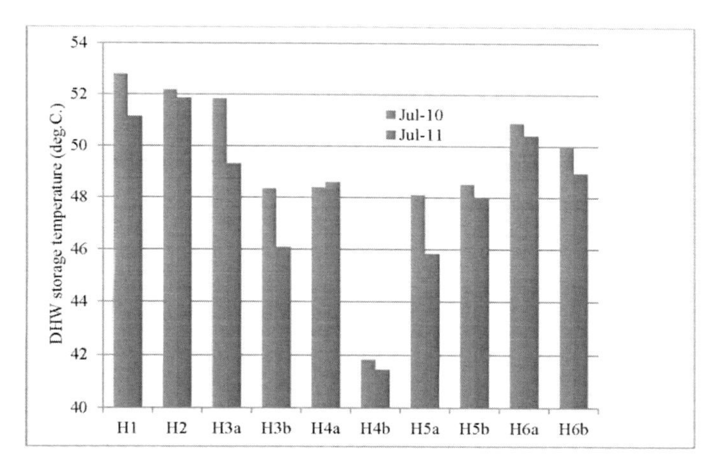

Fig. 3. Comparison of average DHW storage temperatures (top of tank) for July 2010 and 2011

The mean reduction in DHW temperature from July 2010 to July 2011 is 1.125°C, with three dwellings having reductions of over 2°C. This demonstrates the importance of DHW storage temperatures, especially in these dwellings where many of the heat pumps (and integral DHW storage tanks) were located outside the thermal envelope of the dwelling due to space constraints. Although the tanks were well-insulated, this location will inevitably lead to increased tank losses. Tank losses are also dependent to some extent on DHW usage with higher losses for low volume users (Stafford 2011). However, the reductions in storage temperatures clearly resulted in improved summer performances, even though the summer months in 2011 were cooler than 2010 (June and July), or similar (August).

In November 2010, ground loop temperatures were particularly low after the first week of the month, due to rapidly falling external temperatures during that month, and generally lower autumn temperatures than 2011. In 2010, night-time temperatures below freezing were observed in November on several occasions, and temperatures remained almost constantly below freezing from the 26th of the month onwards. The difference in ground-loop temperature is shown in Figure 4.

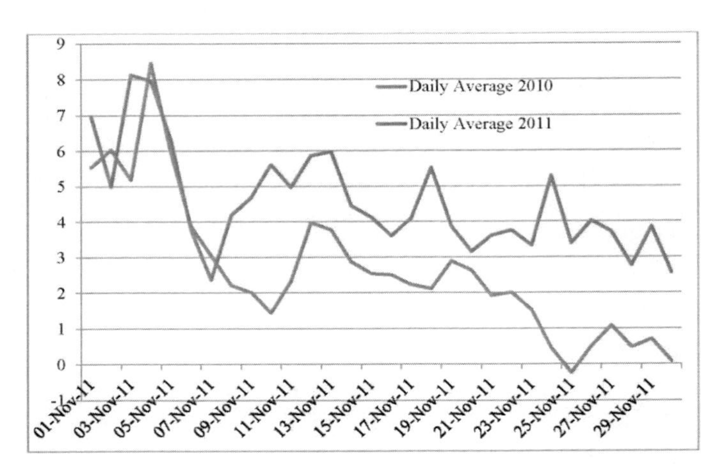

Fig. 4. Temperature of fluid to Ground-loop (averaged over all 10 systems) for November 2010 and 2011

3.2 Relationship of Energy Usage to Degree Days

Perhaps more useful way of assessing the effects of temperature is to plot total heat pump energy usage vs degree days (i.e. the time and temperature-differential measure of heating requirement assuming no space-heating is required above a given baseline temperature, in this case 15.5 °C.) Figure 5 shows that there is an approximately linear relationship between monthly average energy consumption and degree days, with a residual average energy consumption of at least 22.7 kWh per month which accounts for water heating and controls etc. when no space-heating is required.

Separating out the time periods March 2010-December 2010, and Jan 2011-Nov 2011, the line slope and intercept varies as shown in Figure 6.

The slope for 2011 is slightly higher than that for 2010. Interestingly this suggests a slightly poorer general heat pump performance, and may be due to part-loading issues. However, the intercept also changes from 30.165 to 12.458 kWh/month, which over twelve months would represent an additional electricity consumption of around 212 kWh in 2010, compared with 2011. The change in intercept may give an indication of the change in residual DHW heating consumption due to reduction of DHW set-point temperatures. However, the DHW consumption figures derived in this way are indicative only, since the usage of DHW produced by the heat pump in these

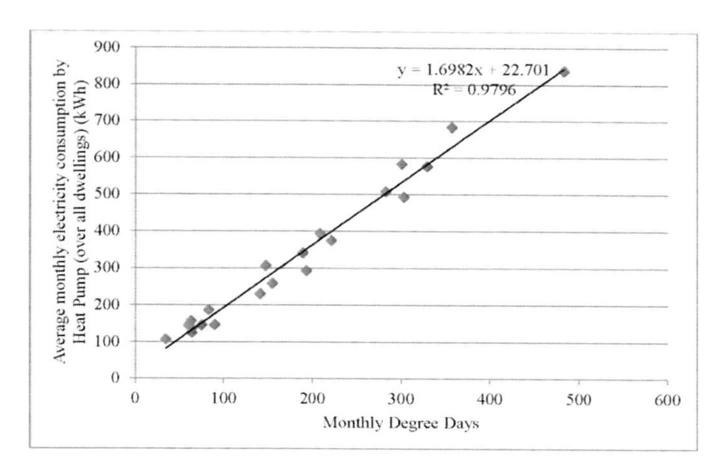

Fig. 5. Monthly energy consumption as a function of degree days

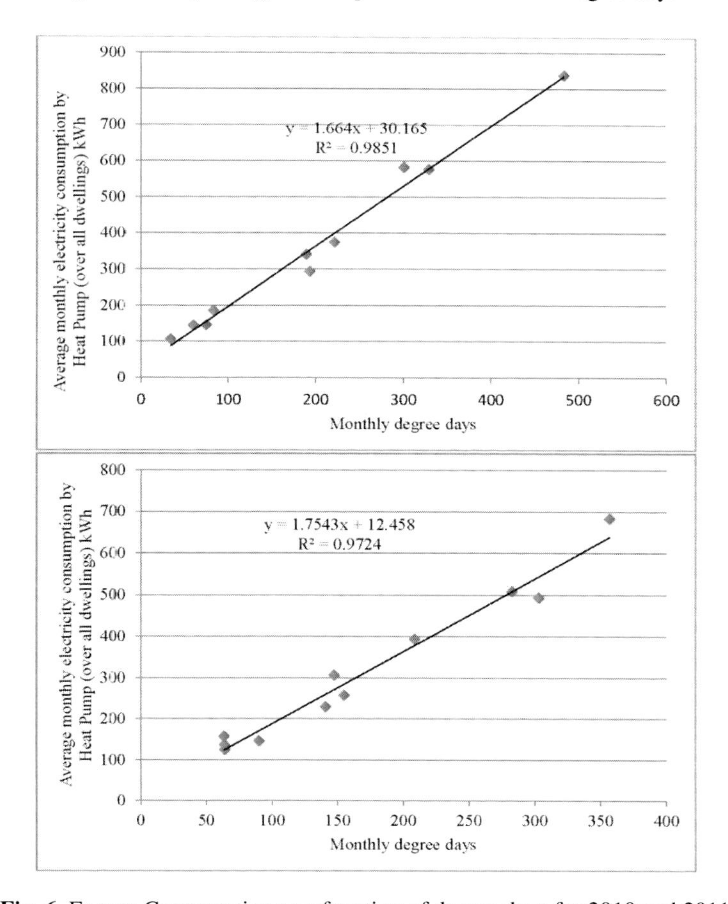

Fig. 6. Energy Consumption as a function of degree days for 2010 and 2011

dwellings tends to be very low, as occupancy is single or two persons, and separate electric showers are present in 9 of the 10 cases. This means that tank and pipework losses become significant. We might expect therefore that there will be a seasonal variation in the DHW element of consumption also.

3.3 Internal Temperatures and Occupant Comfort

Internal conditions in all 10 dwellings were monitored along with heat pump performance. Temperature and relative humidity was measured (at 10 minute intervals) in 4 rooms of each dwelling, i.e. living room, kitchen, occupied bedroom and bathroom, and CO2 levels were monitored as a proxy for air quality in 1 location within the dwelling. In most cases this sensor was located in the occupied bedroom, but in 2 cases it was in the living room.

Internal temperature data shows that the heat pump system was easily capable of maintaining internal demand temperatures up to around 25 °C, and in fact even where high internal temperatures were required, the supplementary electric cassette was never, or hardly ever brought online (except during the weekly pasteurisation cycle). This suggests that the systems may have been somewhat oversized for the requirements, and indeed part-loading may be to some extent responsible for relatively low SPFs.

In November 2010, internal demand temperatures ranged from around 17°C to around 25°C depending upon occupant preferences and heat pump control settings. By November of 2011 however, some alterations to settings had been made, and a narrower range of demand temperatures had evolved (between around 19°C and 23-24°C). When the heat pumps were first installed, many of the tenants were advised not to attempt to alter settings themselves, but to call for assistance if alterations were required. This may have led to tenants tending to tolerate non-optimal conditions at first, because of a reluctance to request minor adjustments.

In most cases the living room temperature is stable within a degree or so. Higher variations are observable in the case of House 6b during 2011, but this is due to a technical intervention (which was installed in March 2011), allowing for the possibility of a limited night-time temperature set-back without energy penalty, via an intelligent control system (Boait et al. 2011).

Occupants sometimes report that late afternoon and evening temperatures are lower than desired, which they attribute to the heat pump's lack of reactivity when temperatures fall rapidly. However, this is not borne out by particularly low late afternoon and evening measured temperatures, and may be a psychological response to a falling temperature profile or diurnal metabolic variations.

There is little evidence of significant overheating on bright summer days for these dwellings. Some morning or afternoon temperature rises can be observed, depending on dwelling orientation, but they are generally modest, remaining within a degree or two of demand temperatures.

4 Discussion and Conclusions

Although this paper concentrates on the effects of variable weather conditions, the results suggest it may be difficult to separate this factor clearly from others. The systems studied here showed overall performances which were broadly in good agreement with the findings of the EST UK major heat pump field trials (EST 2012) and it was evident that the systems were easily capable of supplying sufficient space-heating, even in unusually cold conditions. In itself this suggests that for the most part the systems are running at part-load, and raises the question of system sizing. Part-loading and frequent cycling are both factors which can lead to performance degradation. Given the variability of conditions to be found in the UK, a better solution may be to use a heat pump operating close to optimum loading to provide background space-heating, together with an additional point heat source to cope with peak requirements. All the dwellings in this study had additional electric fires fitted, but these were not used on the grounds of expense, and indeed it may well be that there is at present a tendency for system designers to over-size to ensure adequate heating, especially in the case of vulnerable occupants such as the elderly.

Furthermore, in most of the 10 systems studied, the losses arising from DHW tanks and associated pipework were relatively high. It has been shown that even a modest reduction in DHW storage temperature can improve the overall performance significantly in the summer months. (Savings are to be expected in the winter months also, but of course this represents a smaller fraction of the total energy usage) Once again, however, this could be regarded as a system design issue, in that provision of DHW by the heat pump may not be the optimum solution under these circumstances, despite the fact that the heat pump operates at a higher nominal efficiency than an electric immersion heater.

An effect of weather conditions can be seen clearly in the lower performance resulting from rapid and early onset of cold weather in late 2010 compared with 2011. This was reflected in the system performances and clearly shown in the different ground-loop conditions. This suggests that while there may be a tendency to oversize the heat pumps themselves, this tendency was not necessarily reflected in the external ground-works which may possibly have benefitted from additional heat collection area (though it should be emphasised that the ground-loops were sufficiently sized to enable full recovery of temperatures following both winter periods).

Acknowledgments. The author would like to thank the Engineering and Physical Sciences Research Council (EPSRC) and E.ON UK for providing the financial support for this study as part of the Carbon,Control & Comfort project (EP/G000395/1).

References

Boait, P., Fan, D., Stafford, A.: Performance and control of domestic ground-source heat pumps in retrofit installations. Energy and Buildings 43, 1968–1976 (2011)

DCLG, A Decent Home: Definition and Guidance for Implementation. Department for Communities and Local Government, London, UK (2006)

EST, Getting warmer: a field trial of heat pumps. Energy Saving Trust, London, UK (2010)
EST, Detailed analysis from the first phase of the Energy Saving Trust's heat pump field trial. Department for Communities and Local Government, London, UK (2012)
Huchtemann, K., Müller, D.: Evaluation of a field test with retrofit heat pumps. Building and Environment 53, 100–106 (2012)
IVT. IVT website (2012), http://www.ivt.se (accessed April 30, 2012)
Nordman, R., Andersson, K., Axell, M., Lindahl, L.: Calculation methods for SPF for heat pump systems for comparison, system choice and dimensioning.SP Report No. 49, Energy Technology, SP Technical Research Institute of Sweden (2010)
Stafford, A.: Long-term monitoring and performance of ground source heat pumps. Buildings Research and Information 39(6), 566–573 (2011)
Weather Underground Jennyfields (Harrogate) weather data (2011), http://www.wunderground.com/weatherstation/WXDailyHistory.asp?ID=IENGLAND136 (accessed April 30, 2012)

Chapter 50
Asset and Operational Energy Performance Rating of a Modern Apartment in Malta

Charles Yousif[1], Raquel Mucientes Diez[2], and Francisco Javier Rey Martínez[2]

[1] Institute for Sustainable Energy, University of Malta, Barrakki Street
Marsaxlokk, MXK 1531, Malta
charles.yousif@um.edu.mt
[2] School of Industrial Engineering, University of Valladolid, Valladolid, Spain

Abstract. This paper aims to evaluate the asset and operational energy performance rating of a modern apartment in Malta, by comparing modelling results of DesignBuilder-EnergyPlus and the Energy Performance Rating of Dwellings Malta (EPRDM) software, to actual energy consumption of the apartment. Results showed that EPRDM results compared favourably with the DesignBuilder results, although, the latter one showed higher energy consumption for cooling. This is attributed to the fact that EnergyPlus considers the hottest week in the sizing of cooling systems and the simulation is carried out dynamically for every hour of the day. Actual energy consumption for heating and cooling is generally lower than modelled results, which augurs well for the overall energy consumption in Maltese buildings. The thermal mass of Maltese buildings plays an important role in reducing peak loads.

Keywords: Malta, asset rating, operational rating, DesignBuilder, EPRDM.

1 Introduction

Following Malta's accession to the EU in 2004, the EU Directive on Energy Performance of Buildings has been transposed into local legislation (Legal Notice 261 of 2008), which has also adopted the Minimum Requirements on the Energy Performance of Buildings (Technical Guidance Document F). Eventually, LN 261/2008 will have to be updated to reflect the new requirements of the EU Directive Recast 2010/31/EU (EU-Malta, 2011).

Malta has developed software for calculating the energy performance of domestic single-zone buildings known as the Energy Performance Rating of Residential Dwellings Malta (EPRDM), as other countries have done. For example, Spain has developed its own programmes LIDER and CALENER (EU-Spain, 2011), but has kept the doors open for the use of other auxiliary programmes, such as the well-known DesignBuilder–EnergyPlus software, provided that it fulfils a series of requirements. The UK and Portugal have also adopted DesignBuilder interface for their energy performance rating.

A. Håkansson et al. (Eds.): *Sustainability in Energy and Buildings*, SIST 22, pp. 531–543.
DOI: 10.1007/978-3-642-36645-1_50

This paper mainly deals with the comparison between the actual measured energy consumption (operational energy performance rating) of a recently-built apartment located in the centre of Malta and the modelled outcomes of DesignBuilder-EnergyPlus and EPRDM software (Asset Energy Performance Rating). It also attempts to validate the EPRDM software outputs, given that it is the official software for issuing energy performance certificates for residential buildings in Malta. Within this study, it was also intended to find answers to frequently asked questions such as the real energy consumed for heating and cooling and the energy consumption of specific appliances. Lastly, this work has aimed at developing the first EnergyPlus Weather (EPW) file for Malta, which is necessary to be able to operate DesignBuilder-EnergyPlus software.

2 The EPW File

The EnergyPlus Weather (EPW) file is a weather data format used in DesignBuilder-EnergyPlus software to carry out simulations of energy use in buildings. This file contains headline information written in keywords along eight consecutive lines, followed by hourly weather data in 8760 lines, representing a whole year (Crawley et al., 1999; US DOE, 2010).

The EnergyPlus website has loads of EPW files for different sites but not for Malta. Research has identified alternative sources of EPW files but they are not fully guaranteed, as representing the climate in Malta, since they are either based on interpolation or simply found in internet discussion groups. Hence, a new EPW file for Malta was created using actual data for the year 2010, sourced from the Meteorological Office of the Malta International Airport and the Institute for Sustainable Energy, of the University of Malta. Also, some design data such as the typical hottest and coldest weeks for the year and other typical design data for heating and cooling, including temperatures and degree days, were sourced from ASHRAE (ASHRAE, 2009).

In order to complete the weather information for the EPW file for Malta, other parameters were calculated such as the extraterrestrial direct normal, and extraterrestrial horizontal and total opaque cloud cover. For the headline information, some of the data required was not available in the ASHRAE document, so this was complemented by automatic generation of data by the Auxiliary Energy-Plus "Weather Convert" Programme, in order to calculate ground temperatures, typical/extreme periods as well as the horizontal infrared radiation intensity values. The Weather Convert Editor may be found within the set of auxiliary programmes of EnergyPlus (EnergyPlus, May 2012).

The next step was to amalgamate the data into a file that has the same format as an EPW file. This was carried out by using an existing EPW file from Energy-Plus, converting it to a CSV format file by the "Weather Convert" editor and replacing the data points with Malta's data. Finally, the modified .csv with the information of Malta was introduced again in the programme and reconverted to .epw file to obtain the final EPW file for Malta.

3 Description of the Case Study

The apartment selected for this study is situated in the town of Santa Venera, Malta. It forms part of a block of apartments that was completed in 2009. The building block has three floors, each having three dwellings, left, centre and right. There are also two penthouses on the top of the building. The dwelling under study is situated on the first floor to the right and has an area of around 100 m^2. It was selected because it was the first dwelling to be occupied, or in other words, it was possible to take real data based on the actual behaviour of the occupants. There were three occupants, an elderly but active couple and their adult working daughter.

The dwelling comprises of three bedrooms, combined sitting/kitchen, store room, small ensuite bathroom and main bathroom. There are two terraces, one located in the sitting room and the other in Bedroom 3. At the same time, Bedrooms 1 and 2 and the main bathroom have windows overlooking an internal shaft oriented at 67° NW, while the combined sitting/kitchen room and store room overlook a narrow service shaft. The façade on the main road is oriented at 23° NE.

In order to implement the asset energy performance rating of the apartment under study, the Maltese EPRDM software and the international DesignBuilder-EnergyPlus software were selected to simulate the energy demand for the apartment. Detailed background information of the software has been given in a previous publication (Yousif et al., 2011).

On the other hand, operational energy performance rating requires measurement of the actual energy consumption. For this purpose, a series of data loggers were placed in the apartment to log the individual energy usage in each zone for one year (March 2011-February 2012), as shown in Figure 1.

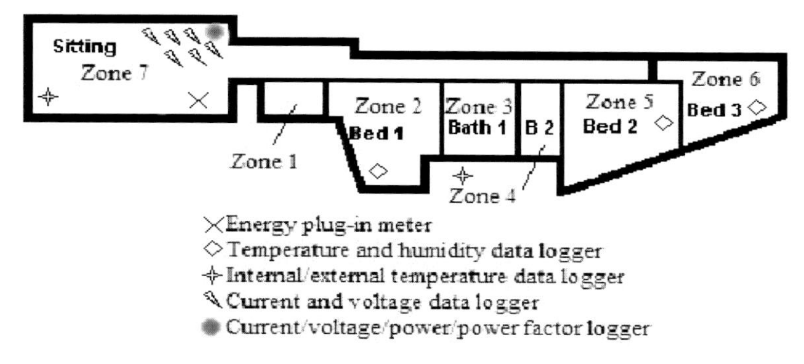

Fig. 1. Type of data loggers used and their position in the different zones within the apartment, as identified in the DesignBuilder-Energy Plus model

4 Results of the Asset Energy Performance Rating

The EPRDM software is able to simulate the energy demand in Maltese buildings with respect to space heating, space cooling, water heating, lighting, other auxiliaries (ventilation, water pumping), as well as energy production from renewable energy sources and savings from the use of second class water. On the other hand, it is not capable of simulating the energy demand of the different electrical appliances normally used in residences. For this reason, the comparison of the asset energy performance rating between EPRDM software and DesignBuilder-EnergyPlus software was carried out by only comparing the main energy demands of energy performance certificates, which are common to both software.

Given that the EPRDM software assumes a single zone of the dwelling, it was necessary to use the "merged zones" option of DesignBuilder, to be able to compare between the two outcomes. Thus, the schedules and the characteristics for the different types of energy consuming appliances, use of rooms in terms of lighting and HVAC systems had to be assumed as common to all.

Due to the versatility and flexibility of DesignBuilder-EnergyPlus, two simulations were carried out. The first case used the default values in the operation timetable of HVAC system of DesignBuilder (07:00-09:00 and 16:00-23:00), while the second simulation was referred to the operation timetable as adopted in EPRDM (06:00-08:00 and 17:00-23:00). Both simulations have a set temperature for cooling of 26.8 °C and for heating of 18.2 °C. Table 1 shows the results obtained and is compared to the actual energy usage in the apartment.

Table 1. Summary results of Asset Energy Performance Rating (kWh/m^2-year)

Software	**Actual**	**EPRDM**	**DB Merged Zones 1st Case**	**DB Merged Zones 2nd Case**
Space Heating	64.4	45.9	75.5	42.5
Space Cooling	3.6	4.8	44.4	27.1
Water Heating	39.0	52.4	49.1	49.1
Lighting	3.2	10.2	4.6	3.4
***EPC**	**110.2**	**113.3**	**172.6**	**122.1**

*EPC is the Energy Performance Rating based on a total floor area of 100 m^2 and primary to electrical energy generation efficiency for Malta of 0.28 (Enemalta, 2006), given that all of the dwelling's energy demand is based on electricity.

It is clear that by matching the timetables for both programmes, better results are obtained from DesignBuilder. Energy for cooling of DesignBuilder was always high for all simulations, since EnergyPlus uses a dynamic approach to calculate the cooling load, while it only uses the minimum external temperature as a sizing limit for the heating load. On the other hand, EPRDM uses monthly average data for calculating the energy demand for heating and cooling. Other limitations of EPRDM included the fact that shading is assumed constant throughout the year and is only caused by horizontal shading elements, while DesignBuilder calculates the effect of shading in all directions every 20 days.

Water heating was simulated in EPRDM considering the number of residents (which is an inbuilt feature based on a ratio of the internal floor area of 60 m^2 per occupant, with a minimum of 2 persons and a maximum of 7 persons) and assuming a constant hot water temperature of 60 °C, while DesignBuilder can take into account the inlet/outlet temperature and be set based on the volume of hot water consumed. This, together with the fact that EPRDM considers a constant heat loss of 15 % per month, while DesignBuilder calculates these losses on an hourly basis, could explain the difference between the two results.

For the case of lighting, the EPRDM over-estimated the requirement for lighting, while DesignBuilder was very close to the actual consumption as will be explained in Section 6.2. DesignBuilder takes into consideration shadows, reflection from floor, ceiling and internal walls. Also, there is a specific option for lighting control, which allows the user to set DesignBuilder to switch on artificial lighting, only when the lux level of daylighting within a zone is below the set minimum value. Such features are not available in the EPRDM software.

5 Results of Operational Energy Performance Rating

In order to implement the operational energy performance rating of the apartment, DesignBuilder-EnergyPlus was used to compare its results to a full year of real data collected from the dwelling, between March 2011 and February 2012.

Different simulations were carried out using different operation schedules for the HVAC system (Table 2) and the time usage of the appliances in each room (Table 3).

Table 2. Updated schedules for simulated HVAC system

	Default Values of DesignBuilder	**Updated Values of DesignBuilder as set in EPRDM**
Heating/Cooling (Operational schedule)	7:00 – 9:00 16:00 – 23:00	6:00 – 8:00 17:00 – 23:00

Table 3. Real time schedules for appliances in each room as input to DesignBuilder-EnergyPlus

Zone	Time of Use	Percentage of Use
Zone 1	11:00 - 12:00, 15:00 - 16:00, 17:00 - 18:00	10 %
Zone 3	5:00 - 8:00, 14:00 - 15:00, 19:00 - 20:00	50 %
Zone 4	11:00 - 12:00, 14:00 - 15:00, 19:00 - 20:00	10 %
Zone 2	6:00 - 7:00, 11:00 - 12:00, 14:00 - 15:00	100 %
Zone 5	5:00 - 6:00, 11:00 - 12:00, 14:00 – 15:00 19:00 - 20:00	100 % 50 %
Zone 6	6:00 - 7:00, 20:00 - 21:00	100 %
Zone 7	6:00 - 8:00, 17:00 - 20:00	50 %

5.1 Heating and Cooling Demand

In the case of energy demand for space heating and cooling, several simulations were carried out with the aim of finding which one would produce results that reflect the real lifestyle of the occupants in the dwelling and the real usage of the HVAC system.

DesignBuilder was used to simulate the apartment using the non-merged zones approach, whereby each room is simulated as a separate thermal zone. Two cases were studied, Case 1 refers to the use of the operational timetable given in Table 2 for the HVAC systems and Case 2 refers to the case when the timetables of Table 3 were used, which reflects the real lifestyle of the users. Moreover, Table 4 indicates separate operational timetable for the remaining appliances and lighting specified for each room.

Table 4. Operation timetable for the appliances in each room

Non-Merged Zones	Table 2
Non-Merged Zones using Actual (New) Timetable	Table 3
Non-Merged Zones using Actual (New) Set Temperatures	Table 2
Non-Merged Zones Actual (New) Timetable & Temperatures	Table 3

The first two simulations used a set temperature of 18.2 ºC for heating and 26.8 ºC for cooling, which are the standard settings of the EPRDM software. The remaining simulations used a set temperature of 19.5 ºC for heating and 26 ºC for cooling, which were the average measured internal temperatures for the coldest and warmest months during one year, respectively, weighted according to the temperatures and areas of each zone.

Figure 2 shows the results of the comparison between the different simulations carried out and the real energy demand in the apartment for space heating and cooling, respectively. DesignBuilder generates less consumption for space heating in the majority of the simulations except for Case 1 of "New Temperatures" and "New Timetable & New Temperatures". This discrepancy could have occurred because the EPW file used in DesignBuilder was for the year 2010, while the actual data gathered in the apartment was between March 2011 and February 2012, with external conditions that were probably different from 2010. This could be seen from Figure 3, when comparing the external temperatures for 2010 and 2011.

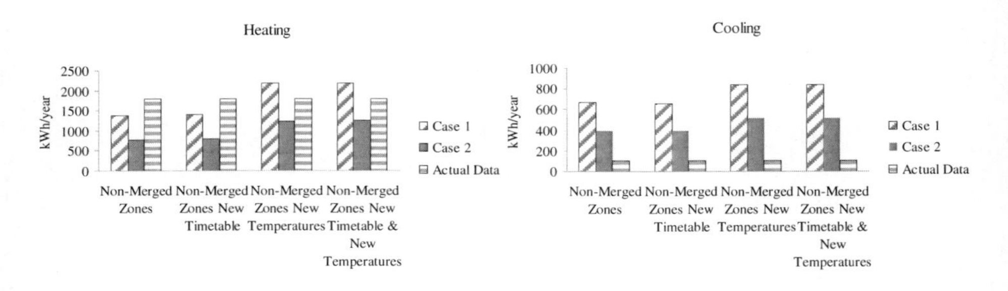

Fig. 2. Comparison between real data and simulated space heating and cooling demands, respectively

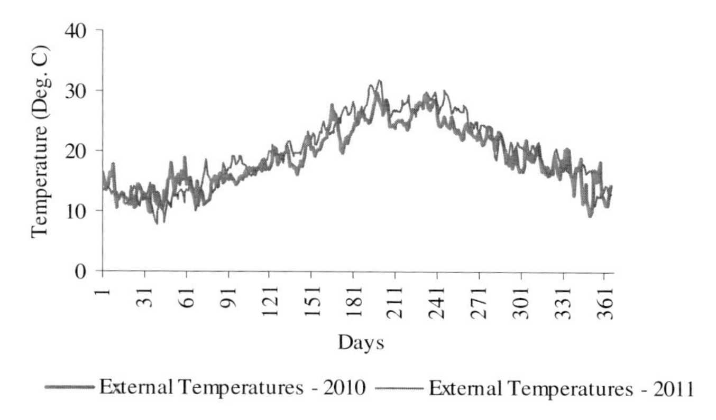

Fig. 3. Comparison of ambient temperatures for the years 2010 and 2011

For this particular apartment, solar gains during winter do not have a strong effect on the energy performance of the building as much as the actual external temperature. This is because fenestration from where direct sunshine can penetrate is only found in Zone 6 (Bedroom 3) and its area is small, while all walls were single brick walls except for the façade.

It is interesting to note that the use of actual timetables and actual set temperatures have a considerable bearing on the results obtained from DesignBuilder and makes them converge towards the actual energy consumption in the apartment, as seen when comparing Case 1 and Case 2 to the actual consumption.

On the other hand, the overestimation in space cooling demand for both simulations with respect to the actual data is also shown in Figure 2. When the operational timetable for the HVAC system and for the other appliances (which generate some heat within the apartment) are updated (Case 2), the difference is still much higher than the actual consumption. One notes that Maltese buildings are categorised as very heavy buildings with large thermal mass, with the result that extreme external temperatures do not fully penetrate the building, but EnergyPlus does not seem to take the thermal mass into consideration. Secondly, it has been noted that users and indeed most Maltese residents, do not resort to air-conditioning until the temperature within the apartment reaches around 28 or 29 °C, since they extensively use electric fans that create evaporative cooling effect. Such a consideration is not considered in DesignBuilder-EnergyPlus.

Once again, one would need to appreciate that the external temperatures for 2011 in summer were different from the EPW file of 2010 and this would also have some bearing on the results of the simulations, as was shown in Figure 3.

5.2 Lighting Demand

The energy consumption for lighting varied according to the seasons, with least artificial lighting used during the months with higher solar intensity and longer days. Exceptionally, November had more lighting demand than December. This is attributed to the fact that November 2011 was particularly cloudy in Malta.

Two simulations were carried out using DesignBuilder. In the first simulation, the operational timetable for lighting was chosen according to the template of "Domestic Family Week Day" (Switch on 7:00 – 9:00 and 16:00 – 23:00) in all the rooms, while in the second simulation, the operational timetable for each room was chosen taking into consideration the real usage of lighting in each room (Table 3).

Figure 4 shows the operational energy performance for the case of lighting, when comparing the real consumption of lighting and the results of the simulations with DesignBuilder. The second simulation provided total results that were closer to the real lighting consumption (90.44 kWh vs. 97.20 kWh, respectively).

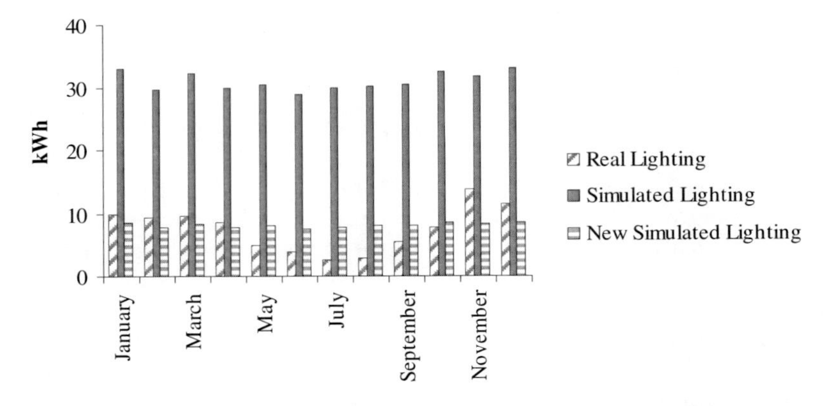

Fig. 4. Comparison between real, simulated and new simulated lighting consumption

5.3 Water Heating Demand

There were two electric boilers in the dwelling that were studied jointly, one of 10-litre capacity situated in the combined kitchen/sitting room and another one of 50-litre capacity in the main bathroom. For this purpose, two simulations were carried out at

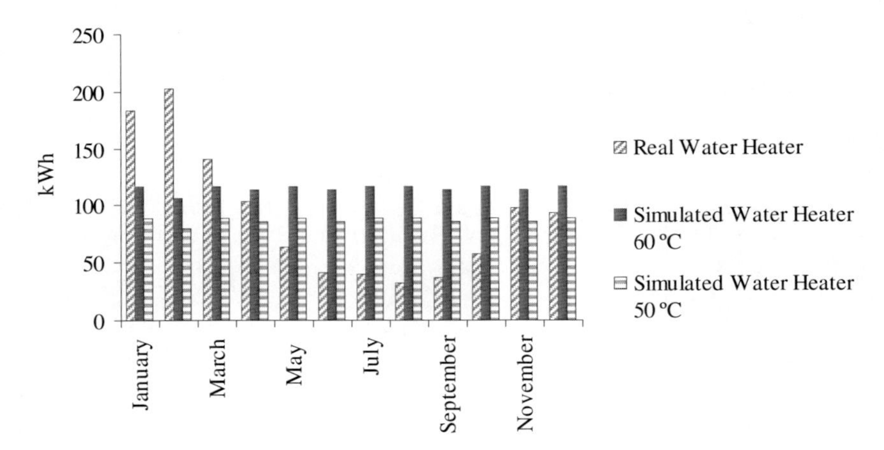

Fig. 5. Comparison between real and simulated water heating consumption (60 ºC and 50 ºC)

the set temperatures of 60 °C and 50 °C. The motive of this decision was that water heaters are normally set at 60 °C to avoid the growth of legionella bacteria in the system and deliver hot water at a temperature of around 50 °C at the outlets. DesignBuilder and EPRDM consider a standard temperature of 60 °C, but the water heaters in the dwelling were set at 50 °C. Figure 5 shows these simulations with the real energy demand for water heating.

In this diagram, it is seen that during the months of January and February 2012, the energy demand for hot water was high. This is attributed to the fact that these months were particularly cold and also the cold water supply was colder than normal, since it comes from a large external storage tank (750 litre capacity), which is subjected to the cold external temperatures on the roof of the building. Placement of mains water storage tanks on top of roofs is a traditional practice in Maltese buildings. The total consumption for water heating was 1153.33 kWh for the actual data, 1381.68 kWh for the simulation at 60 °C and 1041.53 kWh for the simulation at 50 °C. Again, it is seen that DesignBuilder does not take into consideration the seasonal variations in the energy demand for water heating.

5.4 Washing Machine Demand

Washing machines are indispensable in modern homes and could consume significant amounts of energy. Figure 6 shows the comparison between the simulation and the real consumption, taking into consideration a power of 10 W/m^2 with a radiant energy fraction of 0.2 and an operation of one hour every Tuesday, Thursday and Sunday at full load.

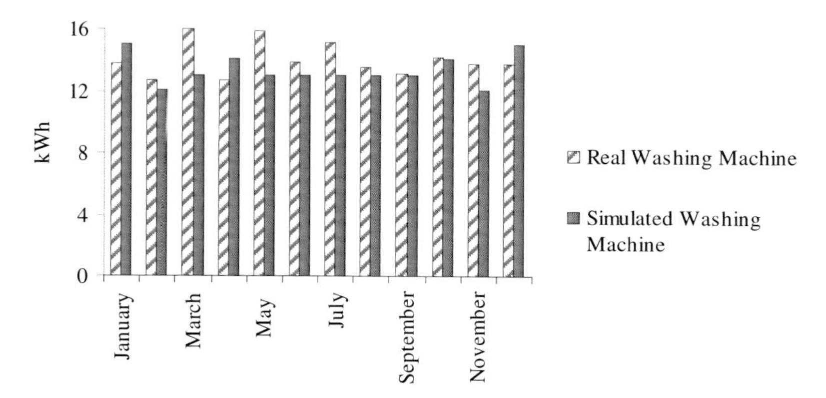

Fig. 6. Comparison between real and simulated washing machine load

5.5 Appliances Demand

The electric oven, extractor fan, fridge/freezer and other appliances were analysed separately and simulated jointly with a power of 2.5 W/m^2 of plan area. Table 5 shows the results of the simulations and the real energy consumption for the appliances.

Table 5. Energy Consumption of the appliances

Appliance	Real Consumption (kWh/year)	Simulated Consumption (kWh/year)
Electric oven	128.34	-
Extractor Fan	8.87	-
Fridge/Freezer	271.35	-
Television	164.25	-
Computer/Printer/Radio	182.50	-
Microwave (since August)	21.00	-
TOTAL	**776.32**	**823.92**

In the case of the fridge, it is interesting to note that the energy demand of this appliance increases in the warmer months due to the extra load on the compressor caused by higher ambient temperature and more frequent opening of the fridge, as seen in Figure 7.

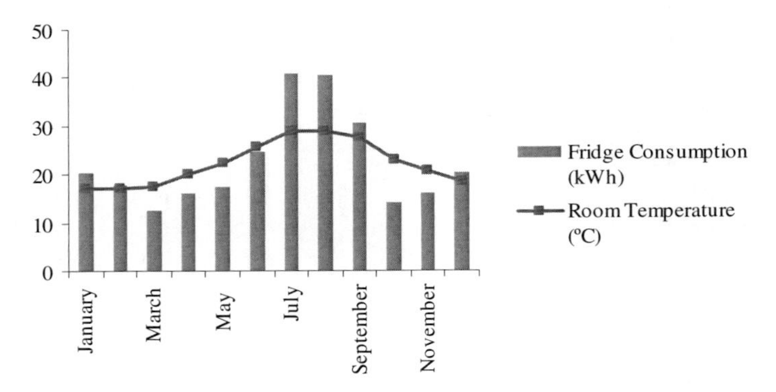

Fig. 7. Seasonal energy consumption of the fridge/freezer

As from October, the energy demand of the fridge started rising again. This was attributed to the accumulation of ice on the evaporator tubes of the freezer, following the humid months of September and October. In January, the freezer was defrosted, since the freezer door could not close properly anymore.

5.6 Overall Summary of Results

Table 6 focuses on the comparison between the actual operational and simulated energy performance rating for the different cases. The over-estimation of DesignBuilder for modelling space cooling demand requires further analysis. Several indications may explain the difference between modelling and actual consumption. First, the thermal mass of the building itself is not modelled in DesignBuilder. Second, natural ventilation and the use of forced evaporative cooling appliances (such as fans) have been extensively applied but these practices are not modelled in DesignBuilder either. One could also add that adaptive behaviour and energy conservation could have played a role in reducing energy demand for cooling.

The effect of using the actual temperatures within the building has a more prominent effect on heating and cooling than using the actual schedule for the appliances. It was also useful to understand the effect of external temperatures on the energy rating of the dwelling, which for this case, had a more marked effect than solar gains.

Table 6. Summary results of energy performance rating (kWh/m^2-year)

Simulations	Actual Data	Non-MZ 1st Case	Non-MZ 2nd Case	Non-MZ 1st Case	Non-MZ 2nd Case	Non-MZ 1st Case	Non-MZ 2nd Case	Non-MZ 1st Case	Non-MZ 2nd Case
Variations				**New (Actual) Templates**		**New (Actual) Temperatures**		**New Templates & New Temperatures**	
Heating	64.4	49.8	27.8	50.2	28.6	77.7	44.0	78.1	45.3
Cooling	3.6	23.9	14.1	23.7	14.1	30.1	18.6	29.9	18.5
Water Heating	39.0	49.3	49.3	49.3	49.3	49.3	49.3	49.3	49.3
Lighting	3.2	3.5	3.5	3.5	3.5	3.5	3.5	3.5	3.5
***EPC**	**110.2**	**126.5**	**94.7**	**126.6**	**95.5**	**160.6**	**115.4**	**160.8**	**116.6**
Washing Machine	6.0	5.7	5.7	5.7	5.7	5.7	5.7	5.7	5.7
Total Appliances	27.2	29.4	29.4	29.4	29.4	29.4	29.4	29.4	29.4

*EPC is the Energy Performance Rating based on a total floor area of 100 m^2 and a primary electrical energy efficiency for Malta of 0.28, given that all of the energy demand is based on electricity.

6 External Temperatures

The external temperatures were measured on the main street (N23°E) and in the internal shaft (N67°W), respectively. Figure 8 shows representative months for the variations between them. The internal shaft exhibits higher external temperatures than the street side during summer. Since the sun's position is higher in the sky during these months, it manages to reach deeper into the shaft and heats up the walls. Since there is no air draft, heat accumulation occurs. Moreover, it is common practice to install the condenser of the split-type air-conditioning units in these shafts, thus adding to the heat generation within the immediate vicinity of the apartment's walls in zones 2, 3 and 5. This brings out the necessity for insulating these walls, which are single brick walls with a high typical U-value of 2.24 W/m^{2}K. The heat trap effect would have been beneficial for winter, but since the elevation of the sun is lower, the internal shaft remains as cold as the ambient temperature. Once again, the insulation of these single walls is required to stop the heat escaping from the rooms.

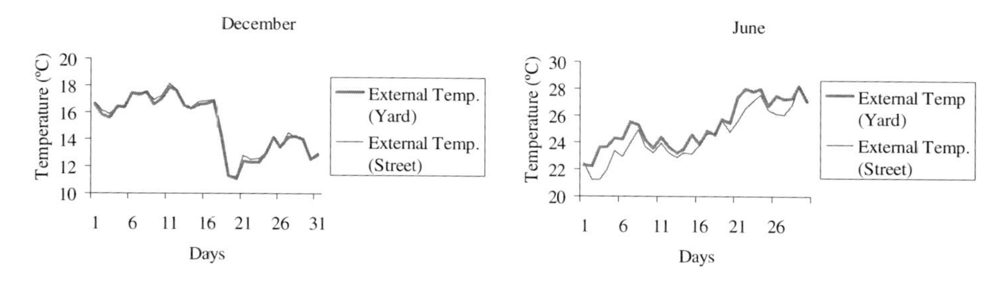

Fig. 8. Comparison between external temperatures on the main street and in the internal shaft

7 Conclusion

This paper has presented the methodology to be used to create an Energy-Plus Weather Data (EPW) file and has generated one for Malta based on data for the year 2010. This same data set is also being adopted by the Building Regulations Office of the Ministry for Resources and Rural Affairs, for the development of national energy performance software for non-residential multi-zone buildings. Since weather variations are to be expected from one year to another, the use of more years would make the EPW file more credible for use in simulation software. However, climate change has affected the global weather and therefore, the typical meteorological year should ideally use recent years of data rather than historical databases of decades ago.

With regards to the asset and operational energy performance rating, the simplicity of the Maltese EPRDM software may be taken as an advantage, as there is less possibility of making errors in the input of data. On the other hand, it is not flexible and the results provide yearly outputs, which are sufficient for the production of the energy performance certificate, but may not provide adequate information for the analysis and improvement of energy performance in buildings. On the other hand, DesignBuilder provides a more powerful tool to analyse any type of building, but it is important to ensure that the input parameters are correctly set by the assessor. Some countries have already adopted DesignBuilder in their scheme of approved software for the production of energy certificates, such as the UK and Portugal. This should also be considered in Malta.

The behaviour of the residents in the dwelling has affected the total energy consumption in the majority of cases. The quantity of times of opening windows and doors, use of artificial lighting and ventilation fans, air-conditioning and water heaters, are all human actions that are hard to predict, as they depend on the particular lifestyle. Nevertheless, the overall simulations showed that this can be quantified to an agreeable level of accuracy.

The operational energy performance rating produced lower results than the asset ratings of EPRDM and DesignBuilder. In practice, this means that energy performance certificates are on the higher rather than the lower end of the energy rating scale, which is commendable.

Acknowledgements. Our thanks go to the Meteorological Office of the Malta International Airport for providing weather data and the University of Malta Scholarship Fund Committee for supporting this study. Also, we acknowledge the support of the Erasmus Programme for giving us the opportunity to share knowledge and research experience between the Universities of Malta and Valladolid, Spain.

References

ASHRAE: ASHRAE Handbook – Fundamentals, Climatic Design Information (ch. 14) (February 2010) ISBN 978-1-933742-55-7

Crawley, D.B., Hand, J.W., Lawrie, L.K.: Improving the Weather Information Available to Simulation Programs. In: Sixth International IBPSA Conference, Kyoto, Japan, cod. P-03 (1999)

Enemalta Corporation: Electricity Generation Plan 2006-2015 (2006), http://www.enemalta.com.mt/enemaltastorage/images/files/archived%20news/generation%20plan%20%2821.06.06%29.pdf
EnergyPlus: Auxiliary EnergyPlus Programs (May 2012), http://apps1.eere.energy.gov/buildings/energyplus/pdfs/auxiliaryprograms.pdf
European Union (Malta Case): Implementing the Energy Performance of Buildings Directive (EPBD) – Implementation of the EPBD in Malta, Brussels, Belgium, pp. III-235–III-244 (April 2011)
European Union (Spain Case), Implementing the Energy Performance of Buildings Directive (EPBD) – Implementation of the EPBD in Spain. Brussels, Belgium, pp. III-111–III-122 (April 2011)
U.S. DoE, Energy Efficiency and Renewable Energy: Auxiliary EnergyPlus Programs. Extra Programs for EnergyPlus (October 2010), http://apps1.eere.energy.gov/buildings/energyplus/energyplus_documentation.cfm
Yousif, C., Gómez Royuela, I., Rey Martínez, F.J.: Energy Performance Rating of Dwellings in Malta. In: Sixth Dubrovnik Conference on Sustainable Development of Energy, Water and Environment Systems, Dubrovnik, Croatia (September 2011)

Chapter 51
Low Carbon Housing: Understanding Occupant Guidance and Training

Isabel Carmona-Andreu[1], Fionn Stevenson[2], and Mary Hancock[3]

[1] CA Sustainable Architecture, Newbury, UK
[2] University of Sheffield, UK
[3] Oxford Brookes University, UK

Abstract. Recent research into occupant behaviour in low carbon housing indicates that for the same type of house, energy and water use can vary by up to fourteen times between different households. This paper assesses the information and training the occupants received in two contrasting building performance evaluation case studies of exemplary low carbon housing. Key findings showed a lack of a coordinated set of guidance for occupants and poor understanding on the trainers' part on specifics of the centralised heating and mechanical ventilation systems. As a consequence occupants were unable to operate or maintain these systems with confidence. Recommendations are made to develop guidance and "hands on" training that keeps usability in mind and empowers occupants to contribute to reductions in carbon emissions.

Keywords: low carbon housing, occupants, guidance, usability, procedures.

1 Introduction

All new-build housing in the UK is required to be "zero-carbon" by 2016 (DCLG, 2007) when all regulated carbon emissions, associated with heating, ventilation and lighting, must be minimised and offset if necessary. Recent research shows that for the same type of low carbon house, energy and water use can vary by up to fourteen times between different households (Pilkington et al, 2011). Why is this happening? One area to examine is how well occupant are guided towards understanding the use of their homes, and how this affects energy consumption.

Current England and Wales Building Regulations require some provision of information to occupants on the efficient operation and maintenance of domestic building services (HMG 2010a and 2010b) (DCLG 2010b and 2011) . The preparation of a guide for occupants containing details for everyday use in a form easily understood is also recommended in the Code for Sustainable Homes (CfSH) (DCLG 2010a). Little guidance, however, is given as to what the preferred format would be, and manuals are generally too complicated (Monaham and Gemmel, 2011). Gill et al (2010) found that only 1 out of 11 households used the guide successfully to find out how to control their heating, and 64% of the households did not use the guide at all. This lack of control capability

A. Håkansson et al. (Eds.): *Sustainability in Energy and Buildings*, SIST 22, pp. 545–554.
DOI: 10.1007/978-3-642-36645-1_51

contributed to above average energy consumption. In 2008, the ground –breaking prototype low carbon Sigma Home at the Building Research Establishment in the UK was inhabited by a family during four fortnight periods. The home introduction tour and home guide proved relatively ineffective with the occupants resorting to a trial and error process while exploring the controls. They still could not use them as intended by the end of their stay (Stevenson and Rijal, 2010).

The Scottish Government reviewed guidance provided by house builders and social landlords in order to derive best practice, which has been adopted as part of their silver sustainability standard (SG 2011a, 2011b). However, the handover process was not reviewed in detail and there is further scope for testing the relationship between the written guidance and the usability of controls. The UK Good Homes Alliance requires developer members to sign up to a sustainability standard. This does not yet cover home guidance but a number of developer members have key personnel involved in the handover process with occupants to familiarise them with their new homes energy efficient features (pers comm., GHA 2012)[1]. Clearly, there is still some way to go in this area.

This paper assesses the guidance given to occupants in two new build housing case studies where extensive building performance evaluation (BPE) studies were carried out. It looks at the roles of different players in providing information and explores how the induction process influences user behaviour. Specifically, it builds on the above precedents and aims to draw out key issues and lessons regarding the organisation, content, accuracy and effectiveness of the home demonstration and home user guidance.

2 The Case Study Housing Developments

The two chosen case studies reflect the extreme scales of development and types of private housing developer. This makes them "paradigmatic" case studies which can be generalised from (Flyvbjerg, 2006). The characteristics of the two case studies are described briefly below:

Large Developer - Crest Nicholson (Photo 1.1 Left)

Operating nationally, their case study development in Kent was designed in 2006 and completed in early 2011. It was based on the winning design entry for an Affordable £60K Home in the UK government's Design for Manufacture competition and designed to achieve an Eco-Homes Excellent level[2]. The low carbon technologies deployed included thermally efficient building envelope using structurally insulated panels (SIPs), a roof lantern, a central services core with mechanical ventilation with heat recovery (MVHR) system and condensing boiler, as well as low energy appliances and lighting The BPE project evaluated one end terrace house in depth together with further evaluations related to the development process.

---

[1] 4 of developer members corresponded with the authors between 20th and 25th April 2012.

[2] Ecohomes was the precursor to the CfSH.

Fig. 1. Left - Crest Nicholson's development, Right - Ecos Homes development

Small Developer - Ecos Homes (Photo 1.1 Right)
Based in the South West of England, their case study development in a small Somerset village was substantially completed in 2009. It comprised two detached houses and three terraced homes situated around a courtyard. Designed to CfSH Level 5, it deployed a substantial number of low carbon technologies including a 2 kWp photovoltaic system, solar thermal panels, a 11.3 kW wood pellet boiler, an indirect, unvented 250 litre domestic hot water cylinder with immersion heater, a mechanical extract ventilation (MEV) system and rainwater harvesting system. The BPE project evaluated one terraced house in depth together with further evaluations related to the development process.

3 Developers Approaches to Occupant Guidance

The large developer organization has a very professional customer care service approach. The dedicated development sales team interacts with each customer and guides them through the purchase. A home demonstration is planned a week prior to legal completion and it is carried out by the sales team together with the site manager to explain more technical details. The information pack with operating manuals and warranties is programmed to be handed over during the home demonstration. After handover, the after sales team provides support and the service includes a 24hour emergency repair service. (Crest Nicholson, 2012).

By contrast, the small developer is relatively inexperienced, having been in existence for only a few years and with only a very small team of staff. For their customer care, they have relied heavily on the installers and a relatively inexperienced project manager to carry out the handover and troubleshoot problems as they emerge.

4 Methodology

The evaluation of handover procedures and guidance in the case studies borrowed from ethnographic observational techniques to gain a 'Thick description' (Geertz 1973). A multi-method approach was used to build up a complex context-rich picture of the homes in use. Six semi structured interviews and walkthroughs with the occupants took place in their homes (four in the large development and two in the small development), which allowed for 'traces' to be gathered that informed the analysis of the occupants' behaviour (Zeisel 1984) and efficacy of the guidance and handover procedures. These were cross-related to a construction audit, a review of design intentions with the design teams, commissioning processes, user questionnaires and a usability study reported elsewhere (Stevenson et al, 2012).

The evaluation of the home demonstration process (Table 1) correlated written guidance and procedures with direct observation by the researcher and the impressions of the home owners. The demonstration was tape recorded and any errors, omissions and deviations from the written procedure noted. Photographs were also taken during the home demonstration and semi structured interviews to record any significant aspects.

Following Bordass et al's (2007) usability criteria, the manuals and other written guidance were evaluated for their *clarity of purpose, usefulness of labelling and annotation and ease of use* (Table 2). The written guidance was effectively viewed as an overall control 'touchpoint' for the dwelling and should therefore follow these usability criteria. The need for congruent natural mapping that Norman (1988) applies to product design also applies to home guidance. One should be able to tell which bit of guidance goes with each part or system of the house and this was also evaluated. The findings of this evaluation are discussed next.

5 Handover Demonstration Process

Generally, there was very little hands on demonstration of controls and the technical accuracy was poor to medium, providing confusion amongst the occupants (Table 1). There were also significant differences in the attitude and clarity of procedure followed by the two developers:

Large Developer

It was observed that the home demonstration team was charismatic and gained the trust of the future occupants. The demonstrators had not used their procedural checklist before and, being unfamiliar with its contents, they missed some items. Both the sales demonstrator and site manager present attempted at times to cover for each other's omission of content. This sounds like good team work but actually led to some unclear messages being communicated to the occupants. Firstly, this was because the sales demonstrator did not fully understand that the roomstat controls the temperature setting and the thermostatic radiator valves (TRVs) are subservient to it. Secondly, the demonstrator thought the MVHR would "balance" the heat of the house, when it cannot do this on its own, as supplementary heating is needed. This led occupants to think that the MVHR was a heating source in itself.

The future occupants handled opening windows and doors and bathroom fittings but there was no demonstration of how to set the boiler programmer or of how to get into MVHR unit for cleaning the heat exchanger filter. Occupants also had no hands on experience of any of the heating or ventilation controls.

Overall, the occupants were content with the handover process, but it became apparent through interviews they did not actually understand some of the environmental control systems in their home, in particular the controls and filter cleaning procedure of the MVHR.

Table 1. Handover demonstration features evaluated

Handover Demonstration feature	Small Developer	Large Developer
Organization	no clear strategy	Very Good ordered impression. Shared- technical and sale staff
Completeness	No	No
Technical accuracy	Poor	Medium - some confused messages
Hands on demonstration of		
Windows/doors	No	Yes - but not rooflights
Taps	No	Yes
Heating controls	Not possible - expected from supplier	No
Heating maintenance/ cleaning	Not possible - expected from supplier	No
Hot water	Not shown - no access to loft	No
Ventilation controls	Not shown - no access to loft	No
Ventilation filters	n/a	No
Rainwater harvesting	Not possible - expected from supplier	n/a
Solar Hot water	Not shown - no access to loft	n/a
Photovoltaic panels	meters shown	n/a
Timeliness	Rushed on moving in day	Handover process part of sales process
Follow up	troubleshooting	Yes by site manager and customer services

Small Developer

The occupants interviewed were dissatisfied with the handover process as it was disorganised and not thorough enough. The timing of the home demonstration was unsuitable, coinciding with their moving day while the builders were still finishing the flooring. It was hurried by the project manager, who simply pointed at items and stated their function. The occupants would have liked a more detailed explanation a week or so after moving in, especially as their home had a lot of relatively unusual kit. They found the written guidance confusing and too technical, and experienced significant troubleshooting when items malfunctioned.

To try and unlock what the problems with the handover were, a "mock up" handover[3] was organised in one of the unoccupied properties. Although the house features were all pointed out, the customer had no hands-on experience of window operation. There was no access to the loft to demonstrate the MEV and solar hot water tank. The demonstration was given by a sales person, who did not know enough about the operation of the installed products. The developer expected the various installers to provide information and demonstrate their products to the customer. This proved unrealistic as tradesmen were often also unfamiliar with the innovative installations provided, and when sales delayed, the installers moved to another job, and were not available to return and provide a demonstration.

6 Home User Guidance

Generally, both developers relied on the manufacturer's technical manuals that were too technical for the occupants and were not interpreted for the specific dwelling (Table 2).

Large Developer

No detail construction drawings were included in the home user guidance box, even though the dwellings were not of traditional construction. This could create problems for occupants in future if they decide to extend their homes. There was no strategic explanation of the heating strategy. Manuals for three items (boiler, programmer and TRVs) were included in the box but in dispersed locations, and the overall heating system operation was left unexplained.

Although the MVHR manual explained what the icons on the control screen display were, it did not explain what buttons to press to change between modes of operation. It had no illustrations to identify and locate the equipment in the specific house. The occupants did not always know how or when to change MVHR filters or what the control screen display symbols meant.

Small Developer

The occupants commented that all the information seemed to be in the guide and handover folder of collated warranties and manuals but it was too technical and not aimed at the end user. In particular, they found it difficult to understand the guidance for the operation of the wood pellet boiler controls and for its cleaning. The document had clearly been translated from Italian and covered different models, even describing cleaning procedures that were not possible for their stove.

The ventilation strategy changed during construction but details of the earlier system still appeared in the folder which was confusing. The mechanical extract ventilation with natural air intake through trickle vents was not adequately explained, with no mention that trickle vents needed to be opened or that the door undercuts were needed to allow air movement through the house.

---

[3] An independent volunteer was recruited to act as home buyer.

Table 2. Home User Guide features evaluated

Home User Guide feature	Small developer	Large Developer
Clarity of purpose		
Presentation	Logical but 2 documents	Good presence: document + box - too precious?
Completeness	Medium	Medium
Certificates and warranties	Yes	Yes
Timeliness	Some late certificates	Part of sales process
Usefulness of labelling and annotation		
Diagrams specific to house?	No	No
Ease of use		
Simplified guidance	No	No
Summary of strategy	No	No
Manufacturer's information	Yes	Yes
Technical accuracy	Raw manufacturer's guidance	Raw manufacturer's guidance
Relevance of information	Fair - Some obsolete items.	Fair
Comprehensive coverage of features		
Construction	Yes	No
Windows/doors	Yes	No
Heating controls	No specific energy efficient advice for product. Mix of installation and user manual for various models. Not specific enough.	User guide and installer manual for boiler - too technical. Programmer controls sheet easily misplaced. No explanation of system (thermostat and TRVs)
Heating maintenance/ cleaning	Mix of installer and user manual for various models. Not specific.	Manufacturer's servicing manual - too technical
Hot water	3 sources of heat described	Installation manual not for user
Ventilation/ controls	Inconsistent between documents	Manufacturer's guide - uncertain how to operate different modes.
Ventilation filters	n/a	Manufacturer's guidance - no visual aids
Rainwater harvesting	no filter maintenance advice	n/a
Photovoltaic panels	Installed system not as per guide	n/a

7 Discussion

Analysis of tables 1 and 2 above highlights three key areas where handover processes and guidance can be improved, taking into consideration the dramatically different sizes of the developers involved.

I. Process Development and Training Needs

A key difference between the case studies lies in the organisation of the handover procedure. The customers of the larger developer were satisfied the handover induction and following customer care arrangement. This is the strength of a larger organisation, with resources to test and evolve these procedures. Smaller developers need to find ways of providing a similar level of customer care, but lack the man power or expertise which adds pressure to their organisations.

To ensure accurate information provision, the demonstrators need a clear technical understanding of all the functions of the systems installed and the design strategy. Good Homes Alliance members have created "champion" roles to help reaffirm the messages and provide training with success (GHA members 2012).

Complexity increases when the low carbon technologies used are unfamiliar to both demonstrators and users. Thus, the level of technical knowledge and training needed by home demonstrators needs to be considered from the start of a development, including adding a usability dimension to briefing and design processes, with questions such as "*How would people actually use it?*" (Way et al 2009) and setting responsibilities within the team to keep usability in mind. Controls' touchpoints and training guidance need testing on real people other than the designers.

II. A Systematic Approach

Repetition of new information is essential for it to be retained and to develop further knowledge (Medina 2008). Retention depends on the conditions of the event when the information was first given, such as intensity of attention and interest (Ebbinghaus 1913).

Home inductions are often that first point of contact with the user and the home guidance becomes the repeated information. However, in the case studies the lack of linkage and coordination between the two parts reduced their effectiveness. Often, the demonstrators did not impart the information but referred the customer to the guide which defeats the purpose of the home induction. A highly visual quick start guide as the initial information source can point to further information when necessary (SG 2011b). Pictures recognition doubles that of text but our brains pay more attention when more than one sense is engaged (Medina 2008), which makes hands-on experience a key part of the learning process. Whatever the starting point, a systematic approach is needed with all the parts of the information "package" fully co-ordinated to promote learning and avoid mixed messages.

III. Clear and Relevant Visual/written Guidance

Congruent mappings (Norman 1988) need to be developed within written guidance to facilitate ease of use. House specific visual diagrams were absent from the case studies manuals but they are essential to provide this clear link between the house and the guidance. The case studies used manufacturers' information to provide user instructions which was often generic. Sometimes various models were covered in the same document which caused confusion. In addition, the manufacturers' information did not show where the installed item was located in the house or when and how the user should interact with the controls for optimum performance. The strategic designers' intentions for efficient use were also missing, as the overall design combined various products, whereas the trade manuals were for single items. This disjunction was typified in the small development, with no guidance on how to optimise three separate modes for heating hot water (biomass boiler, solar hot water or immersion heater); occupants had to develop their own method of decision-making by trial and error.

Understanding the way occupants interact with technology is crucial to be able to give relevant instructions. Pink (2011) argues that as energy use is invisible to ordinary consumers, new ethnographic methods need developing to link the sensory experience of the home to energy use. As guidance manuals are part of the technology, for greater usability, they need also be tested in the real world.

8 Conclusion

These paradigmatic case studies have shown that guidance and handover processes in housing clearly need further development to facilitate occupants understanding of complex low carbon strategies and technologies. Key recommendations include:

1. Clearer handover process within developer organisations that follow a systematic approach focused on usability, from the outset.
2. Training of all home demonstrators so that they are familiar with all technical aspects of the appliances within the home and know how to operate them in an efficient manner.
3. Provision of clearly structured and coordinated home user packages that avoid generic manufacturers information,
4. Evaluation of the effectiveness of all guidance given for usability and consider various modes of presenting the information such as visual diagrams, short films, web links, to simplify the learning process.

Currently there is no legislated standard for home user guidance and processes with regards to accuracy or user friendliness. The above recommendations could be the starting point of a user centred handover and home user guidance initiative leading to a best practice benchmark in the domestic sector. These recommendations need further testing to check their effectiveness and acceptance by occupants. Research is also needed concerning the influence of home user guidance on occupant behaviour, the effectiveness of communication (charts, visual, written, video, hands – on) in providing rapid understanding of key issues and methods for early engagement of occupants in the energy efficient control of homes.

Acknowledgments. The authors gratefully acknowledge the funding provided by the Technology Strategy Board through its Building Performance Evaluation Programme, as well as the generous time given by the developers and occupants involved in these case studies, and. Good Homes Alliance members. With special thanks to Julia Plaskett from Crest Nicholson.

References

Bordass, W., Leaman, A., Bunn, R.: Controls for end-occupants: a guide for good design and implementation, BSRIA, Bracknell (2007)

Crest Nicholson– Customer Charter (2012), `http://www.crestnicholson.com/customercare/default.aspx` (accessed April 19, 2012)

Department of Communities and Local Government (DCLG). Homes for the future. More affordable, more sustainable, Cm 7191, HMSO, London (2007)

Department of Communities and Local Government (DCLG). Code for Sustainable Homes: Technical Guide, DCLG, London (2010a)

Department of Communities and Local Government (DCLG). Domestic Building Services Compliance Guide 2010, DCLG, London (2010b)

Department of Communities and Local Government (DCLG). Domestic Ventilation Compliance Guide, DCLG, London (2011)

Ebbinghaus, H.: Memory: A contribution to experimental psychology - translated by Ruger HA and Bussenius CE (1913) accessed through York University, Ontario, website (1885), http://psychclassics.yorku.ca/Ebbinghaus/index.htm

Flyvbjerg, B.: Five Misunderstandings About Case-Study Research. Qualitative Inquiry 12(2), 219–245 (2006)

Geertz, C.: Thick description: toward an interpretive theory of culture. In: The Interpretation of Cultures: Selected Essays, pp. 3–30. Basic Books, New-York (1973)

Gill, Z.M., Tierney, M.J., Pegg, I.M., Allan, N.: 'Low-energy dwellings: the contribution of behaviours to actual performance. Building Research & Information 38(5), 491–508 (2010)

Good Homes Alliances (GHA) members, e-mail exchange with researcher (2012)

HM Government (HMG) – The Building Regulations 2000 – Approved Document F: Ventilation, 2010 edn. (2010a)

HM Government (HMG) – The Building Regulations 2000 – Approved Document L1A Conservation of fuel and power – New Dwellings, 2010 edn. (2010b)

Medina, J.: Brain rules. Pear Press (2008)

Monahan, S., Gemmel, A.: How occupants behave and interact with their homes - The impact on energy use, comfort, control and satisfaction NHBC Foundation NF35 (2011)

Norman, D.A.: The Design of Everyday Things. Basic Books, New York (1988)

Scottish Government (SG)– Building (Scotland) Regulations 2004 - Technical Domestic handbook (October 2011a)

Scottish Government (SG) – Guidance for Living in a Low Carbon Home – report prepared by MARU and 55NorthArchitecture (2011b)

Pilkington, B., Roach, R., Perkins, J.: Relative benefits of technology and occupant behaviour in moving towards a more energy efficient, sustainable housing paradigm. Energy Policy 39(9), 4962–4970 (2011)

Pink, S.: Ethnography of the invisible: energy in the multisensory home. Ethnologia Europaea: Journal of European Ethnology 41(1), 117–128 (2011)

Stevenson, F., Carmona-Andreu, I., Hancock, M.: Designing for comfort – usability barriers in low carbon housing. In: Proceedings of 7th Windsor Conference: The Changing Context of Comfort in an Unpredictable World, Windsor, UK, London (2012), http://nceub.org.uk

Stevenson, F., Rijal, H.B.: Developing occupancy feedback from a prototype to improve housing production. Building Research and Information 38(5), 550–564 (2010)

Way, M., Bordass, W., Leaman, A., Bunn, R.: The Soft Landings Framework: for better briefing, design, handover and building performance in-use, BSRIA, Bracknell (2009)

Zeisel, J.: Inquiry by Design: Tools for Environment– Behaviour Research. Cambridge University Press, Cambridge (1984)

Printed by Books on Demand, Germany